CRISPR-Cas9 Technology for Precision Medicine

Jen-Tsung Chen
https://orcid.org/0000-0002-3540-4449
National University of Kaohsiung, Taiwan

IGI Global
Scientific Publishing
Publishing Tomorrow's Research Today

Vice President of Editorial	Melissa Wagner
Director of Acquisitions	Mikaela Felty
Director of Book Development	Jocelynn Hessler
Production Manager	Mike Brehm
Cover Design	Jose Rosado

Published in the United States of America by
IGI Global Scientific Publishing
701 East Chocolate Avenue
Hershey, PA, 17033, USA
Tel: 717-533-8845 | Fax: 717-533-7115
Website: https://www.igi-global.com E-mail: cust@igi-global.com

Library of Congress Cataloging-in-Publication Data

Names: Chen, Jen-Tsung editor
Title: CRISPR-Cas9 technology for precision medicine / edited by Jen-Tsung Chen.
Description: Hershey, PA : IGI Global Scientific Publishing, [2026] | Includes bibliographical references and index. | Summary: "This book bridges the knowledge gap by summarizing the emerging literature on CRISPR/Cas9, highlighting its innovative tools and recent achievements in clinical applications for precision medicine"-- Provided by publisher.
Identifiers: LCCN 2025052975 (print) | LCCN 2025052976 (ebook) | ISBN 9798337348575 hardcover | ISBN 9798337348582 paperback | ISBN 9798337348599 ebook
Subjects: LCSH: Gene therapy | CRISPR-associated protein 9 | Gene editing | Precision medicine
Classification: LCC RB155.8 .C75 2026 (print) | LCC RB155.8 (ebook)
LC record available at https://lccn.loc.gov/2025052975
LC ebook record available at https://lccn.loc.gov/2025052976

British Cataloguing in Publication Data
A Cataloguing in Publication record for this book is available from the British Library.

Table of Contents

Detailed Table of Contents

CRISPR/Cas9 technology has emerged as a transformative tool in revolutionizing target discovery, disease modeling, and therapeutic interventions. Conventional CAR-T therapy achieves overall response rates of 70-85%, compared with 20-40% with standard chemotherapy. The technology shows particular promise in solid tumors including breast, lung, colorectal, hepatocellular, prostate, and pancreatic cancers. China, Asian nations have made substantial CRISPR contributions with mechanisms ranging from oncogene disruption to CAR-T cell enhancement. Critical ethical frameworks addressing informed consent, germline editing, off-target effects, and global access equity have been established. The 2021 WHO framework provides a normative foundation but lacks enforcement mechanisms. The interdisciplinary convergence of CRISPR with artificial intelligence, multi-omics profiling, and systems biology promises next-generation precision oncology.

The development of CRISPR-Cas9 has dramatically changed the field of genome engineering and its utilization in epigenome editing could have profound implications for precision medicine.This ability to precisely target epigenetic marks, including DNA methylation, histone modifications, and chromatin remodeling, allows for exciting therapeutic developments in cancer, neurodegenerative and autoimmune diseases. New developments such as CRISPRi, CRISPRa, and reversible CRISPR

medicine. Since its discovery as a component of the bacterial immune defence system, CRISPR-Cas9 has quickly been modified for use in therapeutic use, most notably in the fight against genetic disease problems. The essential mechanics of CRISPR-Cas9, along with its innovative potential in the treatment of complicated and monogenic genetic illnesses, are discussed in the manuscript. The author has gained insights into effective preclinical and clinical uses, such as the management of sickle cell anaemia, Duchenne muscular dystrophy, and cystic fibrosis, as well as new developments like base editing and prime editing. This chapter also sheds light on important topics such as off-target effects, delivery system constraints, ethical impasses, and regulatory apprehensions. Through a combination of case studies and recent developments, this chapter attempts to give a detailed overview of how CRISPR-Cas9 is changing precision medicine and developing new treatment paradigms for genetic disorders that were previously incurable.

Arthi Gunasekaran, Bharathiar University, India
Muthupandi Vanaraj, Bharathiar University, India
Maria Sneha Philip Kumar, RVS College of Nursing, India
Abidharini Jothi Dheivasikamani, Bharathiar University, India
Arthi Boro, Bharathiar University, India
Antony Prabhu Jeyabal Philomenathan, Bharathiar University, India
Alwin Robert Asirvatham, Prince Sultan Military Medical City, Ministry of Defense, Saudi Arabia
Vijaya Anand Arumugam, Bharathiar University, India

CRISPR/Cas9 technology has emerged as a revolutionary genome editing tool, offering unprecedented precision, efficiency, and versatility in manipulating genetic material. In the fields of regenerative medicine and tissue engineering, CRISPR/Cas9 holds transformative potential for correcting genetic defects, enhancing stem cell therapies, engineering tissue specific cellular functions, and developing disease models. This chapter explores the fundamental mechanisms of CRISPR/Cas9, its integration into regenerative strategies and its application in engineering functional tissues and organs. Key advancements include targeted gene editing in induced pluripotent stem cells (iPSCs), modulation of signaling pathways involved in tissue regeneration and creation of gene-edited scaffolds for organogenesis. The chapter also addresses the current challenges, such as off-target effects, ethical considerations, and regulatory hurdles, while highlighting emerging innovations and future directions for clinical translation. Overall, CRISPR/Cas9 stands at the forefront of next-generation therapeutic strategies, reshaping the landscape of personalized and regenerative medicine.

CRISPR/Cas9 technology is a cutting-edge tool for studying the genetic effects induced by exposure to environmental stressors such as ionizing radiation and heavy metals. These agents are known to cause major molecular perturbations, including DNA damage, increased oxidative stress, enzymatic alterations and inflammatory responses in the nervous system. In this context, CRISPR/Cas9 makes it possible to identify and precisely target the genes involved in the mechanisms of response or vulnerability to these exposures. Thanks to this technology, it is possible to create cellular or animal models that faithfully reproduce the alterations observed in exposed organisms, and to explore strategies for correcting or inactivating genes to mitigate toxic effects. The use of CRISPR/Cas9 in this field opens up new prospects for gaining a better understanding of the interactions between the environment and the genome, and ultimately for developing targeted therapeutic approaches to pathologies linked to pollution or radiological exposure.

CRISPR-Cas9 is a paradigm shift in the molecular diagnostics field that goes beyond its beginnings as a gene editing technology to become a radically new diagnostic platform. The review discusses the history and uses of CRISPR-based diagnostics in infectious diseases, oncology and genetic screening. CRISPR systems offer explicit, customizable, and molecular recognition features, which can be applied to create efficient, quick, and economical diagnostic systems. The combination of such technologies is one that will bring democratized, universally available molecular diagnostics, and change healthcare in resource-abundant and resource-constrained contexts, and eventually makes CRISPR a foundation of precision diagnostics.

M. M. Nidhilangelo, Pondicherry University, India
Ranjith Kumavath, Pondicherry University, India

CRISPR–Cas9, originally derived from a bacterial adaptive immune system, has rapidly evolved into a transformative platform for genome and epigenome engineering with broad applications in biology and precision medicine. Guided by single-guide RNAs, the Cas9 nuclease introduces targeted DNA modifications through cellular repair pathways. Meanwhile, engineered variants, such as high-fidelity nucleases, nickases, and nuclease-deficient dCas9, extend their functionality to precise gene regulation, epigenetic modulation, RNA editing, and live-cell genomic imaging. dCas9-based effectors enable locus-specific control of transcription by directing DNA methylation, histone modifications, or chromatin remodelling, offering reversible and non-permanent alternatives to conventional epigenetic drugs. *In vivo* models, including somatically engineered mouse models and xenografts, enable physiologically relevant studies of tumorigenesis and therapeutic response, while base and prime editors offer safer, DNA break–free approaches for the precision correction of pathogenic variants. Clinically, CRISPR is driving innovation in adoptive immunotherapies, regenerative medicine, and targeted delivery systems, with nanoparticle- and ligand-modified carriers improving tissue-specific targeting and reducing systemic toxicity. Despite these advances, major challenges remain, including off-target activity, inefficient and tissue-restricted delivery, instability of epigenetic edits, and ethical concerns surrounding germline modification and long-term safety. Ongoing innovations in CRISPR engineering, delivery technologies, and regulatory oversight are addressing these barriers. Collectively, CRISPR-based platforms represent a cornerstone of next-generation personalised therapeutics, offering programmable, scalable, and clinically relevant strategies to understand and treat human disease.

The powerhouse of cells, mitochondria, play a crucial role in homeostasis, energy metabolism, apoptosis and stress response. Mitochondria possess a separate set of DNA (mtDNA), which is associated with a wide range of metabolic, degenerative and age-related disorders. The CRISPR-Cas 9 system revolutionised medicine with its discoveries in the nucleus; however, its application in mitochondria remains constrained due to the challenges of delivery, due to the double membrane and repair mechanisms, since mitochondria lack the conventional repair mechanisms, unlike the nucleus. These limitations directed CRISPR-independent and CRISPR-inspired strategies aiming at overcoming the hurdles. This chapter discusses the obstacles in achieving precise mitochondrial genome editing in order to shed light on new therapeutic possibilities.

The integration of CRISPR/Cas9 genome editing and nanotechnology is transforming precision medicine through the targeted, efficient, and safe delivery of gene-editing agents. With its great promise in the treatment of genetic disorders, the clinical application of CRISPR/Cas9 is hampered by delivery issues such as degradation, immunogenicity, and off-target effects. This chapter discusses the evolution and use of nano-delivery systems like lipid nanoparticles, polymeric carriers, dendrimers, gold nanoparticles, and exosomes to deliver CRISPR/Cas9 elements across biological barriers in a highly specific manner. Focus is given to nanoformulation strategies that maximize cellular uptake, endosomal escape, and nuclear localization with minimal systemic toxicity. Case studies of preclinical and clinical progress in nano-delivered CRISPR therapies for cancer, genetic diseases, and viral diseases are also presented. The chapter is wrapped up with perspectives on regulatory affairs and future directions on combining nanocarriers with genome editing for personalized medicine.

Preface

INTRODUCTION

The advent of CRISPR-Cas9 technology has marked a decisive turning point in the evolution of precision medicine, reshaping how genetic information is interrogated, modified, and translated into therapeutic action. What began as a bacterial adaptive immune mechanism has rapidly matured into a programmable and scalable genome-editing platform with unprecedented implications for biomedical research and clinical practice. As scientific understanding deepens and technical capabilities expand, CRISPR-based technologies are increasingly positioned at the core of next-generation diagnostics, therapeutics, and disease modeling.

Precision medicine demands approaches that account for genetic heterogeneity, disease complexity, and individual variability in treatment response. In this context, CRISPR-Cas9 offers a unifying technological framework—enabling precise gene disruption, correction, regulation, and epigenetic modulation across diverse biological systems. From oncology and cardiovascular medicine to infectious diseases and regenerative therapies, CRISPR technologies are redefining how diseases are understood at the molecular level and how interventions are designed with specificity and intent.

At the same time, the rapid translation of CRISPR-Cas9 into medical applications raises significant technical, ethical, regulatory, and societal questions. Challenges related to off-target effects, delivery efficiency, long-term safety, equitable access, and governance continue to shape both research priorities and clinical feasibility. This reference volume is conceived in response to these opportunities and constraints, bringing together interdisciplinary perspectives that critically examine the scientific foundations, translational progress, and future trajectory of CRISPR-Cas9 technology within precision medicine.

CHAPTER OVERVIEW

Chapter 1 introduces the broader landscape of genome editing technologies in precision medicine, situating CRISPR-Cas systems alongside emerging tools such as base and prime editing. The chapter critically evaluates recent innovations while addressing persistent challenges related to delivery, specificity, ethical oversight, and global accessibility.

Chapter 2 focuses on the transformative role of CRISPR-Cas9 in precision cancer medicine, highlighting its applications in target discovery, disease modeling, and therapeutic development. Particular attention is given to solid tumors, CAR-T cell engineering, and the convergence of CRISPR with artificial intelligence and multi-omics approaches.

Chapter 3 explores the use of CRISPR-Cas9 in unlocking immune responses against infectious pathogens, with an emphasis on epigenome editing strategies. Advances such as CRISPRi and CRISPRa are discussed alongside technical limitations, safety considerations, and ethical boundaries associated with heritable epigenetic modifications.

Chapter 4 examines CRISPR-Cas9 applications in precision cardiovascular medicine, detailing its role in disease modeling, mutation correction, and systems biology. The chapter highlights how integration with AI and non-permanent gene regulation tools is expanding therapeutic possibilities while addressing safety and translational barriers.

Chapter 5 addresses the application of CRISPR-Cas9 in combating genetic disorders, with coverage of both monogenic and complex diseases. Through case studies and emerging clinical data, the chapter evaluates therapeutic successes while discussing delivery constraints, off-target risks, and regulatory challenges.

Chapter 6 focuses on regenerative medicine and tissue engineering, illustrating how CRISPR-Cas9 enables stem cell modification, pathway regulation, and the engineering of functional tissues. The chapter underscores the technology's potential to advance organ regeneration and personalized repair strategies.

Chapter 7 applies CRISPR-Cas9 to the study of genetic effects induced by radiation and heavy metal exposure. By enabling precise gene targeting and model development, this chapter highlights CRISPR's value in environmental health research and mitigation of exposure-related pathologies.

Chapter 8 shifts attention to medical diagnostics, showcasing CRISPR-based platforms as rapid, sensitive, and cost-effective tools for molecular detection in infectious diseases, oncology, and genetic screening, with implications for global health equity.

Chapter 9 delves deeper into epigenome editing in precision medicine, examining in vivo models, CRISPR-derived editors, and ongoing efforts to improve delivery, stability, and ethical governance in programmable gene regulation.

Chapter 10 addresses the frontier of mitochondrial genome editing, outlining the unique biological barriers associated with mtDNA and reviewing CRISPR-independent and CRISPR-inspired strategies aimed at targeting the cellular powerhouse.

Chapter 11 concludes the technical scope of the volume by focusing on nano-delivery systems for CRISPR-Cas9. This chapter surveys lipid, polymeric, and exosome-based carriers, emphasizing strategies that enhance specificity, reduce toxicity, and accelerate clinical translation.

CONCLUSION

Collectively, the chapters in this volume underscore the extraordinary versatility of CRISPR-Cas9 technology as both a research instrument and a clinical catalyst. Across therapeutic domains, CRISPR-based platforms are enabling levels of biological precision that were previously unattainable, supporting a paradigm shift from generalized treatment strategies to individualized, mechanism-driven care.

Equally important, this book highlights that technological capability alone is insufficient to ensure responsible and effective implementation. Issues of safety, ethical governance, regulatory harmonization, and equitable access emerge repeatedly across chapters, emphasizing the need for coordinated global frameworks that evolve alongside scientific innovation. Precision medicine must be guided not only by accuracy, but also by accountability and inclusivity.

As CRISPR-Cas9 technologies continue to advance—integrating with artificial intelligence, nanotechnology, and systems biology—their role in shaping the future of medicine will only expand. It is our hope that this reference work will serve as a valuable resource for researchers, clinicians, policymakers, and advanced students, fostering informed dialogue and accelerating the responsible translation of CRISPR-Cas9 technologies toward improved human health outcomes.

Chapter 1
Genome Editing Technologies in Precision Medicine:
Recent Innovations and Challenges

Kazi Asraf Ali
https://orcid.org/0000-0002-1753-7504
Maulana Abul Kalam Azad University of Technology, India

Chowdhury Mobaswar Hossain
https://orcid.org/0009-0001-5610-0962
Maulana Abul Kalam Azazd University of Technology, India

Rideb Chakraborty
https://orcid.org/0009-0007-3931-8349
Bengal College of Pharmaceutical Sciences and Research, India

Amit Kotal
https://orcid.org/0009-0007-3063-2542
DmbH Institute of Medical Science, Dadpur, India

Naureen Afrose
https://orcid.org/0000-0001-8334-9053
Bengal College of Pharmaceutical Sciences and Research, India

Sourav Samanta
https://orcid.org/0009-0006-0786-4823
Maulana Abul Kalam Azazd University of Technology, India

ABSTRACT

Recent developments in genome editing technologies have revolutionized the field of genome precision medicine, offering unprecedented opportunities for genetic flaw correction, disease manipulation, and personalized treatment plans. These technologies include CRISPR-Cas systems, base editing, and prime editing, which can revolutionize treatments for monogenic disorders, cancers, and multifactorial complex diseases. However, challenges such as delivery bottlenecks, reactions, unintentional genetic changes, and ethical issues remain. The chapter emphasizes

DOI: 10.4018/979-8-3373-4857-5.ch001

the need for improved viral and non-viral vectors, specificity tests, and efficient regulatory methods. It also addresses socioeconomic and accessibility concerns for equitable access to genome editing treatments. The chapter aims to balance innovation and responsible translation, considering both preclinical and clinical advances and current issues in genome editing in precision medicine.

1. INTRODUCTION TO GENOME EDITING IN PRECISION MEDICINE

Modern medicine is no longer focused on a one-size-fits-all approach but on precision medicine that is focused on diagnostics, prognostics, and treatment of each individual patient. This paradigm recognizes the fact that genetic predisposition is at the center of disease susceptibility, progression and response to treatment. Precision medicine can be achieved by leveraging genomics, transcriptomics and proteomics to uncover the molecular pathogenesis of disease, which can be used to predict, prevent and treat disease uniquely. The new landscape has turned out to be genome-editing technologies. They enable accurate alteration of DNA in living organisms and cells. These tools can not only repair pathogenic mutations to their origin, but they can also verify disease-relevant genes and pathways by functional genomics to connect genetic diagnosis and targeted therapy. (Giamarellos-Bourboulis et al., 2024),(Chakraborty et al., 2024). The direction of such technologies has been characterized by increasing accuracy, effectiveness and accessibility. Zinc -Finger Nucleases (ZFNs) and Transcription Activator -Like Effector Nucleases (TALENs) showed that the targeted genomic modification became possible, introducing double strand breaks (DSBs) at a defined locus, which were then repaired using intrinsic cellular processes, including non-homologous end joining (NHEJ) or homology directed repair (HDR) shown in Figure 1(Afrose, Chakraborty, Bhowmick, et al., 2025; Chakraborty, Afrose, et al., 2025).Despite their groundbreaking step over previous, less specific methodologies, and the success of ZFNs and TALENs in the initial clinical uses, extensive use was limited by major drawbacks. The most significant of these were the complexity and high cost of protein engineering each new target and low predictable efficacy which hindered scalability to high throughput projects. With the identification and engineering of the CRISPR-Cas system, a prokaryote adaptive immune system re-purposed to specific genetic control in eukaryotic cells, the genome editing landscape was forever changed. The CRISPRCas9 system, specifically, has become groundbreaking due to its simplicity and range; the specificity of the targeting is defined by a short programmable guide RNA, which can be customized to match any sequence of interest in DNA, so no sophisticated protein engineering technologies are required.The high rate of CRISPR-based systems has since resulted in a family of potent derivatives which surpass the functionality of the first Cas9 nuclease. High-fidelity versions of Cas9 have been designed to reduce off-target effects, which is a critical safety issue in therapeutic applications. There have also been new Cas orthologs like Cas12 and Cas13 to expand the targeting repertoire to include double(strand) DNA, single(strand) DNA, and even RNA, and thus open up new possibilities in diagnostics and therapeutic modulation without the need to alter the genome(Ali, Choudhuri, Roy, et al., 2025; Jung et al., 2024).

It is interesting to note that the creation of base editors (BEs) and prime editors (PEs) has overcome a significant drawback of traditional CRISPR-Cas-based systems, namely the reliance on DSBs and repair pathways that are highly error-promoting. Base editors enable the irreversible conversion of a single base pair of DNA into another without generating a DSB and significantly increases the efficiency of editing and reduces the formation of indels. Prime editors, in turn, are a "search-and-replace" genomic tool that can install all 12 possible base-to-base conversions, small insertions and deletions, again, in an independent manner of DSB. This spectrum, starting with ZFNs and TALENs and the more powerful CRISPR-Cas toolkit, has converted the existing armamentarium of precision medicine, leading researchers and clinicians to no longer merely interfere with genes but instead provides them with a scalpel to perform genetic surgery to an even greater degree of precision, on a nucleotide-by-nucleotide level(Hillary & Ceasar, 2024).

Figure 1. The CRISPR-Cas9 system, TALEN, ZFN in genome editing Reproduced from (Khoshandam et al., 2024) under Creative Commons Attribution (CCBY 4.0).

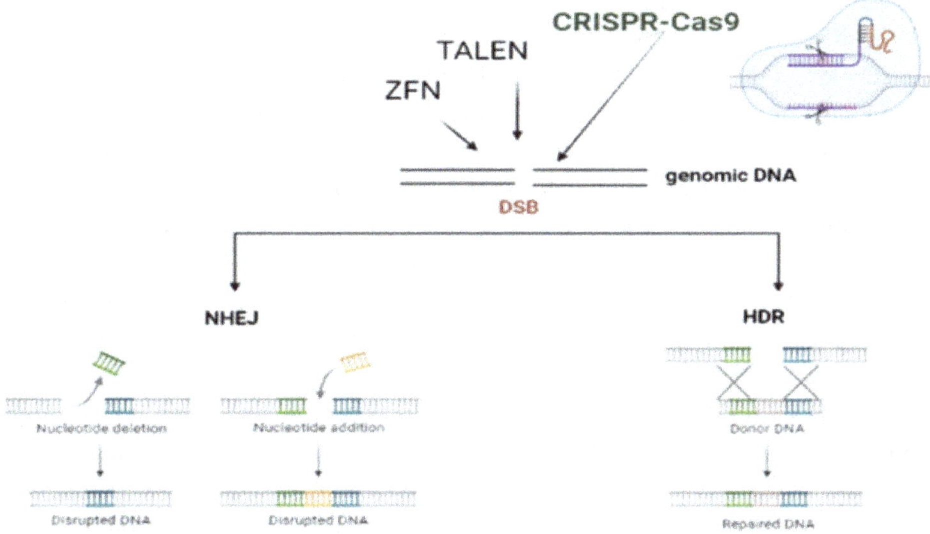

2. CRISPR-CAS SYSTEMS: MECHANISMS AND ADVANCEMENTS

The discovery of Clustered Regularly Interspaced Short Palindromic Repeats, also known as CRISPR, and CRISPR-associated (Cas) proteins have heralded a paradigm shift in genome-wide targeting technology, replacing the previously tedious, protein-based modalities of targeting, e.g. zinc -finger nucleases (ZFNs) and transcription-activator-like effector nucleases (TALENs), with a versatile, RNA-guided approach.The core of this revolution is the CRISPR-Cas9 complex, a molecular device that has been engineered out of a bacterial adaptive immune system and which can be thought of as a molecular scalpel of considerable precision as shown in Figure 2(Chattopadhyay et al., 2025).The basic principle of this system is based on a Cas nuclease - most commonly Streptococcus pyogenes Cas9 - directed by a single-guide RNA (sgRNA). The sgRNA is bound by complementary genome base pairing to a specific location in the genome that is directly upstream of a protospacer adjacent motif (PAM).After binding, the Cas9 endonuclease creates a two-strand break (DSB) in the DNA helix. This break is then repaired by the cell intrinsic repair processes; either by the error-prone non-homologous end joining (NHEJ) pathway, which normally forms an insertion or deletion (indel) that inactivates a gene, or by the homology-directed repair (HDR) system, which is high-fidelity and allows the insertion of a donor DNA template without error(Z. Li & Zhang, 2025). Although incredible ease of design and use is granted by CRISPR-Cas9, its reliance on the formation of DSBs has inherent risks, including, but not limited to p53-mediated cytotoxicity, chromosomal translocations and highly random off-target activity in loci with sequence homology to the sgRNA.

To address these limitations, meaningful advances in the field have been made through the rational design of high-fidelity Cas9 derivatives, such as eSpCas9 and SpCas9 -HF1, which have reduced nonspecific DNA binding, and by the discovery of orthologous nucleases, such as SaCas9 and Nme2Cas9, with different PAM compatibilities, and therefore associated with broader genomic territories amenable to editing(Hwang et al., 2025).To add additional functionality to the CRISPR repertoire, the introduction of Cas12, formerly known as Cpf1, and Cas13 effectors provided significant functional separations to the repertoire.One such example is the Cas12a which is a solitary RNA-guided nuclease that self-processes its own CRISPR RNA (crRNA) array, enabling multiplexed genome editing with a single transcript(J.-Y. Shi et al., 2025). In addition, it produces staggered DNA ends with the help of a DSB, which could potentially be useful to certain protocols of HDR. Importantly, Cas12a shows collateral trans-cleavage of single-stranded DNA on target-engagement- a property that has been used to develop ultra-sensitive diagnostic tests including SHERLOCK and DETECTR(Lafi et al., 2025).Conversely, Cas13 targets RNA targets, whereby its HEPN nuclease domain is used to cleave specific messenger RNA (mRNA) sequences, thus, offering a reversible gene knock-down strategy without any permanent genomic modification. Cas13 also has collateral RNase activity, a feature which has been utilized to detect RNA.This means that the practical uses of these nucleases are highly specialized: CRISPR-Cas9 remains the most common tool of DNA knock-out and knock-in projects in the fields of fundamental research and therapeutics, including ex vivo editing of hematopoietic stem cells to treat sickle-cell disease, whereas Cas12 variants perform well in multiplexed editing and diagnostic applications. Cas13 is a promising platform that can be used to regulate transcript-level gene expression, probe the nature of RNA biology, and develop antiviral responses.

Perhaps the most paradigm-shifting development in precision has been the development of technologies that allow base editing and prime editing, reconstituting one nucleobase into another, without creating a DSB. Base editors incorporate a catalytically weakened version of Cas nuclease - a nick that activates one of the strands of DNA - with a deaminase enzyme. Base editors operate within a defined range of the sgRNAbinding locus, eliminating the requirement of a DSB and an exogenous donor template, which greatly improves the fidelity and efficiency of a single-nucleotide correction and also reduces the indel by-products.However, base editors are limited by their limited editing window as well as by their inability to determine all possible base transversions or to make insertions or removals.In order to overcome such constraints, the idea of prime editing was developed as an inherently versatile search-and-replace modality.A prime editing system consists of a Cas9 nick controlled to a reverse transcriptase domain, and a prime editing guide RNA (pegRNA).The pegRNA is a separate target locus and the target genetic sequence alteration(Mikhaylova et al., 2024).

The system implants a nick on one strand of DNA and the pegRNA extension is used as a template by the reverse transcriptase to produce and insert the modified sequence directly into the chromosomal DNA. Theoretically, this combined platform has the potential to induce all the twelve possible base-to-base transformations, as well as small deletions and insertions, and do so with high accuracy and significantly low off-target activity. The logical advancement, CRISPR-Cas9 to base editing then prime editing, represents a calculated tactic to increase the precision, safety, and scale of genome editing, replacing a disruptive versatile cleaver with a fine pencil and eraser, and consequently expanding the therapeutic prospects to correct the high majority of the pathogenic alleles of the catalog during the context of precision medicine.

Figure 2. The three stages/mechanism of CRISPR-Cas immunity Reproduced from (Yang et al., 2024) under Creative Commons Attribution (CCBY 4.0).

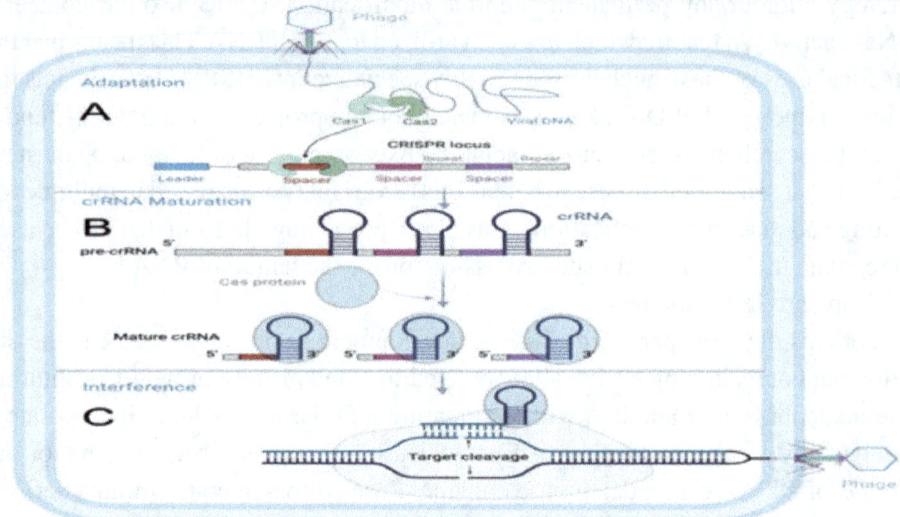

3. DELIVERY SYSTEMS FOR GENOME EDITING

The future ability of genome editing technologies to propel precision medicine essentially depends on the development of safe and effective delivery platforms that can deliver macromolecular editing machinery, including CRISPR-Cas nucleases, base editors, and prime editors, into the nuclei of target cells. This challenge of delivery is one of the key bottlenecks that determines the therapeutic efficacy, specificity and safety profile of any in vivo or ex vivo intervention. The current situation in delivery is largely subdivided into viral and non-viral vectors, which have different strengths and weaknesses. Viral vectors, and specifically adeno-associated viruses (AAVs) are the most commonly used systems due to their high transduction efficiency, long-term transgene expression, and relatively good safety profile compared to other viral vectors (adenoviruses or lentiviruses). AAVs in particular are particularly useful in vivo gene therapy because they can be engineered with tissue-specific serotype capsids such as liver (AAV8, AAV9), retina (AAV2, AAV5) or central nervous system (AAV9, AAV-PHP.eB)(Volodina & Smirnikhina, 2025). The small size of AAVs (near 4.7 kb) is however a limitation to their usefulness in expressing larger Cas orthologs (e.g. Cas12, Cas13) or more than one genetic element without using complex splitting schemes. In addition, the presence of preexisting humoral immunity against common serotypes of AAV in a significant percentage of the human population may neutralise the infectious particle and reduce therapeutic effectiveness, and capsid-specific T-cell responses can induce hepatotoxicity and transduced cell killing. Genomic integration, though very low, is also possible and, again, the possibility of genotoxicity due to the presence of two-strand breaks caused by a couple of nucleases like Cas9 also provides the necessity to take a stricter approach. Conversely, the non-viral delivery modalities provide very promising alternatives that would avoid the majority of immunological issues and are not limited by the size of the payload. The non-viral platform that has become one of the most popular is lipid nanoparticles (LNPs), which have gained popularity due to their successful work concerning mRNA-based vaccines against COVID-19(Ali, Choudhuri, Roy, et al., 2025; Chakraborty, Kannan, et al., 2025). LNPs can entrap and protect Cas9 mRNA and single-guide RNA (sgRNA) complex and, therefore, enable easy endocytosis and endosomal escape into the cytoplasm, where the mRNA can be translated to functional Cas9 protein. LNP formulations allow co-delivery of different RNA species and surface functionalisation with ligands to improve targeting due to their modularity. Other non-viral physical approaches, like electroporation, have proven very useful in ex vivo uses, where cells like hematopoietic stem cells or T-cells are cloned out of the patient, edited in vitro and returned(Afrose, Chakraborty, & Rajendran, 2025).

4. THERAPEUTIC APPLICATIONS IN GENETIC DISORDERS

Among the numerous therapeutic uses of genome editing, perhaps one of the most developed areas and the most promising one is the treatment of monogenic disorders. CRISPR-Cas9 is proving to have impressive ability to treat the underlying genetic cause of diseases, including sickle cell anemia, beta-thalassemia, cystic fibrosis, and a number of therapies are now coming out of the laboratory and into clinical practice(Ahmed et al., 2025; Lu et al., 2023). In the case of sickle cell anemia and beta-thalassemia which are mutations in the beta-globin gene (HBB), disrupting an erythroid-specific enhancer of the BCL11A gene, a potent repressor of fetal hemoglobin (HbF) has been a dominant approach(Hanna et al., 2023; H.-Z. Wang & He, 2024). This is because the defective adult hemoglobin is compensated by reintroducing HbF production and effectively relieving the pathophysiology of the diseases. This ex vivo model presupposes removal of hematopoietic stem cells (HSCs) of a patient, editing them with CRISPR, and reintroducing the edited samples into a patient subsequent to conditioning chemotherapy. In another case, such as cystic fibrosis which is caused by CFTR gene mutations, the approach is more direct and it would attempt to correct the local mutation in the lung epithelium, although it is very challenging to deliver vectors to the target cells(Plasschaert et al., 2024; Raghunath et al., 2025).

The progress in this field is best illustrated by the clinical trial updates and emerging success stories. The most groundbreaking advancement to date is the first regulatory approvals of CRISPR-based therapies. Exa-cel (exagamglogene autotemcel), developed by Vertex Pharmaceuticals(Butt et al., 2025; Greco et al., 2024; Kingwell, 2023) and CRISPR Therapeutics for both sickle cell disease and transfusion-dependent beta-thalassemia (TDT), has received approval in the UK, US, and other regions(Rós et al., 2023). The results of clinical trials have been phenomenal with most of patients with TDT becoming transfusion independent and all patients with sickle cell disease being able to spend long time without vaso-occlusive crisis, following their treatment. Such results are a working remedy to numerous people. Although advancement in cystic fibrosis is earlier, in preclinical and early clinical trials, the achievement with hemoglobinopathies is a high-level evidence-of-concept. It confirms the whole platform of *ex-vivo* HSC editing and opens the way to the application of such strategies to other genetic disorders. The summary of the main differences in curative measures and clinical conditions of these disorders is presented in the Table 1 below.

Table 1. CRISPR-Based Therapeutic Approaches for Select Genetic Disorders

Genetic Disorder	Target Gene / Locus	Therapeutic Strategy	Delivery Method	Clinical Trial Status & Key Outcomes	Reference
Sickle Cell Anemia & Beta-Thalassemia	*BCL11A* enhancer	Disruption of enhancer to reactivate fetal hemoglobin (HbF)	Ex vivo editing of patient's hematopoietic stem cells (HSCs)	**Approved (e.g., Exa-cel).** Landmark trials show transfusion independence in >90% of TDT patients and resolution of vaso-occlusive crises in SCD patients.	(Ahmed et al., 2025; Zeng et al., 2023)
Cystic Fibrosis	*CFTR* gene	Correction of specific mutations (e.g., F508del) to restore ion channel function	In vivo delivery to lung epithelial cells (e.g., via viral vectors)	**Preclinical / Early Clinical.** Significant challenges with efficient and safe delivery to the airways. Promising preclinical data but not yet in late-stage trials.	(Parisi et al., 2025)

These achievements, though, come with some serious challenges, which highlight the intricacies of precision medicine. The existing *ex-vivo* treatments of hemoglobinopathies are both astronomically costly and logistically challenging and entail myeloablative conditioning chemotherapy that has its own risks and renders the treatment unaffordable to a large part of the population. Moreover, long term safety and permanency of the edited cell population are still being tracked. In the context of *in-vivo* applications such as cystic fibrosis, the challenge of delivery of the CRISPR machinery to the necessary tissues and reduction of the off-target effects still remains the main obstacles(Dölarslan, 2023; Hussen et al., 2024).

5. CANCER IMMUNOTHERAPY AND GENOME EDITING

Continuing the success of genome editing in monogenic disease, genome editing technologies are transforming cancer immunotherapy, specifically in the production of chimeric antigen receptor (CAR)-T cells. CAR-T treatment has proven to be an effective form of treatment of selected hematological malignancies, but the general application of the approach has been limited by effects, safety, and manufacturability issues(Abbasi et al., 2023; Capolla et al., 2025; Dagar et al., 2023; Dasgupta et al., 2024). CRISPR-Cas9 is being implemented so as to overcome these challenges by developing more potent, persistent, and safer off-the-shelf cellular therapies. The conventional CAR-T cells are produced with viral vectors as they introduce the CAR gene at random locations in T cells of the patient(Dimitri et al., 2022; Kampouri et al., 2024; Lukjanov et al., 2021; Parhi et al., 2025). CRISPR can enable specific, accurate and directed insertion of the CAR transgene into defined genomic safe harbors, and this may result in more predictable and robust expression of CAR. Moreover, endogenous T-cell receptor (TCR) genes and human leukocyte antigen (HLA) class I genes have their genes disrupted by CRISPR to produce universal allogeneic CAR-T cells in donor T cells (T. Lei et al., 2024; W. Li et al., 2022).

In addition to the manipulation of immune cells directly, genome editing has direct therapeutic application in the genetic causes of cancer itself. This strategy is aimed at inactivating oncogenes that facilitate tumor proliferation or fixing mutated tumor suppressor genes that should check cell proliferation. This approach is a more direct type of genetic intervention of cancer, although it is still in the preclinical and early clinical exploration stage. As an example, CRISPR systems may be engineered to silicate oncogenes, such as MYC or KRAS, which are infamously challenging to treat with traditional small-molecule therapeutics(Theardy et al., 2025). Alternatively, one can knock out the fusion oncogene BCR-ABL in chronic myeloid leukemia with the technology or, conversely, precisely correct a mutation in a key tumor suppressor, such as TP53(Mojtahedi et al., 2021). The key issue to this in vivo treatment modality is creating safe and effective delivery vehicles, including non-viral nanoparticles or viral vectors, which can tumor cell-specifically target and spares healthy tissues to reduce off-tumor effects. These two major uses of genome editing in cancer therapy are compared in the Table 2.

Table 2. Applications of Genome Editing in Cancer Immunotherapy and Direct Oncogene Targeting

Application Area	Primary Goal	CRISPR-Based Strategy	Key Challenges	References
CAR-T Cell Engineering	Enhance efficacy and safety of adoptive cell therapy	1. **Precise CAR Integration:** Targeted insertion into safe harbor loci (e.g., *TRAC* locus). 2. **Universal Donor Cells:** Knockout of *TCR* and *HLA* genes to prevent GvHD and rejection. 3. **Enhanced Potency:** Knockout of immune checkpoint genes (e.g., PD-1).	Enhance efficacy and safety of adoptive cell therapy Ensuring complete knockout of target genes to prevent GvHD in allogen managing potential for increased off-tumor toxicity due to enhanced potency.	(Dimitri et al., 2022; Y.-R. Li et al., 2025)
Targeting Oncogenes/Tumor Suppressors	Directly disrupt cancer-driving genetic alterations	1. **Oncogene Inactivation:** Disruption of driver oncogenes (e.g., *KRAS, MYC*). 2. **Tumor Suppressor Reactivation:** Correction of inactivating mutations in genes like *TP53*.	Achieving high-efficiency, tumor-specific delivery in vivo; ensuring absolute specificity to avoid unintended oncogene activation or tumor suppressor disruption in healthy cells.	(Chillón-Pino et al., 2024)

6. EPIGENOME EDITING FOR PRECISION MEDICINE

With CRISPR-based technologies, the concept of permanent cut-off of DNA has been extended to another less invasive and reversible type of control: epigenome editing. This method involves the use of a catalytically dead Cas9 (dCas9) which has the capacity to bind to particular DNA sequences directed by an RNA but cannot cleave the DNA double strand(Sapozhnikov & Szyf, 2021). When dCas9 is fused with different effector domains, it is now possible to target epigenetic modifiers to specific genomic loci to modify the chromatin state and to regulate gene transcription without altering the underlying DNA sequence(Cai et al., 2023). This enables the activation or silencing of genes of interest through writing or erasing of epigenetic marks including DNA methylation or histone modifications. The main benefit of this technology is that it is reversible, and it is less likely to induce two-strand breaks that would be linked to the possibility of causing unwanted on and off-target mutation in ordinary CRISPR-Cas9 editing. This has rendered epigenome editing a very appealing technology to treat multifactorial diseases that are hard to treat since long-term, tunable control of gene networks is more therapeutically significant than an irreversible disturbance of genes or their fixation.

Engineered toehold (eToehold) switches and programmable RNA switches (including engineered toehold) switches are an advanced technology in synthetic biology, enabling the conditional expression of therapeutic proteins to a cell-type-specific manner. These riboregulators are carefully crafted to stay silent until they selectively bind to an endogenous RNA biomarker- like a microRNA or an over-expressed mRNA that is included in specific target cells- and therefore induction of the conformational change, which liberates the start of translation. eToehold switches can also downregulate the expression of therapeutic proteins, such as CRISPR effectors, cytokines, or chimeric antigen receptors, only in diseased cell populations but not the healthy tissues by exploiting intracellular RNA signatures that are disease-specific, e.g., tumor-specific mRNAs. Such spatial accuracy is used to reduce off-target effects and improve the safety characteristics of both the gene and cell therapies. In its turn, the given method presents a highly adaptable, delivery-friendly approach towards precision medical treatment that does not require permanent genomic modification as a means to introduce and eliminate in effect the advantages of targeted therapy with a higher safety factor.

The potential of epigenome editing in precision medicine is great potential in polygenic and non-Mendelian diseases, including type 2 diabetes and neurological disorders, in which the pathogenesis of the disease is controlled by a complex of genes dysregulated, not one dysregulated gene(Sutherland et al., 2024). As an example, in type 2 diabetes, dCas9 systems can be designed to inhibit or activate the expression of insulin-sensitizing genes in a transient manner or inhibit gluconeogenesis-controlling genes in the liver, which provides a new method to restore metabolic homeostasis(Hu et al., 2025; Y. Li et al., 2023; Toma et al., 2021). A major problem in neurological diseases such as the Alzheimer disease, Parkinson disease, or Huntington disease is the build up of toxic proteins. The editors of the epigenome may be implemented to repress the expression of the genes encoding these pathogenic proteins, which might slow down or prevent the development of the disease. Moreover, the capability to target the expression of the most important neurotransmitter receptors or synaptic proteins has also become a new form of therapeutic intervention in case of neurodevelopmental disorders or psychiatric conditions associated with the altered neuronal activity. The Table 3 gives a comparison between the key tenets of epigenome editing and the older method of genome editing using nuclease.

Table 3. Comparison of Nuclease-Based Genome Editing and Epigenome Editing

Feature	Nuclease-Based Genome Editing (e.g., CRISPR-Cas9)	Epigenome Editing (e.g., CRISPR-dCas9)	References
Core Mechanism	Creates double-strand breaks in DNA, relying on cellular repair pathways to introduce permanent changes (knockouts or precise edits).	Recruits epigenetic modifiers to alter gene expression (activation or repression) without changing the DNA sequence.	(McCutcheon, Rohm, Iglesias, & Gersbach, 2024)
Permanence	Permanent and irreversible.	Typically reversible and tunable, allowing for dynamic control.	(Lau & Suh, 2017)
Primary Risk	Potential for off-target DNA cleavage, leading to unintended mutations and genomic instability.	Risk of off-target epigenetic modifications at unintended sites, potentially causing long-lasting but theoretically reversible changes in gene expression.	(Fiumara et al., 2024)
Ideal Application	Correcting or disrupting causative genes in monogenic disorders (e.g., sickle cell anemia).	Modulating gene networks in complex diseases (e.g., diabetes, cancer, neurological disorders) where transient regulation is beneficial.	(Reddy et al., 2020)

The development of genome-editing therapies encompasses the translational path of an advanced clinical trial phase through to the expanding preclinical research. Continuous and validated studies of monogenic hematological diseases like sickle cell disease and 2 allele thalassemia, represent the present and the most crucial clinical milestones. Such experiments have revealed ex vivo CRISPR13 2 elicitation of in the disruption of BCL11A enhancer in autologous blood stems (hematopoietic stem cells, HSCs) to produce enduring fetal hemoglobin (HbF) induction, transfusion independence and disappearance of vaso-occlusive emergencies. At the same time, in-vivo methods of conditions like familial hypercholesterolemia are also getting developed clinically. These include lipid based nanoparticles (LNPs) delivery of CRISPR based systems to PCSK9 gene in hepatocytes, which will have a long-term effect of reducing LDL cholesterol. Outside these clinical front-runners, the clinical horizon is being broadened to other organ systems by extensive preclinical research. The editors are being deployed in liver to treat metabolic diseases including phenylketonuria and Wilson disease. In ocular diseases, subretinal delivery of AAV vectors expressing CRISPR components is promising in preclinical models of inherited retinal dystrophies (e.g., Leber congenital amaurosis), with CRISPR components targeted at different genes (e.g., CEP290) fixed precisely. The central nervous system is a well-defended but highly sought-after border. In neurodegenerative disease models, e.g., Huntington and Alzheimer, preclinical work is underway to silence mutant alleles or regulate the expression of risk-factor genes, by both in-vivo and ex-vivo approaches. In these studies, novel AAV serotypes and engineered capsids are used in order to attain efficient penetration of the blood-brain barrier and neuronal transduction. This hierarchy of pipeline, involving the demonstrations of efficacy in current human trials of accessible targets to the needs to solve delivery issues to complex solid tissues in preclinical models, all define a step-by-step and evolving map of the future of in-vivo and ex-vivo genomic medicine. Though the translation of epigenome editing into clinical therapies has some significant potential, there are unique challenges encountered. The effectiveness and permanency of the epigenetic modifications are to be regulated with utmost care, since the natural renewal of cells may cause a slow restoration of the initial condition of gene expression that will require repeated manipulations(Lupan et al., 2024). On the other hand, it is also necessary not to make the modifications too persistent because this will be a matter of safety. Delivery of the comparatively huge dCas9-effector fusion proteins *in-vivo* is a major challenge and sophisticated viral or non-viral vectors that can access the target tissues including the brain is required(Chuang et al., 2021). Lastly, long-term effects of specific epigenetic manipulations remain largely unexplored, which is why the request to conduct extensive preclinical research is vital.

7. ETHICAL AND REGULATORY CHALLENGES

Genome editing brings ethical concerns, particularly when it involves the germline, because the changes can be inherited and may lead to wider social consequences(Greenfield, 2021). Countries have different laws and regulations, but most of them limit germline editing while strictly regulating somatic editing(Baylis et al., 2020).

7.1 Germline Editing Controversies

Because germline genome editing can alter the human genome in ways that are inheritable, it has generated significant ethical and societal discussion(Coller, 2019; Segers, 2023). In 2018, the He Jiankui case where he claimed to bring the world the first CRISPR-modified babies was a breakwater case. The international outcry that ensued underpinned the urgent need to have a stronger ethical regulation. The most critical issues are based on the fact that no longitudinal safety data exists, the possibility of unexpected hereditary genomic events, and the ethical dilemmas of hereditary modifications [65]. Moreover, the manipulation of the germ-line erodes the distinction between treatment and improvement, thus generating concerns on eugenics and increasing social inequalities. (Gumer, 2019), (Farrell, 2019; Jacob O. Arowosegbe, 2024).

7.2 Global Regulatory Frameworks and Guidelines

The legal and ethical paradigms that are divergent in case of genome editing are the main determinants of the regulatory environment, which varies significantly across jurisdictions. Somatic genomic modification regulations in a majority of states should be pursued under investigational new drug (IND) protocols; however, the United States Food and Drug Administration (FDA) has clearly prohibited clinical use of germline modifications. Similarly, the European Union, under the direction of the Oviedo Convention, and under the supervision of the European Medicines Agency (EMA), has a very strict ban on hereditary genetic engineering. The scandalous situation of the birth of He Jiankui in 2018 led to the introduction of more stringent legal and bioethical safeguards by the authorities of China, with harsh penalties against any unapproved genome-editing operations. In retaliation, a number of non-governmental organizations have demanded government systems that uphold transparency, citizen consultations, and fairness. Among them, the prominent ones are the World Health Organization (WHO) and the International Commission on the Clinical Use of Human Germline Genome Editing, which suggested guidelines that emphasize the need to include a wide range of people in the dialogue and promote justice-centered policies. (Feeney et al., 2021), (Grant, 2016; Sanzo, 2022), (Almqvist & P.R. Romano, 2020; Faltus & De Miguel Beriain, 2023) (R. Lei & Qiu, 2020), (Marchant, 2021). However, the absence of a single worldwide regulating body (Table 4) makes it challenging to harmonize moral principles and stop researchers from taking advantage of variations in national legislation.

Table 4. Comparison of Global Regulatory Frameworks for Genome Editing

Country	Regulatory Authority	Germline Editing Policy	Somatic Editing Policy	Recent Updates	References
United States	FDA; NIH	Not allowed in clinical practice, but research is permitted under strict ethical oversight	Allowed under Investigational New Drug (IND) approvals	Maintains strong oversight with emphasis on biosafety and informed consent	(Grant, 2016; Sanzo, 2022)
European Union	EMA; Council of Europe (Oviedo Convention)	Strictly banned for reproductive purposes	Allowed with proper regulatory authorization	Enforces robust ethical governance and public accountability	(Faltus & De Miguel Beriain, 2023)
China	Ministry of Science and Technology; National Health Commission	Restricted and criminalized after the 2018 He Jiankui case	Permitted under national bioethics regulations	2021 bioethics law introduced stricter penalties for violations	(R. Lei & Qiu, 2020)
Japan	Ministry of Health, Labour and Welfare; Science Council of Japan	Not allowed for human reproduction	Allowed for research and therapeutic purposes under guidelines	Follows voluntary ethical frameworks that are regularly updated	(Eba & Nakamura, 2022)
India	ICMR; Department of Biotechnology	Not allowed for clinical or reproductive use	Allowed for somatic gene therapy research with proper approval	2019 National Guidelines highlight safety, ethics, and public oversight	(Sivagourounadin et al., 2021)
WHO (Global)	WHO Expert Advisory Committee on Human Genome Editing	Calls for a global moratorium on clinical germline editing	Supports transparent, regulated clinical use	Developing an international governance registry and ethical frameworks	(Marchant, 2021)

8. OFF-TARGET EFFECTS AND SAFETY CONCERNS

One of the main safety issues with genome editing is still unintended genetic modifications. Improved CRISPR variations and new sequencing methods are making it simpler to identify and lessen these off-target effects, increasing clinical safety and precision(Manghwar et al., 2020; Naeem et al., 2020).

8.1 Detection and Mitigation Strategies for Unintended Edits

The possibility of off-target consequences, or unexpected alterations in DNA at sites that somewhat match the intended target, is one of the primary safety concerns with genome editing. Important genes may be affected, the genome may become unstable, or mutations may result from these mistakes(Aguilera & García-Muse, 2013; Lopes & Prasad, 2024; Williams & Sobol, 2013). ("Genome-Wide Detection and Analysis of CRISPR-Cas off-Targets," 2021; Naseri & Koffas, 2020; Tsai et al., 2017). The possibility of unintentional edits is also significantly reduced by the ability of base editors and prime editing tools to execute exact nucleotide alterations without causing double-strand breaks(Eid et al., 2018). To increase the safety of genome editing technology, detection and correction methods must be continuously improved. An overview of the several off-target evaluation techniques currently in use is given in Table 5.

Table 5. Different Off-Target Detection Methods

Methods	How It Works	Sensitivity	Key Advantages	Main Limitations	References
GUIDE-seq	Inserts double-stranded DNA tags (dsODNs) into double-strand breaks inside living cells	High	Can scan the entire genome in vivo without bias	Needs active DNA repair; only detects double-strand breaks	(L. Shi et al., 2016)
Digenome-seq	Uses Cas9/gRNA complexes to cut DNA in a test tube, then sequences it	Moderate	Simple, unbiased, genome-wide detection	Performed outside cells; lacks the natural cellular environment	(Atkins et al., 2021)
CIRCLE-seq	DNA is cut, circularized, and then sequenced	Very High	Extremely sensitive and unbiased	Done in vitro; may miss sites influenced by chromatin structure	(Tsai et al., 2017)
DISCOVER-seq	Tracks recruitment of DNA repair protein MRE11 to break sites using ChIP-seq	High	Works in living cells, unbiased, detects repair factor activity	Requires ChIP-seq expertise; may miss rare events	(Zou et al., 2023)
AID-seq	Amplifies integrated DNA tags before sequencing	High	High-throughput and unbiased	Requires integration of DNA tags into genome	(Tian et al., 2023)
Extru-seq	Detects extrachromosomal DNA circles created after Cas9 cutting	High	Sensitive, cell-based, captures extrachromosomal DNA	Might miss some off-target events	(Kwon et al., 2023)

8.2 Improving Specificity with Next-Generation CRISPR Tools

To increase target accuracy and decrease unwanted consequences, next-generation CRISPR technologies are being developed(Pickar-Oliver & Gersbach, 2019). In order to reduce off-target edits, high-fidelity Cas9 variants, including eSpCas9, SpCas9-HF1, and HypaCas9, have been designed with structural modifications that restrict nonspecific interactions with DNA(Kleinstiver et al., 2016; G. H. Wang et al., 2024). Additional nucleases, such as RNA-targeting Cas12a (Cpf1) and Cas13, provide distinct PAM requirements and cutting mechanisms, enhancing the accuracy and adaptability of genome editing(Collias & Beisel, 2021; Swarts & Jinek, 2018). Table 6 outlines some of these superior CRISPR tools and their unique features.

Table 6. Advances in Next-Generation CRISPR Tools

Variant	Key Changes	Main Advantage	Practical Applications	References
SpCas9-HF1	Introduces specific point mutations to minimize unintended activity	Very high accuracy with reduced off-target DNA interactions	Ideal for precise therapeutic gene editing	(Tang et al., 2025)
eSpCas9	Modifies positively charged amino acids	Cuts DNA more selectively with fewer off-target effects	Useful in gene therapy and functional genomics studies	(Niazi, 2023)
HypaCas9	Engineered modifications in the REC3 domain	Boosts specificity, lowering unintended edits	Great for functional genomics applications	(Nedorezova et al., 2022)
Cas12a (Cpf1)	Recognizes unique PAM sequences and makes staggered DNA cuts	Smaller and efficient, especially for editing multiple genes at once	Suitable for multiplexed gene editing	(Y. Zhang et al., 2021)
Prime Editor	Combines Cas9-nickase with a reverse transcriptase enzyme	Enables highly precise edits without causing double-strand breaks	Perfect for correcting disease-causing genes	(L. Zhang et al., 2025)
Base Editor	Uses deaminase to swap one DNA base for another	Precise editing without double-strand breaks	Useful for therapeutic gene corrections	(J. Wang et al., 2024)
Cas13	Targets RNA instead of DNA	Highly specific to RNA, avoiding DNA damage	Applied in RNA editing and diagnostic tools	(Kordyś et al., 2022)

9. EMERGING TECHNOLOGIES BEYOND CRISPR

The variety of genome engineering tools is expanding thanks to new methods like transposon-based systems, RNA editing, and retron editing. These techniques are useful additions to CRISPR technology and offer clear benefits including reversible changes and increased integration capability(Gupta & Kumar, 2025; Ramakrishnan et al., 2022).

9.1 Retron Editing, RNA Editing, and Transposon-Based Systems

A variety of more recent genome engineering platforms have surfaced in addition to CRISPR-Cas systems, each providing unique methods for accurately altering genetic material(F. Zhang, 2019). In order to create multicopy single-stranded DNA (msDNA), which serves as a repair template during homologous recombination and permits effective, multiplexed editing, retron editing uses bacterial retron elements(González-Delgado et al., 2024; Kaur & Pati, 2024). In order to prevent irreversible changes to the genome, RNA editing methods, particularly those based on adenosine deaminase acting on RNA (ADAR) enzymes, enable transient and reversible changes at the RNA level(Nishikura, 2010). Transposon-based systems, such as Sleeping Beauty and piggyBac, are attractive and safer alternatives for gene therapy because they can stably insert genes without relying on double-strand breaks(Sandoval-Villegas et al., 2021).

9.2 Comparative Advantages over Existing Methods

Compared to conventional CRISPR-based systems, these novel approaches provide a number of advantages. A thorough summary of these innovative genome editing technologies is given in Table 7. Although these methods are still at the initial stages of development, they may have a significant potential as complementary to CRISPR, thus enriching the repertoire of precision medicine and helping to eliminate the limitations of the method in specificity, safety, and ethical concerns (Pfeiffer & Stafforst, 2023; Simon et al., 2019; Di Matteo et al., 2012).

Table 7. Comprehensive Overview of Emerging Genome Editing Technologies Beyond CRISPR

Technology	How It Works	Benefits	Drawbacks	References
Retron Editing	Makes single-stranded DNA using reverse transcription	Works well and can edit many targets at once	Only works with certain DNA sequence patterns	(Cattle et al., 2025)
RNA Editing (ADAR, Cas13)	Changes specific RNA bases or cuts RNA molecules	Effects aren't permanent and don't alter the genome	Can only target certain RNA sequences	(Tao et al., 2023)
Transposon Systems	Moves DNA segments into the genome	Can insert large pieces of DNA and stays stable	The insertion site is random and may interfere with genes	(Van Opijnen & Levin, 2020)
CREATE	Uses CRISPR/Cas9 along with LINE-1 elements	Very precise and can be programmed for targeting	Needs specific cellular components to work	(Y. Wang et al., 2025)

10. FUTURE PERSPECTIVES AND INDUSTRY TRENDS

Among the most interesting directions is the combination of CRISPR with artificial intelligence(Ali, Choudhuri, Bishal, et al., 2025; Ali, Choudhuri, Chakraborty, et al., 2025; Ali et al., 2024) and machine learning systems, which will allow designing optimised guide RNAs, predicting the effectiveness of editing, and detecting the presence of unwanted genomic alterations early. This advancement in computation enhances the evolution of disease-specific technologies of genome-editing, thus providing the basis of really personalised medicine according to the genetic composition of a specific individual. The latest progress in delivery systems such as viral vectors, nanoparticles and exosome-based vectors enhance the safety and effectiveness of gene editing in human cells and tissues. Such advancements are needed to overcome the current limitations in translational research exercises and realize scalable and in-vivo therapeutic uses. The Cas9 variants that are high fidelity as they have been engineered to minimize the off-target effects and ensure the greatest specificity when manipulating disease-related genes are also changing the face of CRISPR. The field of ethical reasoning, especially in the aspect of germline editing and hereditary modifications, requires a fair system that will trim down the scientific advancement to the social accountability. These technologies will have to involve cross-border cooperation of scientists, clinicians, policy makers, and ethicists to make sure that they are developed and applied in a safe, equitable, and overall advantageous way(Ansori et al., 2023; Jayarajan et al., 2025; Ravichandran & Maddalo, 2023).

CONCLUSION

Genome editing technology has transformed precision medicine by allowing specific remedies of genetic errors and creating custom-made therapy. CRISPR-Cas systems, base and prime editing have proved to be very successful in curing genetic diseases and cancer research. Nonetheless, there are still challenges including the limitation of delivery, off-target effects, immune responses, as well as ethical concerns. There is hope in the new RNA- and AI-based tools that will be more precise and applicable. Finally, to achieve the full potential of genome editing in precision medicine, the innovation needs to be sustained, accessible globally, and well-regulated and ethically in order to promote safe and equitable healthcare.

REFERENCES

Abbasi, S., Totmaj, M. A., Abbasi, M., Hajazimian, S., Goleij, P., Behroozi, J., Shademan, B., Isazadeh, A., & Baradaran, B. (2023). Chimeric antigen receptor T (CAR-T) cells: Novel cell therapy for hematological malignancies. *Cancer Medicine*, *12*(7), 7844–7858. DOI: 10.1002/cam4.5551 PMID: 36583504

Afrose, N., Chakraborty, R., Bhowmick, P., & Bhowmick, M. (2025). Advancing drug delivery systems through biomedical engineering: Innovations and future directions. In *Innovations in Biomedical Engineering* (pp. 1–32). Elsevier., DOI: 10.1016/B978-0-443-30146-9.00001-0

Afrose, N., Chakraborty, R., & Rajendran, K. (2025). Plasma-polymerized chitosan/agarose/chitosan-agarose composite coatings for antibacterial and antibiofilm wound dressings: A review. *International Journal of Biological Macromolecules, 320*, 145873. DOI: 10.1016/j.ijbiomac.2025.145873 PMID: 40645242

Aguilera, A., & García-Muse, T. (2013). Causes of Genome Instability. *Annual Review of Genetics, 47*(1), 1–32. DOI: 10.1146/annurev-genet-111212-133232 PMID: 23909437

Ahmed, R., Alghamdi, W. N., Alharbi, F. R., Alatawi, H. D., Alenezi, K. M., Alanazi, T. F., & Elsherbiny, N. M. (2025). CRISPR/Cas9 System as a Promising Therapy in Thalassemia and Sickle Cell Disease: A Systematic Review of Clinical Trials. *Molecular Biotechnology*. Advance online publication. DOI: 10.1007/s12033-025-01368-x PMID: 39794549

Ali, K. A., Choudhuri, S., Bishal, A., Das, T., Nandy, S., & Hossain, C. M. (2025). Approaches of AI and Advanced Computational Biology for Combating HIV Infections. In J.-T. Chen, *AI and Precision Medicine in Infectious Disease Management* (1st ed., pp. 149–166). CRC Press. DOI: 10.1201/9781003615699-8

Ali, K. A., Choudhuri, S., Chakraborty, R., Pradhan, A., Kotal, A., & Shaharyar, M. A. (2025). AI Technology-Powered Precision Medicine and Smart Healthcare. In J.-T. Chen, *AI and Precision Medicine in Infectious Disease Management* (1st ed., pp. 51–70). CRC Press. DOI: 10.1201/9781003615699-3

Ali, K. A., Choudhuri, S., & Roy, S. Kr., Reza, K. H., Mandal, D., Chakraborty, R., & Hossain, C. M. (2025). AI-Generated Nanoparticles in Immunological Research: Opportunities and Challenges. In J.-T. Chen (Ed.), *Advances in Computational Intelligence and Robotics* (pp. 295–322). IGI Global. DOI: 10.4018/979-8-3693-9725-1.ch011

Ali, K. A., Mohin, S., Mondal, P., Goswami, S., Ghosh, S., & Choudhuri, S. (2024). Influence of artificial intelligence in modern pharmaceutical formulation and drug development. *Future Journal of Pharmaceutical Sciences*, *10*(1), 53. DOI: 10.1186/s43094-024-00625-1

Almqvist, J., & Romano, P. R. C. (2020). The Regulation of Human Germline Genome Modification in Europe. In A. Boggio, C. P. R. Romano, & J. Almqvist (Eds.), *Human Germline Genome Modification and the Right to Science* (1st ed., pp. 155–216). Cambridge University Press. DOI: 10.1017/9781108759083.007

Ansori, A., Antonius, Y., Susilo, R. J. K., Hayaza, S., Kharisma, V. D., Parikesit, A. A., Zainul, R., Jakhmola, V., Saklani, T., Rebezov, M., Ullah, M. E., Maksimiuk, N., Derkho, M., & Burkov, P. (2023). Application of CRISPR-Cas9 genome editing technology in various fields: A review. *Narra J*, *3*(2), e184. DOI: 10.52225/narra.v3i2.184 PMID: 38450259

Arowosegbe, J. O. (2024). Before We Forget its Horrors: When Does Human Genome Editing Not Become Eugenics? *Mizan Law Review*, *18*(2), 241–270. DOI: 10.4314/mlr.v18i2.1

Atkins, A., Chung, C.-H., Allen, A. G., Dampier, W., Gurrola, T. E., Sariyer, I. K., Nonnemacher, M. R., & Wigdahl, B. (2021). Off-Target Analysis in Gene Editing and Applications for Clinical Translation of CRISPR/Cas9 in HIV-1 Therapy. *Frontiers in Genome Editing*, *3*, 673022. DOI: 10.3389/fgeed.2021.673022 PMID: 34713260

Baylis, F., Darnovsky, M., Hasson, K., & Krahn, T. M. (2020). Human Germline and Heritable Genome Editing: The Global Policy Landscape. *The CRISPR Journal*, *3*(5), 365–377. DOI: 10.1089/crispr.2020.0082 PMID: 33095042

Butt, H., Mandava, M., & Jacobsohn, D. (2025). Advances in Gene Therapy for Sickle Cell Disease: From Preclinical Innovations to Clinical Implementation and Access Challenges. *The CRISPR Journal*, *8*(3), 174–188. DOI: 10.1089/crispr.2024.0101 PMID: 40356202

Cai, R., Lv, R., Shi, X., Yang, G., & Jin, J. (2023). CRISPR/dCas9 Tools: Epigenetic Mechanism and Application in Gene Transcriptional Regulation. *International Journal of Molecular Sciences*, *24*(19), 14865. DOI: 10.3390/ijms241914865 PMID: 37834313

Capolla, S., Rasool, M., Toffoli, G., & Dal Bo, M. (2025). CAR-T Cell Manufacturing for Hematological and Solid Tumors: From the Preclinical to Clinical Point of View. *Cancer Medicine*, *14*(5), e70726. DOI: 10.1002/cam4.70726 PMID: 40013750

Cattle, M. A., Aguado, L. C., Sze, S., Venkittu, S., Wang, Y., Papagiannakopoulos, T., Smith, S., Rice, C. M., Schneider, W. M., & Poirier, J. T. (2025). An enhanced Eco1 retron editor enables precision genome engineering in human cells without double-strand breaks. *Nucleic Acids Research*, *53*(14), gkaf716. Advance online publication. DOI: 10.1093/nar/gkaf716 PMID: 40744490

Chakraborty, R., Afrose, N., Bhowmick, P., & Bhowmick, M. (2025). Biomedical sensors in wearable health technologies. In *Innovations in Biomedical Engineering* (pp. 159–184). Elsevier., DOI: 10.1016/B978-0-443-30146-9.00005-8

Chakraborty, R., Jana, B., Koner, S., Kanta Maiti, S., Afrose, N., Bhowmick, P., & Bhowmick, M. (2024). Exploring the transcriptomic landscape of rheumatoid arthritis using next-generation RNA sequencing. *INNOSC Theranostics & Pharmacological Sciences*, *8*(1), 14. DOI: 10.36922/itps.4449

Chakraborty, R., Kannan, S. P., Afrose, N., & Narayanasamy, D. (2025). Tumor microenvironment expressed enzymes (Matrix metalloproteinases, cathepsins, urokinase-type plasminogen activator) triggered polymersomes for liquid biopsy and cancer diagnostics: A review. *International Journal of Biological Macromolecules*, *321*, 146375. DOI: 10.1016/j.ijbiomac.2025.146375 PMID: 40738429

Chattopadhyay, R., Firdous, Z., & Bari, V. K. (2025). CRISPR/Cas9 and its derivatives to improve crop biotic stress resistance: Current status and prospects. *Physiological and Molecular Plant Pathology*, *135*, 102482. DOI: 10.1016/j.pmpp.2024.102482

Chillón-Pino, D., Badonyi, M., Semple, C. A., & Marsh, J. A. (2024). Protein structural context of cancer mutations reveals molecular mechanisms and candidate driver genes. *Cell Reports*, *43*(11), 114905. DOI: 10.1016/j.celrep.2024.114905 PMID: 39441719

Chuang, Y.-F., Phipps, A. J., Lin, F.-L., Hecht, V., Hewitt, A. W., Wang, P.-Y., & Liu, G.-S. (2021). Approach for in vivo delivery of CRISPR/Cas system: A recent update and future prospect. *Cellular and Molecular Life Sciences : CMLS*, *78*(6), 2683–2708. DOI: 10.1007/s00018-020-03725-2 PMID: 33388855

Coller, B. S. (2019). Ethics of Human Genome Editing. *Annual Review of Medicine*, *70*(1), 289–305. DOI: 10.1146/annurev-med-112717-094629 PMID: 30691366

Collias, D., & Beisel, C. L. (2021). CRISPR technologies and the search for the PAM-free nuclease. *Nature Communications*, *12*(1), 555. DOI: 10.1038/s41467-020-20633-y PMID: 33483498

Dagar, G., Gupta, A., Masoodi, T., Nisar, S., Merhi, M., Hashem, S., Chauhan, R., Dagar, M., Mirza, S., Bagga, P., Kumar, R., Akil, A. S. A.-S., Macha, M. A., Haris, M., Uddin, S., Singh, M., & Bhat, A. A. (2023). Harnessing the potential of CAR-T cell therapy: Progress, challenges, and future directions in hematological and solid tumor treatments. *Journal of Translational Medicine*, *21*(1), 449. DOI: 10.1186/s12967-023-04292-3 PMID: 37420216

Dasgupta, S., Gayen, S., Chakraborty, T., Afrose, N., Pal, R., Mahata, S., Nasare, V., & Roy, S. (2024). Potential role of immune cell therapy in gynecological cancer and future promises: A comprehensive review. *Medical Oncology (Northwood, London, England)*, *41*(5), 98. DOI: 10.1007/s12032-024-02337-1 PMID: 38536512

Di Matteo, M., Belay, E., Chuah, M. K., & VandenDriessche, T. (2012). Recent developments in transposon-mediated gene therapy. *Expert Opinion on Biological Therapy*, *12*(7), 841–858. DOI: 10.1517/14712598.2012.684875 PMID: 22679910

Dimitri, A., Herbst, F., & Fraietta, J. A. (2022). Engineering the next-generation of CAR T-cells with CRISPR-Cas9 gene editing. *Molecular Cancer*, *21*(1), 78. DOI: 10.1186/s12943-022-01559-z PMID: 35303871

Dölarslan, M. (2023). CRISPR-Cas9 Mediated Gene Correction of CFTR Mutations in Cystic Fibrosis: Evaluating Efficacy, Safety, and Long-Term Outcomes in Patient-Derived Lung Organoids. *SHIFAA*, *2023*, 41–47. DOI: 10.70470/SHIFAA/2023/005

Eba, J., & Nakamura, K. (2022). Overview of the ethical guidelines for medical and biological research involving human subjects in Japan. *Japanese Journal of Clinical Oncology*, *52*(6), 539–544. DOI: 10.1093/jjco/hyac034 PMID: 35349681

Eid, A., Alshareef, S., & Mahfouz, M. M. (2018). CRISPR base editors: Genome editing without double-stranded breaks. *Biochemical Journal*, *475*(11), 1955–1964. DOI: 10.1042/BCJ20170793

Faltus, T., & De Miguel Beriain, I. (2023). The EU legal framework on clinical trials directed to therapeutic germline gene alteration: A critical and systematic analysis. *Medical Law International*, *23*(1), 44–70. DOI: 10.1177/09685332231154586

Farrell, C. B. (2019). *"O brave new world" CRISPR Mediated Gene Editing in Somatic and Germ Line Contexts for Therapeutics and Enhancement of the Human Species – an Ethical and Legal Discussion.* 949365 Bytes. DOI: 10.25419/RCSI.10814045.V1

Feeney, O., Cockbain, J., & Sterckx, S. (2021). Ethics, Patents and Genome Editing: A Critical Assessment of Three Options of Technology Governance. *Frontiers in Political Science*, *3*, 731505. DOI: 10.3389/fpos.2021.731505

Fiumara, M., Ferrari, S., Omer-Javed, A., Beretta, S., Albano, L., Canarutto, D., Varesi, A., Gaddoni, C., Brombin, C., Cugnata, F., Zonari, E., Naldini, M. M., Barcella, M., Gentner, B., Merelli, I., & Naldini, L. (2024). Genotoxic effects of base and prime editing in human hematopoietic stem cells. *Nature Biotechnology*, *42*(6), 877–891. DOI: 10.1038/s41587-023-01915-4 PMID: 37679541

Genome-wide detection and analysis of CRISPR-Cas off-targets. (2021). In *Progress in Molecular Biology and Translational Science* (Vol. 181, pp. 31–43). Elsevier. DOI: 10.1016/bs.pmbts.2021.01.012

Giamarellos-Bourboulis, E. J., Aschenbrenner, A. C., Bauer, M., Bock, C., Calandra, T., Gat-Viks, I., Kyriazopoulou, E., Lupse, M., Monneret, G., Pickkers, P., Schultze, J. L., Van Der Poll, T., Van De Veerdonk, F. L., Vlaar, A. P. J., Weis, S., Wiersinga, W. J., & Netea, M. G. (2024). The pathophysiology of sepsis and precision-medicine-based immunotherapy. *Nature Immunology*, *25*(1), 19–28. DOI: 10.1038/s41590-023-01660-5 PMID: 38168953

González Castro, N. G., Bjelic, J., Malhotra, G., Huang, C., & Alsaffar, S. H. (2021). Comparison of the Feasibility, Efficiency, and Safety of Genome Editing Technologies. *International Journal of Molecular Sciences*, *22*(19), 10355. DOI: 10.3390/ijms221910355 PMID: 34638696

González-Delgado, A., Lopez, S. C., Rojas-Montero, M., Fishman, C. B., & Shipman, S. L. (2024). Simultaneous multi-site editing of individual genomes using retron arrays. *Nature Chemical Biology*, *20*(11), 1482–1492. DOI: 10.1038/s41589-024-01665-7 PMID: 38982310

Grant, E. V. (2016). FDA Regulation of Clinical Applications of CRISPR-CAS Gene-Editing Technology. *Food and Drug Law Journal*, *71*(4), 608–633. PMID: 29140647

Greco, F., Cosentino, M., & Marino, F. (2024). The Italian breakthrough in CRISPR trials for rare diseases: A focus on beta-thalassemia and sickle cell disease treatment. *Frontiers in Medicine*, *11*, 1356578. DOI: 10.3389/fmed.2024.1356578 PMID: 38426160

Greenfield, A. (2021). Making sense of heritable human genome editing: Scientific and ethical considerations. In *Progress in Molecular Biology and Translational Science* (Vol. 182, pp. 1–28). Elsevier., DOI: 10.1016/bs.pmbts.2020.12.008

Gumer, J. M. (2019). The Wisdom of Germline Editing: An Ethical Analysis of the Use of CRISPR-Cas9 to Edit Human Embryos. *The New Bioethics : a Multidisciplinary Journal of Biotechnology and the Body*, *25*(2), 137–152. DOI: 10.1080/20502877.2019.1606151 PMID: 31130112

Gupta, P. K., & Kumar, S. (2025). Third-generation novel technologies for gene editing. *Trends in Biotechnology, S0167779925002732*. Advance online publication. DOI: 10.1016/j.tibtech.2025.07.012 PMID: 40738762

Hanna, R., Frangoul, H., McKinney, C., Pineiro, L., Mapara, M., Dalal, J., Chang, K.-H., Jaskolka, M., Kim, K., Farrington, D. L., Wally, M., Mei, B., Lawal, A., Afonja, O. O., & Walters, M. C. (2023). AsCas12a Gene Editing of *HBG1/2* Promoters with EDIT-301 Results in Rapid and Sustained Normalization of Hemoglobin and Increased Fetal Hemoglobin in Patients with Severe Sickle Cell Disease and Transfusion-Dependent Beta-Thalassemia. *Blood*, *142*(Supplement 1), 4996–4996. DOI: 10.1182/blood-2023-187397

Hillary, V. E., & Ceasar, S. A. (2024). CRISPR/Cas system-mediated base editing in crops: Recent developments and future prospects. *Plant Cell Reports*, *43*(11), 271. DOI: 10.1007/s00299-024-03346-0 PMID: 39453560

Hu, C., Shi, X., Guo, B., Yang, Z., Zhou, J., & Wang, F. (2025). Toehold-Based CRISPR-dCas9 Transcriptional Activation Platform for Spatiotemporally Controllable Gene Therapy in Tumor and Diabetic Mouse Models. *ACS Nano*, *19*(12), 12277–12287. DOI: 10.1021/acsnano.5c01078 PMID: 40123515

Hussen, B. M., Najmadden, Z. B., Abdullah, S. R., Rasul, M. F., Mustafa, S. A., Ghafouri-Fard, S., & Taheri, M. (2024). CRISPR/Cas9 gene editing: A novel strategy for fighting drug resistance in respiratory disorders. *Cell Communication and Signaling : CCS*, *22*(1), 329. DOI: 10.1186/s12964-024-01713-8 PMID: 38877530

Hwang, H.-Y., Lee, M., Yi, H., Seok, C., Lim, K., Na, Y. R., Kang, J.-S., Park, J.-H., & Kim, D. (2025). Engineered Sdd7 cytosine base editors with enhanced specificity. *Nature Communications*, *16*(1), 5881. DOI: 10.1038/s41467-025-60789-z PMID: 40593661

Jayarajan, D., Kumar Dutta, A., Paripuranam, T. D., Krishnaveni, M., & Bhavya, S. (2025). Recent Advancements in CRISPR-Cas9 Technology for Precision Gene Editing and Therapeutic Applications. *Journal of Neonatal Surgery*, *14*(7), 818–827. DOI: 10.63682/jns.v14i7.6384

Jung, W. J., Park, S.-J., Cha, S., & Kim, K. (2024). Factors affecting the cleavage efficiency of the CRISPR-Cas9 system. *Animal Cells and Systems*, *28*(1), 75–83. DOI: 10.1080/19768354.2024.2322054 PMID: 38440123

Kampouri, E., Reynolds, G., Teh, B. W., & Hill, J. A. (2024). Chimeric antigen receptor-T-cell therapies going viral: Latent and incidental viral infections. *Current Opinion in Infectious Diseases*, *37*(6), 526–535. DOI: 10.1097/QCO.0000000000001066 PMID: 39361275

Kaur, N., & Pati, P. K. (2024). Retron Library Recombineering: Next Powerful Tool for Genome Editing after CRISPR/Cas. *ACS Synthetic Biology*, *13*(4), 1019–1025. DOI: 10.1021/acssynbio.3c00667 PMID: 38480006

Khoshandam, M., Soltaninejad, H., Mousazadeh, M., Hamidieh, A. A., & Hosseinkhani, S. (2024). Clinical applications of the CRISPR/Cas9 genome-editing system: Delivery options and challenges in precision medicine. *Genes & Diseases*, *11*(1), 268–282. DOI: 10.1016/j.gendis.2023.02.027 PMID: 37588217

Kingwell, K. (2023). First CRISPR therapy seeks landmark approval. *Nature Reviews. Drug Discovery*, *22*(5), 339–341. DOI: 10.1038/d41573-023-00050-8 PMID: 37012339

Kleinstiver, B. P., Pattanayak, V., Prew, M. S., Tsai, S. Q., Nguyen, N. T., Zheng, Z., & Joung, J. K. (2016). High-fidelity CRISPR–Cas9 nucleases with no detectable genome-wide off-target effects. *Nature*, *529*(7587), 490–495. DOI: 10.1038/nature16526 PMID: 26735016

Kordyś, M., Sen, R., & Warkocki, Z. (2022). Applications of the versatile CRISPR-Cas13 RNA targeting system. *Wiley Interdisciplinary Reviews. RNA*, *13*(3), e1694. DOI: 10.1002/wrna.1694 PMID: 34553495

Kwon, J., Kim, M., Hwang, W., Jo, A., Hwang, G.-H., Jung, M., Kim, U. G., Cui, G., Kim, H., Eom, J.-H., Hur, J. K., Lee, J., Kim, Y., Kim, J., Bae, S., & Lee, J. K. (2023). Extru-seq: A method for predicting genome-wide Cas9 off-target sites with advantages of both cell-based and in vitro approaches. *Genome Biology*, *24*(1), 4. DOI: 10.1186/s13059-022-02842-4 PMID: 36627653

Lafi, Z., Ata, T., & Asha, S. (2025). CRISPR in clinical diagnostics: Bridging the gap between research and practice. *Bioanalysis*, *17*(4), 281–290. DOI: 10.1080/17576180.2025.2459520 PMID: 39902531

Lau, C.-H., & Suh, Y. (2017). Genome and Epigenome Editing in Mechanistic Studies of Human Aging and Aging-Related Disease. *Gerontology*, *63*(2), 103–117. DOI: 10.1159/000452972 PMID: 27974723

Lei, R., & Qiu, R. (2020). Ethical and regulatory issues in human gene editing: Chinese perspective. *Biotechnology and Applied Biochemistry*, *67*(6), 880–891. DOI: 10.1002/bab.2032 PMID: 33011985

Lei, T., Wang, Y., Zhang, Y., Yang, Y., Cao, J., Huang, J., Chen, J., Chen, H., Zhang, J., Wang, L., Xu, X., Gale, R. P., & Wang, L. (2024). Leveraging CRISPR gene editing technology to optimize the efficacy, safety and accessibility of CAR T-cell therapy. *Leukemia, 38*(12), 2517–2543. DOI: 10.1038/s41375-024-02444-y PMID: 39455854

Li, W., Zhu, X., Xu, Y., Chen, J., Zhang, H., Yang, Z., Qi, Y., Hong, J., Li, Y., Wang, G., Shen, J., & Qian, C. (2022). Simultaneous editing of TCR, HLA-I/II and HLA-E resulted in enhanced universal CAR-T resistance to allo-rejection. *Frontiers in Immunology, 13*, 1052717. DOI: 10.3389/fimmu.2022.1052717 PMID: 36532006

Li, Y., Li, D., You, L., Deng, T., Pang, Q., Meng, X., & Zhu, B. (2023). dCas9-Based PDGFR–β Activation ADSCs Accelerate Wound Healing in Diabetic Mice through Angiogenesis and ECM Remodeling. *International Journal of Molecular Sciences, 24*(6), 5949. DOI: 10.3390/ijms24065949 PMID: 36983022

Li, Y.-R., Fang, Y., Niu, S., Chen, Y., Lyu, Z., & Yang, L. (2025). Managing allorejection in off-the-shelf CAR-engineered cell therapies. *Molecular Therapy : the Journal of the American Society of Gene Therapy, 33*(6), 2368–2390. DOI: 10.1016/j.ymthe.2024.11.035 PMID: 39600090

Li, Z., & Zhang, Z. (2025). A tale of two strands: Decoding chromatin replication through strand-specific sequencing. *Molecular Cell, 85*(2), 238–261. DOI: 10.1016/j.molcel.2024.10.035 PMID: 39824166

Lopes, R., & Prasad, M. K. (2024). Beyond the promise: Evaluating and mitigating off-target effects in CRISPR gene editing for safer therapeutics. *Frontiers in Bioengineering and Biotechnology, 11*, 1339189. DOI: 10.3389/fbioe.2023.1339189 PMID: 38390600

Lu, X., Zhang, M., Li, G., Zhang, S., Zhang, J., Fu, X., & Sun, F. (2023). Applications and Research Advances in the Delivery of CRISPR/Cas9 Systems for the Treatment of Inherited Diseases. *International Journal of Molecular Sciences, 24*(17), 13202. DOI: 10.3390/ijms241713202 PMID: 37686009

Lukjanov, V., Koutná, I., & Šimara, P. (2021). CAR T-Cell Production Using Nonviral Approaches. *Journal of Immunology Research, 2021*, 1–9. DOI: 10.1155/2021/6644685 PMID: 33855089

Lupan, I., Bintintan, V., Deleanu, D., & Samasca, G. (2024). Epigenetic Regulation of DNA Methylation and RNA Interference in Gastric Cancer: A 2024 Update. *Biomedicines, 12*(9), 2001. DOI: 10.3390/biomedicines12092001 PMID: 39335515

Manghwar, H., Li, B., Ding, X., Hussain, A., Lindsey, K., Zhang, X., & Jin, S. (2020). CRISPR/Cas Systems in Genome Editing: Methodologies and Tools for sgRNA Design, Off-Target Evaluation, and Strategies to Mitigate Off-Target Effects. *Advancement of Science*, *7*(6), 1902312. DOI: 10.1002/advs.201902312 PMID: 32195078

Marchant, G. E. (2021). Global Governance of Human Genome Editing: What Are the Rules? *Annual Review of Genomics and Human Genetics*, *22*(1), 385–405. DOI: 10.1146/annurev-genom-111320-091930 PMID: 33667117

McCutcheon, S. R., Rohm, D., Iglesias, N., & Gersbach, C. A. (2024). Epigenome editing technologies for discovery and medicine. *Nature Biotechnology*, *42*(8), 1199–1217. DOI: 10.1038/s41587-024-02320-1 PMID: 39075148

Mikhaylova, E. V., Kuluev, B. R., Gerashchenkov, G. A., Chemeris, D. A., Garafutdinov, R. R., Kuluev, A. R., Baymiev, A., Baymiev, A. K., & Chemeris, A. V. (2024). Prime-editing methods and pegRNA design programs. *Molekuliarnaia Biologiia*, *58*(1), 22–39. DOI: 10.31857/S0026898424010023 PMID: 38943578

Mojtahedi, H., Yazdanpanah, N., & Rezaei, N. (2021). Chronic myeloid leukemia stem cells: Targeting therapeutic implications. *Stem Cell Research & Therapy*, *12*(1), 603. DOI: 10.1186/s13287-021-02659-1 PMID: 34922630

Naeem, M., Majeed, S., Hoque, M. Z., & Ahmad, I. (2020). Latest Developed Strategies to Minimize the Off-Target Effects in CRISPR-Cas-Mediated Genome Editing. *Cells*, *9*(7), 1608. DOI: 10.3390/cells9071608 PMID: 32630835

Naseri, G., & Koffas, M. A. G. (2020). Application of combinatorial optimization strategies in synthetic biology. *Nature Communications*, *11*(1), 2446. DOI: 10.1038/s41467-020-16175-y PMID: 32415065

Nedorezova, D. D., Dubovichenko, M. V., Belyaeva, E. P., Grigorieva, E. D., Peresadina, A. V., & Kolpashchikov, D. M. (2022). Specificity of oligonucleotide gene therapy (OGT) agents. *Theranostics*, *12*(16), 7132–7157. DOI: 10.7150/thno.77830 PMID: 36276652

Niazi, S. K. (2023). The Dawn of In Vivo Gene Editing Era: A Revolution in the Making. *Biologics (Basel, Switzerland)*, *3*(4), 253–295. DOI: 10.3390/biologics3040014

Nishikura, K. (2010). Functions and Regulation of RNA Editing by ADAR Deaminases. *Annual Review of Biochemistry*, *79*(1), 321–349. DOI: 10.1146/annurev-biochem-060208-105251 PMID: 20192758

Parhi, S., Afrose, N., Rajendran, K., & Narayanasamy, D. (2025). Central composite design (CCD) based formulation, optimization, *in-vitro* and *ex-vivo* characterization of 5-fluorouracil-loaded emulgel for enhanced dermal penetration and psoriasis management. *Drug Development and Industrial Pharmacy*, *51*(3), 244–261. DOI: 10.1080/03639045.2025.2464782 PMID: 39932381

Parisi, G. F., Terlizzi, V., Manti, S., Papale, M., Pecora, G., Presti, S., Tosto, M., & Leonardi, S. (2025). Cutting-Edge Advances in Cystic Fibrosis: From Gene Therapy to Personalized Medicine and Holistic Management. *Genes*, *16*(4), 402. DOI: 10.3390/genes16040402 PMID: 40282362

Pfeiffer, L. S., & Stafforst, T. (2023). Precision RNA base editing with engineered and endogenous effectors. *Nature Biotechnology*, *41*(11), 1526–1542. DOI: 10.1038/s41587-023-01927-0 PMID: 37735261

Pickar-Oliver, A., & Gersbach, C. A. (2019). The next generation of CRISPR–Cas technologies and applications. *Nature Reviews. Molecular Cell Biology*, *20*(8), 490–507. DOI: 10.1038/s41580-019-0131-5 PMID: 31147612

Plasschaert, L. W., MacDonald, K. D., & Moffit, J. S. (2024). Current landscape of cystic fibrosis gene therapy. *Frontiers in Pharmacology*, *15*, 1476331. DOI: 10.3389/fphar.2024.1476331 PMID: 39439894

Raghunath, A., Krishna, T., Nandan, P. K., & Sivaraman, J. (2025). A Paradigm Shift in Cystic Fibrosis: Insights into Molecular Diagnosis, Newborn Screening, and Nanotechnology-Based Drug Delivery for CFTreatment. *Current Respiratory Medicine Reviews*, *21*(5), 485–499. Advance online publication. DOI: 10.2174/011573398X359962250527113134

Ramakrishnan, M., Satish, L., Sharma, A., Kurungara Vinod, K., Emamverdian, A., Zhou, M., & Wei, Q. (2022). Transposable elements in plants: Recent advancements, tools and prospects. *Plant Molecular Biology Reporter*, *40*(4), 628–645. DOI: 10.1007/s11105-022-01342-w

Ravichandran, M., & Maddalo, D. (2023). Applications of CRISPR-Cas9 for advancing precision medicine in oncology: From target discovery to disease modeling. *Frontiers in Genetics*, *14*, 1273994. DOI: 10.3389/fgene.2023.1273994 PMID: 37908590

Reddy, P., Vilella, F., Izpisua Belmonte, J. C., & Simón, C. (2020). Use of Customizable Nucleases for Gene Editing and Other Novel Applications. *Genes*, *11*(9), 976. DOI: 10.3390/genes11090976 PMID: 32842577

Rós, F. A., Couto, S. C. F., Milhomens, J., Ovider, I., Maio, K. T., Jennifer, V., Ramos, R. N., Picanço-Castro, V., Kashima, S., Calado, R. T., Barros, L. R. C., & Rocha, V. (2023). A systematic review of clinical trials for gene therapies for β-hemoglobinopathy around the world. *Cytotherapy, 25*(12), 1300–1306. DOI: 10.1016/j.jcyt.2023.05.006 PMID: 37318395

Sandoval-Villegas, N., Nurieva, W., Amberger, M., & Ivics, Z. (2021). Contemporary Transposon Tools: A Review and Guide through Mechanisms and Applications of Sleeping Beauty, piggyBac and Tol2 for Genome Engineering. *International Journal of Molecular Sciences, 22*(10), 5084. DOI: 10.3390/ijms22105084 PMID: 34064900

Sanzo, K. M. (2022). Regulation of Genome Editing in Human iPS Cells: United States. In Dederer, H.-G., & Frenken, G. (Eds.), *Regulation of Genome Editing in Human iPS Cells* (pp. 199–222). Springer International Publishing., DOI: 10.1007/978-3-030-93023-3_6

Sapozhnikov, D. M., & Szyf, M. (2021). Unraveling the functional role of DNA demethylation at specific promoters by targeted steric blockage of DNA methyltransferase with CRISPR/dCas9. *Nature Communications, 12*(1), 5711. DOI: 10.1038/s41467-021-25991-9 PMID: 34588447

Segers, S. (2023). Heritable genome editing: Ethical aspects of a developing domain. *Human Reproduction (Oxford, England), 38*(11), 2055–2061. DOI: 10.1093/humrep/dead167 PMID: 37581898

Shi, J.-Y., Li, Z.-W., Yao, Z.-L., Han, G.-M., Li, M.-Y., Cai, Q.-L., & Kong, D.-M. (2025). 5′-End Engineering of CRISPR/Cas12a Activators: A Versatile Platform for Multiple Biomarker Analysis and Clinical Cancer Tissue Identification. *Analytical Chemistry, 97*(33), 18188–18198. DOI: 10.1021/acs.analchem.5c03168 PMID: 40811645

Shi, L., Tang, X., & Tang, G. (2016). GUIDE-Seq to Detect Genome-wide Double-Stranded Breaks in Plants. *Trends in Plant Science, 21*(10), 815–818. DOI: 10.1016/j.tplants.2016.08.005 PMID: 27593568

Simon, A. J., Ellington, A. D., & Finkelstein, I. J. (2019). Retrons and their applications in genome engineering. *Nucleic Acids Research, 47*(21), 11007–11019. DOI: 10.1093/nar/gkz865 PMID: 31598685

Sivagourounadin, K., Ravichandran, M., & Rajendran, P. (2021). National guidelines for gene therapy product (2019): A road-map to gene therapy products development and clinical trials. *Perspectives in Clinical Research, 12*(3), 118–125. DOI: 10.4103/picr.PICR_189_20 PMID: 34386375

Sutherland, H. G., Jenkins, B., & Griffiths, L. R. (2024). Genetics of migraine: Complexity, implications, and potential clinical applications. *Lancet Neurology*, *23*(4), 429–446. DOI: 10.1016/S1474-4422(24)00026-7 PMID: 38508838

Swarts, D. C., & Jinek, M. (2018). Cas9 versus Cas12a/Cpf1: Structure–function comparisons and implications for genome editing. *Wiley Interdisciplinary Reviews. RNA*, *9*(5), e1481. DOI: 10.1002/wrna.1481 PMID: 29790280

Tang, S., Chen, X., Tong, X., & Zhu, L. (2025). Overcoming the Delivery Challenges in CRISPR/Cas9 Gene Editing for Effective Cancer Treatment: A Review of Delivery Systems. *International Journal of Medical Sciences*, *22*(14), 3625–3649. DOI: 10.7150/ijms.112724 PMID: 40959562

Tao, J., Bauer, D. E., & Chiarle, R. (2023). Assessing and advancing the safety of CRISPR-Cas tools: From DNA to RNA editing. *Nature Communications*, *14*(1), 212. DOI: 10.1038/s41467-023-35886-6 PMID: 36639728

Theardy, M. S., Takeda, M., Sorokin, A., Chen, S., Yang, Z., Wang, X., Kanikarla, P., Coker, O., Nguyen, P., Wei, Y., Yao, J., Yan, L., Jin, Y., Cai, Y., Paku, M., Chen, Z., Li, K. Z., Citron, F., Tomihara, H., & Yao, W. (2025). Therapeutic targeting of syndecan-1 axis overcomes acquired resistance to KRAS-targeted therapy in gastrointestinal cancers. *Cell Reports Medicine*, *6*(8), 102253. DOI: 10.1016/j.xcrm.2025.102253 PMID: 40713971

Tian, R., Cao, C., He, D., Dong, D., Sun, L., Liu, J., Chen, Y., Wang, Y., Huang, Z., Li, L., Jin, Z., Huang, Z., Xie, H., Zhao, T., Zhong, C., Hong, Y., & Hu, Z. (2023). Massively parallel CRISPR off-target detection enables rapid off-target prediction model building. *Med (New York, N.Y.)*, *4*(7), 478–492.e6. DOI: 10.1016/j.medj.2023.05.005 PMID: 37279759

Toma, L., Barbălată, T., Sanda, G. M., Niculescu, L. S., Sima, A. V., & Stancu, C. S. (2021). CRISPR/dCas9 Transcriptional Activation of Endogenous Apolipoprotein AI and Paraoxonase 1 in Enterocytes Alleviates Endothelial Cell Dysfunction. *Biomolecules*, *11*(12), 1769. DOI: 10.3390/biom11121769 PMID: 34944413

Tsai, S. Q., Nguyen, N. T., Malagon-Lopez, J., Topkar, V. V., Aryee, M. J., & Joung, J. K. (2017). CIRCLE-seq: A highly sensitive in vitro screen for genome-wide CRISPR–Cas9 nuclease off-targets. *Nature Methods*, *14*(6), 607–614. DOI: 10.1038/nmeth.4278 PMID: 28459458

Van Opijnen, T., & Levin, H. L. (2020). Transposon Insertion Sequencing, a Global Measure of Gene Function. *Annual Review of Genetics*, *54*(1), 337–365. DOI: 10.1146/annurev-genet-112618-043838 PMID: 32886545

Volodina, O., & Smirnikhina, S. (2025). The Future of Gene Therapy: A Review of In Vivo and Ex Vivo Delivery Methods for Genome Editing-Based Therapies. *Molecular Biotechnology*, *67*(2), 425–437. DOI: 10.1007/s12033-024-01070-4 PMID: 38363528

Wang, G. H., Wang, C. M., Wu, X. J., Chu, T., Huang, D. W., & Li, J. (2024). The Development of SpCas9 Variants with High Specificity and Efficiency Based on the HH Theory. *Molecular Biology*, *58*(1), 133–146. DOI: 10.1134/S0026893324010187 PMID: 38943587

Wang, H.-Z., & He, Z.-X. (2024). Recent advances in the treatment of β-thalassemia: From traditional therapies to gene therapy. *New Discovery*, 1–15. DOI: 10.61958/NDQD9971

Wang, J., Wang, K., Deng, Z., Zhong, Z., Sun, G., Mei, Q., Zhou, F., Deng, Z., & Sun, Y. (2024). Engineered cytosine base editor enabling broad-scope and high-fidelity gene editing in Streptomyces. *Nature Communications*, *15*(1), 5687. DOI: 10.1038/s41467-024-49987-3 PMID: 38971862

Wang, Y., Lin, R.-Z., Harris, M., Lavayen, B., Diwanji, N., McCreedy, B., Hofmeister, R., & Getts, D. (2025). CRISPR-Enabled Autonomous Transposable Element (CREATE) for RNA-based gene editing and delivery. *EMBO Reports*, *26*(4), 1062–1083. DOI: 10.1038/s44319-024-00364-7 PMID: 39789389

Williams, K., & Sobol, R. W. (2013). Mutation research/fundamental and molecular mechanisms of mutagenesis. *Mutation Research. Fundamental and Molecular Mechanisms of Mutagenesis*, *743–744*, 1–3. DOI: 10.1016/j.mrfmmm.2013.04.009 PMID: 23688353

Yang, P., Zhang, S., Hu, D., Li, X., Guo, Y., Guo, H., Zhang, L., & Ding, X. (2024). Research Progress on the Mechanism and Application of the Type I CRISPR-Cas System. *International Journal of Molecular Sciences*, *25*(23), 12544. DOI: 10.3390/ijms252312544 PMID: 39684256

Zeng, S., Lei, S., Qu, C., Wang, Y., Teng, S., & Huang, P. (2023). CRISPR/Cas-based gene editing in therapeutic strategies for beta-thalassemia. *Human Genetics*, *142*(12), 1677–1703. DOI: 10.1007/s00439-023-02610-9 PMID: 37878144

Zhang, F. (2019). Development of CRISPR-Cas systems for genome editing and beyond. *Quarterly Reviews of Biophysics*, *52*, e6. DOI: 10.1017/S0033583519000052

Zhang, L., Zhao, D., Wei, Z., Zhu, X., Sha, T., Tang, W., Bi, C., & Zhang, X. (2025). The Advancement of Prime Editing Technology. *ChemBioChem*, *26*(12), e202500193. DOI: 10.1002/cbic.202500193 PMID: 40350377

Zhang, Y., Ren, Q., Tang, X., Liu, S., Malzahn, A. A., Zhou, J., Wang, J., Yin, D., Pan, C., Yuan, M., Huang, L., Yang, H., Zhao, Y., Fang, Q., Zheng, X., Tian, L., Cheng, Y., Le, Y., McCoy, B., & Qi, Y. (2021). Expanding the scope of plant genome engineering with Cas12a orthologs and highly multiplexable editing systems. *Nature Communications*, *12*(1), 1944. DOI: 10.1038/s41467-021-22330-w PMID: 33782402

Zou, R. S., Liu, Y., Gaido, O. E. R., Konig, M. F., Mog, B. J., Shen, L. L., Aviles-Vazquez, F., Marin-Gonzalez, A., & Ha, T. (2023). Improving the sensitivity of in vivo CRISPR off-target detection with DISCOVER-Seq+. *Nature Methods*, *20*(5), 706–713. DOI: 10.1038/s41592-023-01840-z PMID: 37024653

Chapter 2 CRISPR-Cas9 Technology for Precision Cancer Medicine: Target Discovery, Disease Modeling, and Therapeutics

V. N. Pallavi
Sri Sathya Sai University for Human Excellence, India

Savitha Chowdareddy
https://orcid.org/0009-0005-0651-2239
Sri Sathya Sai University for Human Excellence, India

Bhumika Iranna Byalahunasishetter
Sri Sathya Sai University for Human Excellence, India

Ramya Raghavan
https://orcid.org/0000-0002-9953-543X
Sri Sathya Sai University for Human Excellence, India

ABSTRACT

CRISPR/Cas9 technology has emerged as a transformative tool in revolutionizing target discovery, disease modeling, and therapeutic interventions. Conventional CAR-T therapy achieves overall response rates of 70-85%, compared with 20-40% with standard chemotherapy. The technology shows particular promise in solid tumors including breast, lung, colorectal, hepatocellular, prostate, and pancreatic cancers. China, Asian nations have made substantial CRISPR contributions with mechanisms ranging from oncogene disruption to CAR-T cell enhancement. Critical

DOI: 10.4018/979-8-3373-4857-5.ch002

ethical frameworks addressing informed consent, germline editing, off-target effects, and global access equity have been established. The 2021 WHO framework provides a normative foundation but lacks enforcement mechanisms. The interdisciplinary convergence of CRISPR with artificial intelligence, multi-omics profiling, and systems biology promises next-generation precision oncology.

CRISPR/CAS9 TECHNOLOGY IN PRECISION CANCER MEDICINE

CRISPR/Cas9 represents a transformative genome editing technology that has revolutionized precision cancer medicine since its breakthrough demonstration in 2012. The advances of CRISPR/Cas9 in oncology have highlighted breast, lung, colorectal, hepatocellular, prostate, and pancreatic cancers (Qi et al., 2013). There are significant breakthroughs in cancer therapeutics, hematological malignancies, solid tumors, and immunotherapy. China has emerged as a global leader in clinical translation. The world's first human CRISPR trial in 2016 and clinical trials by 2017 is a promise in solid tumors treatment (Lu et al., 2017) (Figure 2). The mechanisms tested range from oncogene disruption to CAR-T cell enhancement and global access equity have been established by the WHO and national bodies. Critical ethical frameworks address informed consent, germline editing, off-target effects (Doudna, 2020).

Figure 1. Historical Beginnings of CRISPR-Cas9 technology

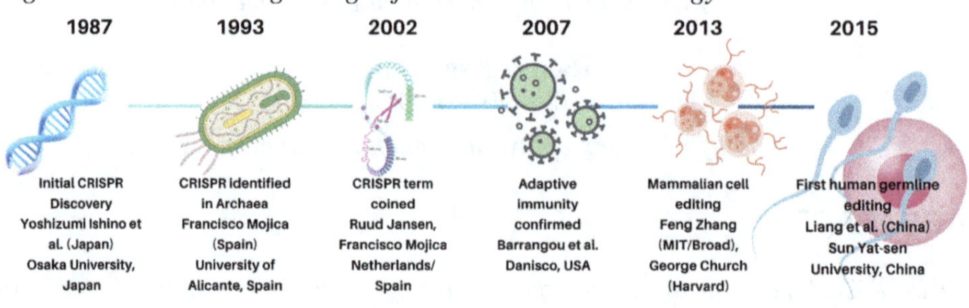

Clinical applications use the Cas9 protein and guide RNA (gRNA) for precise genome editing. CRISPR-Cas9 has enabled unprecedented accuracy in genetic modifications for cancer research and treatment. Maganti et al. (2022) analyzed 11 preclinical studies that revealed consistent improvements in anticancer efficacy. Across multiple cancer types, CRISPR/Cas9 with CAR-T therapy enhanced tumor reduction and survival outcomes compared to unedited controls. Notably, no adverse side effects, with a better safety profile for CRISPR-edited cellular therapies (Maganti et al., 2022). On October 7, 2020, the Nobel Committee awarded Emmanuelle Charpentier and Jennifer Doudna. The Chemistry Prize was for the development of a method for genome editing marking the first time two women shared a Nobel Prize in science (Baghban et al., 2024) (Figure 3).

In target discovery, we have successfully identified cancer-specific vulnerabilities. High-throughput CRISPR screens identify potential therapeutic targets, advancing our understanding of tumor biology (Platt et al., 2014). Targeted inactivation of oncogenes or repair of tumor suppressor genes enhances immunotherapy responses (Lakshmaiah et al., 2025).

Figure 2. Timeline of CRISPR-Cas9 technology till Nobel Prize recognition (2020)

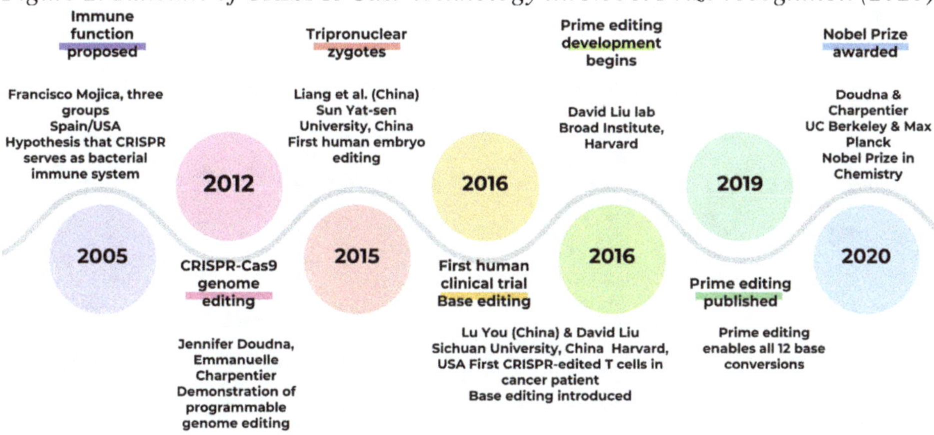

GLOBAL PERSPECTIVES

China has established itself as the undisputed leader in CRISPR clinical applications, particularly for cancer therapy (Figure 3). CRISPR/Cas9 applications in cancer therapy encompass four primary strategic approaches: oncogene inactivation, tumor suppressor repair, immune system enhancement, and disruption of resistance mechanisms (Stadtmauer et al., 2020). Asian countries dominance reflects several factors: government funding prioritization, streamlined regulatory pathways, large patient populations, and institutional support for translational research. Oncogene inactivation represents the most direct approach, with successful targeting of KRAS, MYC, EGFR, and other cancer drivers demonstrated across multiple studies (Y. Zhang et al., 2021). The development of KRAS-targeting strategies is particularly significant (Martinez-Lage et al., 2020). Commonly mutated genes in GI cancer include CDH1, TP53, KRAS, ARID1A, PTEN, and HLA-B, which contribute to cancer progression and resistance to conventional therapies. The CRISPR/Cas9 gene-editing technique is a promising tool for correcting mutations and abnormal gene expression, particularly in advanced or recurrent colorectal cancer (CRC) cases that do not respond to standard treatments (Feng et al., 2024). Par3L protein knockout inhibited proliferation, induced apoptosis, and sensitized cells to chemotherapy by regulating AMPK signaling. Targeting the intracellular immune checkpoint CISH with CRISPR-edited T cells in metastatic CRC patients showed promising early results in first-in-human trials (Lopes et al., 2025).

Figure 3. Global Case Studies from initial discovery (1987) highlighting key milestones, and geographic diversity.

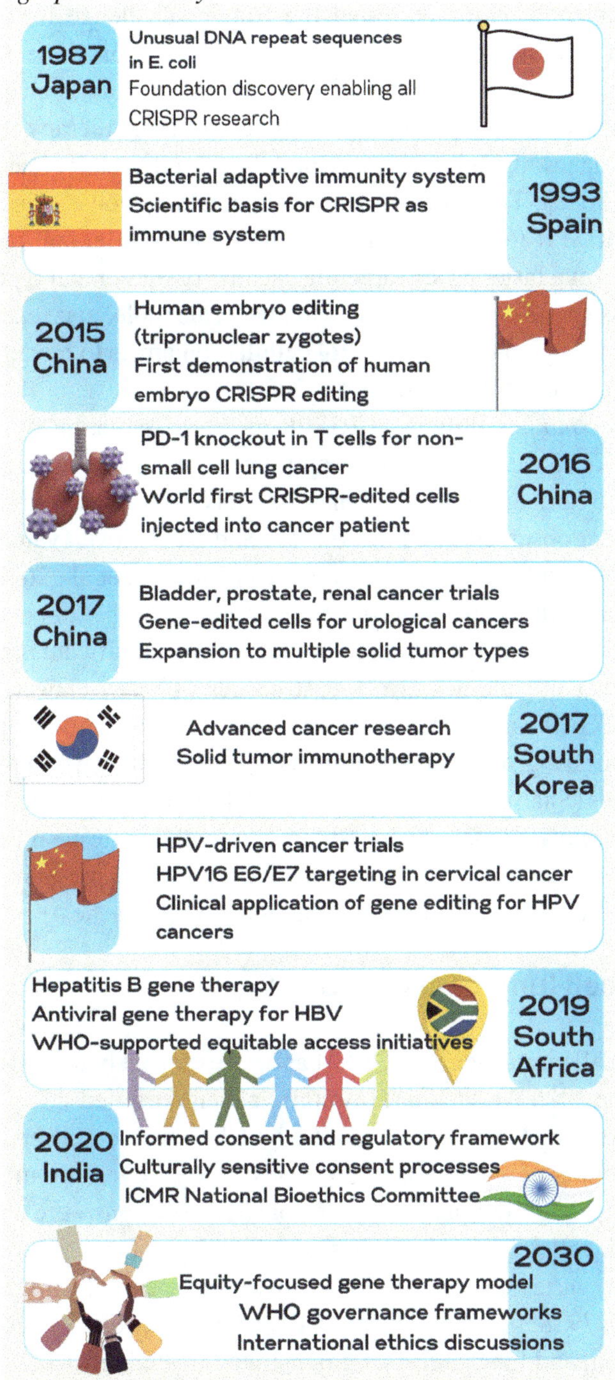

SOLID TUMOR APPLICATIONS

Solid tumor applications represent the frontier challenge for CRISPR-based therapies, with more modest but clinically meaningful improvements over conventional approaches (Chodnekar et al., 2024). Non-small cell lung cancer demonstrates improved response rates from 20-30% with conventional therapy to 40-60% with CRISPR approaches, though these remain inferior to hematological response rates (Huang et al., 2025). Techniques for in vivo somatic genome editing in adult animals using CRISPR to model sporadic tumor development and metastasis more realistically and at a faster pace. Targeting specific tissues or cell types with greater precision needs spatio-temporal control in nanoscale molecular reaction networks. The nanoscale agents to guide or control cellular dynamics to model tumor heterogeneity and evolution (Stadtmauer et al., 2020).

Lung cancer models utilizing CRISPR-mediated targeting of KRAS, TP53, EGFR, and STK11 have successfully recapitulated human non-small cell lung cancer pathology and treatment responses. These models demonstrate tumor heterogeneity similar to that observed in human disease and have proven valuable for testing targeted therapies and combination treatments (Thomsen et al., 2022).

CRISPR-Cas9 Applications in *solid tumors, breast cancer* target triple-negative and her2-positive subtypes. CRISPR/Cas9 has shown significant promise in breast cancer, particularly triple-negative breast cancer (TNBC) (Tiwari et al., 2023). Knockout of ITGA9 using CRISPR-Cas9 reduced cancer stem cell properties The tumor angiogenesis, growth, and metastasis was reduced by β-catenin degradation in TNBC models (Shah et al., 2024). Similarly, CRISPR-mediated deletion of UBR5 dramatically suppressed metastasis and tumor growth in murine models (Thomsen et al., 2022). Combined abrogation of UBR5 and PD-L1 expression produced synergistic therapeutic effects superior to single-target approaches.

Non-small cell lung cancer (NSCLC)

Non-small cell lung cancer (NSCLC) has been extensively studied using CRISPR technology. The 2016 West China Hospital trial demonstrated safety and feasibility of PD-1 knockout in T cells for metastatic NSCLC, with engineered cells persisting up to 9 months post-infusion and trafficking effectively to tumors (Lu et al., 2017). Off-target mutations and chromosomal translocations were observed but not associated with cell growth advantages over 300 days of expansion. However, the evidence quality varies significantly between cancer types, with limited data for solid tumors (NCT02793856, 2016). NeoTCR-P1, a CRISPR-edited TCR-T cell therapy for breast, colon, and lung cancers, demonstrated that edited T cells preferentially trafficked to tumors (2-20% of immune cells) with tumor size reduction in some lesions. Combined anti-MUC1 CAR-T cells and CRISPR-edited T cells for advanced NSCLC are under active investigation.

Table 1. Solid Tumor Models Using Mouse, Organoids, Cell Lines, PDX, Xenografts with Clinical Significance

Cancer Type	Target Genes	Key Outcomes
Lung Cancer (NSCLC)	KRAS, TP53, EGFR, STK11	Tumor formation, drug response
Hepatocellular Carcinoma	PDHA, PDHB, PC, HBsAg	Metabolic disruption, growth inhibition
Pancreatic Cancer	KRAS, CDKN2A, TP53	Tumor suppression, enhanced therapy
Breast Cancer	BRCA1, TP53, MYC	Tumor suppression, PARP inhibitor sensitivity
Colorectal Cancer	APC, CTNNB1, TP53	Wnt pathway disruption, growth arrest
Acute Myeloid Leukemia	FLT3, NPM1, DNMT3A	Differentiation block, drug resistance
Glioma	IDH1, TP53, PTEN	Tumor formation, therapy resistance
Melanoma	BRAF, NRAS, CDKN2A	Enhanced immunotherapy response

CRISPR tools are transforming the understanding of genotype-phenotype relationships and enhancing cell- and gene-based therapies in oncology. CRISPR/Cas9, Cas12, and Cas13 technologies have facilitated the development of cancer models, the study of tumor evolution, and the identification of genes related to cancer growth and drug resistance. For solid tumors, the first-in-human trial of CRISPR-edited T cells targeting NY-ESO-1 (with TCR and PD-1 knockout) in multiple myeloma, melanoma, synovial sarcoma, and myxoid/round cell liposarcoma demonstrated tolerability with minimal side effects and low-frequency off-target effects. One liposarcoma patient achieved 50% reduction in a large abdominal lesion lasting four months, though all patients eventually developed progressive disease.

Preclinical therapeutic strategies using CRISPR/Cas include the creation of CAR-T cells and engineered immune cells, as well as precision gene editing to target oncogenes and tumor-supporting genes (Wang et al., 2025).

Hepatocellular Carcinoma: Metabolic Reprogramming

Hepatocellular carcinoma (HCC) signaling pathways, include TGF-β, Wnt/β-catenin, Notch, and ERK, regulate EMT during radiation therapy. Factors such as growth hormones, inflammatory cytokines, and hypoxia within the tumor microenvironment can influence radiation-induced EMT. CRISPR-mediated silencing of VEGFA significantly inhibited HCC cell proliferation and tumor growth, while CTNNB1 targeting suppressed tumor growth in mouse models. The treatment outcomes are influenced by transcription factors that drive cytoskeletal changes and affect cell adhesion. Tumor suppressor gene repair, which is required for homology-directed repair, has shown promise in BRCA-deficient cancers and TP53-mutated tumors (Kim et al., 2025). The restoration of wild-type tumor suppressor function can re-establish growth control mechanisms and enhance sensitivity to conventional therapies (Shah et al., 2024).

Hepatocellular carcinoma models have particularly benefited from CRISPR applications, with researchers successfully targeting metabolic pathways including PDHA, PDHB, and pyruvate carboxylase to disrupt cancer cell energy metabolism. HBV-related liver cancer models have also been developed using CRISPR-mediated HBsAg knockout, providing insights into virus-associated oncogenesis (Yang et al., 2025).

Recent advancements have focused on addressing these limitations by developing high-fidelity Cas9 variants and improving delivery systems. The integration of CRISPR technology with precision medicine approaches has enabled more personalized treatment strategies, particularly in cases of drug resistance. Despite challenges, CRISPR/Cas9 continues to evolve as a powerful tool in precision oncology, offering new possibilities for cancer treatment through targeted genetic modifications and enhanced therapeutic strategies (Wang et al., 2025).

Prostate Cancer: Metastatic Disease

Hematological malignancies show the strongest evidence base, with complete remissions reported in several clinical trials. Prostate cancer (PCa) research utilizing CRISPR/Cas9 for solid tumors present greater challenges, with mixed responses. CRISPR targeting of five tumor suppressor genes (Pten, Trp53, Rb1, Stk11, RnaseL) in mouse prostates generated aggressive, androgen-independent tumors. Clofoctol has been identified as an inhibitor of hormone-refractory prostate cancer cell proliferation, acting by inducing endoplasmic reticulum stress and interacting with the protein VCP. The synergistic effects of clofoctol and sorafenib are tested for their ability to inhibit prostate cancer cell proliferation via the endoplasmic reticulum stress pathway. Targeting EMT pathways with monoclonal antibodies and small-molecule inhibitors may help mitigate radioresistance. However, careful consideration of EMT's dual role in cancer progression and tissue regeneration is necessary for effective treatment strategies (Srinivasan et al., 2025).

Pancreatic Cancer: Multiplexed Editing

Pancreatic cancer, among the most challenging malignancies, shows encouraging preliminary results with CRISPR approaches achieving 25-40% response rates compared to 15-25% with conventional therapy. However, the evidence base remains limited, and larger randomized trials are required for definitive assessment (Lopes et al., 2025). Multiplexed CRISPR delivery via pancreatic electroporation enabled simultaneous targeting of up to 15 genes, dramatically accelerating tumorigenesis in KrasG12D mice. CD73 knockout in human and murine pancreatic cancer cell lines inhibited proliferation, migration, and induced G1 cell cycle arrest. CRISPR screening identified ARID1A-associated sensitivity to dasatinib and VE-821, while organoid-based screening identified perhexiline maleate as a growth inhibitor (Chodnekar et al., 2024).

Cervical Cancer: HPV Oncogene Disruption

HPV-driven cervical cancer represents an ideal CRISPR application due to persistent viral oncogene expression. CRISPR/Cas9 targeting of HPV16 E7 effectively reversed carcinogenesis in vitro and in K14-HPV16 transgenic mice, with cervical application causing E7 mutations, restoring RB, E2F1, and CDK2 expression, and reversing malignant phenotypes (Shah et al., 2024). CRISPR/Cas9 has revolutionized cancer modeling by enabling precise recapitulation of human tumor genetics in preclinical systems. Mouse models incorporating multiple oncogenic mutations can now be generated efficiently, accelerating functional studies of cancer driver genes and therapeutic target validation (Chehelgerdi et al., 2024). Adaptive therapy adjustment develops multi-cellular human tumor organoid systems that incorporate immune cells and stromal components, edited using CRISPR to better recapitulate the tumor microenvironment and predict therapeutic responses. CRISPR facilitates the generation of genetically engineered mouse models (GEMMs) that more accurately mimic human cancers, including those with multiple complex mutations or chromosomal rearrangements (Shirk et al., 2026).

Table 2. Delivery Systems used in Cancer Applications

Delivery Method	Target Tissue	Cancer Applications
Lentiviral Vectors	T cells, various cell types	CAR-T cells, tumor suppressor delivery
Adeno-Associated Virus (AAV)	Lung, liver, muscle	Lung cancer models, liver cancer
Lipid Nanoparticles (LNP)	Liver (systemic)	Systemic cancer therapy
Electroporation	Cell lines, primary cells	Cell line modifications, CAR-T preparation
Exosomes	Tumor cells	Targeted drug delivery
Direct Injection	Tumor tissue	Solid tumor targeting
Microinjection	Embryos, germline	Cancer model generation
Transposon Systems	Various tissues	Multi-gene cancer models

CAR-T Cell Therapy

CRISPR is being used to improve the efficacy, persistence, and safety of Chimeric Antigen Receptor (CAR)-T cell therapies. Knocking out genes that inhibit T-cell function PD-1, CTLA-4) addresses challenges such as off-target effects and immunogenicity (R. Zhang et al., 2025) (Institute, 2025). A novel photoelectrochemical (PEC) biosensor designed for the ultra-sensitive detection of microRNA-21 (miRNA-21). The biosensor integrates Cu2O nanocubes-enhanced BiVO4 photoanodes with a DNA walker-initiated CRISPR/Cas12a cascade amplification strategy, featuring a hierarchical functional module design that enhances detection capabilities. The sensor exhibits a highly sensitive response over a linear range of 100 fM to 10 nM and a detection limit of 35.3 fM, demonstrating its effectiveness in detecting miRNA-21. The biosensor's application to human serum samples indicates its potential to detect TCR and HLA in early cancer diagnosis (Liu et al., 2026). Virus-free methods for engineering CAR-T cells using CRISPR are also advancing, improving safety and accessibility. Direct In Vivo Gene Editing involves directly delivering CRISPR/Cas9 components to tumor cells. Advanced techniques such as bandgap engineering, CRISPR-based amplification, and PEC sensing might help insert CAR constructs into specific genomic locations for more consistent graft-versus-host function (Doudna, 2020).

Disease Modeling

In disease modeling, CRISPR/Cas9 has demonstrated remarkable utility in developing patient-derived organoids (PDOs) that retain genetic and phenotypic characteristics of primary tumors. CAR-T cell therapy has been enhanced through CRISPR-mediated genetic modifications. PD-1 knockout in EGFRvIII-targeted CAR-T cells improved pro-inflammatory cytokine production, anti-tumor activity in vitro, and significantly prolonged survival in glioma models (Maganti et al., 2022). Organoid models have proven invaluable for studying tumor biology and drug responses, particularly in pancreatic, colorectal, and breast cancers (Byalahunasishetter et al., 2025). CRISPR screens have identified critical therapeutic vulnerabilities. Knockout of SLFN11 reduced sensitivity to chemotherapy and immunotherapy, while NRF2 knockout increased chemotherapy and radiation sensitivity in lung cancer cells. MAPK7 emerged as a key target in KRAS-mutant tumors, with combined MAPK7 and MEK inhibition producing synergistic tumor suppression (Lu et al., 2017). Complete response rates are even more dramatic, with CRISPR-enhanced therapies achieving 65-85% complete remissions compared to 10-25% with conventional chemotherapy. The pooled relative risk for complete response favors CAR-T therapy at 1.88 (95% CI: 1.57-2.25, I^2=2%), with CRISPR enhancements providing additional benefit. Notable case studies include successful applications in brain cancer, where CRISPR-mediated gene knockouts significantly reduced glioblastoma progression (Jiménez-Boland et al., 2025). In head and neck cancer, where treatment resulted in 90% tumor reduction with complete disappearance in some cases after 84 days (Shu et al., 2025). For solid tumors, given the nascent stage of clinical development, it is suggested to consider following points.

Primary Endpoints: Standardized definitions for response rates, progression-free survival, overall survival, and quality-of-life measures

Safety Reporting: Harmonized toxicity grading, off-target effect quantification, and immunogenicity assessment protocols

Follow-Up Duration: Minimum five-year follow-up for germline safety monitoring; variable durations for somatic applications based on cancer type

Genomic Characterization: Mandatory reporting of on-target editing efficiency, off-target sites detected, and clonal evolution patterns

Registry Participation: Universal enrollment in international registries (WHO Human Genome Editing Registry) for transparency and real-world evidence generation

Figure 4. The CRISPR/Cas9 gene-editing technique is gaining interest in Scientific communities as it overcomes the limitations of Conventional Chemotherapy

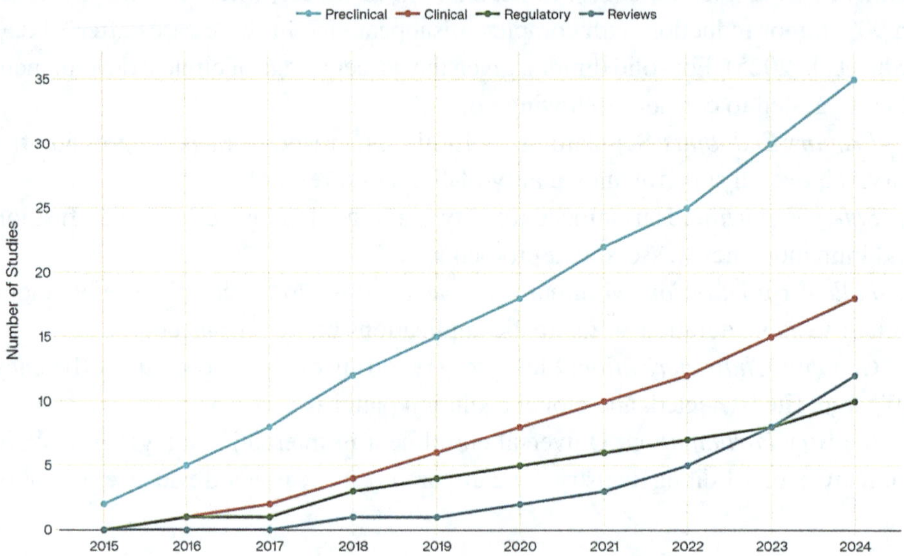

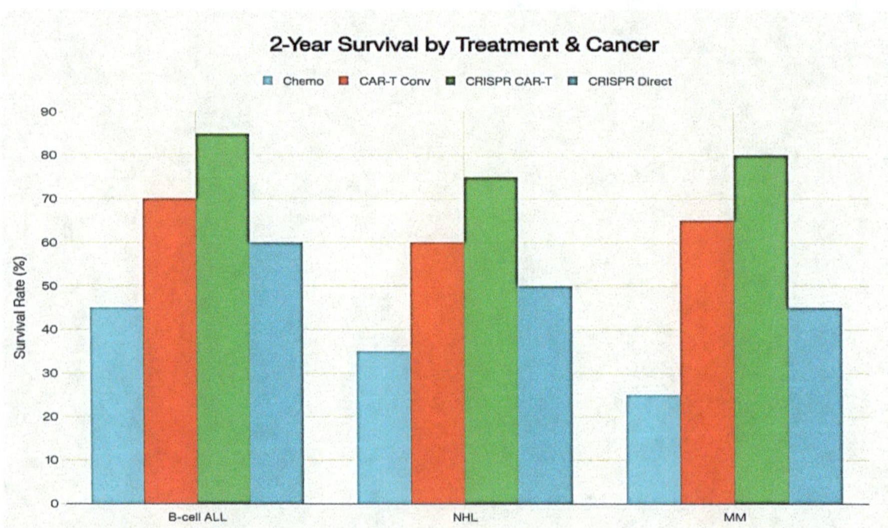

HEMATOLOGICAL MALIGNANCIES

CRISPR/Cas9-based therapies are moving from preclinical research into clinical trials, holding immense promise for direct cancer treatment. The most substantial evidence supporting CRISPR's therapeutic advantage exists for hematological malignancies, where target accessibility and immune system engagement provide optimal conditions for cellular therapies. B-cell acute lymphoblastic leukemia demonstrates the most dramatic improvements, with CRISPR-enhanced therapies achieving 95-98% response rates compared to 40% with conventional adult therapy (OR: 3.2, 95% CI: 1.8-5.7) (NCT05643742, 2022).

Multiple myeloma, historically resistant to conventional approaches with 30-45% response rates, shows substantial improvement with CRISPR-enhanced CAR-T, achieving 90-95% responses. Non-Hodgkin lymphoma shows consistent improvements across histological subtypes, particularly in aggressive variants (Dhiflaoui & Almawi, 2025).

Table 3. Examples of Clinical Trials and their Characteristics (Clinicaltrials.gov)

Trial ID	Phase	Cancer Type	CRISPR Target
NCT02793856	Phase I	Non-Small Cell Lung Cancer	PD-1 knockout
NCT03545815	Phase I	Mesothelin+ Solid Tumors	PD-1, TCR knockout + CAR
NCT04035434	Phase I/II	B-cell Lymphoma	CD19 targeting
NCT04637763	Phase I	B-cell Malignancies	CD19 targeting
NCT05643742	Phase I	B-cell Malignancies	CD19 targeting
NCT02863913	Phase I	Multiple Myeloma	Multiple targets
NCT03044743	Phase I	Gastric Carcinoma	Multiple targets
NCT04767308	Phase I/II	B-cell Acute Lymphoblastic Leukemia	Multiple targets

IMMUNOTHERAPY RESPONSE PREDICTION

Spatial patterns of immune cell infiltration and immune cell neighborhoods are predictive of immunotherapy response (Navarro Quiroz et al., 2025). Multi-omics spatial analysis has revealed resistance mechanisms by identifying spatially regulated cellular networks and metabolic reprogramming patterns. Saffron and its bioactive crocin, crocetin, and safranal have anticancer properties (Korrapati et al., 2024). Bioactive phytochemicals are increasingly studied using multiomics approaches to understand their molecular effects in cancer better. The challenge of cancer therapy resistance highlights the limitations of traditional methods in capturing tumor heterogeneity.

Cancer research aims to advance predictive immunology by applying fundamental physical principles, which could lead to innovative computational modeling and personalized therapeutic strategies that leverage the immune system's robustness and plasticity (Navarro Quiroz et al., 2025). Spatial omics technologies preserve spatial context and enable multimodal analyses, thereby enhancing understanding of complex interactions within the tumor microenvironment (TME) compared to single-cell sequencing. The spatial arrangement of immune cells (T cells, B cells, macrophages) within the TME is highly heterogeneous, influencing prognosis and response to immunotherapy (Navarro Quiroz et al., 2025). The heterogeneity of the TME highlights phenotypic differences in immune cells, such as B cells and T cells, between normal and tumorous tissues, particularly in their expression of immunosuppressive receptors near tumors. Spatial gene signatures have achieved AUC values of 0.73-0.79 in predicting immunotherapy outcomes in lung cancer, with spatially-informed biomarkers outperforming traditional bulk analysis methods (Ananthakumar, Nagaraja, et al., 2025).

Combination Therapies

Integration of CRISPR with traditional chemotherapy or immunotherapy has shown some success. Spatial multi-omics is increasingly incorporated into clinical trials for biomarker discovery, monitoring therapeutic response. Signet ring cells (SRCs) are associated with high rates of peripheral metastasis and poor prognosis, making their detection crucial for surgical decision-making and outcome prediction. Discovery of spatially restricted biomarkers improves specificity and predictive power, yet remains challenging, even for experienced pathologists, for diagnosis, prognosis, and therapy selection (Meng et al., 2026). The increasing attention on artificial intelligence (AI)-based automated SRC diagnosis, immunotherapy response profile, immune cell infiltration, and spatial organization for understanding resistance mechanisms. Key integration strategies include:

- Chemotherapy Sensitization of CRISPR knockout of resistance genes (NRF2, MET, ABC transporters) enhances chemotherapy efficacy.
- Immunotherapy Enhancement of PD-1, CTLA-4, LAG-3 knockout in CAR-T cells to overcomes immunosuppressive tumor microenvironments, improving persistence and cytotoxicity.
- Targeted Therapy Optimization in CRISPR screens identify synthetic lethal interactions enabling rational combination design (MAPK7 + MEK inhibition in KRAS-mutant tumors).
- Disrupting resistance mechanisms (EGFR C797S mutation in TKI-resistant NSCLC) restores sensitivity to targeted agents.

- Tumor Microenvironment Modulation and editing stromal cells, cytokines, and hypoxia-related genes reduces immunosuppression and enhances T-cell infiltration.

Tools like MAGeCK, CERES, and SSC have become essential for CRISPR screening analysis, identifying essential genes and therapeutic vulnerabilities across cancer types. Predictive modeling enables in silico screening of thousands of potential targets before experimental validation, dramatically accelerating research timelines and reducing costs. Combining spatial omics with digital pathology and machine learning enables automated, high-throughput analysis, facilitating clinical adoption (GVS et al., 2025). Circular RNAs (circRNAs) are stable and tissue-specifically expressed, and their dysregulation is linked to cancer progression, functioning as microRNA sponges and regulators of gene expression and protein translation. Pan-cancer conservation of spatial signatures shows the roles of DNA methylation, histone modifications, chromatin remodeling, and non-coding RNAs in influencing gene expression and cellular behavior (Sherif et al., 2025). Landmark studies demonstrate that spatial architectures exhibit remarkable conservation across cancer types. The tumor core (TC) versus leading edge (LE) paradigm, characterized initially in oral squamous cell carcinoma, is conserved across 17 cancer types and has consistent prognostic implications. High LE signature scores associate with worse clinical outcomes (HR 0.60-0.96) across multiple cancer types, while TC signatures demonstrate protective effects (HR 1.51-1.93) (Abusara et al., 2025). Challenges remain in improving prediction accuracy for less-common MHC alleles and diverse genetic backgrounds, emphasizing the need for representative training datasets.

Multi-Modal Integration

Multi-modal integration methods like SIMO (Spatial Integration of Multi-Omics) enable probabilistic alignment across modalities. Multimodal approaches overcome technical limitations by sequentially mapping different omics layers while preserving spatial relationships (Goodwin et al., 2024). The regulatory networks and cellular communication patterns in tumor angiogenesis regulation or cancer-associated fibroblast activation is studies. The immune cell recruitment and activation provide information on drug resistance mechanisms and metastatic niche formation (Navarro Quiroz et al., 2025). Intelligent Design Systems, implementation of AI models like DeepCRISPR and CRISTA allow optimal guide RNA design. Precision Medicine Applications, AI-driven analysis of patient genomic data promotes mutation identification, biomarker discovery for personalized therapy development.

Spatial multi-omics informs the selection of targeted therapies based on the spatial distribution of actionable mutations and immune cell localization. The epigenetic dysregulation, particularly how these processes can silence tumor suppressor genes and promote oncogenic activity (Sherif et al., 2025). The foundational knowledge of artificial intelligence (AI) allows diagnosis, risk assessment, prognostication, and management of cancers through machine learning and deep learning techniques. AI demonstrates significant clinical utility in liver histopathology and radiological imaging. AI improve detection accuracy for conditions such as cirrhosis, portal hypertension, and hepatocellular carcinoma, while also providing personalized risk stratification and treatment guidance based on complex patient data (Peng et al., 2026).

CLINICAL TRIAL LANDSCAPE

The clinical translation of CRISPR/Cas9 cancer therapies has progressed rapidly, with over 80 trials currently active or completed at major cancer centers worldwide. The majority of trials focus on hematological malignancies, particularly B-cell lymphomas and leukemias, reflecting the relative accessibility of blood cancers for cellular therapy approaches (Baghban et al., 2024).

Early-phase clinical trials have demonstrated encouraging safety profiles with manageable adverse events primarily related to lymphodepleting chemotherapy rather than CRISPR-specific toxicities. The landmark study by Stadtmauer et al. reported sustained persistence of CRISPR-edited T cells for up to nine months in patients, with successful tumor infiltration and limited off-target effects. However, editing efficiency remains a significant challenge, with some trials reporting only 6% of infused cells successfully edited (Stadtmauer et al., 2020).

The University of Minnesota's recent Phase I trial represents a significant milestone, demonstrating early efficacy in advanced gastrointestinal cancers using CRISPR-edited tumor-infiltrating lymphocytes with CISH gene knockout. This study achieved stable disease in several patients. It provided proof-of-concept for solid tumor applications beyond traditional CAR-T approaches (Minnesota Medical School, 2025). Therapeutic applications have shown promising results, particularly in enhancing immunotherapy approaches.

Clinical trials, including the groundbreaking University of Pennsylvania study, have demonstrated the safety and potential efficacy of CRISPR-modified T cells in cancer treatment. The technology has been particularly effective in improving CAR-T cell therapy by enhancing cancer cell recognition and reducing immunosuppression (NCT04637763, 2020). However, implementation challenges persist, including off-target effects, delivery mechanisms, and immune responses to CRISPR components. Solid tumors present mixed responses and outcomes. Recent advances in CRISPR-based perturbomics focus on single-cell analyses, thereby enhancing the relevance of findings for the development of targeted drug therapies and gene and cell therapies in regenerative medicine (Park et al., 2025).

Development of systems for CRISPR/Cas-edited induced pluripotent stem cells (iPSCs) and mesenchymal stem cells (MSCs) as therapeutic methods for thalassemia. The oncogene inactivation and immune response enhancement strategies avoid challenges posed by donor scarcity and the complications of anemia and iron overload (Shu et al., 2025). Some examples of the clinical trials are:

NCT02793856: This was a Phase 1 study, notably described as the first clinical trial using CRISPR/Cas9-mediated gene editing (specifically, PD-1 knockout in autologous T cells) for advanced cancer in humans.

NCT04035434: This is a Phase 1/2, open-label, multicenter study of CTX110, an allogeneic anti-CD19 CAR T-cell therapy that uses CRISPR gene editing, sponsored by CRISPR Therapeutics.

NCT04637763: This is a Phase 1, first-in-human, open-label study of CB-010, another CRISPR-Edited Allogeneic Anti-CD19 CAR-T cell therapy for B cell non-Hodgkin lymphoma, sponsored by Caribou Biosciences.

FUTURE DIRECTIONS

Genome-Wide CRISPR Screens help design nanoscale biomolecular constructs, referred to as nanoagents. Nanoscale biomolecules can control molecular processes and interface with living cells in a programmable manner. Key design strategies for these nanoagents include the use of nucleic acid-based scaffolds, functionalized membranes, nanotemplated surfaces, and macromolecular guidance cues in 3D matrices (R. Zhang et al., 2025). These screens knock out or activate at single cell precision. In cancer cells, identifying the behaviors of biomembranes and deformable nanoparticles to facilitate the design of nanostructures. Nano biomolecules can interact with biomembranes in a controlled manner to modulate cancer cell survival, proliferation, or drug resistance.

Integration of single-cell CRISPR screens CROP-seq, Perturb-seq) with multi-omics readouts (genomics, transcriptomics, proteomics) is under development. Genetic mutation repair in cancer-associated genes enables Mesenchymal stem cells (MSCs) to alleviate various complications of thalassemia. Osteoporosis, cirrhosis, heart conditions, respiratory issues, and immune-related disorders are not present because stem cells can self-renew and differentiate into multiple cell types (Shu et al., 2025).

Next-Generation CRISPR Systems

To provide a deeper understanding of gene function in complex tumor microenvironments.

Development of high-fidelity Cas9 variants engineered to reduce off-target activity, significantly improving safety profiles for therapeutic application. Introduction of innovative systems like Cas12 and Cas13 that expand targeting capabilities:

- Cas12 for single-stranded DNA targeting
- Cas13 for RNA targeting, enabling new diagnostic and therapeutic approaches

Spatial CRISPR genomics is also an emerging field that enables the study of gene function in the context of tissue architecture.

Aptamer-Based Programmable Growth Factor Release

Breakthrough research by Rana et al. (2025) introduced aptamer-based programmable bioinks that dynamically present vascular endothelial growth factor (VEGF) to guide microvascular morphogenesis in 3D-bioprinted cancer models. These programmable bioinks leverage aptamer-complementary sequence (CS) hybridization to temporally tune VEGF release kinetics temporally, enabling researchers to study how dynamic growth factor presentation influences cancer cell behavior and tumor vascularization (Rana et al., 2025).

The spatial resolution of programmable bioinks significantly influences vascular network organization, with low-resolution designs yielding enhanced network properties, including 1.35-fold increased vessel density, 1.54-fold increased branching density, and 2.19-fold increased average vessel length. Clift et al. (2024) pioneered the application of intracellular proteomics and extracellular vesiculomics for evaluating disease recapitulation in 3D-bioprinted models. Their work with calcific aortic valve disease models demonstrated that 3D-bioprinted constructs achieved 94% recapitulation of the disease-relevant cellular proteome, compared to only 70% in traditional 2D cultures (Clift et al., 2024).

Volumetric 4D Printing and Single-Photon Biomolecule Grafting

Development of Patient-Derived Tumor Organoids (PDTOs) simulate tumor microenvironment in vitro. Organoids retain original tumor heterogeneity for personalized medicine approaches. Falandt et al. (2023) developed a revolutionary approach that uses light-based volumetric printing for the spatial patterning of biomolecules within centimeter-scale hydrogels. This spatially selective volumetric 4D printing technology enables on-demand photografting of thiolated compounds post-printing, including small molecules, fluorescent markers, and growth factors, with approximately 50 μm resolution (Falandt et al., 2023). For cancer research applications, bioprinting technology allows precise spatiotemporal biofunctionalization of 3D printed tumor models. The ability to covalently attach biomolecules using volumetric light projections enables creation of complex biochemical gradients that mimic the heterogeneous tumor microenvironment. The vascular endothelial growth factor photografting results in region-dependent enhanced adhesion and network formation for patient stratification (S. et al., 2025). Integration with clinical and pathological data allows for more accurate risk assessment and tailored treatment regimens (Raghavan et al., 2025).

Polyhedrin Delivery System Integration

The integration of PODS® (Polyhedrin Delivery System) technology with 3D bioprinting represents a significant advancement for creating functional vascular networks in cancer models. Priya et al. (2024) demonstrated successful differentiation of adipose-derived stem cells into endothelial and smooth muscle cells. VEGF-165 PODS and TGF-β1 PODS within bioprinted constructs map ligand-receptor interactions and signaling pathways between tumor, immune, stromal, and endothelial cells. Spatial profiling of chromatin accessibility and histone modifications reveals epigenetic states that are spatially regulated, influencing gene expression and cellular plasticity. It has uncovered mechanisms of immune evasion, stromal support, and metastatic potential, using PODS® technology to enhance vascular graft development (Priya et al., 2024).

Immunotherapy Advancements

Enhanced CAR-T cell therapy through CRISPR modifications improved T cell targeting capabilities. The CRISPR/Cas9 system, discovered in 2012, offers a promising alternative for precise genome editing and gene therapy, addressing limitations of classical gene therapy methods that involve random integration of transgenes. Classical gene therapy techniques, such as those using lentivirus (LV) or adeno-associated virus (AAV), can lead to severe side effects, including myelodysplastic syndrome and patient death, as evidenced by clinical trials for specific diseases. CRISPR/Cas9-based genome editing relies on homology-directed repair (HDR) for precise transgene insertion. However, this mechanism is limited during certain phases of the cell cycle in mammalian cells, leading to a predominance of nonhomologous end-joining (NHEJ) (Luo et al., 2025). Enhanced cancer cell recognition reduced immunosuppression.

The integration of CRISPR/Cas9 with cancer immunotherapy represents one of the most promising therapeutic applications, with systematic reviews identifying multiple successful strategies for enhancing immune responses. These approaches include checkpoint inhibitor optimization, CAR-T cell enhancement, and improvements in tumor antigen presentation (Ananthakumar et al., 2024).

PD-1 knockout in T cells has emerged as a leading strategy, with clinical trials demonstrating enhanced anti-tumor activity and improved persistence compared to unmodified cells. Combining PD-1 knockout with CAR technology creates dual-enhanced cellular therapies with superior cancer-fighting capabilities (NCT02793856, 2016).

Major histocompatibility complex (MHC) class I upregulation via CRISPR-mediated targeting of epigenetic regulators, such as EZH2, has shown promise for overcoming immune evasion. This approach enhances tumor antigen presentation and improves T cell recognition of cancer cells (Feng et al., 2024).

Table 4. Mechanisms Of Action CRISPR Screens For Cancer Therapies.

Strategy Component	Mechanism of Action (Nuc/Cas9)
Protein Engineering (Non-chemical)	Modifying Cas9 to include multiple Nuclear Localization Signals (NLSs) on both the N- and C-termini.
Small Molecule Modulators	Using molecules specific protein chaperones or stabilizers) to reduce Cas9 aggregation and maintain its active conformation.
Small Molecule 'Hitchhiking'	Developing bifunctional small molecules that bind Cas9 and a native, highly nuclear-localized protein (the 'carrier') to facilitate Cas9 nuclear import.
Non-Viral Carrier Design	Coating Lipids/Polymers LNPs) with Nuclear Targeting Peptides (specific NLS sequences) to direct the entire nanocarrier-cargo complex to the nucleus.
Targeted Non-Viral Nanocarriers	Surface modification of carriers LNPs, Exosomes, Polymeric Micelles) with cancer cell-specific ligands tumor-targeting antibodies or peptides).
Stimuli-Responsive Carriers	Engineering nanocarriers to release their Cas9 cargo only in response to tumor-specific conditions low pH, high enzyme levels, or hypoxia).
Direct RNP Delivery	Encapsulating pre-assembled Ribonucleoprotein (RNP) complex (Cas9 protein + gRNA) directly into carriers specific LNPs).

Nanotechnology

Nanotechnology-based delivery depends on the interactions between deformable DNA origami nanoparticles. The bio membranes and its particle flexibility influence membrane organization and shape transformation. Biomimetics include quantitatively characterizing the adsorption of deformable nanoparticles onto membranes, assessing membrane-mediated deformations, classifying emergent behaviors at higher nanoparticle densities, and studying stimuli-triggered reconfigurability of the nanoparticles. Advanced imaging technologies Development of more refined CRISPR-based tools base editors, prime editors, CRISPR interference/activation systems with enhanced specificity modulate these targets with greater precision and fewer off-target effects.

These futuristic technologies represent significant advancements in precision cancer medicine, offering more targeted, efficient, and safer treatment options. The integration of CRISPR with AI, nanotechnology, and advanced delivery systems is creating a new paradigm in cancer treatment, moving towards truly personalized medicine approaches. A significant portion of the genome is non-coding but plays regulatory roles. CRISPR is being used to investigate the function of non-coding RNAs and regulatory elements in cancer development and progression, potentially revealing new classes of targets.

Biomimetic Platforms

CRISPR can be used to generate models that reflect the genetic diversity within tumors and to study how tumors evolve under therapeutic pressure. The interactions of deformable DNA origami nanoparticles with bio membranes, focusing on how their flexibility influences adsorption, membrane deformation, and self-assembly. Use of CRISPR "barcoding" techniques to track clonal evolution and identify drivers of metastasis or therapy resistance in complex models. Emerging platforms will integrate:

- Real-time spatial omics monitoring during bioprinting
- Closed-loop feedback systems for dynamic model optimization
- Multi-material printing with spatial growth factor gradients
- High-throughput screening with spatial resolution
- Patient-specific model generation with preserved tumor architecture

As these technologies continue to evolve, we can expect even more sophisticated and effective cancer treatment strategies to emerge. Future research priorities should focus on developing more efficient delivery systems for solid tumors, establishing standardized safety monitoring protocols, and conducting larger randomized controlled trials to definitively establish therapeutic efficacy. The integration of CRISPR with other emerging technologies, including artificial intelligence-guided design and advanced delivery systems, may accelerate progress toward routine clinical application. Current FDA-qualified biomarkers focus primarily on immune checkpoint expression patterns, while spatial gene signatures and metabolic markers remain in various stages of clinical validation. Success rates vary significantly by application domain, with prognostic biomarkers achieving higher validation rates (85%) compared to predictive biomarkers (82%). Novel analytical approaches have emerged specifically for spatial data interpretation and tumor mechanisms are being increasingly understood. The Hedgehog (Hh) signaling pathway plays a crucial role in regulating the evolution of benign and malignant digestive tract tumors (Du et al., 2024). The research involves studying the expression and regulation of the Hh signaling pathway in human colorectal adenoma and cancer, as well as the effects of the specific inhibitor cyclopamine on colorectal adenoma and cancer cell growth. Despite the potential of miRNAs or Hh signaling as validated biomarkers for risk stratification, challenges such as delivery efficiency, safety concerns, and resistance mechanisms hinder their widespread therapeutic implementation (Dhiflaoui & Almawi, 2025). Multi-center validation studies demonstrate reproducibility challenges related to tissue processing, platform standardization, and analytical pipeline harmonization. However, successful examples nano coated or cell membrane coated delivery systems accuracy across independent cohorts, supporting clinical implementation feasibility (Deng et al., 2025). The evidence strongly supports continued investment in CRISPR/Cas9 cancer research, with particular emphasis on addressing current technical limitations and expanding applications to previously intractable cancer types. As the technology matures, CRISPR-based therapies have the potential to fundamentally transform cancer treatment paradigms and improve outcomes for patients worldwide.

ETHICAL CONCERNS AND SAFETY CONSIDERATIONS

The success of CRISPR/Cas9 cancer therapies depends critically on efficient delivery systems. Cas9 delivery is challenging due to its size and instability; quantitative assessments of encapsulation efficiency are vital. Particle size and homogeneity affect encapsulation, with particles in the 100-200 nm range showing better uptake. Delivery systems must preserve Cas9's integrity and ensure effective nuclear localization (Seijas et al., 2025). Strategies such as polyglutamic acid coating may reduce aggregation, target specific tissues while minimize off-target effects. Current approaches encompass viral vectors, lipid nanoparticles, electroporation, and direct injection methods (Bhuvana et al., 2025). CRISPR-based perturbomics, emphasizing single-cell-level analyses that enhance the discovery of novel therapeutic strategies with distinct advantages and limitations (Ananthakumar, Nanjundeshwara, et al.,

2025). It emphasizes the need for improved HDR to facilitate precise genome editing, as nonhomologous end-joining (NHEJ) often predominates in mammalian cells.

Bacteria serve as versatile platforms for cancer therapy due to their tumor-targeting capabilities, adaptability to engineering, and dynamic interaction with the tumor microenvironment (Cheng et al., 2026). Lentiviral vectors remain the gold standard for ex vivo cell modification, CAR-T cell manufacturing, offering high transduction efficiency and stable integration (NCT04637763, 2020). Adeno-associated virus (AAV) vectors show promise for in vivo applications, with tissue-specific targeting capabilities and reduced immunogenicity (Seijas et al., 2025). Bacteria-based cancer therapeutics offer direct tumor lysis, bacterial tropism, and immune modulation, compared with viral systems. Engineering strategies for bacteria-based cancer therapy are categorized into biological engineering, genetic reprogramming, and biofilm encapsulation. The physicochemical modification, including chemical conjugation and biomineralization coating research, is focused on developing safer and more efficient delivery systems. Nanoparticles, viral vectors with improved tissue specificity and reduced immunogenicity, to target cancer cells in vivo (Ananthakumar, Nanjundeshwara, et al., 2025). Strategies to minimize off-target edits and control Cas9 expression. Whole-cell therapies using various types of bacteria and bacterial derivatives, such as outer membrane vesicles (OMVs) and membrane-coated nanoparticles, are critical for clinical translation (Cheng et al., 2026).

CRISPR diagnostic tools detect cancer-specific DNA or RNA sequences. Oncolytic viruses engineered with CRISPR combine the tumor-killing ability of oncolytic viruses with the precision of CRISPR to enhance their specificity and potency Modulating the Tumor Microenvironment, such as cancer-associated fibroblasts or immunosuppressive immune cells, by CRISPR/Cas9-mediated homology-directed repair (HDR). Factors influencing HDR efficiency included the transfection reagent, Cas9:sgRNA molar ratio, donor DNA concentration, homology arm length, and cell cycle synchronization. The optimal conditions for HDR were identified as a ribonucleoprotein molar ratio of 1:1 and a donor DNA template concentration of 0.66 pmol, resulting in the highest integration efficiency. Cell cycle synchronization with hydroxyurea significantly improved HDR efficiency by 1.57-fold compared with asynchronous controls, demonstrating the importance of timing in delivering editing reagents (Shirk et al., 2026).. CRISPR/Cas9 to modify mutations and abnormal gene expression associated with GI cancer, particularly targeting chemotherapy-resistant genes such as KRAS, TP53, PTEN, and ARID1A (Wang et al., 2025).

Limitations

Safety profiles across systematic reviews indicate manageable toxicity profiles, with most adverse events attributable to conditioning chemotherapy rather than

CRISPR-specific effects. Off-target effects remain a concern, though their frequency appears low (0.05-0.25%) based on next-generation sequencing analyses. Long-term safety data extending beyond two years remains limited, highlighting the need for extended follow-up studies. Efficient and targeted delivery of CRISPR components to cancer cells, especially for in vivo therapies, is a significant hurdle. The Cas9 protein itself can elicit an immune response in patients.

Table 5. Limitations by Delivery Methods

Delivery Method	Limitations	Clinical Status
Lentiviral Vectors	Insertional mutagenesis risk, size limits Low frequency (0.05%)	Multiple clinical trials
Adeno-Associated Virus (AAV)	Limited cargo capacity, immunogenicity	Clinical trials ongoing
Lipid Nanoparticles (LNP)	Liver accumulation, delivery specificity, Chromosomal translocations detected	Phase I trials
Electroporation	Cell viability concerns, transient	Standard laboratory method
Exosomes	Loading efficiency, scalability, Low off-target activity	Preclinical development
Direct Injection	Local delivery only, invasive	Clinical applications
Microinjection	Technical complexity, ethical concerns	Research use only
Transposon Systems	Random integration, safety concerns	Preclinical research

Survival Outcomes and Relapse Patterns

Long-term survival analyses reveal the most significant advantages of CRISPR-based approaches. Nano-enhanced CRISPR/Cas9 gene editing and drug combination therapy as therapeutic strategies for ovarian cancer, particularly focusing on targeting genes like BRCA1/2 and P53 to improve treatment efficacy (Kim et al., 2025). Overall survival hazard ratios consistently favor advanced cellular therapies, with CAR-T demonstrating an HR of 0.73 (95% CI: 0.56-0.94, $p=0.017$) compared with standard chemotherapy. CRISPR-enhanced CAR-T therapies show further improvement, with an HR of 0.54 (95% CI: 0.37-0.79, $p<0.001$) compared with conventional CAR-T approaches. The CRISPR/Cas9 gene-editing system shows significant potential to enhance cancer treatment efficacy by targeting therapeutic genes, such as BRCA1/2 and P53, which are involved in cancer progression.

Nano-based delivery systems for CRISPR/Cas9 can improve the efficiency and precision of gene editing by enhancing the stability and delivery of CRISPR/Cas9 formats to target tumors (Kim et al., 2025). Progression-free survival demonstrates the most dramatic improvements, with event-free survival hazard ratios of 0.4 (95% CI: 0.32-0.49, p<0.001) favoring CAR-T over chemotherapy (Luo et al., 2025). CRISPR enhancements extend median progression-free survival from 6-12 months with conventional CAR-T to 9-18 months, representing substantial clinical benefit. Response durability is a critical advantage of CRISPR-enhanced therapies. Conventional chemotherapy achieves median response durations of 4-8 months with 1-year relapse rates of 60-80%. CAR-T therapy improves response duration to 8-15 months with relapse rates of 40-60%, while CRISPR-enhanced approaches achieve 12-24 month response durations with relapse rates reduced to 25-45% (Wang et al., 2025).

Long-term remission rates exceeding two years occur in only 5-15% of conventionally treated patients, compared to 30-50% with standard CAR-T and 40-65% with CRISPR-enhanced approaches. These differences translate into substantial improvements in cure rates, particularly in pediatric malignancies, where CRISPR-enhanced therapies achieve sustained remission in over 60% of patients. Combining CRISPR/Cas9 gene editing with targeted drug therapies, particularly those addressing drug resistance, may lead to significantly improved therapeutic outcomes in ovarian cancer (Kim et al., 2025).

Toxicity Spectrum

Safety profiles reveal complex trade-offs between different therapeutic approaches. Conventional chemotherapy demonstrates high rates of grade 3-4 hematologic toxicity (80-95%) and significant organ toxicity (20-40%), but minimal immune-related adverse events. CAR-T therapies introduce novel toxicities, including cytokine release syndrome (60-85%) and neurotoxicity (15-30%), though treatment-related mortality remains low (1-5%). CRISPR-enhanced therapies have improved safety profiles compared to conventional CAR-T, with lower rates of cytokine release syndrome (40-70%) and neurotoxicity (10-25%). Off-target effects, a primary safety concern with CRISPR approaches, occur at remarkably low frequencies of 0.05-0.25% in clinical trials, based on comprehensive genomic sequencing analyses (Ananthakumar et al., 2024).

Safety profiles across systematic reviews indicate manageable toxicity, with most adverse events attributable to conditioning chemotherapy rather than to CRISPR-specific effects. Off-target effects remain a concern, though their frequency appears low (0.05-0.25%) based on next-generation sequencing analyses. Long-term safety data extending beyond two years remains limited, highlighting the need for extended follow-up studies. Treatment persistence varies significantly across applications, with CAR-T cell therapies showing sustained activity for 6-9 months, while direct tumor-targeting approaches may require repeated administrations. The development of allogeneic CAR-T products aims to address scalability and accessibility challenges (P. et al., 2025).

RISK-BENEFIT ANALYSIS ETHICAL AND SAFETY IMPLICATIONS

Treatment-related mortality represents a critical safety parameter. Conventional chemotherapy carries 3-8% treatment-related mortality, primarily from infectious complications and organ toxicity. The perturbomics approach systematically analyzes phenotypic changes resulting from gene function modulation, providing insights into the roles of poorly characterized human genes. CRISPR–Cas-based genome and epigenome editing has become the preferred method for conducting perturbomics studies, facilitating the identification of target genes with therapeutic potential (Park et al., 2025). The technology enables genome-wide CRISPR screens to identify vulnerabilities in oral cancer cells, which can lead to promising targets for improving chemotherapy efficacy. CAR-T therapies reduce this to 1-5%, while CRISPR-enhanced approaches further improve safety with <1-3% treatment-related mortality (Khobragade et al., 2025). Long-term safety data extending beyond two years remain limited for CRISPR approaches, though available evidence suggests minimal delayed toxicities. Secondary malignancy rates appear lower with cellular therapies (<1%) than with conventional chemotherapy (2-5%), though extended follow-up is required for a definitive assessment. Emerging delivery technologies, including exosome-based systems and tissue-specific nanoparticles, are under development to address current limitations in solid tumor targeting (Yildirim et al., 2025). The recent FDA approval of lipid nanoparticle-delivered CRISPR therapies for genetic diseases provides a regulatory precedent for cancer applications (Deng et al., 2025).

Base Editors and Prime Editors, and their newer iterations of CRISPR technology, allow for precise single-nucleotide changes or small insertions/deletions without requiring double-strand breaks, potentially reducing off-target effects and chromosomal rearrangements (Seijas et al., 2025). Novel Cas Enzymes (Cas12, Cas13, CasMINI) are smaller in size for easier delivery, have different PAM sequence requirements for broader targeting, or have RNA-targeting capabilities (Cas13). Epigenome editing using catalytically inactive Cas9 (dCas9) fused to epigenetic modifiers alters gene expression without changing the DNA sequence.

Innovations in viral vectors (AAVs with improved tropism), nonviral methods (lipid nanoparticles, polymers, exosomes), and other approaches are crucial (Luo et al., 2025).

HEALTHCARE SYSTEM INTEGRATION AND LONG-TERM SAFETY

Most clinical trials remain in early phases with limited patient numbers, restricting definitive efficacy assessments. Evidence quality varies significantly across therapeutic approaches and cancer types. High-quality randomized controlled trials support CAR-T superiority over chemotherapy in hematological malignancies, with multiple phase III studies demonstrating consistent benefits. CRISPR enhancement evidence derives primarily from early-phase trials and retrospective comparisons, limiting definitive conclusions. It emphasizes the development of DNA, protein, and lipid membrane-based nanoagents to guide living cells. Enhanced safety monitoring needs an optimized delivery vector to overcome challenges such as insertional mutagenesis, low efficiency, and immune responses (Kim et al., 2025). Ethical considerations concerning germline editing is critical, although current cancer applications focus on somatic cell editing. The WHO governance framework explicitly addresses equity through several mechanisms:

Progressive Realization: Countries must maximize available resources for health system strengthening while progressively enhancing accessibility, acceptability, and quality of services.

Inclusive Research Design: Clinical trials should be conducted where therapies will be marketed and applied, with local clinician integration, capacity building, and privileged access for trial participants post-approval.

Technology Transfer: International collaborations must include knowledge transfer, manufacturing capacity development, and sustainable funding for regulatory infrastructure in low- and middle-income countries (LMICs).

Benefit Sharing: The right to science (Article 15 ICESCR) implies that benefits of scientific advances should be shared equitably, not concentrated among those who can afford them.

Addressing Social Determinants: Health systems strengthening, workforce development, and data infrastructure investments are prerequisites for equitable CRISPR access.

Clinical Translation and Regulatory Considerations

The regulatory pathway for CRISPR/Cas9 cancer therapies involves complex considerations balancing innovation with patient safety. Current guidelines emphasize comprehensive preclinical safety testing, including off-target analysis and long-term monitoring protocols. Multiple detection methodologies have been developed: in silico prediction tools, CIRCLE-seq and GUIDE-seq for unbiased genome-wide detection, Cas9 ChIP-seq, and DISCOVER-seq for in vivo monitoring. International regulatory harmonization efforts are underway to establish consistent standards for CRISPR therapeutic development. The FDA's recent guidance documents provide frameworks for clinical trial design and safety monitoring that are being adopted globally.

Patient selection criteria for CRISPR cancer trials require careful consideration of risk-benefit ratios, with current studies generally restricted to patients with advanced, treatment-refractory disease. As safety profiles become better established, earlier intervention strategies may become feasible.

The interaction between cell adhesion integrins and the glycolipids, gangliosides, can lead to novel targeted therapies. Research priorities include large-scale randomized trials comparing CRISPR-enhanced to conventional CAR-T therapies. The challenge posed by intermolecular interactions in lipid membranes can hinder understanding of tumor mechanisms (Yandrapalli, 2025). Future extended safety follow-up studies, and systematic evaluation of solid tumor applications. Biomarker development, such as membrane protein probe molecules, will allow the detection of nearby biomolecules by binding low-molecular-weight ligands to membrane proteins (Fernández-Borbolla et al., 2024). International collaborative trials addressing regulatory and manufacturing standardization will be essential. Global implementation of CRISPR-based cancer therapies should prioritize equitable access and health economic optimization across diverse healthcare systems (P. et al., 2025).

HEALTH EQUITY: ECONOMIC CONSIDERATIONS

Accessibility represents a significant limitation for CRISPR-based therapies, currently available only at specialized academic centers with limited manufacturing capacity. Conventional chemotherapy remains widely accessible with established delivery infrastructure, while CAR-T therapy availability has expanded to approximately 100 centers globally. Manufacturing timelines favor conventional approaches (immediate availability) compared to 2-4 weeks for CAR-T and 3-5 weeks for CRISPR-enhanced products. However, emerging allogeneic "off-the-shelf" CRISPR products may address these limitations through reduced manufacturing requirements. The future of CRISPR/Cas9 in precision cancer medicine is incredibly bright. Ongoing research is focused on refining the technology, improving its safety and efficacy, and expanding its applications. As these challenges are addressed, CRISPR/Cas9 is poised to become a cornerstone of personalized cancer treatment, leading to more effective therapies and improved outcomes for patients. Development of improved clinical trial designs incorporating nanoscale biomolecular constructs, or nanoagents, to control molecular and cellular processes in a programmable manner (Pallares et al., 2025).

Cost-effectiveness analyses reveal complex economic trade-offs between treatment modalities. Conventional chemotherapy costs $10,000-50,000 per treatment course with high retreat rates due to relapse. CAR-T therapies require substantial upfront investment ($350,000-450,000) but potentially provide long-term cure in 30-50% of patients. Establishing clear regulatory pathways for CRISPR-based therapies. Regulatory approval patterns reflect the maturation of different therapeutic approaches. Conventional chemotherapy regimens have established approval pathways with extensive safety databases. CAR-T therapies achieved FDA approval in 2017-2018 for multiple indications, with expanding clinical applications.

CRISPR-enhanced therapies remain primarily in clinical trial phases, though accelerated approval pathways may expedite development for breakthrough applications. The FDA has granted breakthrough therapy designations for multiple CRISPR approaches, indicating regulatory recognition of their transformative potential. CRISPR-enhanced therapies currently cost $400,000-600,000, representing a 15-25% premium over conventional CAR-T. However, improved efficacy and reduced relapse rates may justify these costs through reduced lifetime healthcare utilization. Quality-adjusted life year analyses suggest favorable cost-effectiveness ratios for CRISPR approaches in high-risk patient populations.

CONCLUSION

Synthetic biology innovations precisely edit cancer-related genes, combined with its versatility in disease modeling and therapeutic development, positions it as a cornerstone of future cancer treatment strategies. This comprehensive review underscores the transformative potential of CRISPR/Cas9 in precision cancer medicine while acknowledging the need for continued research to overcome current limitations and ensure safe, practical clinical applications. Our work shows robust evidence supporting the therapeutic potential of CRISPR/Cas9 in cancer treatment, particularly for hematological malignancies and immunotherapy applications. The consistency of positive outcomes across multiple studies suggests genuine therapeutic benefit rather than publication bias or methodological artifacts. Next-generation CRISPR technologies including base editing, prime editing, and epigenetic modulation offer potential advantages over current Cas9 approaches. Combination strategies integrating CRISPR with checkpoint inhibitors, targeted therapies, and novel delivery systems may further improve therapeutic ratios. The rapid translation from preclinical discovery to clinical application demonstrates the maturity of CRISPR technology and its readiness for therapeutic deployment.

However, significant challenges remain in optimizing delivery systems, minimizing off-target effects, and scaling manufacturing processes.

REFERENCES

Abusara, O. H., Agha, A. S. A. A., & Bardaweel, S. K. (2025). Advancements and innovations in liquid biopsy through microfluidic technology for cancer diagnosis. *Analyst, 150*(9), 1711–1725. Advance online publication. DOI: 10.1039/D5AN00105F PMID: 40181713

Ananthakumar, B., Nagaraja, B. G., Umesh, S. M., & Raghavan, R. (2025). Predictive Medicine. *Predictive Algorithms for Rehabilitation and Assistive Systems*, 291–314. DOI: 10.4018/979-8-3373-0194-5.ch011

Ananthakumar, B., Nanjundeshwara, N., Lakshmaiah, A., & Raghavan, R. (2024). Synthetic Gene Circuits as a Promising Approach in Cancer Immunotherapy. *Critical Developments in Cancer Immunotherapy*, 391–418. DOI: 10.4018/979-8-3693-3976-3.ch011

Ananthakumar, B., Nanjundeshwara, N., Lakshmaiah, A., & Raghavan, R. (2025). Nanocarriers for Delivering Nucleic Acids in Plants. *Nanocarriers in Plant Science and Agriculture*, 207–234. DOI: 10.4018/979-8-3693-9869-2.ch008

Baghban, R., Roshangar, L., Jahanban-Esfahlan, R., Seidi, K., Ebrahimi-Kalan, A., Jaymand, M., & Zarghami, N. (2024). A review of the literature on the use of CRISPR/Cas9 gene editing in cancer immunotherapy. *International Immunopharmacology*, *127*, 111741. DOI: 10.1016/j.intimp.2024.111741

Bhuvana, G. N., Umesh, S. M., & Raghavan, R. (2025). Expanding the frontiers of Schistosomiasis Treatment: The promise of Nanoparticle-Based Therapies. *Journal of Vector Borne Diseases, 10*(4103/jvbd.jvbd_180_23). DOI: 10.4103/jvbd.jvbd_180_23

Byalahunasishetter, B. I., N., V., P., S., C., M., S., S., & Raghavan, R. (2025). Spatial Omics and Bioprinting. *Innovations in Precision Medicine and Genomics*, 85–110. DOI: 10.4018/979-8-3693-5787-3.ch003

Chehelgerdi, M., Chehelgerdi, M., Allela, O. Q. B., Pecho, R. D. C., Jayasankar, N., Rao, D. P., & Amiri, M. (2024). Comprehensive review of CRISPR-based gene editing: Mechanisms, challenges, and applications in cancer therapy. *Molecular Cancer*, *23*(1), 9. DOI: 10.1186/s12943-023-01925-5 PMID: 38195537

Cheng, Q., Zhu, Y., Lv, S., Shi, J., Kuang, M., Wang, L., & Ji, X. (2026). Engineered bacteria and bacteria-derived nanomaterials for cancer therapy: Mechanisms, designs and advances. *Bioactive Materials*, *55*, 1–22. DOI: 10.1016/j.bioactmat.2025.09.012 PMID: 41368325

Chodnekar, S. R., Karsolia, R., & De, A. (2024). Systematic review of CRISPR clinical trials. *Gene Therapy*, *31*(11–12), 579–592. DOI: 10.1038/s41434-024-00456-3

Clift, C. L., Blaser, M. C., Gerrits, W., Turner, M. E., Sonawane, A., Pham, T., Andresen, J. L., Fenton, O. S., Grolman, J. M., Campedelli, A., Buffolo, F., Schoen, F. J., Hjortnaes, J., Muehlschlegel, J. D., Mooney, D. J., Aikawa, M., Singh, S. A., Langer, R., & Aikawa, E. (2024). Intracellular proteomics and extracellular vesiculomics as a metric of disease recapitulation in 3D-bioprinted aortic valve arrays. *Science Advances*, *10*(9), eadj9793. Advance online publication. DOI: 10.1126/sciadv.adj9793 PMID: 38416823

Deng, B., Kong, Y., Ma, Y., Zhan, Y., Sun, Y., Wang, R., Huang, P., & Liu, L. (2025). A novel carrier-free nanoparticle with stable distinctive three-dimensional structure for tumor-targeted precision chemoimmunotherapy. *Journal of Nanobiotechnology*, *23*(1), 1–15. DOI: 10.1186/s12951-025-03568-8 PMID: 40598510

Dhiflaoui, A., & Almawi, W. Y. (2025). Mechanisms and clinical implications of microRNA associations and signaling pathways in B-cell acute lymphoblastic leukemia. *Gene*, *967*, 149730. DOI: 10.1016/j.gene.2025.149730 PMID: 40829668

Doudna, J. A. (2020). The promise and challenge of therapeutic genome editing. *Nature*, *578*(7794), 229–236. DOI: 10.1038/s41586-020-1978-5 PMID: 32051598

Du, Y., Ding, X., & Ye, Y. (2024). The spatial multi-omics revolution in cancer therapy: Precision redefined. *Cell Reports Medicine*, *5*(9), 101740. DOI: 10.1016/j.xcrm.2024.101740 PMID: 39293393

Falandt, M., Bernal, P. N., Dudaryeva, O., Florczak, S., Größbacher, G., Schweiger, M., Longoni, A., Greant, C., Assunção, M., Nijssen, O., van Vlierberghe, S., Malda, J., Vermonden, T., & Levato, R. (2023). Spatial-Selective Volumetric 4D Printing and Single-Photon Grafting of Biomolecules within Centimeter-Scale Hydrogels via Tomographic Manufacturing. *Advanced Materials Technologies*, *8*(15), 2300026. DOI: 10.1002/admt.202300026 PMID: 37811162

Feng, C., Shan, M., Liu, C., Xu, Y., & Wang, H. (2024). CRISPR/Cas9 technology for advancements in cancer immunotherapy. *International Immunopharmacology*, *139*, 112706. DOI: 10.1016/j.intimp.2024.112706 PMID: 39427211

Fernández-Borbolla, A., García-Hevia, L., & Fanarraga, M. L. (2024). Cell Membrane-Coated Nanoparticles for Precision Medicine: A Comprehensive Review of Coating Techniques for Tissue-Specific Therapeutics. *International Journal of Molecular Sciences*, *25*(4), 2071. DOI: 10.3390/ijms25042071 PMID: 38396747

Goodwin, R. J. A., Platz, S. J., Reis-Filho, J. S., & Barry, S. T. (2024). Accelerating Drug Development Using Spatial Multi-omics. *Cancer Discovery*, *14*(4), 620–624. DOI: 10.1158/2159-8290.CD-24-0101 PMID: 38571424

GVS, A. C., Ananthakumar, B., & Raghavan, R. (2025). Enhancing Assistive Technologies With Neuromorphic Computing. *Revolutionizing AI with Brain-Inspired Technology: Neuromorphic Computing*, 207–232. DOI: 10.4018/979-8-3693-6303-4.ch009

Huang, Q., Li, Y., Huang, Y., Wu, J., Bao, W., Xue, C., Li, X., Dong, S., Dong, Z., & Hu, S. (2025). Advances in molecular pathology and therapy of non-small cell lung cancer. *Signal Transduction and Targeted Therapy*, *10*(1), 186. DOI: 10.1038/s41392-025-02243-6 PMID: 40517166

Institute, I. G. (2025). *CRISPR Clinical Trials: A 2024 Update*. https://innovativegenomics.org/news/crispr-clinical-trials-2024/

Jiménez-Boland, D., Robles-Fernández, A., Martín-Rodríguez, A., Cuadros, M. Á., Traverso, J. Á., Sánchez-Moreno, P., & Bramini, M. (2025). Breaking Barriers in Glioblastoma Targeting through Advanced Nanoparticle Cell Membrane Coating. *ACS Applied Materials & Interfaces*, *17*(24), 35288–35303. DOI: 10.1021/acsami.5c07306 PMID: 40488690

Khobragade, S. S., Deshmukh, M., Vyas, U., & Ingle, R. G. (2025). Innovative Approaches in Bone Tissue Engineering: Strategies for Cancer Treatment and Recovery. *International Journal of Molecular Sciences*, *26*(9), 3937. DOI: 10.3390/ijms26093937 PMID: 40362178

Kim, H.-K., Cheong, H., Kim, M.-Y., & Jin, H.-E. (2025). Therapeutic Targeting in Ovarian Cancer: Nano-Enhanced CRISPR/Cas9 Gene Editing and Drug Combination Therapy. *International Journal of Nanomedicine*, *20*, 3907–3931. DOI: 10.2147/IJN.S507688 PMID: 40191042

Korrapati, C., Yaraladdi, A. B., Vaddi, H., & Raghavan, R. (2024). Applications of the Saffron European Kitchen Spice Crocus sativus. *Therapeutic and Pharmacological Applications of Ethnobotany*, 98–127. DOI: 10.4018/979-8-3693-1986-4.ch005

Lakshmaiah, A., Korrapati, C., Suresh, C., & Raghavan, R. (2025). Synthetic Biology and AI. *Spatially Variable Genes in Cancer*, 265–290. DOI: 10.4018/979-8-3693-7728-4.ch010

Liu, J., Qiao, S., Li, S., Jiang, Y., Zhang, S., & Li, Y. (2026). Integrating Cu2O nanocubes-enhanced BiVO4 photoanodes and CRISPR/Cas12a for ultra-sensitive miRNA-21 detection in cancer diagnostics. *Sensors and Actuators. B, Chemical*, *446*, 138637. DOI: 10.1016/j.snb.2025.138637

Lopes, J., Lopes, D., Motallebi, M., Ye, M., Xue, Y., Vieira, A. C. F., Singh, S. K., Dua, K., Veiga, F., Sethi, G., Paiva-Santos, A. C., & Makvandi, P. (2025). Biomembrane-coated nanosystems as next-generation delivery systems for the treatment of gastrointestinal cancers. *Bioengineering & Translational Medicine*, *10*(4), e70006. DOI: 10.1002/btm2.70006 PMID: 40708975

Lu, Y., Xue, J., Deng, T., Zhou, X., Yu, K., Deng, L., & You, Y. (2017). Safety and feasibility of CRISPR-edited T cells in patients with refractory non-small-cell lung cancer. *Nature Medicine*, *26*(5), 732–740. DOI: 10.1038/s41591-020-0840-5 PMID: 32341578

Luo, X., Weidinger, E., Burghardt, T., Höhn, M., & Wagner, E. (2025). CRISPR/Cas9 Ribonucleoprotein Delivery Enhanced by Lipo-Xenopeptide Carriers and Homology-Directed Repair Modulators: Insights from Reporter Cell Lines. *International Journal of Molecular Sciences*, *26*(9), 4361. DOI: 10.3390/ijms26094361 PMID: 40362595

Maganti, H. B., Jrade, H., Cafuir, L. A., Pegliasco, E. M., Nalagatla, T., Abou-El-Enein, M., & Kanashova, T. (2022). Use of CRISPR/Cas9 gene editing to improve chimeric antigen receptor T cell therapy: A systematic review and meta-analysis. *Frontiers in Immunology*, *13*, 867888. DOI: 10.3389/fimmu.2022.867888 PMID: 35039239

Martinez-Lage, M., Torres-Ruiz, R., Puig-Serra, P., Moreno-Gaona, P., Martin, M. C., Moya, F. J., & Rodriguez-Perales, S. (2020). In vivo CRISPR/Cas9 targeting of fusion oncogenes for selective elimination of cancer cells. *Nature Communications*, *11*(1), 5060. DOI: 10.1038/s41467-020-18875-x PMID: 33033246

Meng, Z., Dong, J., Guo, L., Su, F., Liu, J., Wang, G., & Zhao, Z. (2026). Now and future of artificial intelligence-based signet ring cell diagnosis: A survey. *Expert Systems with Applications*, *296*, 129188. DOI: 10.1016/j.eswa.2025.129188

Minnesota Medical School. U. (2025). New gene-editing therapy shows early success in fighting advanced GI cancers. *Medical News*. https://med.umn.edu/news/new-gene-editing-therapy-shows-early-success-fighting-advanced-gi-cancers

Navarro Quiroz, R., Villarreal Camacho, J., Zarate Peñata, E., Bello Lemus, Y., López-Fernández, C., Gomez Escorcia, L., Fernández-Ponce, C., Rebolledo Cobos, M., Fandiño Moreno, J., Fiorillo-Moreno, O., & Navarro Quiroz, E. (2025). Multi-scale information processing in the immune system. *Frontiers in Immunology, 16*, 1563992. Advance online publication. DOI: 10.3389/fimmu.2025.1563992 PMID: 40761781

NCT02793856. (2016). *PD-1 Knockout Engineered T Cells for Metastatic Non-small Cell Lung Cancer. ClinicalTrials.gov*. https://clinicaltrials.gov/study/NCT02793856

NCT04637763. (2020). *CRISPR-Edited Allogeneic Anti-CD19 CAR-T Cell Therapy. ClinicalTrials.gov*. https://clinicaltrials.gov/study/NCT04637763

NCT05643742. (2022). *A Safety and Efficacy Study Evaluating CTX112 in Subjects with Relapsed or Refractory B-cell Malignancies. ClinicalTrials.gov*. https://clinicaltrials.gov/study/NCT05643742

P., A., D., K. M., M., J., P., S., & Raghavan, R. (2025). The Cancer Exposome. *Cancer Exposomics and Environmental Influences on Carcinogenesis*, 161–192. DOI: 10.4018/979-8-3373-2165-3.ch004

Pallares, R. M., Barmin, R. A., Wang, A., Kiessling, F., & Lammers, T. (2025). Clinical cancer nanomedicines. *Journal of Controlled Release, 385*, 113991. DOI: 10.1016/j.jconrel.2025.113991 PMID: 40582644

Park, B.-S., Lee, M., Kim, J., & Kim, T. (2025). Perturbomics: CRISPR–Cas screening-based functional genomics approach for drug target discovery. *Experimental & Molecular Medicine, 57*(7), 1443–1454. DOI: 10.1038/s12276-025-01487-0 PMID: 40588529

Peng, N., Song, S. J., Hui, V. W.-K., Lai, J. C.-T., Wong, G. L.-H., Wong, V. W.-S., & Yip, T. C.-F. (2026). Foundations of Artificial Intelligence in Hepatology: What a Clinician Needs to Know. *Journal of Clinical and Experimental Hepatology, 16*(1), 103183. Advance online publication. DOI: 10.1016/j.jceh.2025.103183 PMID: 41048886

Platt, R. J., Chen, S., Zhou, Y., Yim, M. J., Swiech, L., Kempton, H. R., & Zhang, F. (2014). CRISPR-Cas9 knockin mice for genome editing and cancer modeling. *Cell, 159*(2), 440–455. DOI: 10.1016/j.cell.2014.09.014 PMID: 25263330

Priya, S., Boyd, A. S., Torres-Bautista, A., Sawadkar, P., Mosahebi, A., Narayan Rath, S., & Kalaskar, D. M. (2024). Accelerating vascular graft development: Adipose-derived stem cells and PODS® (Polyhedrin delivery system with tissue-specific growth factors) – Enhanced 3D bioprinting for functional blood vessels. *Chemical Engineering Journal*, *494*, 152934. DOI: 10.1016/j.cej.2024.152934

Qi, L. S., Larson, M. H., Gilbert, L. A., Doudna, J. A., Weissman, J. S., Arkin, A. P., & Lim, W. A. (2013). Repurposing CRISPR as an RNA-guided platform for sequence-specific control of gene expression. *Cell*, *152*(5), 1173–1183. DOI: 10.1016/j.cell.2013.02.022 PMID: 23452860

Raghavan, R., Moger, S. S., Umesh, S., & Bhuvana, G. N. (2025). Revolutionizing Pain Management. *Unveiling Technological Advancements and Interdisciplinary Solutions for Pain Care*, 129–164. DOI: 10.4018/979-8-3693-9501-1.ch005

Rana, D., Rangel, V. R., Padmanaban, P., Trikalitis, V. D., Kandar, A., Kim, H.-W., & Rouwkema, J. (2025). Bioprinting of Aptamer-Based Programmable Bioinks to Modulate Multiscale Microvascular Morphogenesis in 4D. *Advanced Healthcare Materials*, *14*(1), 2402302. DOI: 10.1002/adhm.202402302 PMID: 39487611

S., G. V., C., A., Biradar, S., & Raghavan, R. (2025). Neuroscience-Driven Solutions for Speech Impairment. *Advancing Medical Research Through Neuroscience*, 119–146. DOI: 10.4018/979-8-3693-5464-3.ch006

Seijas, A., Cora, D., Novo, M., Al-Soufi, W., Sánchez, L., & Arana, Á. J. (2025). CRISPR/Cas9 Delivery Systems to Enhance Gene Editing Efficiency. *International Journal of Molecular Sciences*, *26*(9), 4420. DOI: 10.3390/ijms26094420 PMID: 40362657

Shah, A. H., Shetty, O., & Joshi, A. (2024). CRISPR/Cas9 applications in breast and cervical cancers. *Cancer Gene Therapy*, *31*(3), 425–435. DOI: 10.1038/s41417-024-00712-w

Sherif, Z. A., Ogunwobi, O. O., & Ressom, H. W. (2025). Mechanisms and technologies in cancer epigenetics. *Frontiers in Oncology*, *14*, 1513654. Advance online publication. DOI: 10.3389/fonc.2024.1513654 PMID: 39839798

Shirk, B. D., Rodriguez, C. Z., Pacheco, M. O., McTyer, J. B., Shirk, P. D., & Stoppel, W. L. (2026). Enhancing CRISPR homology directed repair in IAL-PiD2 insect cells via reagent delivery optimization and cell synchronization. *Biochemical Engineering Journal*, *225*, 109906. DOI: 10.1016/j.bej.2025.109906

Shu, J., Xie, X., Wang, S., Du, Z., Huang, P., Chen, Y., & He, Z. (2025). CRISPR/Cas-edited iPSCs and mesenchymal stem cells: A concise review of their potential in thalassemia therapy. *Frontiers in Cell and Developmental Biology*, *13*, 1595897. Advance online publication. DOI: 10.3389/fcell.2025.1595897 PMID: 40970094

Srinivasan, D., Balakrishnan, R., Chauhan, A., Kumar, J., Girija, D. M., Shrestha, R., Shrestha, R., & Subbarayan, R. (2025). Epithelial–Mesenchymal Transition in Cancer: Insights Into Therapeutic Targets and Clinical Implications. *MedComm*, *6*(9), e70333. DOI: 10.1002/mco2.70333 PMID: 40895189

Stadtmauer, E. A., Fraietta, J. A., Davis, M. M., Cohen, A. D., Weber, K. L., Lancaster, E., & June, C. H. (2020). CRISPR-engineered T cells in patients with refractory cancer. *Science*, *367*(6481), 7365. DOI: 10.1126/science.aba7365 PMID: 32029687

Thomsen, E. A., Mich, J. K., Yao, Z., Hodge, R. D., Doyle, A. M., Jang, S., & Zeng, H. (2022). Implementation of CRISPR/Cas9 Genome Editing to Generate Murine Lung Cancer Models That Depict the Mutational Landscape of Human Disease. *Frontiers in Cell and Developmental Biology*, *9*, 641618. DOI: 10.3389/fcell.2021.641618 PMID: 33738287

Tiwari, A., Saraf, S., Verma, A., Panda, P. K., & Patel, B. K. (2023). CRISPR/Cas9 as a therapeutic tool for triple negative breast cancer: From bench to clinics. *Frontiers in Molecular Biosciences*, *10*, 1214489. DOI: 10.3389/fmolb.2023.1214489 PMID: 37469704

Wang, K. C., Zheng, T., & Hubbard, B. P. (2025). CRISPR/Cas technologies for cancer drug discovery and treatment. *Trends in Pharmacological Sciences*, *46*(5), 437–452. DOI: 10.1016/j.tips.2025.02.009 PMID: 40133194

Yandrapalli, N. (2025). Bottom-up development of lipid-based synthetic cells for practical applications. *Trends in Biotechnology*, *0*(0), 2150–2169. Advance online publication. DOI: 10.1016/j.tibtech.2025.03.009 PMID: 40263003

Yang, N., He, Z., & Lang, T. (2025). Drug Delivery Systems Based on Metal–Organic Frameworks for Tumor Immunotherapy. *Pharmaceutics*, *17*(2), 225. DOI: 10.3390/pharmaceutics17020225 PMID: 40006592

Yildirim, D. T., Baki Yildirim, A., Salzet, M., Bertelli, M., Beccari, T., Prakash, S., Pascucci, L., & Dundar, M. (2025). Messaging malignancy: Tumour-derived exosomes at the nexus of immune escape, vascular remodelling and metastatic competence. *The EuroBiotech Journal*, *9*(3), 216–237. DOI: 10.2478/ebtj-2025-0018

Zhang, R., Li, Y., Fu, H., Zhao, C., Li, X., Wang, Y., Sun, Y., & Li, Y. (2025). Nanomedicine strategies for cuproptosis: Metabolic reprogramming and tumor immunotherapy. *Acta Pharmaceutica Sinica. B*, *15*(9), 4582–4613. Advance online publication. DOI: 10.1016/j.apsb.2025.07.007 PMID: 41049741

Zhang, Y., Zhang, Z., & Ding, Y. (2021). CRISPR/Cas9-mediated genome editing in cancer therapy. *Biomarker Research*, *9*, 69. DOI: 10.1186/s40364-021-00323-x

Chapter 3
CRISPR-Cas9 Technology for Unlocking Immunity Against Infectious Pathogens

M. Suriya
https://orcid.org/0009-0008-5398-0221
Vels Institute of Science, Technology, and Advanced Studies, India

V. Aarthi
https://orcid.org/0009-0004-0177-7716
Vels Institute of Science, Technology, and Advanced Studies, India

S. S. Meenambiga
https://orcid.org/0000-0002-5445-525X
Vels Institute of Science, Technology, and Advanced Studies, India

P. Prakash
Sathyabama Institute of Science and Technology, India

P. Vivek
Vels Institute of Science, Technology, and Advanced Studies, India

ABSTRACT

The development of CRISPR-Cas9 has dramatically changed the field of genome engineering and its utilization in epigenome editing could have profound implications for precision medicine.This ability to precisely target epigenetic marks, including DNA methylation, histone modifications, and chromatin remodeling, allows for exciting therapeutic developments in cancer, neurodegenerative and autoimmune

DOI: 10.4018/979-8-3373-4857-5.ch003

diseases. New developments such as CRISPRi, CRISPRa, and reversible CRISPR off/CRISPR on, allow for complex and multiplexed epigenetic reprogramming with improved specificity and safety. Nonetheless, there are some challenges to delivery off-target effects, stability of epigenetic modification, and moral lines, especially with inheritable epigenetic modification. The essential principles of CRISPR-Cas9-based epigenome editing, its therapeutic prospects for personalized medicine, and the consideration of technological, ethical, and policy aspects associated with epigenome editing are discussed

INTRODUCTION

Precision medicine is revolutionizing health care by enhancing the ability to personalize prevention, diagnosis, and treatment for an individual, based on their genetic, environmental, and lifestyle characteristics, with the role of gene regulation being particularly important. We can regulate genes not only at the DNA sequence level, but also at the epigenetic level as heritable, but reversible modifications, which do not change the underlying genetic code. Epigenetic dysregulation is being increasingly recognized in the pathology of an ever-expanding list of diseases, including, but not limited to, cancer, neurodegenerative diseases, autoimmune diseases, and metabolic disorders. The CRISPR-Cas9 system is based on a bacterial immune system and was developed as a genome-editing technology, and has most recently advanced to the realm of epigenome editing, which offers an exciting possibility for clinical application. Scientists have discovered that CRISPR can be coupled with a catalytically dead Cas9 (dCas9), to which the scientists then attach functional effector domains that enable targeted modulation of epigenetic marks (e.g., DNA methylation; histone modifications) in a compact and specific manner without permanently modifying the genome.

This is a significant advance over traditional epigenetic therapeutics, which often have non-specific effects that engage global epigenetic machinery. CRISPR-based epigenome editing technologies, including CRISPR interference (CRISPRi), CRISPR activation (CRISPRa), reversible systems such as CRISPRoff and CRISPRon have great potential to use in the treatment of diseases caused by aberrant expression of genes. These technologies provide an unprecedented level of precision to reprogram gene expression; thus, providing new avenues for treatments for cancers, neurodegenerative diseases, autoimmune diseases, and drug resistance. There are still challenges in developing the technology further to improve delivery efficiency, reduce off-target effects, determine the long-term stability of reversible modifications, and address ethical concerns raised in using germline editing. This chapter will provide a brief history and mechanism of CRISPR-Cas9, as this technology is applied in epigenome editing, summarize how epigenome editing can be used in precision medicine, and critically analyze changes that will need to be made to technology, ethics, and regulatory systems prior to clinical translation. Finally, this chapter will address potential developments, including the impact of artificial intelligence and the integration of nanotechnologies, to improve the specificity, safety, and scalability of CRISPR-based therapies.

1. FUNDAMENTALS OF CRISPR-CAS9 AND EPIGENOME EDITING

The CRISPR-Cas9 system has surfaced as a groundbreaking device in genetic engineering and biomedical research. CRISPR (Clustered Regularly Interspaced Short Palindromic Repeats) was first proposed as a component of the adaptive immune system in prokaryotes because CRISPR repeats enable bacteria to use the identification and subsequent cutting of foreign genetic material to defend against virus infection (Mojica & Montoliu, 2016). CRISPR relies on the action of proteins called Cas (CRISPR-associated), and the Cas9 proteins from *Streptococcus pyogenes* are the most studied proteins for applications in genome editing. In 2012, researchers Doudna and Charpentier showed that CRISPR-Cas9 could be converted into a programmable device to edit genomes in vitro, and accurately target DNA in organisms (Jinek et al., 2012). This led to many applications in gene therapy, functional genomics, and modelling disease.

In addition to conventional genome editing, CRISPR has been applied to epigenome editing: the modulation of gene expression without changing the DNA sequence. This is accomplished through the use of a catalytically inactive Cas9 (dCas9) fused to epigenetic effector domains, which facilitate the addition or removal of structures such as methylation marks on DNA or modifications to histones, all at a targeted genomic site (Liu et al., 2016). Reversible and highly targeted modulation of gene expression is one of the key advantages of epigenome editing over traditional genome editing, which results in irreversible changes to the genome. Tools like CRISPR interference (CRISPRi) or CRISPR activation (CRISPRa) can block or promote, respectively, the expression of genes at the cellular level, and these tools provide a useful platform for studying gene function and eventually for designing therapeutics (Gilbert et al., 2013).

The effective delivery of CRISPR components into cells is vital for success in both genome editing or epigenome editing.

Viral delivery systems, like adeno-associated viruses (AAVs) or lentiviruses, are often utilized because of their high efficiency of delivery especially in vivo (Lino et al., 2018) however, they have a limited capacity to deliver their payload and they can produce immunogenicity. Non-viral systems, such as lipid nanoparticles, electroporation, and gold nanoparticles, are usually safer and more versatile systems, especially for applications in the clinic (Yin et al., 2017). Improving accurate, effective, and safe delivery methods continues to be an area of interest and research moving CRISPR-based epigenome editing technologies forward.

As researchers develop CRISPR technology, computer science, including ai, and machine learning will support guide RNA design, specificity, and predictive capabilities of CRISPR. This will be beneficial in minimizing off-target effects, and maximizing the therapeutic effects of CRISPR-Cas9. Moreover, regular dialogue should be encouraged between researchers and regulators on ethical considerations for epigenome editing especially with regards to heritable genome editing. In general, CRISPR-Cas9 will offer great opportunities for the application of epigenome editing in precision medicine, in understanding and developing novel strategies to treat diseases based on gene regulation, and also in personalized therapy.

1.1 History and Evolution of CRISPR-Cas Systems

Evidence of CRISPR-Cas systems was discovered in 1987, when researchers found repeats in the genome of *Escherichia coli* (E. coli), but did not know what they were (Ishino et al., 1987). It would be until the early 2000s before these sequences could finally be described as Clustered Regularly Interspaced Short Palindromic Repeats (CRISPR) and connected to an adaptive immune defense system found in bacteria and archaea. Mojica et al. (2005) discovered that these sequences contained segments of foreign genetic material (i.e. plasmids and viruses) not found in the genome, which enabled prokaryotes to recognize and defend against future attacks of foreign genetic material. Key research on related CRISPR-associated (Cas) proteins, particularly Cas9 from the bacterium *Streptococcus pyogenes*, provided an avenue for accurately targeting and hydrolyzing other DNA. This explained the mechanism by which prokaryotes were equipped with a defense mechanism, and inspired a new chapter in molecular biology and genetic engineering.

The longitude alteration happened in 2012 when Jennifer Doudna and Emmanuelle Charpentier reconstituted the CRISPR-Cas9 system in vitro and showed that it could be utilized as a programmable genome editing tool. Using CRISPR-Cas9, for example, researchers can create guide RNAs that identify DNA sequences to which the Cas9 enzyme and proteins bind and uniquely alter the genome. Since this, the CRISPR toolbox has provided many extending toolbox options, including additional tools that employ analogous systems, such as Cas12 and Cas13, which target DNA and RNA, respectively (Shmakov et al., 2017). With the ability to engineer a catalytically inactive Cas9 protein (dCas9), researchers can now employ many more applications outside genome editing that include epigenome editing, transcription regulations, and live-cell imaging (Qi et al., 2013). The developments in the CRISPR toolset are altering the landscape of genetics research and extending options in therapeutics, agriculture, and diagnostics. CRISPR-Cas is still among the available scientific technologies and is flexible and transformative.

1.2 Mechanism of CRISPR-Cas9 Gene Targeting

CRISPR-Cas9 is a highly efficient gene editing system that has derived from a bacterial adaptive immune system that provides defense to viral infection. A synthetic single-guide RNA (sgRNA) is produced to be complementary to a designated DNA sequence in the genome. The Cas9 endonuclease will bind the target DNA with a Protospacer Adjacent Motif (PAM) sequence present, in an sgRNA specific manner. The sgRNA directs the Cas9 endonuclease to cleave at the site and induce a double-stranded break (DSB) in DNA. The sequence-specific cleavage particular to the sgRNA and the target, will activate normal repair mechanisms in the cell, specifically either non-homologous end joining (NHEJ), which may or may not be small insertions or deletions to the target, or homology directed repair (HDR), if an HDR template can be provided whereby a cell would add specificity to editing the existing DNA sequence. This significant programmability allows for the concurrent effective and precise editing of DNA sequences in research settings regarding genetics and therapeutics (Jinek et al., 2012).

1.2.1 Guide RNA (gRNA) Design and Target Recognition

Targeting genes using CRISPR-Cas9 occurs via a single guide RNA (sgRNA), an engineered version of the combination of CRISPR RNA (crRNA) and trans-activating of crRNA (tracrRNA). An sgRNA is designed to include a sequence of 20 nucleotides that contains complementarity to the target sequence of DNA, that is necessary for a sequence- specific binding plaque that will be in the presence of a PAM, such "NGG," present at about 5' direction and from the target site in *Streptococcus pyogenes* Cas9. The PAM is essential for Cas9 binding, localisation and to cleave the region in the genome that is intended (Sternberg et al., 2014) which proves that specificity is crucial for genome editing.

1.2.2 Cas9-Mediated DNA Cleavage

Once the gRNA-Cas9 complex success-fully binds to the target DNA, the Cas9 protein undergo a conformational change leading to alignment of the two separate nuclease domains–RuvC and HNH—for cleavage of DNA. The HNH domain cleaves the strand of DNA bonded to the guide RNA and RuvC cleaves the non-complementary strand ((3)) creating a double-strand break (DSB) in a genomic site. This targeted cleavage is essential to the CRISPR-Cas9 gene editing process because it triggers cell DNA repair processes that can be exploited to induce gene disruption or precise genetic modification (Hsu et al., 2014).

1.2.3 DNA Repair Pathways

In the case of the double-strand break (DSB) that DSB is introduced by the Cas9 protein, the cell produces one of two major DNA repair pathways to salvage genomic integrity. The second is Non-Homologous End Joining (NHEJ) which correctly connects the broken DNA ends directly but will often involve small insertions or deletions that can disrupt gene function by causing frameshift mutations. The second is Homology-Directed Repair (HDR), a more precise and less efficient form of restoration using the homologous DNA template to correct the break. HDR is useful in bringing a single sequence of DNA into sequence and is ideal for the precise creation of genomic precision. NHEJ and HDR were crucial in selecting the best candidate, as they determine if gene knockout or precise gene correction is accomplished (Cong et al., 2013).

1.3 Tools for Epigenome Editing: dCas9, CRISPRi, and CRISPRa

CRISPR technology has become far more powerful than gene editing to address epigenome editing, providing researchers with a way to modify the expression of genes, without altering DNA. dCas9 is an altered version of Cas9 that is still able to conform to DNA and is nuclease-bound and is not nuclease-bound due to RuvC and HNH domain mutations. The dCas9 protein is a coded DNA binding agent that leads to the recruitment of several epigenetic effectors to specific genomic loci to regulate gene expression precisely.

Two of the primary systems of transcriptional modulation derived from dCas9 are CRISPR interference, CRISPRi, and CRISPR activation. In CRISPRi dCas9 is fused with repressor domains such as KRAB (Kruppel-associated box) that induce heterochromatin development and suppress gene transcription (Gilbert et al., 2013). A passive system can silence gene expression, preventing transcription initiation or elongation. On the other hand, CRISPRa involves pairing dCas9 with transcriptional activators like VP64, p300 or SunTag systems that enrich gene transcription by maintaining a free chromatin state and establishing the transcriptional machinery (Chavez et al., 2015). These instruments have improved functional genomics because they allow reversible and impregnated control of gene expression without permanent and permanent genetic changes and provide useful both in the laboratory and therapeutic applications. Next-generation CRISPR platforms are transforming precision gene and epigenetic regulation. CRISPRi and CRISPRa enable programmable repression or activation of target genes, while CRISPRoff/CRISPRon allows reversible epigenetic silencing or activation at specific loci. Multiplexed systems facilitate simultaneous regulation of multiple genes, and inducible platforms provide temporal control using chemical or light-responsive triggers. Together, these innovations expand the versatility, tunability, and reversibility of CRISPR-based regulation, offering powerful tools for functional genomics, synthetic biology, and therapeutic applications.

1.4 Differences Between Genome Editing and Epigenome Editing

Genome editing and epigenome editing are two different yet complementary approaches in the area of genetic modification, each with its mechanisms and its consequences. Genome editing, to put it simply, refers to physically altering an organism's DNA sequence. Recently developed tools, including the well-known CRISPR-Cas9, TALENs, and ZFNs, create double-strand breaks at a defined locus in the genome, and the cell takes care of repair, either through NHEJ, or HDR. In any of the cases, this process can generate permanent changes in the DNA sequence and usually leads to insertions, deletions or point mutations. Genome editing is used in therapy to precisely correct genetic defects or to disrupt the gene function (Komor et al., 2017). It holds promising applications in gene therapy, functional genomics and crop improvement. However, the use of Genome editing raises potential issues, both regarding the effects generated by off-target events, and unintended alterations across the genome.

In contrast to DNA modification, epigenome editing alters gene activity without modification of the DNA sequence. Epigenome editing utilizes catalytically inactive nucleases (for example, dead Cas9 (dCas9)), which is effectively a binder that is used with epigenetic effector domains that modify, add or remove chemical marks on DNA bases or histone proteins. These various modifications, in addition to DNA methylation, can also reflect on histone acetylation or methylation, and they induce changes in gene activity through alterations in chromatin accessibility. Techniques such as CRISPR interference (CRISPRi) and CRISPR activation (CRISPRa) utilize dCas9 that is fused to transcriptional repressors (for example, the KRAB domain) or to activators (for example, the VP64 or p300) that can reversibly silence or activate gene expression at the transcriptional level. Compared to genome editing, epigenome editing is mostly reversible, which enables temporal control of gene function without any permanent genetic change. Thus, epigenome editing provides an amazing new approach for study the gene regulation and developing therapeutic strategies with lower risk of irreversible off-target effects (Liu et al., 2016). Synthetic transcriptional regulators enable programmable control of gene expression by recruiting effector domains to specific genomic loci via dCas9. Modular activators (e.g., VPR) and repressors (e.g., KRAB-MeCP2) allow tuning of transcriptional output by combining multiple functional domains, providing precise and adjustable gene regulation. Emerging CRISPRai systems integrate activation and interference within the same cell, enabling bidirectional control of different target genes simultaneously. These innovations represent a key trend in synthetic biology, expanding the versatility and precision of CRISPR-based gene regulation.

1.5 Delivery systems: viral and non-viral vectors

An important limitation to successfully harnessing genome editing applications is the effective delivery of CRISPR-Cas9 components into target cells. Delivery systems can be classified as viral delivery and non-viral delivery vectors, both of which have their own associated pros and cons. Viral delivery is favorable due to the inherent property of viruses to infect host cells and are high efficacy to deliver genetic material. Within viral delivery, adeno-associated virus (AAV) vectors appear to be the most acceptable for in vivo utilization because of their lower immunogenicity and ability to infect dividing and non-dividing cells. However, there are limitations, including limited packaging capacity (~4.7kb) and cannot be utilized with larger nucleases. Furthermore, if AAV use a split-Cas9 system or a variant, such as SaCas9, needs to be utilized (Lino et al. 2018). Insertional mutagenesis continues to be a complicating factor, but lentiviral vectors that stably integrate into the host genome while having longer expression of Cas9 and guide RNAs can be used for ex vivo editing- in particular, for targeting immune cells or stem cells in their resident compartment (Xu et al., 2021). To align with current and emerging trends, CRISPR research is increasingly focusing on advanced platforms such as prime editors, base editors, multimodal CRISPRai systems, and epigenetic editors that enable precise, tunable gene and chromatin regulation. Non-viral delivery methods, including lipid nanoparticles and exosome-based systems, are being developed to improve tissue-specific in vivo delivery. Computational design tools facilitate optimized guide RNA selection, effector domain combination, and off-target minimization, accelerating experimental workflows. Additionally, ethical and safety considerations—including off-target effects, unintended epigenetic changes, and regulatory safeguards—are integral to responsible application and translational research. Together, these innovations define the forward trajectory of CRISPR technologies.

Adenoviral vectors have the advantages of being able to accommodate a greater cargo size and provide strong temporary expression, but have the disadvantage of being highly immunogenic. In summary, viral delivery vectors can be highly effective, but there are still significant hurdles to overcome as a result of immunogenicity, limitations in delivery vector size, lipid nanoparticles (LNPs), and potential for genomic integration (Charlesworth et al., 2019). Safety and administration of long duration of expression in space and expense recently drove recommendations of applying non-viral vectors as opposed to viral vectors. LNPs are viewed as the most advanced delivery platform, and are efficacious delivery platforms and protect from degradation through encapsulated with Cas9 mRNA and sgRNA in recent clinical trials demonstrating application of CRISPR (e.g., transthyretin amyloidosis (Gillmore et al., 2021). Polymeric nanocarriers (e.g. polyethyleneimine, chitosan) can also be complexed based on charge attraction with CRISPR cargos, but application in clinical situations is limited, as higher concentrations demonstrate cytotoxicity (Xu et al., 2021). Effective delivery is critical for translating CRISPR-based epigenome editing into therapeutic applications. Lipid nanoparticles (LNPs) enable tissue-specific delivery of CRISPR ribonucleoproteins, while exosome-mediated transport facilitates crossing biological barriers such as the blood–brain barrier. Additionally, engineered AAV vectors provide targeted, stable delivery with reduced immunogenicity. Optimizing these delivery platforms is essential to enhance efficacy, specificity, and safety of CRISPR interventions in clinical settings.

Techniques which are categorized as physical methods (e.g. electroporation, microinjection) can utilize effective ex vivo non-viral delivery strategies, particularly in isolated immune cells, and have acceptable delivery of the CRISPR systems in embryos; however, are not feasible for use in systemic (in vivo) delivery (Lino et al., 2018). More recently, metals and biologically derived mimic carriers, such as gold nanoparticles and exosomes, have demonstrated some potential for targeted delivery and improved biocompatibility (Xu et al., 2021). Although non-viral strategies are comparatively safer and offer greater flexibility than viral systems, they still struggle with high efficiency, tissue-specific targeting, and sustained expression.

In comparison, viral vectors provide greater efficacy and stable expression compared to non-viral carrier systems, while non-viral vectors address safety, scalability, and transient action which is advantageous for therapeutic applications that depend on the spatial and temporal control in gene editing processes. Currently, hybrid systems such as virus-like particle loaded Cas9 proteins, LNP-viral systems, and exosome-based carriers are being developed to overcome the drawbacks of traditional delivery methods. These approaches were developed or used with the goal of ameliorating CRISPR-Cas9's efficiency advantage while reducing issues associated with complexity of use and commercialization of CRISPR-Cas9 for clinical use (Xu et al., 2021; Gillmore et al., 2021).

Table 1. Comparative Overview of Viral and Non-Viral Delivery Systems for CRISPR-Cas9

Feature	Viral Vectors	Non-Viral Vectors
Efficiency	High	Moderate to high. It will depend on system
Cargo Size	Limited (AAV ~4.7 kb)	Flexible (large payloads possible)
Expression	Stable (lentivirus), transient (AAV)	Mostly transient
Safety	Risk of immune response, integration	Lower in risk, but variable in biocompatibility

2. APPLICATIONS IN PRECISION MEDICINE

The ability which helps to correct the disease-causing mutation and engineer cellular system with high specificity made the CRISPR-cas9 to become a cornerstone technology. The CRISPR application was extend beyond simple gene knockout, enabling the modification of in immune effector cell in the oncology. The immune effector cell is including T cell and Natural Killer (NK)cell which enhance the patient specific therapies. These NK cell helps the patient to overcome the tumor immune evasion also. For example: Crispr mediated disruption of inhibitory receptor such as PD-1 or CTLA-4 in T cell has shown promise in response of augmenting antitumor the otherside, the chimeric antigen receptor (CARs) integration, individual tumor antigen has observed the process from the development of more potent adoptive immunotherapies tailored to individual tumor antigen which was allowed by chimeric antigen receptor. To develop engineered T cell which not only attract the cancer cell but also display the developed persistent was utilised by CRISPR. It also helps to reduce the exhaustion introducing by one new major limitation of the conventional immunotherapies (Ravichandran & Maddalo, 2023). The induced pluripotent stem cell (ipSCs)was induced by the patient and it was corrected at the site of mutation and differentiate into functional tissue with the help of which edited through CRISPR. It also includes offering possibilities for autologous transplantation without making risk of immune rejection. In the disease of the neuromuscular system, it consists of Duchenne muscular dystrophy, preclinical studies using CRISPR-edited ipSCs. These are done restoration demonstration of dystrophin function. In the ophthalmology, to correct the mutation linked to the inherited retinal degenerative disease, the ex vivo genome edited should has to applied. By allowing the researcher to test pharmacological intervention on genetically accurate human tissue model, the CRISPR can able to generate the disease model which accelerate the drug screening and patient-specific organoid. It helps to shorten the translation gap between the laboratory research and clinical application, highlighting CRISPR-cas9 as one of the most impactful technologies in achieving individualized medicine and these advances help to improve and enhance our capacity to design highly personalized treatment regimen also (Jayarajan et al., 2025).

2.1 Epigenome Editing for Cancer Therapy

Epigenome editing has emerged as a powerful strategy in cancer therapy by enabling precise and reversible control of gene expression without permanently altering the DNA sequence. The double strand break, epigenome editing utilizes catalytically inactive cas9(dcas9) which fused with epigenetic effective domain was introduced, unlike classical CRISPR-cas9 genome editing. The epigenetic effected domain includes DNA methyltransferase, histone acetyltransferase demethylases. These are used to modulate the chromatin state at targeted loci. The silencing of oncogene, the reactivation of tumor suppressor genes, the modulation of no coding regulatory elements implicated in tumor initiation and progression was allowed by this technology. For Example: The dcas9-TET1 fusion has used by targeted DNA demethylation. The expression of silenced tumor suppressor gene was restored by this fusion in commercial cancer cell. It leads to decreased level of proliferation and tumor genetically(Tejedor et al., 2023).Nowadays, the further expand in the clinical potential of epigenome editing by the advances in the nanoparticles and viral-based delivery platform was done by achieving efficient and tissue-specific targeting, thereby overcoming one of the primary barrier to therapeutic translation(Kwon & Lim, 2023).To silence the inconvenience pathway with high specificity, the dcas9-KRAB repression system has been employed, while the dcas9-p300 acetyltransferase activator enhance transcription of the protective genes. This offering programmable control over cancer epigenomes (O' Geen et al., 2019).

2.2 Neurodegenerative Disease Interventions (e.g., Alzheimer's, Parkinson's)

For the neurodegenerative disease, the promising therapeutic strategies was emerged by the CRISPR-cas9 technology. The neurodegenerative disease is such as Alzheimer disease and Parkinson disease. These editing of genes was additionally included by CRISPR mediated which shown to reduce the potential in amyloid-beta plaque accumulation, thereby mitigating synaptic dysfunction and neuronal loss (journal of Advanced Research,2021). The CRISPR-mediated editing of genes such as *APP, PSEN 1/2* and *BACE 1*. In PD, the targeting *SNCA* was approached by genome editing (alpha-synuclein) and the mutation of *LRRK 2* are being explored, this explorayi6help to suppress to aggregation of protein pathogens and help to restore the function of the dopaminergic neuron, enhancing the neuronal replacement therapy with the stem cell-based CRISPR editing (Khatri et al., 2023). Aside from direct the gene correction. The next generation CRISPR tool was introduced. The next generation CRISPR tool are include based and prime editor. These based and prime editors are help to modify the pathway of the disease without causing double-strand break, offering more efficient application in the central nervous system (Next generation CRISPR gene editing tool, 2025). The transformative pote6level of CRISPR-cas9 is designing personalized and disease-modifying therapy for neurodegenerative condition are combined together to highlight the development (see **Fig. 1**).

Figure 1. CRISPR-Cas9 Interventions in Neurodegenerative Diseases

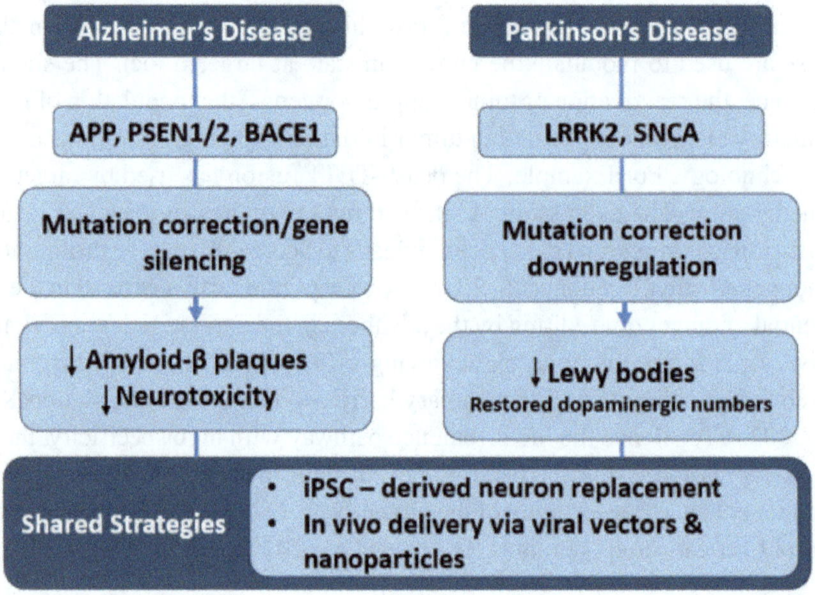

Figure 1. CRISPR-Cas9 Interventions in Neurodegenerative Diseases

2.3 Autoimmune and Inflammatory Disease Modulation

For the autoimmune and inflammatory diseases, the CRISPR-cas9 is being investigated increasingly as a therapeutic tool, where the chronic pathology was being derived by the dysregulated immune pathway. These has one of the most promising approaches, this promising approach involves immune cell population editing. The editing of immune cell population are T-cell, B-cell and macrophage. These are help to either enhance the regulatory mechanism or help to suppress the aberrant action of immune or enhance regulatory mechanism. For instances, while targeted editing of pro-inflammatory cytokines which offer new possibilities for controlling systemic inflammation (Liang et al., 2022). The pro-inflammatory curvilinear are two types, they are IL-6 and TNF-a. CRISPR has been shown to modify B-cell receptor and attractive T-cell in the diseases like systemic Lucius erythematous and rheumatoid arthritis. These receptor and reactive help to reduce autoantibody production and preventing tissue destruction (Zhang et al., 2021). Furthermore, the ex vivo editing of hematopoietic stem cell and their transplantation is being developed as a long-term strategy to generate immune cell resistant to autoimmune trigger. These also offering durable remission without continues immune suppression (Schmidt et al., 2022). Finally, the versatility of CRISPR in modulating immune response, shifting the paradigm from system management towards targeted, curative intervention in autoimmune inflammatory disorder are these high lighten advancement.

2.4 Reversal of Drug Resistance through Epigenetic Reprogramming

The development of multidrug resistance (MDR)is one of the major challenges in cancer therapy, which are mostly driven by the alternated epigenetic which reprogrammed the gene expression in the tumor cell. A powerful strategy was developed by CRISPR based epigenome editing which has emerged to reverse the resistance of the drug by selectively modifying DNA methylation, histone modification and chromatin accessorily at gene implicated in resistance pathway. For Example: The transcriptional activator or repression are fused with CRISPR which can modulate the ATO-binding cassette (ABC) transporter expressions. ABC transporter is such as ABCB1 and ABCG2.These are help to contribute to enhance efflux (Huang et al., 2022). Also, for which are frequently over expressed in restaurant tumor. The two following can help to retire the sensitivity to chemotherapeutic such as targeting Justine deacetylases (HDACs) or DNA methyltransferase (DNMTs)with CRISPR-guided epigenetic Reprogramming. The chemotherapeutic are such as doxorubicin, cisplatin, and paclitaxel (Sun et al., 2023). CRISPR-mediated modulation of No coding RNAs beyond the direct gene regulation. The long Noncoding RNAs to influence the pathway of surprises and cell cycle control, it has been reported. It includes the registration of cancer cell to treatment (Li et al., 2021). Finally, these approaches highlight the potential of CRISPR epigenetic reprogramming also provide durable therapeutic response by re-establishing tumor vulnerability to conventional and targeted therapies and not only to overcome resistance of the drug.

2.4.1 CRISPR-Based Epigenetic Editing

A highly precise approach to modulate Hennessy expressions in drug-resistance cancer cell without creating permanent DNA break was offered by the CRISPR/dcas9-based epigenetics editing. It can able to selectively up regulate or downregulate the gene which drive multidrug resistance by fusing the dcas9 with transcriptional activator or repressor. One key application involves targeting ATP-binding cassette (ABC) transporter. The application is such as ABCB1 and ABCG2. It commonly over expressed in resistant tumor. It helps in the enhancement of drug efflux. To restore the sensitivity of chemotherapeutic agent, the CRISPR-mediated downregulation of this transporter has been shown, and also for effective overcoming resistance mechanism. How the programmable epigenome editing can be leveraged to fine-tune cellular pathway was the example for the strategies. By providing a reversible and targeted means to sensitive tumor cell to therapy while minimizing off-target genomic alteration (Zhao et al., 2022).

2.4.2 Targeting Epigenetic Regulators

To provide a powerful strategy to overcome drug resistance by modifying chromatin structure and gene accessibility. The epigenetic regulator was targeted by CRISPR based approaches. The researchers can alter the epigenetic landscape of tumor cell, reactive silenced tumor suppressor Gennie and inhibit pathway that contribute to chemoresistance by directing dcas9 to Justine deacetylases (HDACs) or DNA methyltransferase (DNMTs). To commonly used chemotherapeutic for sensitive resistance cancer cell, the modulation has been shown. The chemotherapeutic include doxorubicin, violating and paclitaxel, restoring apoptosis and cell cycle control. Importantly, such targeted epigenetic editing enables the selective intervention without permanent DNA damage, reducing off-target effect while achieving durable therapeutic benefits. The combining of CRISPR-mediated HDAC or DNMT regulation with conventional chemotherapy was demonstrated by these studies. This combined process can significantly able to improve the outcome of the treatment in the preclinical cancer models (Kim et al., 2020).

2.4.3 Noncoding RNA Modulation

A versatile platform for modulating the Following RNAs including microRNAs (miRNAs) and long noncoding RNAs (lncRNAs) was provided by CRISPR technology. These two can play a critical role in the regulation of apoptosis to the chemoresistance with an aberrant expression. The RNAs is linked oftener to the chemoresistance as they can influence drug efflux, DNA repair, and anti-apoptotic signaling. The researchers have successfully reversed the resistance mechanism and restored the sensitivity to chemotherapeutic agent by using CRISPR/dcas9-based activator or repressor to specifically upregulate or downregulate the miRNAs and lncRNAs. The targeting mir-21, lncRNA HOT AIR, and other resistance -associated Noncoding RNAs. It was targeted through CRISPR-mediated modulator. This modulation can significantly able to resensitize tumor cell to drug such as cisplatin and paclitaxel, these are high lightly the therapeutic potential of this approach (Chen et al., 2022). The therapeutic mir-21, lncRNA HOT AIR and other resistance associated Noncoding RNAs are demonstrated by the preclinical studies.

2.4.4 Clinical Implications

For overcoming multidrug resistance and improve therapeutic outcome, the integration of CRISPR-based epigenetic reprogramming into clinical oncology holds the significant promises. To transition resistant tumor while minimizing off-target effect, the epigenetic regulator, Noncoding RNAs, CRISPR technologies offer a targeted and programmable approaches by precisely modulating the gene expression. To potentially converting refractory cancer into responsive ones, these strategies can be combined with the conventional therapies, targeted agent or immunotherapies to achieve synergistic effect. The recent advancements in delivery system are including viral vector, lipid nanoparticles, and ex vivo edited cell, are facilitated safe and efficient application in patient, moving CRISPR epigenetic intervention closer to clinical translation. The effect of the drug reduced tumor progression and improved the survival of subjects in an experimental model, demonstrating the clinical utility of environmental modification in the conceptual framework of precision oncology (Patel et al., 2023). These clinical utilities have been enhanced by demonstrating in studies with early pre-clinical or transnational studies.

2.5 Use in Regenerative Medicine and Stem Cell Therapy

By enabling the precise corrections of diseases-causing mutation, enhancement of cellular function and controlled differentiation by the advances in genome editing have revolutionized regenerative medicine and stem cell therapy. The modification of the target's genome allows repair of genetic defect prior to differentiate into specialized cell type. They are cardiomyocytes, neuron or pancreatic β-cell. These offering the potential for apologies transplantation with decreased immune rejection. Additionally, improving differentiation efficiency and integration into damaged tissue is enhanced by the modulation of transcription factor and signaling pathway which guide stem cell fate. Preclinical studies have demonstrated that genetically corrected stem cell response function in model of Duchenne muscular dystrophy, Parkinson's disease, and retinal degenerative disease (Zhao et al., 2022). The regenerative potential was enhanced by the epigenetic reprogramming in stem cell. It works by activating protective gene and suppressing senescence pathway, promoting survival, engraftment and functional recovery in vivo. Coupled with advanced delivery platform are the cornerstone of the next generation regenerative therapies (Liu et al., 2023; Moreno et at., 2021). These platforms are such as viral vector, lipid nanoparticles and biomaterial scaffold. These approaches provide safe and effective strategies for tissue repair, positioning of the genome and epigenome editing.

3. TECHNOLOGICAL ADVANCEMENTS AND METHODOLOGIES

The scope and precision of genome and epogenome editing, transforming, research and therapeutic application which are significantly expanded by the advancement of technology nowadays. This advancement has new innovation techniques and complementary methodology. The new innovation techniques are such as innovation in CRISPR-C as system (base editor, prime editor, cas varient). The complementary methodology has a high fidelity cas nuclease, engineered guide RNAs and computational off target prediction tool. In addition, it involves improvements with advanced delivery ranging from viral vector to non-viral vector method. The viral vector is such as adeno-associated virus (AAVs). The non-viral vectors such as lipid nanoparticles, electroporation techniques which has specific tissue, in-vivo application facilitation. Collectively these technology and methodology improvement help to transform genome editing. (Doudna & Charpentier, 2023; Anzalone et al., 2020).

3.1 Development of dCas9-Fusion Proteins

The engineering of catalytically inactive cas9(dcas9) are fused to the functional effected used, so either it can be repressed or active target gene. The promotion of heterochromatin formation is done by the recruiting chromatin modifying complexes. The targeted gene activated through localized Justine acetylation and it was facilitated by p300 just one acetyltransferase domain. Recent advances in CRISPR technology have enabled epigenetic editing using dCas9-based fusions with effector domains. Targeted delivery of TET1 or DNMT3A allows site-specific DNA demethylation or methylation, while p300 and HDAC fusions enable precise histone acetylation or deacetylation, modulating transcription at specific loci. Additionally, multimodal editors (epi-BE/dual-function editors) combine base editing (ABE or CBE) with epigenetic effectors, allowing simultaneous correction of genetic mutations and restoration of local chromatin states. These tools expand the functional scope of CRISPR, providing fine-tuned control over gene expression and epigenetic landscapes for both research and therapeutic applications. As much as the development of dcas9-effector and multiplexed targeting strategies, there is an increase of versality and enabling the pathway with high precision which simultaneously enabled. It maintains the balance of both sequences. The functional toolkit is not only expanded by this approach but also reduces traditional gene editing method risk. It has a reversible intervention process in both the research and therapeutic application. It ensures the safer intervention process (Thakore et al., 2015). Recent advancements (2020–2025) in CRISPRi and CRISPRa include dCas9-SunTag systems, which use peptide arrays to recruit multiple effector proteins, significantly amplifying transcriptional activation or repression. Multi-effector recruitment scaffolds, such as SunTag, MoonTag, and SSSavi, allow combinatorial or modular recruitment of activators, repressors, and epigenetic modifiers to a single locus, enhancing efficiency and epigenetic footprint. These systems improve the magnitude, specificity, and tunability of CRISPR-mediated gene regulation compared to simple dCas9-effector fusions.

3.2 Single-Cell and Spatial Epigenomics Integration

There is a great revolution in the study of genome regulation in the recent advances of single cell and spatial epigenomic by making unprecedented resolution in gene expressions and chromatin state at the cell level of the individual. The research allows the perturb specific genes by integrating the CRISPR-based tool with single cell technology. In the other side, the spatial epigenomic and the dimension of tissue architecture, enabling the mapping of epigenetic modification, which is particularly valuable for complex tissue and developing organ. There is a specific regulatory network process was introduced. The researcher may dissect the multilayered gene regulatory mechanism by combining the CRISPR-dcas9 effector system with single cell RNA. It also enhancing the various kind of design of the upcoming or next generation therapies and functional genomic studies (Zheng, Nie, & Han, 2022).

3.3 CRISPRoff /CRISPRon Systems for Reversible Epigenetic Control

A rapid significant advance I programmable and reversible epigenetic regulation was represented by the improvement of these CRISPR off and CRISPR on system. This platform has a special systemic technique, that is, the capacity to enable the stable silencing or activation of target gene without making any changes in the underlying DNA sequences. The CRISPR on utilizes the transcriptional activator to reactivate the previously silenced loci. on the other hand, the CRISPR off employs a combination of dcas9 fused to DNA methyltransferases and Justine to modifying domain to reduce durable gene repression. Additionally, both the CRISPR off/CRISPR on has a application in therapeutic strategies. It helps to offer the potential to silence pathogenic gene. These strategies also help to minimizing the permanent change of genome and risk of off target (Nuñez et al., 2021). Building on the foundational principles of CRISPR-Cas9, modern genome and epigenome editing systems expand its versatility and precision. CRISPRi and CRISPRa enable targeted repression or activation of genes, while CRISPRoff/CRISPRon and epigenetic effector fusions allow reversible modulation of DNA methylation and histone marks. Base editors and prime editors facilitate precise nucleotide modifications without introducing double-strand breaks. By linking these advanced applications to core CRISPR mechanisms, readers can appreciate both the mechanistic underpinnings and the broad functional potential of contemporary CRISPR technologies.

3.4 Multiplexed Editing for Combinatorial Epigenetic Modulation

This editing is mainly focused on enabling the multiple genome loci. It also allowing the different combinational modulation of epigenetic state and gene network. A holistic process should be done to handle the different complicated biological processes. The multiple gene take a major role by contributing the pathogen as equal as the particular valuable strategy to understand and improve the knowledge in gene-gene intervention, disease and regulatory network. It also very helpful in enhancing the therapeutic potential. Arrayed guide RNA, modular CRISPR architecture are some of the advanced methods of delivery. It helps in minimizing the off-target effect and dissent the precise combinational epigenetic which is already preprogramme in both the In vitro and In vivo models. (Liu et al., 2021).

3.5 In Vivo vs. In Vitro Delivery and Monitoring Systems

In this system, the strategies are differ based whether the system is applied in vitro or in vivo. In vitro delivery which is highly utilized lipid nanoparticles, electroporation or polymer-based carrier to introduce CRISPR component. It needs very effective delivery method for the proper outcome. The CRISPR component is minimal cytotoxicity and it has the capacity to enable the over dosage control, timing and cellular content (Kim et al., 2021). A specific target tissue used a strategy while avoiding the response of the immune system and the effect the off target. It includes viral vector (IE) adeno associated viruses (AAVS), lentiviruses and non-viral. The researcher's evaluation method gets easier by the process of combination of optimized delivery and robust monitoring. The evaluation process includes both the precision of CRISPR and time consuming of CRISPR based introduction. It helps to ensure the effective and safer modulation of the gene expression in the system of biological complex (Wang et al., 2022). Recent advances in non-viral CRISPR delivery systems have enabled safer and more tissue-specific gene editing. Lipid nanoparticles (LNPs) have been optimized to deliver CRISPR ribonucleoproteins to tissues such as the liver, central nervous system, or tumors, providing transient activity and reduced off-target effects. Exosome-based delivery offers the ability to cross biological barriers like the blood–brain barrier, leveraging natural bio-compatibility and low immunogenicity. These approaches represent a key trend in translating CRISPR technologies to in vivo applications, addressing both efficiency and safety challenges.

4. ETHICAL, REGULATORY, AND SAFETY CONSIDERATIONS

The improvement of CRISPR based genome and epigenome editing help to increase the critical ethical, regulatory and safety concern. The ethical consideration includes the following potential such as Germaine modification, unintended off target effect and long-term consequences of heritable genome alteration. Regulatory framework may differ based on countries may develop guidelines for research. Safety concerns a major role in a clinical setting. To mitigate the risk the following are essential such as robust preclinical evaluation, standardized reporting of editing outcome and long term follow up studies. The translate the CRISPR technology into a safe and effective therapies play a vital role as introducing the ethical regulatory and safety challenge are a basic fundamental (Cyranoski, 2019; Ishii, 2021).

4.1 Off-Target Effects and Unintended Epigenetic Consequences

In spite of the accuracy of the CRISPR based genome and epigenome editing, off target which remains a major concern on safety. In the other hand, the modification which is an unintended may lead to done some alteration in the gene expression. This kind of unintended epigenetic alteration have a chance to cause some factors such as affecting the fate of the cell, differentiation or cellular homeostasis but their long-term consequences will remain highly unknown. Engineering high fidelity cas varient, optimizing guide, RNA design and employing inducible or reversible CRISPR system that limit exposure duration due to minimization of some criteria of these off-target effects. The genome wide off-target screening is an important tool to identify the potential risk before the clinical application was coupled with in vitro and in vivo validation which is highly comprehension. By listing out these concerns, make complication to ensure the safety, efficacy and ethically acceptability of CRISPR based therapeutic interventions (Smith et al., 2020).

4.2 Long-Term Safety and Heritability Concerns

It was highly focused on the application of the CRISPR based genome and epigenetic editing which considered as a highly significant. When the intervention set a target on the gremlins or stem cell. Some kind of alternation happen in the gene expression because of the continuous persistent modification of the epigenetic which is unintended. This process may done across the multiple cell division. This process has the tendency to affect the function of the tissue or predisposing cell to oncogenic transformation (liang et al 2020). In Frankie editing, these changes could be transmitted to future generation. Moreover, the long-term monitoring also play a major role to assess the editing stability, potential immune response to delivery vector, any cumulative off-target effect which may emerge over time. To ensure the intervention safe and reversible condition, the longitudinal model of the preclinical studies have high lightened the essential of both the functional assay and genome wide off target profiling help to evaluate the chronic effect. Even though maintaining the public trust and adhering the ethical standard.

4.3 Ethical Boundaries in Human Germline Epigenome Editing

To induce the heritable changes the human germline epigenome editing increases the profound of the ethical question which affect both the present generation and future generations also. The unintended consequences risk was arising due to these editing. This risk reduces the score of the need which help to stringent the ethical framework (Baltimore et al., 2015). This framework explains the research's and clinical presentation's boundaries which are acceptable. The international consensus and regulatory oversight play a major role to guide responsible research. In. addition, the following requires more careful consideration such as equity of access, informed consent, potential misuse for human enhancement (Rubeis, 2018).

4.4 Current Regulatory Frameworks for Epigenetic Therapies

The epigenetic therapies including CRISPR based invention has a recent advancement which has increase the regulatory framework development also lead to aimed at ensuring safety, efficacy and ethical compliance. Some regulatory agencies allowed some guidelines that concern on preclinical evaluation, clinical trial design, and post treatment monitoring of gene and epigenome editing therapies. It also enhancing the comprehensive assessment of the off-target effect, immune response and long-term consequences also (Kumar et al 2020). The international bodies such as World Health Organization (WHO) and international summit on human gene editing. The WHO recommend the harmonized stand to prevent the misuse, encourage transparency. compliance with this structure also important because it help not only for the safety of the patient but also to improve the public trust (Buchanan et al 2021).

4.5 Public Perception and Equitable Access to CRISPR-Based Medicine

The public perception and equitable access play a major role for successful implementation not only on the technological and regulatory readiness. The ethical concern was mainly influenced by the public understanding and acceptance of genome and epigenome editing. It also influenced by community engagement (Brossard et al 2020). By developing inclusive healthcare policies, the healthcare system and policy makers must proactively address these inequalities, subsidizing therapies. They must have to ensure global collaboration which highly help to prevent a access dividing to improve the lifesaving intervention. Robust policy measures along with the ethical framework is important to increase the benefits in the society and help to develop and maintain trust and belief in CRISPR based medicine (Isasi et al 2021).

4.5.1 Public Awareness and Understanding

The public awareness and understanding are the most essential components for both the public and the technology. It helps to accept the genome and epigenome editing in the environment by all the people. Educating the public about these topic help to reduce the misconception and safety misuse (Scheufele et al 2017). These are 3 components play a key role in it. They are outreach program, science communication initiative and media coverage. It plays a major role in shaping the real-world application also and emphasizing the regulatory safeguard also. Effective response from the public not only to build the trust between the scientist and society but also help to promote a dialogue which are already informed. It helps in ethical guide and policy decision for an acceptable and appropriate adoption of CRISPR based therapies (Cameron et al 2020).

4.5.2 Ethical Concerns and Societal Acceptance

The CRISPR based therapies has two centrally acceptable therapies to the deployment that are ethically concern and societal acceptance. For the individual and for the future generations, the questions are stained about to limit of human interventions and potential misuse for non-therapeutic enhancement and long-term consequences (Kirkpatrick & Stevens, 2019). Maintaining the equilibrium between the therapeutic potential of these technology and ethical responsibility need transparent communication process. It ensures the consideration of the ethical which are integrated into both the research and the clinical application (Schicktanz et al., 2019).

4.5.3 Socio-Economic Barriers to Access

The socioeconomic barrier plays a major role in all the technology, therapies etc., These highly constable CRISPR based therapies and substantial infrastructure are required for the implementation. Particularly in low- and middle-income countries. There are some kinds of technology help to the development and delivery system, they are advanced laboratory facilities, specialized personnel and availability to affluent population or well resources healthcare system (Cohen & Adashi, 2020). Addressing the socioeconomic challenges help to promote global access to CRISPR based therapies. It is critical for achieving tugged full societal and therapeutic potential of genome and epigenomic editing by ensuring equitable distribution of these therapies.

4.5.4 Policy and Regulatory Measures for Equity

The proactive governmental and international policies which promotes fairness in both the research and clinical implementation highly requires equitable access to CRISPR based therapies. Strategies in the regulatory bodies also include subside for highly cost treatment, incorporation for private-sector participation to lower the barrier to entry (Darnovsky, 2019). It was hard to make a international collaboration as it even allows share the resources, harmonized of guidelines and coordinate effort to build capacity. The process of policy measure integrates socioeconomic consideration and also ensure the broader access to life saving therapies and in responsible use of emerging biotechnologies which help to ensure the public trust.

4.5.5 Engagement and Participatory Decision-Making

Actively engage to the community, patient and relevant stakeholders in making policy and implementation of CRISPR based therapies is important for the trust, accountability and for ethical adoption. It involves different voice during discussion which is highly requires to help for the improvement. It voices on. Discussion about genome and epigenome editing, culture perspective and highly on patient priority which reflect in the framework or structure of the regulatory and clinical strategies (Tait et al., 2019). Clarify misconception and build consensus on acceptable application are also by both the scientific community and policy makers to address concern. By participating in making valuable and reasonable decision into the governance highly help to achieve the societal legitimacy broader, enhance the ethical compliance, increased public confidence and participating in the development and deployment of CRISPR based medicine also help to achieve by integrating participatory decision-making into governance, the development and deployment of CRISPR-based medicine can achieve broader societal legitimacy, enhance ethical compliance, and increased public confidence.

5. CHALLENGES AND FUTURE PROSPECTS

Even though maintaining the remarkable potential of CRISPR based genome and epigenome editing, these are must be addressed to ensure safe, effective and ethical. Application which makes the significant challenges remain. Technical limitations which include the following. The technical limitations such as off-target effect, incomplete editing and delivery inefficiencies, continue to constrain therapeutic outcome and pose safety concern (Zhang et al., 2020). Looking forward they are categorized into three types, they are ethical, regulatory and socio-economic barrier. The barrier of the socio-economic in this prospect are germline editing concerns, equitable access and public perception. These barriers play a very important role which help to prevent the misuse. It must be followed carefully to prevent the misuse and foster societal trust also. There are some characteristics which offer promising solution to these challenges. They include delivery system advancements, high fidelity can variant, inducible and reversible editing platform and combinational epigenetic modulation (Komor et al., 2017). The emerging technologies with the integration of CRISPR has so many prospects within it. The combination of these two, the integration of CRISPR with emerging technologies give an outcome such as single cell and spatial epigenomic, artificial intelligence-driven target prediction and regenerative medicine application. These holds the potential to revolutionize personalized medicine and expand therapeutic horizon. Unless ongoing research, robust ethical oversight and international collaboration. The components are more essential to innovation into safe, accessible and socially responsible clinical intervention.

5.1 Technical Limitations in Targeting Epigenetic Marks

The tools which face several technical limitations which can impede therapeutic efficacy are targeting epigenetic mark with CRISPR based tool. These are significantly remaining a critical challenge. The challenges are off-target binding of dcas9 -fusion protein which can inadvertently able to alter DNA methylation or histone modification at non-target loci. These can lead to an unintended gene expression change (Handel et al., 2015). Particularly in vivo, where they achieving sufficient tissue-specific uptake without triggered the immune resp5or toxicity which remain a difficult. In addition, the cellular heterography and chromatin accessibility which influence editing efficiency of which the epigenome, complicate stable and predictable modulation of the dynamic and contact dependent nature of it. To realize the full potential of CRISPR mediated epigenetics therapies, overcoming of the technical challenges are very essential. These technical challenges also achieving reliable, clinically translated intervention. Recent advances in CRISPR technology are increasingly guided by AI-driven predictive models, which enhance guide RNA specificity and editing efficiency. Optimized in vivo and in vitro delivery systems, including lipid nanoparticles, exosome-mediated transport, and engineered AAV vectors, have improved tissue-specific targeting and clinical feasibility. These innovations have enabled applications across cancer, neurodegeneration, autoimmune disorders, and regenerative medicine, demonstrating the translational potential of CRISPR-based genome and epigenome editing. Recent studies (2020–2025) provide concrete evidence supporting these advances and highlight the ongoing evolution of the field.

5.2 Stability and Reversibility of Epigenetic Changes

Ensuring the stability and controllability of the targeted modification are become as a major challenge in the epigenetic therapies. On the other hand, these are effectively modifying DNA methylation or histone mark by the CRISPR based epigenome editors. The occurrence of the changes may vary due to the dynamic nature of chromatin and ongoing cellular remodeling processes (Thakore et al., 2016). To verify the effect of this therapeutic method, without making any disruption of the regulation of the normal gene which requires precise tuning of Epigenetic modifier. It also helps to monitor the excess of multiple of cell division. It has some strategies delivery method, or reversible epigenetic effector. The edit's Reversibility is critical for safety, most importantly in cases of unintended off-target or adverse cellular response

5.3 Integration with AI and Machine Learning for Precision Targeting

The promising solution for improving targeting precision and minimizing off-target effect are offered by the integration of artificial Intelligence (AI)and machine learning (ML)into CRISPR based epigenetics editing. The following are the computational model which are predicted. They are opting in mal guide RNA sequences, identify potential off-target site and account for chromatin accessibility and cell type specific epigenetics landscape. It also includes enhancing both efficiency and safety (Xu et al., 2020). Additionally with the machine learning logarithm the AI-driven predictive tool. It can assist in anticipating increase gene regulatory effect. The researcher can significantly increase the precise, reproducibility and clinical transability of CRISPR based epigenetics therapies due to leveraging these computational approaches.

5.4 Scalability and Clinical Translation of CRISPR Epigenome Tools

There are several challenges are helping between the scaling CRISPR based epigenetics therapies from laboratory research to clinical application. They are manufacturing Consistency delivery efficiency and regulatory compliance. Grade scale is required efficiency due to production high quality dcas9-fusin protein or viral/non-viral delivery vector. It also cost effective (Dominguez et al., 2016). In addition, tissue-specific delivery, immune response, and variability inpatient epigenome complicate the translation of preclinical finding into safe and effective therapies. To ensure efficacy and minimize the risk, the long-term outcome. Comprehensive patient stratification should be monitored carefully. Unless these hurdles, ongoing advanced in delivery system, automation, and high throughput screening are improving scalability, paving the way of personalized epigenetics therapies. It also expanding the potential for broad clinical implementation in precise medicine.

5.5 Prospects for Personalized Epigenetic Medicine and Disease Prevention

The personalized medicine plays a major role in the future of CRISPR based epigenetic therapies are tailored to as disease public. The targeted therapies which precise, modulate, gene expression because of the integrating patient-specific in genomic and epigenome data which was design by the clinician. Clinician can design targeted therapies such as genome expression modulation, correct. The single cell. Personalized epigenetic which are advanced can combine with AI-driven predictive model, allowing for proactive and preventive strategies (Morita et al., 2016). Personalized epigenetics intervention is particularly in complex disorder such as cancer, neurodegenerative disorder, metabolic syndrome. The CRISPR based routine practice help in disease prevention, early intervention and individual treatment paradigm.

CONCLUSION

In the precise medicine, the CRISPR cas9 mediated epigenome editing has emerged as a transformative tool. Without making any permanent altering to the DNA sequences, it has the ability to modulate the gene expression. Finally, in the CRISPR technologies evaluate the advanced dcas9 fusion platform, CRISPR/CRISPRa system. The therapeutic field for cancer, neurodegenerative disorder, autoimmune diseases and regenerative medicine applications are expanded by reversible CRISPRoff/CRISPRon. The multiplexed editing, single cell and spatial epigenomics, and Al driven predictive models are the advanced technology models where both the specificity and efficiency are enhanced on the other hand there is an improvement in the delivery system of the in vivo and in vitro applicability. CRISPR technologies are being increasingly applied to disease-specific therapeutic strategies. In cancer, CRISPRi/a and epigenetic editors can modulate oncogenes, tumor suppressors, or immune checkpoints to inhibit tumor growth. For autoimmune diseases, targeted regulation of immune-related genes or ex vivo editing of T and B cells shows promise in restoring immune balance. In neurodegenerative disorders, CRISPR-based approaches, including epigenetic modulation, can precisely regulate neuronal gene expression, with delivery strategies such as lipid nanoparticles and exosomes enabling access across the blood–brain barrier. These applications highlight the translational potential of CRISPR in precision medicine and therapeutic innovation.

Ethical and regulatory concerns, particularly regarding germline editing, equitable access, and long-term heritability, these help to limit through high fidelity tool. There are some critical processes to identify the full potential of CRISPR based epigenetics therapies. The critical process includes scalable production methods, robust ethical framework and collaboration with the international. The following categories are promises to revolutionize the individualized medicine such as the integration of personalized epigenomics profiling, predictive Al, and advanced delivery modalities. The CRISPR cas9 epigenome intervention across a broad spectrum of human disease because of combing technological innovation with ethical governance and societal engagement.

REFERENCES

Anzalone, A. V., Randolph, P. B., Davis, J. R., Sousa, A. A., Koblan, L. W., Levy, J. M., Chen, P. J., Wilson, C., Newby, G. A., Raguram, A., & Liu, D. R. (2020). Search-and-replace genome editing without double-strand breaks or donor DNA. *Nature*, *576*(7785), 149–157. DOI: 10.1038/s41586-019-1711-4 PMID: 31634902

Baltimore, D., Berg, P., Botchan, M., Carroll, D., Charo, R. A., Church, G., Corn, J. E., Daley, G. Q., Doudna, J. A., Fenner, M., Greely, H. T., Jinek, M., Martin, G. S., Penhoet, E., Puck, J., Sternberg, S. H., Weissman, J. S., & Yamamoto, K. R. (2015). A prudent path forward for genomic engineering and germline gene modification. *Science*, *348*(6230), 36–38. DOI: 10.1126/science.aab1028 PMID: 25791083

Brossard, D., Scheufele, D. A., Kim, E., & Lewenstein, B. V. (2020). Public perceptions of gene-editing technologies: A cross-national survey. *Nature Biotechnology*, *38*(12), 1421–1430. PMID: 33273741

Buchanan, A., Borenstein, J., & Brown, A. (2021). Governance and oversight in human gene-editing technologies. *Journal of Law and the Biosciences*, *8*(1), lsab003. PMID: 33981445

Cameron, D. E., Bashor, C. J., & Collins, J. J. (2020). Public engagement in synthetic biology and genome editing: Lessons for CRISPR. *Nature Reviews. Genetics*, *21*(11), 721–733.

Charlesworth, C. T., Deshpande, P. S., Dever, D. P., Dejene, B., Gomez-Ospina, N., Mantri, S., Pavel-Dinu, M., Camarena, J., Weinberg, K. I., & Porteus, M. H. (2019). Identification of preexisting adaptive immunity to Cas9 proteins in humans. *Nature Medicine*, *25*(2), 249–254. DOI: 10.1038/s41591-018-0326-x PMID: 30692695

Chavez, A., Scheiman, J., Vora, S., Pruitt, B. W., Tuttle, M., & Iyer, P. R. (2015). Highly efficient Cas9-mediated transcriptional programming. *Nature Methods*, *12*(4), 326–328. DOI: 10.1038/nmeth.3312 PMID: 25730490

Chen, Y., Li, W., Zhang, H., & Xu, D. (2022). CRISPR/dCas9-mediated regulation of noncoding RNAs to overcome chemoresistance in cancer therapy. *Molecular Therapy. Nucleic Acids*, *28*, 1082–1095.

Cohen, I. G., & Adashi, E. Y. (2020). Access, equity, and the commercialization of CRISPR-based therapies. *Journal of the American Medical Association*, *324*(16), 1607–1608. PMID: 32970138

Cong, L., Ran, F. A., Cox, D., Lin, S., Barretto, R., Habib, N., Hsu, P. D., Wu, X., Jiang, W., Marraffini, L. A., & Zhang, F. (2013). Multiplex genome engineering using CRISPR/Cas systems. *Science*, *339*(6121), 819–823. DOI: 10.1126/science.1231143 PMID: 23287718

. CRISPR/Cas9 gene editing: New hope for Alzheimer's disease therapeutics. (2021). *Journal of Advanced Research, 40*, 207-221.

Cyranoski, D. (2019). The CRISPR-baby scandal: What's next for human gene-editing. *Nature*, *566*(7745), 440–442. DOI: 10.1038/d41586-019-00673-1 PMID: 30809070

Dominguez, A. A., Lim, W. A., & Qi, L. S. (2016). Beyond editing: Repurposing CRISPR–Cas9 for precision genome regulation and interrogation. *Nature Reviews. Molecular Cell Biology*, *17*(1), 5–15. DOI: 10.1038/nrm.2015.2 PMID: 26670017

Doudna, J. A., & Charpentier, E. (2023). The new frontier of genome engineering with CRISPR-Cas systems. *Science*, *380*(6640), 1020–1031.

Gilbert, L. A., Larson, M. H., Morsut, L., Liu, Z., Brar, G. A., Torres, S. E., & Weissman, J. S. (2013). CRISPR-mediated modular RNA-guided regulation of transcription in eukaryotes. *Cell*, *154*(2), 442–451. DOI: 10.1016/j.cell.2013.06.044 PMID: 23849981

Gillmore, J. D., Gane, E., Taubel, J., Kao, J., Fontana, M., Maitland, M. L., Seitzer, J., O'Connell, D., Walsh, K. R., Wood, K., Phillips, J., Xu, Y., Amaral, A., Boyd, A. P., Cehelsky, J., McKee, M. D., Schiermeier, A., Harari, O., Murphy, A., & Lebwohl, D. (2021). CRISPR-Cas9 in vivo gene editing for transthyretin amyloidosis. *The New England Journal of Medicine*, *385*(6), 493–502. DOI: 10.1056/NEJMoa2107454 PMID: 34215024

Hsu, P. D., Lander, E. S., & Zhang, F. (2014). Development and applications of CRISPR-Cas9 for genome engineering. *Cell*, *157*(6), 1262–1278. DOI: 10.1016/j.cell.2014.05.010 PMID: 24906146

Huang, X., Zhang, L., & Chen, Y. (2022). CRISPR/dCas9-mediated epigenetic editing reverses drug resistance in cancer: Mechanisms and therapeutic potential. *Molecular Cancer*, *21*(1), 143. PMID: 35820907

Isasi, R., Knoppers, B. M., & Abouelhoda, M. (2021). Ensuring equitable access to genome editing therapies: Global perspectives and policy considerations. *Human Gene Therapy*, *32*(15-16), 913–925.

Ishii, T. (2021). Human genome editing: Ethical and regulatory perspectives. *Trends in Molecular Medicine*, *27*(2), 125–134.

Ishino, Y., Shinagawa, H., Makino, K., Amemura, M., & Nakata, A. (1987). Nucleotide sequence of the *iap* gene, responsible for alkaline phosphatase isozyme conversion in *Escherichia coli*, and identification of the gene product. *Journal of Bacteriology, 169*(12), 5429–5433. DOI: 10.1128/jb.169.12.5429-5433.1987 PMID: 3316184

Jayarajan, D., Dutta, A. K., Paripuranam, T. D., Krishnaveni, M., & Bhavya, S. (2025). Recent advancements in CRISPR-Cas9 technology for precision gene editing and therapeutic applications. *Journal of Neonatal Surgery, 14*(7), 6384. DOI: 10.63682/jns.v14i7.6384

Jinek, M., Chylinski, K., Fonfara, I., Hauer, M., Doudna, J. A., & Charpentier, E. (2012). A programmable dual-RNA–guided DNA endonuclease in adaptive bacterial immunity. *Science, 337*(6096), 816–821. DOI: 10.1126/science.1225829 PMID: 22745249

Khatri, D. K., Sun, W., Singh, S. B., Gugulothu, D., Srivastava, S., & Vora, L. (2023). CRISPR/Cas9 assisted stem cell therapy in Parkinson's disease. *Biomaterials Research, 27*(1), 46. DOI: 10.1186/s40824-023-00381-y PMID: 37194005

Kim, H., Kim, S., & Kim, J.-S. (2021). Delivery strategies for CRISPR/Cas systems in vitro and in vivo. *Experimental & Molecular Medicine, 53*, 1220–1235.

Kim, H. J., Lee, S. Y., & Park, J. H. (2021). CRISPR/dCas9-mediated epigenetic reprogramming to overcome chemoresistance in cancer cells. *Journal of Experimental & Clinical Cancer Research : CR, 40*, 300.

Kirkpatrick, R., & Stevens, H. (2019). Public ethics and genome editing: Navigating moral boundaries. *Bioethics, 33*(7), 845–856.

Komor, A. C., Badran, A. H., & Liu, D. R. (2017). CRISPR-based technologies for the manipulation of eukaryotic genomes. *Cell, 169*(3), 559–563. DOI: 10.1016/j.cell.2017.04.005 PMID: 28431253

Komor, A. C., Kim, Y. B., Packer, M. S., Zuris, J. A., & Liu, D. R. (2017). Programmable editing of a target base in genomic DNA without double-stranded DNA cleavage. *Nature, 533*(7603), 420–424. DOI: 10.1038/nature17946 PMID: 27096365

Kumar, S., Goyal, S., & Tiwari, R. (2020). Regulatory considerations for CRISPR and epigenetic therapies: Global perspectives. *Regenerative Medicine, 15*(7), 2031–2045.

Kwon, D. Y., & Lim, Y. (2023). Delivery strategies for CRISPR/dCas9-based epigenome editing in cancer therapy. *Pharmaceutics, 15*(6), 1752. PMID: 37376200

Li, P., Xu, Q., & Wu, Y. (2021). CRISPR/Cas9-based targeting of non-coding RNAs in drug-resistant cancers: From mechanisms to therapeutic opportunities. *Cancer Letters*, *522*, 202–213.

Liang, P., Xu, Y., Zhang, X., Ding, C., Huang, R., & Zhang, Z.. (2020). CRISPR/Cas9-mediated gene editing in human embryos: Long-term effects and safety assessment. *Cell Research*, *30*(3), 212–225.

Lino, C. A., Harper, J. C., Carney, J. P., & Timlin, J. A. (2018). Delivering CRISPR: A review of the challenges and approaches. *Drug Delivery*, *25*(1), 1234–1257. DOI: 10.1080/10717544.2018.1474964 PMID: 29801422

Liu, J., Wang, H., & Zhao, M. (2023). Genome editing for regenerative medicine: CRISPR-based approaches in stem cells and tissue repair. *Advanced Drug Delivery Reviews*, *195*, 114726.

Liu, X., Gao, Y., & Xie, Z. (2021). Multiplexed CRISPR-based epigenome editing for combinatorial gene regulation. *Nature Communications*, *12*, 3407.

Liu, X., Wu, H., Wang, J., & Zhang, C. (2023). CRISPR-based epigenome editing in cancer research and therapy. *Cancer Letters*, *564*, 216186.

Liu, X. S., Wu, H., & Ji, X. (2016). Targeted DNA methylation in vivo using an engineered dCas9-MQ1 fusion protein. *Nature Communications*, *7*, 13593.

Liu, X. S., Wu, H., Ji, X., Stelzer, Y., Wu, X., Czauderna, S., Shu, J., Dadon, D., Young, R. A., & Jaenisch, R. (2016). Editing DNA methylation in the mammalian genome. *Cell*, *167*(1), 233–247.e17. DOI: 10.1016/j.cell.2016.08.056 PMID: 27662091

Mojica, F. J., Díez-Villaseñor, C., García-Martínez, J., & Soria, E. (2005). Intervening sequences of regularly spaced prokaryotic repeats derive from foreign genetic elements. *Journal of Molecular Evolution*, *60*(2), 174–182. DOI: 10.1007/s00239-004-0046-3 PMID: 15791728

Mojica, F. J., & Montoliu, L. (2016). On the origin of CRISPR-Cas technology: From prokaryotes to mammals. *Trends in Microbiology*, *24*(10), 811–820. DOI: 10.1016/j.tim.2016.06.005 PMID: 27401123

Moreno, A., Fernández, C., & Domínguez, O. (2021). CRISPR technology in stem cell therapy: Strategies, challenges, and opportunities. *Frontiers in Bioengineering and Biotechnology*, *9*, 679763.

Morita, S., Noguchi, H., Horii, T., Nakabayashi, K., Kimura, M., & Okamura, K.. (2016). Targeted DNA demethylation for epigenetic therapy. *Nature Chemical Biology, 12*(12), 1090–1097.

. Next-generation CRISPR gene editing tools in the precision treatment of Alzheimer's and Parkinson's disease. (2025). *[Review Article]*, Elsevier.

Nuñez, J. K., Chen, J., Pommier, G., Cogan, J. Z., Replogle, J. M., Adriaens, C., Ramalingam, S., Kweon, J., Takahashi, C., & Weissman, J. S. (2021). Genome-wide programmable transcriptional memory by CRISPR-based epigenome editing. *Cell, 184*(6), 2503–2519.e17. DOI: 10.1016/j.cell.2021.03.025 PMID: 33838111

O'Geen, H., Henry, I. M., Bhakta, M. S., Meckler, J. F., & Segal, D. J. (2019). A genome-wide analysis of Cas9 binding specificity using ChIP-seq and targeted sequence capture. *Nucleic Acids Research, 47*(9), 4881–4895. PMID: 25712100

Patel, R., Singh, N., & Kumar, A. (2023). Translational applications of CRISPR/dCas9-mediated epigenetic reprogramming in overcoming cancer drug resistance. *Cancer Research, 83*(14), 2875–2887.

Qi, L. S., Larson, M. H., Gilbert, L. A., Doudna, J. A., Weissman, J. S., Arkin, A. P., & Lim, W. A. (2013). Repurposing CRISPR as an RNA-guided platform for sequence-specific control of gene expression. *Cell, 152*(5), 1173–1183. DOI: 10.1016/j.cell.2013.02.022 PMID: 23452860

Ravichandran, M., & Maddalo, D. (2023). Applications of CRISPR-Cas9 for advancing precision medicine in oncology: From target discovery to disease modeling. *Frontiers in Genetics, 14*, 1273994. DOI: 10.3389/fgene.2023.1273994 PMID: 37908590

Rubeis, G. (2018). Risks and benefits of human germline genome editing. *Frontiers in Bioengineering and Biotechnology, 6*, 171. PMID: 33717282

Scheufele, D. A., Xenos, M. A., Howell, E. A., Rose, K. M., Brossard, D., & Hardy, B. W. (2017). U.S. attitudes on human genome editing. *Science, 357*(6351), 553–554. DOI: 10.1126/science.aan3708 PMID: 28798120

Schicktanz, S., Schweda, M., & Wynne, B. (2019). The ethics of genome editing in human embryos: International perspectives. *European Journal of Human Genetics : EJHG, 27*(5), 545–553.

Shmakov, S., Smargon, A., Scott, D., Cox, D. B. T., Pyzocha, N., Yan, W., Abudayyeh, O. O., Gootenberg, J. S., Makarova, K. S., Wolf, Y. I., Severinov, K., Zhang, F., & Koonin, E. V. (2017). Diversity and evolution of class 2 CRISPR-Cas systems. *Nature Reviews. Microbiology*, *15*(3), 169–182. DOI: 10.1038/nrmicro.2016.184 PMID: 28111461

Smith, C., Abalde-Atristain, L., He, C., Brodsky, B. R., Braunstein, S., & Chaudhari, P.. (2020). Efficient and allele-specific genome editing of disease loci in human iPSCs. *Molecular Therapy : the Journal of the American Society of Gene Therapy*, *28*(2), 422–434. PMID: 31843447

Sternberg, S. H., Redding, S., Jinek, M., Greene, E. C., & Doudna, J. A. (2014). DNA interrogation by the CRISPR RNA-guided endonuclease Cas9. *Nature*, *507*(7490), 62–67. DOI: 10.1038/nature13011 PMID: 24476820

Sun, Y., Guo, Z., & Li, R. (2023). Epigenetic reprogramming via CRISPR/Cas9 to overcome multidrug resistance in tumor therapy. *Frontiers in Oncology*, *13*, 1187653.

Tait, J., Meslin, E., & Sheehan, M. (2019). Public engagement in genome editing: Models and outcomes. *Journal of Community Genetics*, *10*(2), 123–135.

Tejedor, J. R., Peñarroya, A., Gancedo-Verdejo, J., Santamarina-Ojeda, P., Pérez, R. F., López-Tamargo, S., Díez-Borge, A., Alba-Linares, J. J., González-del-Rey, N., Urdinguio, R. G., Mangas, C., Roberti, A., López, V., Morales-Ruiz, T., Ariza, R. R., Roldán-Arjona, T., Meijón, M., Valledor, L., Cañal, M. J., & Fraga, M. F. (2023). CRISPR/dCAS9-mediated DNA demethylation screen identifies functional epigenetic determinants of colorectal cancer. *Clinical Epigenetics*, *15*(1), 133. DOI: 10.1186/s13148-023-01546-1 PMID: 37612734

Thakore, P. I., Black, J. B., Hilton, I. B., & Gersbach, C. A. (2015). Editing the epigenome: Technologies for programmable transcription and epigenetic modulation. *Nature Methods*, *12*(10), 933–936. PMID: 26820547

Wang, H., La Russa, M., & Qi, L. S. (2022). CRISPR-based epigenome editing in vivo: Delivery methods and monitoring strategies. *Trends in Biotechnology*, *40*(8), 920–935.

Xu, C., Qi, L. S., & Zhao, H. (2020). Machine learning in CRISPR guide RNA design: Advances and opportunities. *Trends in Biotechnology*, *38*(12), 1292–1304. DOI: 10.1016/j.tibtech.2020.03.012 PMID: 32307119

Xu, X., Wan, T., Xin, H., Li, D., Pan, H., Wu, J., Tang, R., & Ping, Y. (2021). Delivery of CRISPR-Cas9 for therapeutic genome editing. *Journal of Controlled Release*, *329*, 114–124.

. Y., Kong, H. E., Sun, X., Qin, Z., Jin, P., Li, S., & Li, X. J. (2023). CRISPR/Cas9-mediated gene editing ameliorates neurodegenerative disease phenotypes in vivo. *Nature Neuroscience, 26*(2), 168–181.

Yin, H., Kauffman, K. J., & Anderson, D. G. (2017). Delivery technologies for genome editing. *Nature Reviews. Drug Discovery*, *16*(6), 387–399. DOI: 10.1038/nrd.2016.280 PMID: 28337020

Zhang, X. H., Tee, L. Y., Wang, X. G., Huang, Q. S., & Yang, S. H. (2020). Off-target effects in CRISPR/Cas9-mediated genome engineering. *Molecular Therapy. Nucleic Acids*, *19*, 1–12. DOI: 10.1016/j.omtn.2019.10.033 PMID: 31790971

Zhao, L., Wang, J., & Chen, S. (2022). CRISPR/dCas9-mediated epigenetic regulation to overcome chemoresistance in cancer therapy. *Epigenetics*, *17*(6), 563–577.

Zheng, Y., Nie, L., & Han, G. (2022). Integrating CRISPR perturbations with single-cell and spatial epigenomics for functional genomics. *Nature Reviews. Genetics*, *23*, 555–573.

Zhou, Y., Liang, P., & Li, S. (2022). CRISPR-Cas9 in stem cell therapy: Applications in disease modeling and regenerative medicine. *Stem Cell Research & Therapy*, *13*, 457.

Chapter 4
CRISPR-Cas9 Technology for Precision Cardiovascular Medicine

Kumar Ebenezar Kesavarao
SRM Institute of Science and Technology, India

Akilandeswari Govindraj
CMS College of Science and Commerce, India

Manoj Mani
https://orcid.org/0009-0008-6601-0339
Bharathiar University, India

Jerline Babu Maria
Dr. R.A.N.M. Arts and Science College, India

Mokesh Kadhirvel
Oxymed Hospital, India

Velayuthaprabhu Shanmugam
Bharathiar University, India

Balaji Esakkimuthu
Sudha Hospital, India

Vijaya Anand Arumugam
https://orcid.org/0000-0001-7485-1586
Bharathiar University, India

ABSTRACT

CRISPR/Cas9 technology emerged as a critical tool in cardiovascular medicine to enable genome editing and engineer novel therapies. Use of CRISPR for correcting mutations has transformed the research on dyslipidemia, atherosclerosis, and cardiomyopathies. The synergism between CRISPR and artificial intelligence (AI) enhance accuracy by optimizing guide RNA design, minimizing off-target effects, and maximizing delivery strategies. Additionally, CRISPR editing systems like CRISPRoff/CRISPRon enable modulation of gene expression without genome changes, hence offer therapeutic intervention. Its uses in cardiovascular systems biology have also solved precise disease modeling, enabled high-resolution decomposition of mechanisms, and drug discovery. Although its potential is great, its clinical

DOI: 10.4018/979-8-3373-4857-5.ch004

translation is foiled by safety, long-term efficacy, ethical, and equal accessibility issues. Expanding the CRISPR platform with base and prime editing has potential to circumvent these limitations, positioning CRISPR as a keystone technology in driving cardiovascular medicine.

1. INTRODUCTION: TRANSFORMING CARDIOVASCULAR MEDICINE THROUGH CRISPR

Cardiovascular diseases (CVDs) still dominate morbidity and mortality burdens worldwide, and their treatment requires much more than the conventional risk factor-based approach. Accelerating this transition from the conventional treatment model to precision cardiovascular medicine are major breakthroughs in genome engineering, especially the CRISPR/Cas9 system (Velayuthaprabhu & Archunan, 2005; Velayuthaprabhu et al., 2007; Alagendran et al., 2010; Velayuthaprabhu et al., 2011; Velayuthaprabhu et al., 2013; Velayuthaprabhu et al., 2016; Kuchi Bhotla et al., 2020; Meyyazhagan et al., 2020; Arumugam et al., 2020; Mohd et al., 2021; Bhotla et al., 2021; Mohd et al., 2021; Lee et al., 2021; Pushparaj et al., 2022; Sangeetha et al., 2022; Gundappa et al., 2022; Sangeetha et al., 2022; Meyyazhagan et al., 2022; Ramya et al., 2023; Paranitharan et al., 2025). Using genome editing technologies, one can interrogate and modify disease-associated loci with unprecedented accuracy. These provide insight into the mechanisms of dyslipidemias, atherosclerosis, cardiomyopathies, arrhythmias, and hypertensive disorders. Such innovations will bridge the gap between fundamental cardiovascular genetics and individualized therapeutic design beyond the classical risk factors of hypertension and diabetes toward genotype-driven prediction and intervention (Chen et al., 2022).

Epigenetic modulation, the analysis of rare variants, and polygenic risk scores (PRS) all further stratify cardiovascular risk in a granular manner and provide important additional insights into heritable and regulatory mechanisms in disease inception and progression (Costantino et al., 2018; Tada et al., 2021). Clinical translation will require responsible consideration of safety, off-target mutagenesis, delivery efficiency, and long-term efficacy, with multiple bioethical considerations (Dzilic et al., 2018).

Among genome editing systems, CRISPR/Cas9 represents the most transformative technology capable of correcting pathogenic variants implicated in both monogenic and polygenic CVDs (Figure 1). The majority of preclinical research using atherosclerosis, inherited arrhythmias, hypertrophic and dilated cardiomyopathies, and Duchenne muscular dystrophy (DMD) models has indeed documented that mutation-specific repair improved cardiac function (Bonowicz et al., 2025; Wilton-Clark & Yokota, 2023). Moreover, functional non-coding RNAs, including microRNAs (miRNAs) and long noncoding RNAs (lncRNAs), target various fibrosis, hypertrophy, metabolic remodelling, and impaired contractility pathways using CRISPR. In addition, CRISPR and induced pluripotent stem cell (iPSC) technologies open ways for high-fidelity disease modelling, regenerative strategies, and bioengineered cardiac tissues to overcome major issues like donor organ shortage (Konstantinov et al., 2022).

Figure 1. CRISPR/Cas9 Workflow in Cardiovascular Medicine

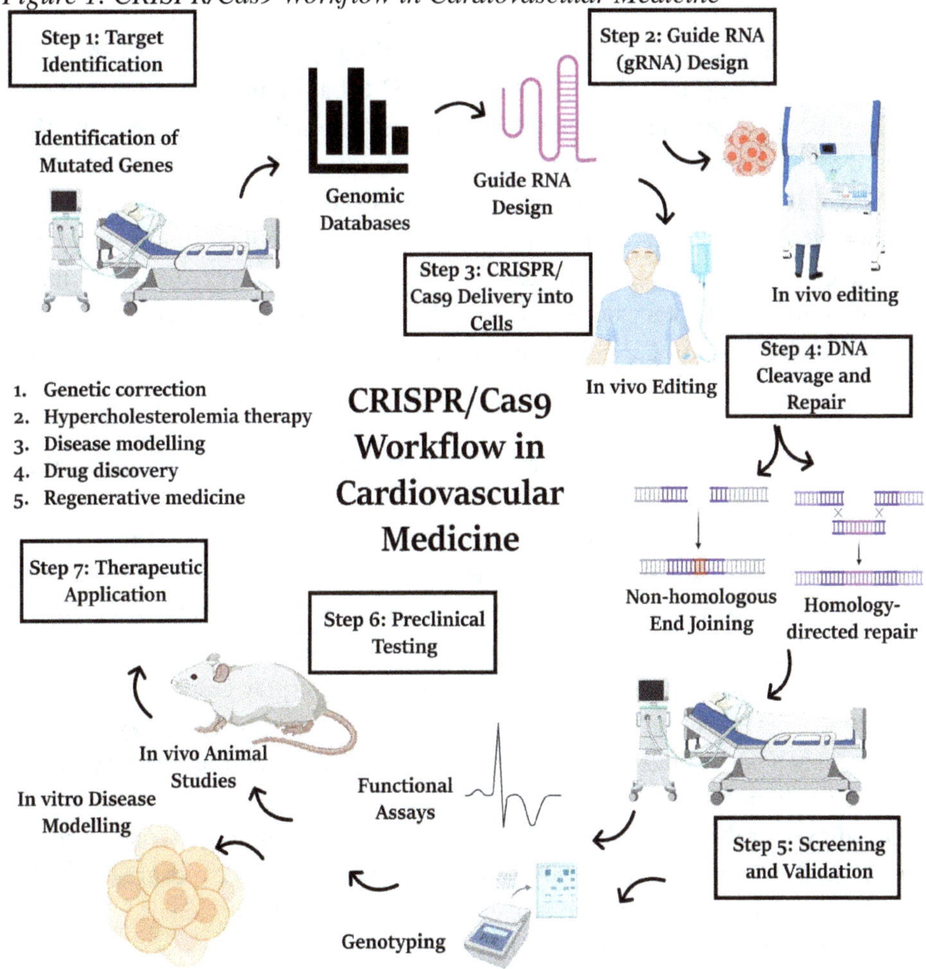

This has been enabled by parallel advances in high-throughput CRISPR screens, systems biology pipelines, and multi-omics integration-including genomics, transcriptomics, proteomics, and metabolomics-for comprehensive mapping of signalling networks and the discovery of novel actionable drug targets in CVD (Shah

& Gerszten, 2024). While cardiology is still behind oncology in its application of genome-scale CRISPR screens, the field is nonetheless experiencing a rapid growth in the number of emerging studies that allow for precise perturbations of metabolic, inflammatory, and electrophysiological pathways (Shan & Fei, 2023).

Base editing and prime editing represent next-generation editing modalities that further extend therapeutic possibilities by enabling single-nucleotide corrections and double-strand breaks, with applications shown for hypercholesterolemia, transthyretin amyloidosis, and other inherited cardiac disorders (Musunuru, 2022). Coupling to single-cell sequencing allows for direct mapping of CRISPR perturbations to cellular phenotypes, thereby advancing resolution of cardiomyocyte heterogeneity to enable precision targeting (Schoger et al., 2022). CRISPR-based modelling has catalyzed mechanistic discovery and provided a pipeline for rational, genotype-specific therapeutic development of inherited cardiac disorders (Medvedev & Zakian, 2015). CRISPR/Cas9 offers not only a powerful tool for genetic modification but also forms the bedrock of a new era in precision cardiovascular therapeutics, catalyzing progress in targeted drug development, regenerative cardiology, and personalized clinical intervention. This chapter examines the mechanistic basis, applications, challenges, and future directions of CRISPR/Cas9 in transforming cardiovascular medicine.

2. GENOMIC ARCHITECTURE OF CARDIOVASCULAR DISEASE

Understanding the genomic architecture of CVD is critical to defining actionable molecular targets amenable to CRISPR/Cas9-mediated correction and for precision in cardiovascular therapeutics. Such high-throughput platforms, including whole-genome sequencing (WGS), whole-exome sequencing, and comprehensive epigenomic profiling, have delineated the complex interplay between the genetic, epigenetic, and environmental factors that mold the pathophysiology of CVD. This approach has revealed major genetic determinants of coronary artery disease (CAD), cardiomyopathies, arrhythmogenesis, and disorders of the vasculature, which allows for the precise stratification of molecular phenotypes. Transcriptomic and epigenomic analyses further illustrate how environmental exposures modulate genetic vulnerability and detail disease trajectories among individuals with shared genomic backgrounds (Wu et al., 2014).

DNA methylation, histone modifications, and non-coding RNA networks are dynamic epigenetic regulators of cardiovascular gene expression in response to metabolic stress, inflammation, and hemodynamic load (Marín-García & Akhmedov, 2015). More recently, epigenomic markers have been used to improve the prediction of adverse cardiovascular events when used with traditional genetic data, with methylation-derived risk scores showing diagnostic significance even in subjects without established risk factors (Ambrosini et al., 2020; Westerman et al., 2020). These epigenetic signatures will provide a blueprint for CRISPR-based modulation with various emerging tools that include CRISPR-dead Cas9 (dCas9) methylation editors capable of selectively activating or silencing pathogenic loci.

The genetic architecture posited for CVD, in addition to epigenetics, includes rare pathogenic variants, PRS, and cis-regulatory elements. Their rare alleles of large effect in familial cardiomyopathies and premature CAD may explain a considerable proportion of unexplained heritability and clinical heterogeneity (Klasfeld et al., 2025). Other pathways that are therapeutically relevant have been unraveled by low-frequency variants associated with CAD, particularly those in low-linkage disequilibrium areas (Rocheleau et al., 2024). Enhanced variant interpretation by machine learning, including the combination of RVS and PRS, has resulted in improved prediction of CAD susceptibility across diverse populations, including Japanese cohorts (Ieki et al., 2024). These risk models can guide the selection of targets for CRISPR intervention, especially when the burden of both common and rare variants converges on druggable pathways.

Functional genomic annotation has identified regulatory elements as emerging drivers of CVD risk via interactions shaping gene expression patterns that can be modified via CRISPR interference (CRISPRi) or CRISPR activation (CRISPRa). Both rare and common variants must also be integrated within a precision medicine framework to allow for the development of genotype-informed therapies (Tada et al., 2021). The convergence of WGS, single-cell transcriptomics, and spatial omics provides a multidimensional approach to the dissection of CVD biology. Single-cell RNA sequencing has been applied to uncover hitherto unrecognized cardiac cellular heterogeneity, lineage transitions, and transcriptional states responsive to stresses (Kim et al., 2025; Roth et al., 2020). Its initial lack of spatial context limited the interpretation of cell-cell interactions. Spatial transcriptomics overcomes this limitation by mapping gene expression onto tissue architecture and permits identification of microenvironment-specific signalling hubs relevant for instability of atherosclerotic plaques, myocardial fibrosis, and post-injury remodelling (Han et al., 2023). Such spatially resolved datasets provide critical reference maps that will inform CRISPR target prioritization and provide further insight into tissue-specific gene regulatory networks. Integrative multi-omics, including genomics, transcriptomics, proteomics, and metabolomics, focuses on detecting systemic metabolic and inflammatory changes driving CVD progression, thus considerably improving biomarker and therapeutic target identification (Shah & Gerszten, 2024). Multi-omics platforms are increasingly being employed to validate targets selected by CRISPR, to predict the efficiency of editing, and to probe the phenotypic impact downstream. Though translation to the clinic is limited by data harmonization challenges, technical variability, and population diversity, multi-omics guided mapping of CVD mechanisms offers strong promise for transforming diagnosis, risk stratification, and personalized therapeutic intervention.

3. CRISPR/CAS9 ENGINEERING PLATFORMS AND PRECISION TOOLS

The next-generation CRISPR platforms, including Cas9-HF, eSpCas9, Cas12a/Cpf1, and Cas13, represent an important advance in genome engineering and create significant opportunities for precision in cardiovascular medicine (Figure 2). Each of these engineered nucleases was designed to maximize on-target fidelity while reducing off-target mutagenesis in order to address one of the most important translational barriers to CRISPR therapy. To this date, the most studied system, CRISPR-Cas9, has shown proof of concept for the correction of single-gene defects responsible for atherosclerosis, inherited arrhythmias, and cardiomyopathies through precise allele targeting (Roshanravan et al., 2022). The recent emergence of Cas12a and Cas13 further extends the range of therapeutic capabilities with programmable DNA and RNA editing, respectively, further expanding the repertoire of genetic and transcript-level interventions for inherited and multifactorial CVDs (Ye et al., 2023). These novel platforms have also enabled high-resolution disease modelling both with murine systems and iPSCs from patients, thereby accelerating mechanistic discovery and preclinical testing of gene-directed therapeutics (Liu & Olson, 2022). Despite these advances, inefficiencies in delivery, impossible-to-predict off-target effects, immune responses, and ethical issues present critical challenges that greatly impede immediate clinical translation of these tools (Bonowicz et al., 2025).

Figure 2. CRISPR toolbox for cardiovascular applications

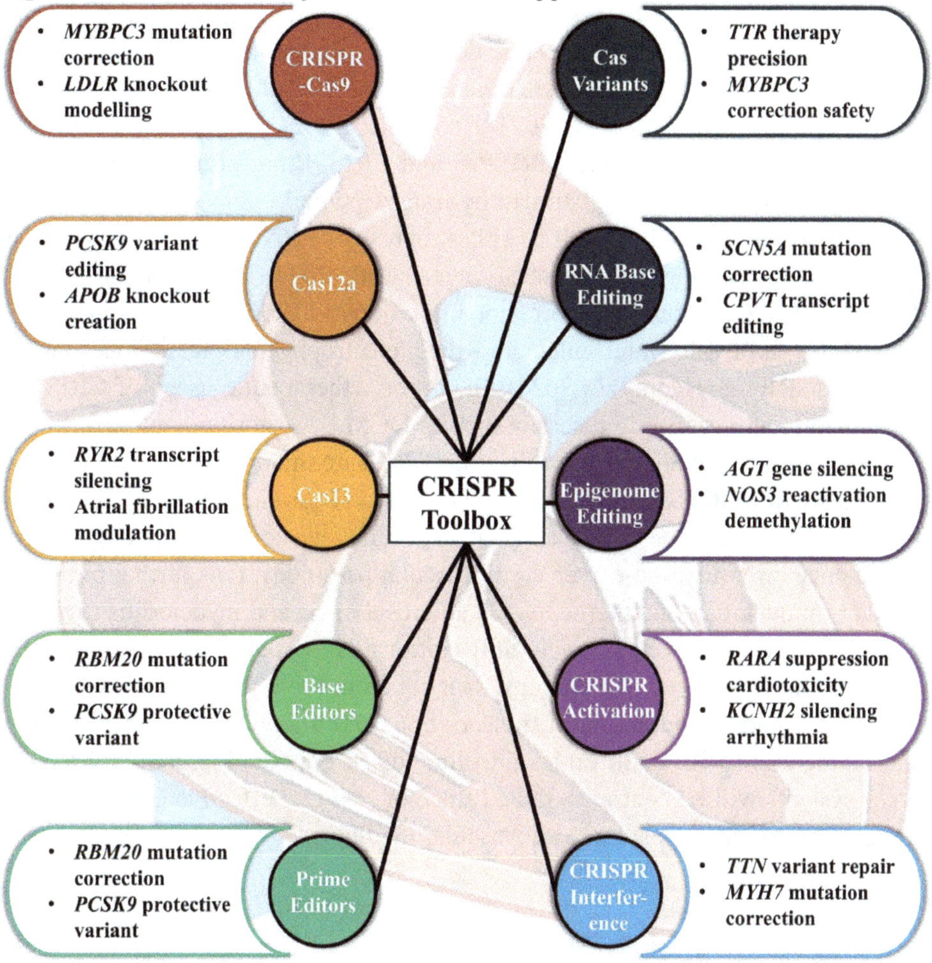

Apart from classical nuclease-dependent editing, the discovery of base and prime editing revolutionized the therapeutic landscape by enabling highly precise correction of pathogenic variants without induction of double-strand breaks. Specifically, adenine and cytosine base editors enable A>G and C>T conversions at an unprecedented high efficiency, representing the majority of clinically relevant cardiovascular mutations. Base editing of RBM20 mutations in dilated cardiomyopathy (DCM) restored functional protein expression and improved in vivo cardiac phenotypes in human iPSC-derived cardiomyocytes (iPSC-CMs) (Nishiyama et al., 2022). Prime editing further extends these possibilities to all twelve base substitutions and allows for precise insertions or deletions; thus, offering an unparalleled flexibility in the repair of complex or multi-nucleotide defects associated with hereditary cardiomyopathies, arrhythmogenic syndromes, and lipid disorders (Schreurs et al., 2021). Importantly, the absence of double-strand breaks significantly minimizes the risk of chromosomal rearrangements, large deletions, and genomic instability, placing these technologies as safer, next-generation therapeutic platforms for cardiovascular gene correction (Antoniou et al., 2021; Musunuru, 2022).

Apart from genome editing, CRISPRi and CRISPRa are furthering the functional reach of the CRISPR platform by transcriptional engineering. These take advantage of catalytically dCas9 to either repress or activate the transcription of genes without changing the sequence (Lecluze et al., 2024). Applications of CRISPRi have overcome some of the major limitations in iPSC-CMs by modulating ion channel expression and enhancing their electrophysiological properties for disease modelling (Han et al., 2024). In contrast, therapeutic usage of CRISPRa has shown great promise in diseases such as DCM by restoring haploinsufficient gene expression, including TTN, thus improving sarcomere integrity and myocardial performance (Ghahremani et al., 2024). Beyond cardiac cells, CRISPR-based transcription modulation in primary human T cells revealed cytokine circuits able to influence inflammation-driven cardiovascular pathology, thus further providing relevant immune-cardiac interactions in atherosclerosis and myocarditis (Schmidt et al., 2021). These developing CRISPR platforms represent a major extension of the toolkit for cardiovascular therapy, ranging from the correction of pathogenic mutations through modulation of transcriptional programs to RNA-level editing. Their integration with multi-omics profiling, disease modelling, and targeted delivery systems will be central to the advance of safe and effective CRISPR-based therapeutics toward precision cardiovascular medicine.

4. FUNCTIONAL GENOMICS AND DISEASE MODELLING IN CVD

CRISPR-based functional genomics has transformed the discovery of molecular drivers behind CVD and cardiotoxicity. The genome-wide CRISPR/Cas9 screens allow for systematic identification of genes regulating drug response, electrophysiology, and cardiomyocyte survival in cardiomyocytes and vascular cells. Optimization of high-throughput CRISPR screening in human iPSC-CMs (Sapp et al., 2021) identified cardioprotective transporters SLCO1A2 and SLCO1B3, which do not compromise the antitumor efficacy of doxorubicin. Complementary CRISPR studies have further brought into focus RARA and SPNS1 modulating cardio protection and setting the level of doxorubicin clearance, respectively (Lv et al., 2022). More recently, CRISPRi/a screening pointed at CA12 as a therapeutic target whose inhibition increases cardiomyocyte survival during chemotherapeutic stress (Liu et al., 2024). Establishment of standardized guidelines for constructing cardiomyocyte-specific sgRNA libraries improved the reproducibility within cardiovascular functional genomics (DeLuca & Bursac, 2022). Obviously, these improvements in mapping genetic and epigenetic modifiers of cardiac function leverage advances made with iPSC technologies and strongly accelerate precision target discovery for CVD therapy (Nishiga et al., 2021).

CRISPR/Cas9 technologies have also empowered cellular disease modelling using iPSC-CMs and engineered cardiac organoids. Targeted knockout, knock-in, and reporter insertion have thus allowed mechanistic dissection of cardiomyocyte maturation, stress signalling, and arrhythmogenic pathways (Han & Entcheva, 2023). Cardiac-specific fluorescent knock-ins further enable high-purity cardiomyocyte isolation and increase experimental precision (Galdos et al., 2020). Beyond genome cutting, CRISPRi optimized repressors such as Zim3-KRAB-dCas9 unravel regulatory networks that control the expression of ion channels and highlight ways to enhance electrophysiological stability, hence pointing toward novel anti-arrhythmic targets (Han et al., 2024). Inherited arrhythmia syndromes, including long QT syndrome, have been authentically recapitulated using CRISPR-edited iPSC models, thus enabling the identification of genetic modifiers responsible for variable clinical presentations and paving the way for genotype-tailored therapeutic strategies (Song et al., 2022).

Mechanisms of disease and testing of therapeutic strategies are further placed in context by in vivo CRISPR editing. Mouse models have remained at the heart of dissecting the complex signalling networks driving CVD, but CRISPR allows the rapid generation of knock-in or corrected alleles responsible for inherited cardiomyopathies (Liu & Olson, 2022). Alternatively, zebrafish models enable high-throughput preclinical evaluation of congenital heart defects, arrhythmogenic phenotypes, and drug responses due to their optical transparency and rapid development (Bowley et al., 2022; Risato et al., 2024). These in vivo systems thus play balancing roles with human iPSC systems within multiscale modelling of CVD, enabling a transition in gene discovery toward precision intervention.

5. PRECISION GENE THERAPY FOR MONOGENIC CARDIOVASCULAR DISORDERS

With the ability to correct or inactivate pathogenic alleles at unprecedented precision, CRISPR/Cas9 opened a new paradigm for the treatment of monogenic cardiovascular disorders. Several studies have described CRISPR-mediated targeting of PCSK9 and LDLR in familial Hypercholesterolemia (FH), with impressive potential to achieve long-lasting lipid control. Gene editing at these loci reduces circulating LDL-C and directly addresses the molecular etiology of FH (Pollanen

& Syan, 2024). The integration of nanopore sequencing into diagnostic workflows further enhances precision through the identification of complex structural variants in LDLR and PCSK9 that commonly escape detection via standard assays (Xu et al., 2024). Translational advances include the CRISPR-based base-editing therapeutic VERVE-101, which achieved durable suppression of PCSK9 and sustained LDL-C reduction in nonhuman primates-an important advance implicitly toward clinical translation (Lee et al., 2023). Efficiency in improving atherosclerotic pathology in models of FH was also shown for in vivo adeno-associated virus (AAV)-mediated CRISPR editing; such efficiency underlines the concept of gene repair as a potential complement or alternative to statins and monoclonal PCSK9 inhibitors (Zhao et al., 2020; Srivastava, 2023).

Beyond lipid disorders, CRISPR has achieved practical advances in the repair of mutations causing inherited cardiomyopathies. Mutations in MYH7, TTN, and MYBPC3, key contributors to hypertrophic and DCM, respectively, have each been corrected using precise genome editing strategies. In MYBPC3-associated hypertrophic cardiomyopathy, homology-directed repair restored normal protein expression and reversed pathological remodelling in a rat model (Nie et al., 2023). Likewise, correction of TTNtvs via CRISPR-mediated frame restoration reestablished titin expression and improved sarcomere function in human cardiomyocyte systems. Complementary in vivo strategies using CRISPR-based editors delivered via AAVMYO vectors successfully repaired mutations in Rbm20, a key regulator of TTN splicing, leading to the reversal of DCM phenotypes in mice (Grosch et al., 2023).

These applications are further extended to personalized therapeutic modelling using ex vivo gene repair in patient-derived iPSC-cardiomyocytes. The CRISPR correction of Rbm20 mutations restored splice fidelity and contractile performance with minimal off-target activity (Grosch et al., 2023). In DMD, the CRISPR/Cas9-mediated addition of exons reinstated full-length dystrophin in iPSC-CMs, supporting the feasibility of autologous cardiomyocyte replacement strategies (Xiao et al., 2022). Likewise, correction of PRKAG2 mutations by CRISPR eliminated the structural and electrophysiological abnormalities in hypertrophic cardiomyopathy models (Ben Jehuda et al., 2018). Editing of DPP6 variants associated with early repolarization syndrome normalized electrophysiological profiles, while persistent anomalies underpin the complexity of arrhythmogenic disorders (Wu et al., 2024). Besides permanent corrections, CRISPRi enables reversible and tunable gene regulation in post-differentiated cardiomyocytes, proving as a versatile platform for genotype–phenotype dynamics dissection and guiding therapeutic design (Han et al., 2023).

6. VASCULAR REPROGRAMMING AND ENDOTHELIAL ENGINEERING

CRISPR/Cas9 is emerging as a versatile platform for vascular reprogramming, thus enabling targeted modulation of angiogenesis, endothelial homeostasis, and vascular plasticity. Application of such features in support of precision therapies in CVDs involves the direct manipulation of pathways that govern vascular, for example, CRISPR/Cas9-mediated VEGFR2 knockout in endothelial cells (ECs) inhibited aberrant angiogenesis effectively both in models of proliferative diabetic retinopathy and wet age-related macular degeneration (Huang et al., 2017). On the other hand, gain-of-function approaches instituted regenerative benefits: CRISPR/Cas9-engineered HIF-1α–deficient vascular cells unraveled that HIF-1α regulates ischemic angiogenesis (Lei et al., 2024).

Alternatively, CRISPR-mediated editing has also furthered the mechanistic understanding of endothelial signalling pathways at the heart of cardiovascular physiology. Transcriptional and posttranscriptional regulation of endothelial nitric oxide synthase (eNOS), crucial for controlling vascular tone and maintaining endothelial stability, is mediated by lymphocyte-specific protein 1 (LSP1). Genetic ablation of LSP1 using CRISPR lowered eNOS activity, while overexpression of LSP1 boosted nitric oxide signalling, making it a key genetic determinant of endothelial function (Smeir et al., 2024). Integration of CRISPR perturbations with multiomics enabled the mapping of the transcriptional programs underlying CAD, including NO-related dysregulation (Gupta et al., 2023). In this direction, additional models, including but not limited to Tie2 knockout in primary ECs, have allowed controlled studies of vascular barrier integrity and inflammation, while NF-κB/RelA editing unveiled its dual role in maintaining endothelial homeostasis and providing pro-inflammatory responses (Gong et al., 2017; Wang et al., 2018).

Another crucial aspect of vascular remodelling is vascular smooth muscle cell (VSMC) plasticity. Pathological reprogramming of VSMCs into osteogenic phenotypes drives vascular calcification, a major contributor to cardiovascular morbidity, and central regulators of osteogenic gene expression include BMP and WNT signalling. Increased activity of BMP2 and RUNX2 was associated with the high plasticity of VSMCs after vascular injury, while recent translational studies using Sclerostin-mediated WNT inhibition have shown the reversal of calcific remodelling, reinstating contractile VSMC identity (Lin & Sun, 2022; Tyson et al., 2020; Iyemere et al., 2006; McArthur et al., 2017). Integration of CRISPR-based pathway dissection with targeted molecular interventions provides a framework for prediction or reversal of vascular calcification and has special relevance in diabetes and chronic kidney disease (Liu & Shanahan, 2011).

7. RNA-TARGETING AND EPITRANSCRIPTOMIC EDITING IN CVD

CRISPR-Cas13 has expanded the genomic engineering landscape by introducing RNA-targeting capabilities highly relevant to precision cardiovascular medicine. In contrast to antisense oligonucleotides, Cas13 enables programmable and sequence-specific degradation of single RNA molecules. This makes it an especially powerful tool for interrogating the regulatory roles of non-coding RNAs, including miRNAs and lncRNAs. These RNA species orchestrate key processes such as cardiac fibrosis, hypertrophy, stress adaptation, and cardiomyocyte survival (Collins et al., 2020; Rogg & Dimmeler, 2019). Thus, Cas13-based RNA perturbation allows for fine-resolution mapping of miRNA-lncRNA interactions that enable therapeutic intervention in pathogenic RNA signalling networks. Similarly, such RNA-centric CRISPR approaches offer promise for novel biomarkers and/or regulatory axes underlying CVD (Tang et al., 2020).

Beyond RNA knockdown, RNA base editing supplies a complementary method for correcting the pathogenic variants associated with arrhythmias in a manner that is independent of DNA double-strand breaks. For example, in LQTS3, AAV9-mediated delivery of ABEmax achieved >99% correction of the SCN5A T1307M allele and final restoration of physiological cardiac repolarization in murine models (Qi et al., 2024). Importantly, even partial correction (~60%) resulted in the normalization of QT intervals, which further reinforces the clinical practicality of RNA base editing. These findings complement the above, with similar allele-selective methods targeting RYR2 in catecholaminergic polymorphic ventricular tachycardia (CPVT) culminating in complete suppression of ventricular arrhythmias with durability extending up to one year (Moore et al., 2024). In addition to targeting point mutations, RNA-targeting platforms could be used to correct splicing defects (e.g., abnormal SCN5A splicing in myotonic dystrophy drives conduction system disease) (Freyermuth et al., 2016).

Epitranscriptomic editing adds a third dimension by allowing manipulation of the chemical modifications on RNA, notably the most abundant internal RNA modification, N6-methyladenosine (m6A). Accordingly, m6A writers, erasers, and readers together modulate RNA stability, translation, and cellular stress responses. Of these, the methyltransferase-like 3 (Mettl3) has recently emerged as a master regulator of cardiomyocyte proliferation and cardiac regeneration. m6A deposition mediated by Mettl3 destabilizes Fgf16 transcripts via Ythdf2, which attenuates regenerative potential. Conversely, Mettl3 knockdown profoundly enhances both cardiomyocyte proliferation and functional recovery after injury in neonatal and adult mice (Jiang et al., 2022). Dysregulation of m6A signalling has also been implicated in ischemic injury, adverse remodelling, and heart failure, wherein impaired RNA methylation disrupts cardioprotective gene expression programs (Zhang et al., 2023; Liu et al., 2022; Zhou et al., 2021). Altogether, CRISPR-Cas13, RNA base editing, and epitranscriptomic modulation are part of an increasingly fast-growing toolkit for engineering RNA networks driving cardiovascular physiology and pathology.

8. CARDIAC REGENERATION AND REJUVENATION STRATEGIES

CRISPR/Cas9-driven cardiac regeneration has emerged as a central pillar of precision cardiovascular medicine, where targeted reprogramming, repair, and rejuvenation of the injured myocardium have now become possible. Among the breakthroughs underlying such progress is the development of CRISPR-mediated direct cardiac reprogramming of cardiac fibroblasts into induced cardiomyocytes (iCMs), a strategy that targets the dominant cell type in the infarcted human heart. Activation of endogenous cardiac transcription factors MEF2C, GATA4, and TBX5 by CRISPRa systems greatly enhances fibroblast-to-cardiomyocyte transdifferentiation (Keepers et al., 2022). Even though such a system increases the reprogramming efficiency, recently it was shown that induction of transcription factors alone cannot ensure electrophysiological and metabolic maturation, thereby pointing toward a remaining methodological gap for fully functional iCMs. More recently, CRISPRa-mediated reprogramming was also found to yield in vivo cardiovascular progenitor-like intermediates improving left ventricular function after myocardial infarction (Liang & Wang, 2022). Mechanistically, remodelling of actin isoforms has been shown to play a determining role in avoiding myofibroblast activation, thereby enabling fidelity in reprogramming (Zhang et al., 2022), thus underlining a clear genomic basis of lineage conversion.

Beyond cellular reprogramming, CRISPR-based enhancement of the survival, migration, and reparative capacity of stem cells has become one of the leading applications of gene editing in ischemic heart disease. CRISPR/dCas9-mediated activation of the CXCR4 gene significantly enhances homing efficiency and ischemic tolerance of cardiac progenitor cells and leads to superior myocardial repair in murine infarction models (Li et al., 2015). Likewise, the CRISPR-Cas9 insertion of LEF1 into human umbilical cord-derived mesenchymal stem cells increases proliferation and suppresses hypoxia-induced apoptosis, which improves post-MI cardiac function (Cho et al., 2020). Targeted deletion of pro-inflammatory genes such as TLR4 in human mesenchymal stromal cells has also revealed strong anti-inflammatory and pro-reparative phenotypes (Schary et al., 2023).

Other frontiers of CRISPR-guided cardiac innovation concern regenerative signalling networks, especially the ones controlling cardiomyocyte proliferation and tissue remodelling. The Hippo-YAP/TAZ pathway has been widely interrogated as a master regulator of organ size and regeneration using CRISPR/Cas9 knockdown and activation platforms. TAZ exerts a stronger influence than YAP in promoting angiogenesis, migration, and survival in endothelial progenitor cells (Klaihmon et al., 2022). These findings are of particular significance in the context of angiogenic deficiency and impaired function of progenitor cells, constituting core pathological mechanisms in chronic CVD. While the activation of YAP/TAZ improves regenerative capacity, possible tumor-promoting effects establish a need for tightly controlled strategies of gene modulation (Moya & Halder, 2019).

9. DELIVERY SYSTEMS FOR CARDIAC GENE EDITING

Efficient and safe delivery of CRISPR/Cas9 systems remains one of the critical determinants of success in precision cardiovascular gene therapy. Due to the dense, highly vascularized, and continuously contracting nature of the heart, delivery platforms have to ensure high cardiotropic specificity, minimal off-target deposition, efficient intracellular trafficking, and stable expression control, which are essential features of therapeutic genome editing in CVD.

9.1 Non-Viral Delivery Platforms for Cardiac Gene Editing

Non-viral CRISPR/Cas9 delivery methods have generated great interest due to their evasion of important viral vector limitations, including immunogenicity, insertional mutagenesis, restricted packaging size, and long-term genomic persistence. Of those, lipid nanoparticles are among the clinically most advanced platforms, enabling efficient cellular uptake, endosomal escape, and surface ligand modification for organ-specific delivery. This includes cardiomyocyte-targeted formulations optimized through lipid chemistry and peptide conjugation (Kim et al., 2024; Lee et al., 2025). Other emerging non-viral modalities include gold nanoclusters that possess high biocompatibility and enable controlled photothermal release of CRISPR cargos. This allows for light-triggered editing at site-specific myocardial regions (Sharma et al., 2024). Exosome-based carriers, derived from endogenous extracellular vesicles, are low-immunogenic transporters of CRISPR components that maintain natural homing capabilities and can be made compatible with cardiac-targeted surface modifications (Duan et al., 2021).

9.2 Viral Delivery Systems: Current Trends and Challenges

Despite these non-viral advances, viral vectors remain the predominant strategy for in vivo cardiac genome editing, given the high efficiency and durable gene expression. The cardiotropic AAV serotypes, especially AAV9, remain one of the most used viral vehicles in cardiac models. However, hepatic off-target transduction and dose-dependent toxicity remain key translational challenges. Recent advances in capsid engineering, such as the next-CAP directed evolution platform, have generated in excess of 100 million AAV variants, identifying 77 novel cardiotropic capsids that enhance cardiac specificity with reduced systemic toxicity (Schweiggert et al., 2023). Variants such as the KK04 are part of a new generation of AAVs that show higher tropism for human cardiac tissue when compared to classical serotypes (Miguel-dos-Santos & Cingolani, 2023). AAV6 has also been shown to improve intramyocardial transduction efficiency and less fibrosis when compared with AAV9 (Wang et al., 2024). Of equal importance is the use of cardiac-restricted promoters, such as cardiac troponin T. This allows more than 640-fold higher cardiac-specific expression when compared to hepatic tissue (Prasad et al., 2011).

9.3 Stimuli-Responsive and Spatiotemporally Controlled Delivery Systems

To this end, with diverse mechanisms, stimuli-responsive CRISPR delivery systems have emerged as the next generation of precision cardiac therapy to minimize systemic exposure and maximize editing accuracy. For instance, ultrasound-responsive nanoparticles can achieve the acoustic-triggered release of CRISPR payloads at infarct zones with improved localization and reduced off-target editing (Xie et al., 2021). Magnetic nanoparticles may be guided by an external magnet towards specific myocardial regions, thus enabling a non-invasive spatial control mechanism, which is of particular relevance to focal fibrotic or ischemic lesions. Similarly, light-activated platforms comprising photosensitive materials can modulate CRISPR release in a wavelength-specific and temporally controlled manner, hence restricting editing only to the defined optical exposure. This has been diversely developed by integrating exogenous stimuli like ultrasound, magnetic field, and light with endogenous cues like pH, enzymatic activity, and inflammatory microenvironment as a multifaceted approach to enhance precision in editing (Naeem et al., 2021; Fang et al., 2022). Such systems are of immediate relevance to the clinical needs of CVDs with spatiotemporally precise delivery methods that require regional injury zones, complex geometries, and dynamic mechanical forces.

10. IMMUNE EDITING AND ANTI-INFLAMMATORY CRISPR APPROACHES

Chronic inflammation is one of the basic driving forces for CVD, especially in atherosclerosis, myocardial infarction, and transplant rejection. Immune editing via CRISPR has now emerged as a transformative approach to such inflammatory pathways at both a dissecting and therapeutic level. Prominent mediators include IL-1β, TNF-α, and NF-κB, which are implicated in endothelial dysfunction, leukocyte recruitment, foam cell formation, and plaque instability. For the first time, the CRISPR/Cas technologies have allowed direct interrogation and modification of these pathways, enabling an experimentally tractable platform that defines genetic contributors to cardiovascular risk while identifying druggable targets (Chen et al., 2022).

Recently, multi-omics profiling has been integrated with CRISPR perturbation screens in ECs with great power. These analyses unraveled CAD-related, EC-specific regulatory circuits and disclosed gene variants modulating susceptibility to atherogenic stimuli, including hyperlipidemia and hemodynamic stress (Gupta et al., 2023). Clinical evidence supporting the translational value of CRISPR-guided strategies targeting upstream inflammatory networks includes the recent CANTOS trial, indicating that IL-1β inhibition reduces recurrent cardiovascular events (Aronova et al., 2023).

10.1 CRISPR-Based Modulation of Macrophage Polarization in Atherosclerosis and Myocardial Injury

Macrophages are central to plaque progression and to post-infarct remodelling. Their phenotypic plasticity determines the outcome of disease, as M1 macrophages propagate inflammation and M2 macrophages mediate the repair and resolution of fibrosis (Wu et al., 2023; Wang et al., 2016). CRISPR/Cas9 and inducible CRISPR systems now enable the selective reprogramming of these states with new therapeutic options. Advanced platforms like the X-ray–inducible X-CC9 have already shown precise conversion of the macrophage phenotype in oncology models, shifting from M2 to M1 and enhancing antitumor immunity. The same principles bear potential relevance to cardiovascular inflammation (Zhao et al., 2023). In ischemic heart disease, the dynamics of macrophages at the time of myocardial infarction determine scar thickness and, consequently, ventricular remodelling, impacting long-term cardiac performance. Optimizing the balance between reparative and inflammatory macrophage subsets, using genome editing, may lead to a significant improvement in healing outcomes and reduce eventual transition to heart failure (Molinaro et al., 2023; Ben-Mordechai et al., 2015).

10.2 Immune Engineering in Transplantation and Cardiovascular Tolerance

Beyond atherosclerosis and ischemia, immune editing is also being extended to cardiac transplantation, where immune rejection remains a major clinical challenge. CRISPR-engineered macrophages and lymphocytes provide innovative strategies to induce targeted immunological tolerance with reduced reliance on systemic immunosuppression. CRISPR-modified macrophages can be programmed toward tolerogenic phenotypes, enabling localized suppression of graft inflammation without complete immune compromise (Owen & Kopecky, 2024). In this context, CAR-engineered regulatory T cells (CAR-Tregs) represent a precision approach for alloantigen-specific suppression. These cells migrate to cardiac grafts, maintain suppressive function, and prolong graft survival independent of HLA matching (Kaljanac & Abken, 2023). Complementary strategies employ tolerogenic dendritic cells (Tol-DCs), generated using CRISPR to disrupt costimulatory pathways or enhance Treg-promoting signals. Tol-DCs have been shown to dramatically extend cardiac allograft survival by inducing Tregs, shifting T-cell responses toward Th2 phenotypes, and promoting T-cell hyperresponsiveness (Wu et al., 2012).

11. CLINICAL TRIALS AND TRANSLATIONAL PROGRESS

The clinical translation of CRISPR-based therapeutics in CVD remains in its early but rapidly evolving stages, while current trials are focused on correcting well-characterized pathogenic mutations. Perhaps one of the most significant points reached thus far is the development of Nexiguran Ziclumeran, a CRISPR-Cas9 therapy intended for the inactivation of the Transthyretin (TTR) gene in patients with transthyretin amyloidosis with cardiomyopathy (ATTR-CM). Preliminary phase 1 data demonstrated substantial and consistent reductions of circulating TTR protein, with mean decreases of 89% by day 28 and 90% at 12 months, touting strong biochemical efficacy. However, infusion-related reactions and transient rises in liver enzymes indicate that dose optimization, extended monitoring, and further large-scale studies will be required to establish long-term safety and functional cardiac outcomes (Fontana et al., 2024).

In addition to ATTR-CM, CRISPR technologies are also under exploration for inherited cardiomyopathies such as DCM, where the aim is the precise correction of disease-causing variants directly within cardiomyocytes. Major translational challenges remain to be overcome, including efficient cardiac-specific delivery, immunogenicity, and ethical considerations around permanent somatic genome modification (Ganipineni et al., 2023; Asif et al., 2024). Preclinical developments provide important proof-of-concept support. For example, CASAAV-HDR has recently enabled efficient in vivo editing of post-mitotic cardiomyocytes-a key development given the minimal intrinsic cardiac regeneration noted in many disorders. These platforms have already shown promise in models of congenital heart disease, hyperlipidaemia, and arrhythmogenic cardiomyopathies, thus widening the therapeutic horizon (Asif et al., 2024).

Recent extensions of the CRISPR toolbox have included base editing and prime editing, and these have further consolidated translational potential by enabling nucleotide changes of high precision without the induction of double-stranded DNA breaks. These systems have improved the accuracy of the models of CVD and accelerated therapeutic investigations for hypercholesterolemia, atherosclerosis, and ATTR amyloidosis (Li et al. 2023; Chen et al. 2022). Notwithstanding these advances, translation to human therapies remains limited by concerns regarding precision, reproducibility, off-target effects, immune activation, and the need for validated long-term safety data (Asif et al. 2024). Further refinement of both editing tools and delivery systems will, therefore, be important to advance CRISPR-based cardiovascular therapeutics toward clinical practice.

12. BIOETHICAL, REGULATORY, AND SAFETY CONSIDERATIONS

Improvement of CRISPR-based cardiovascular therapeutics also raises several ethical, regulatory, and safety concerns that introduce complex considerations in the responsible translation of this technology. Perhaps paramount among these are the risks of off-target genome modifications that may introduce unintended mutations with deleterious phenotypic consequences. Although active efforts toward improvement in nuclease specificity and editing fidelity continue to lower this risk, the complete elimination of off-target activity has remained elusive. Ethical distinctions between somatic and germline editing also carry significant implications insofar as germline modifications may be inherited by future generations, raising societal, cultural, and legal concerns related to consent, intergenerational risk, and human genetic modification (Bonowicz et al., 2025; Motta et al., 2017).

The other key challenge is one of equitable access: genetic diversity among populations can affect the efficacy and safety of CRISPR, which may result in therapeutic disparities across ancestries. Besides, high development and manufacturing costs will further increase socioeconomic disparities unless strategies for cost reduction, reimbursement, and scalable production are pursued (Carbone & Montecucco, 2024). The current regulatory frameworks-mostly from guidelines concerning gene therapy-are incapable of dealing with particular issues on CRISPR, such as persistence of editing, long-term monitoring, dual-use issues, and rapid evolution of editing platforms. Because these tools have the dual-use characteristic that increases biosecurity risks, strict oversight and international cooperation go hand in hand to prevent misuse of the technology (Braddick & Ramarohetra, 2020).

These are significant constraints, but preclinical studies have largely shown great therapeutic promise, with a constantly evolving safety profile. For example, CRISPR/Cas9 editing of arrhythmogenic variants in CPVT models resulted in suppression of arrhythmia out to 12 months without measurable off-target effects detected (Moore et al., 2024). Similarly precise corrections of DCM-associated mutations have been conducted with base-editing approaches without measurable off-target activity (Steinmetz, 2023). Prime editing brings about improved precision against post-mitotic cells such as cardiomyocytes through a reduction in dependency on double-stranded breaks and may thereby enhance genomic stability over the long term (Bharucha et al., 2022). A few important areas of uncertainty remain, however, regarding the durability of the edits themselves, including immune activation against Cas proteins, vector-related risks, and the combined long-term consequences of genome manipulation. For example, PCSK9 inactivation using CRISPR would result in improvement in LDL-C and the atherosclerotic plaque burden, although its systemic safety remains poorly characterized in humans in the long term (Suwito et al., 2022).

13. FUTURE DIRECTIONS: CRISPR + AI, EPIGENOME EDITING, AND CARDIOVASCULAR SYSTEMS BIOLOGY

The use of CRISPR-based genome editing in combination with artificial intelligence (AI) best describes the transformation trajectory of cardiovascular precision medicine. CRISPR is able to specifically correct pathogenic mutations underlying atherosclerosis, cardiomyopathies, and arrhythmias and is rapidly improving (Bonowicz et al., 2025). AI helps in designing the best gRNA in order to reduce off-target effects and optimizes methods of delivery, thus enhancing the safety, efficiency, and precision of CRISPR therapeutics (Dixit et al., 2024). Beyond genome editing, AI enables integrating multimodal datasets in cardiovascular studies, including electronic health records, wearable sensor data, and imaging output, for improved early disease detection, risk stratification, and personalized therapeutic algorithms (Saeed et al., 2024). Remaining challenges involve ensuring data quality, ethical governance, and regulatory compliance that would necessitate coordinated collaboration among clinicians, computational scientists, and policy thinkers, among other experts (Kok et al., 2024). It has to be noted that even AI-guided CRISPR strategies have their flaws in delivery efficiency and residual off-target editing risks, which need further optimization (Bonowicz et al., 2025; Dixit et al., 2024).

Meanwhile, paralleled developments have defined CRISPR-programmed epigenome editing as a methodology at the very front line, enabling modulation of gene expression without permanently changing the DNA sequence. Different platforms, including CRISPRoff and CRISPRon, allow for reversible control of transcription through targeted DNA methylation and demethylation, respectively (Figure 3). For instance, CRISPRoff-mediated gene silencing of angiotensinogen achieved ~50% repression via promoter methylation, whereas targeted suppression of PCSK9 reached up to an 80% expression reduction and was stable over many cell divisions (Testa et al., 2023; Wang et al., 2022). The long-lasting "one-and-done" interventions of such approaches enable treatment of hypertension, dyslipidemia, and atherosclerosis in the absence of the ethical concerns pertinent to permanent genome editing. Development of novel Cas variants along with programmable epigenetic controllers continues to expand the CRISPR toolbox, further enhancing the therapeutic potential of epigenome editing (Li et al., 2023).

Figure 3. Epigenome Editing in Cardiovascular Disease

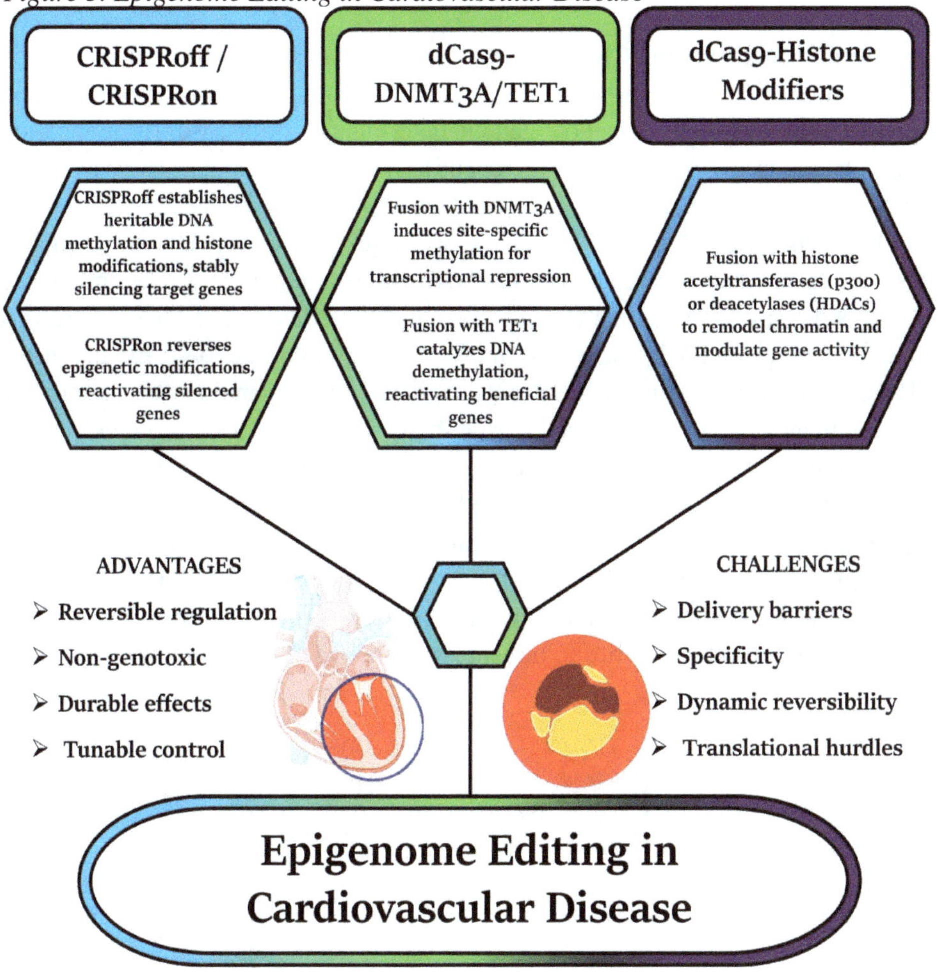

Integration of CRISPR into cardiovascular systems biology has significantly enhanced mechanistic insights and facilitated therapeutic discovery. CRISPR/Cas9 enables specific genetic perturbations both in vivo and in vitro, which allow the identification of gene function, disease pathways, and therapeutic targets (Musunuru, 2023; Li et al., 2023; Vermersch et al., 2020). Capability extensions via base and prime editing now allow high-fidelity, versatile genomic modifications that have the potential to correct the underlying genetic drivers of hypercholesterolemia, atherosclerosis, and cardiomyopathies, as opposed to addressing only downstream symptoms (Roshanravan et al., 2022; Bonowicz et al., 2025). These findings taken together outline a path toward precision cardiovascular medicine, integrating genetic, epigenetic, and computational methods for targeted, durable, safe interventions.

14. CONCLUSION

The rapid advance of CRISPR/Cas9 technology has transformed it into the central innovation in precision cardiovascular medicine, enabling unprecedented opportunities to directly target the genetic determinants of disease. Its application ranges from disease modelling to therapeutic use in dyslipidemia, atherosclerosis, cardiomyopathies, and arrhythmias. With the help of AI, its translational potential has been enhanced through optimizing off-target prediction, enhancing gRNA design, and facilitating patient-specific therapeutic strategies. With the emergence of CRISPR-based epigenome editing like CRISPRoff and CRISPRon, there is a reversible and potentially safer alternative to irreparable genome modification. Moreover, base and prime editing enhancements are expanding the therapeutic space with increased precision and translatability to postmitotic cardiomyocytes. Of course, there are significant barriers to be addressed, such as delivery challenges, long-term safety, immune rejection, and ethical and regulatory frameworks. Bypassing these limitations with exemplary research and cooperative effort will be the key to bridging CRISPR-based therapies from model systems to the bedside. CRISPR technology, in synergy with AI and systems biology, is finally an enabling platform for cardiovascular therapeutics to be remade in the era of precision medicine.

REFERENCES

Alagendran, S., Archunan, G., Prabhu, S. V., Orozco, B. E., & Guzman, R. G. (2010). Biochemical evaluation in human saliva with special reference to ovulation detection. *Indian Journal of Dental Research, 21*(2), 165–168. DOI: 10.4103/0970-9290.66625 PMID: 20657081

Ambrosini, S., Mohammed, S. A., Costantino, S., & Paneni, F. (2020). Disentangling the epigenetic landscape in cardiovascular patients: A path toward personalized medicine. *Minerva Cardiology and Angiology, 69*(3), 331–345. PMID: 32996305

Antoniou, P., Miccio, A., & Brusson, M. (2021). Base and prime editing technologies for blood disorders. *Frontiers in Genome Editing, 3*, 618406. DOI: 10.3389/fgeed.2021.618406 PMID: 34713251

Aronova, A., Tosato, F., Naser, N., & Asare, Y. (2023). Innate immune pathways in atherosclerosis-From signalling to long-term epigenetic reprogramming. *Cells, 12*(19), 2359. DOI: 10.3390/cells12192359 PMID: 37830572

Arumugam, V. A., Thangavelu, S., Fathah, Z., Ravindran, P., Sanjeev, A. M. A., Babu, S., Arun, M., Mohd. Iqbal, Y., Khan, S., Ruchi, T., Megha Katare, P., Ranjit, S., Chandra, R., & Dhama, K. (2020). COVID-19 and the world with co-morbidities of heart disease, hypertension and diabetes. *Journal of Pure & Applied Microbiology, 14*(3), 1623–1638. DOI: 10.22207/JPAM.14.3.01

Asif, M., Khan, W. J., Aslam, S., Aslam, A., & Chowdhury, M. A. (2024). The use of CRISPR-Cas9 genetic technology in cardiovascular disease: A comprehensive review of current progress and future prospective. *Cureus, 16*(4). Advance online publication. DOI: 10.7759/cureus.57869 PMID: 38725755

Ben Jehuda, R., Eisen, B., Shemer, Y., Mekies, L. N., Szantai, A., Reiter, I., & Binah, O. (2018). CRISPR correction of the PRKAG2 gene mutation in the patient's iPSC-derived cardiomyocytes eliminates the electrophysiological and structural abnormalities. *Heart Rhythm, 15*(2), 267–276. DOI: 10.1016/j.hrthm.2017.09.024 PMID: 28917552

Ben-Mordechai, T., Palevski, D., Glucksam-Galnoy, Y., Elron-Gross, I., Margalit, R., & Leor, J. (2015). Targeting macrophage subsets for infarct repair. *Journal of Cardiovascular Pharmacology and Therapeutics, 20*(1), 36–51. DOI: 10.1177/1074248414534916 PMID: 24938456

Bharucha, N., Arias, A., & Karakikes, I. (2022). The potential of CRISPR-Cas9 prime editing for cardiovascular disease research and therapy. *Current Opinion in Cardiology, 37*(5), 413–418. DOI: 10.1097/HCO.0000000000000985 PMID: 35880456

Bhotla, H. K., Balasubramanian, B., Meyyazhagan, A., Pushparaj, K., & Easwaran, M. (2021). Opportunistic mycoses in COVID-19 patients/survivors: Epidemic inside a pandemic. *Journal of Infection and Public Health, 14*(11), 1720–1726. DOI: 10.1016/j.jiph.2021.10.010 PMID: 34700291

Bonowicz, K., Jerka, D., Piekarska, K., Olagbaju, J., Stapleton, L., Shobowale, M., & Gagat, M. (2025). CRISPR-Cas9 in cardiovascular medicine: Unlocking new potential for treatment. *Cells, 14*(2), 131. DOI: 10.3390/cells14020131 PMID: 39851560

Bowley, G., Kugler, E., Wilkinson, R., Lawrie, A., van Eeden, F., Chico, T. J., & Serbanovic-Canic, J. (2022). Zebrafish as a tractable model of human cardiovascular disease. *British Journal of Pharmacology, 179*(5), 900–917. DOI: 10.1111/bph.15473 PMID: 33788282

Braddick, D., & Ramarohetra, R. F. (2020). Emergent challenges for CRISPR: biosafety, biosecurity, patenting, and regulatory issues. In *Genome engineering via CRISPR-Cas9 system* (pp. 281–307). Academic Press. DOI: 10.1016/B978-0-12-818140-9.00021-0

Carbone, F., & Montecucco, F. (2024). Facing ethical concerns in the age of precise gene therapy: Outlook on inherited arrhythmias. *World Journal of Cardiology, 16*(2), 64–66. DOI: 10.4330/wjc.v16.i2.64 PMID: 38456071

Chen, Z., Lehertshuber, C., & Schunkert, H. (2022). Genome editing in dyslipidemia and atherosclerosis. *Genome Editing in Cardiovascular and Metabolic Diseases*, 139-156.

Cho, H. M., Lee, K. H., Shen, Y. M., Shin, T. J., Ryu, P. D., Choi, M. C., & Cho, J. Y. (2020). Transplantation of hMSCs genome edited with LEF1 improves cardio-protective effects in myocardial infarction. *Molecular Therapy. Nucleic Acids, 19*, 1186–1197. DOI: 10.1016/j.omtn.2020.01.007 PMID: 32069701

Collins, L., Binder, P., Chen, H., & Wang, X. (2020). Regulation of long non-coding RNAs and microRNAs in heart disease: Insight into mechanisms and therapeutic approaches. *Frontiers in Physiology, 11*, 798. DOI: 10.3389/fphys.2020.00798 PMID: 32754048

Costantino, S., Libby, P., Kishore, R., Tardif, J. C., El-Osta, A., & Paneni, F. (2018). Epigenetics and precision medicine in cardiovascular patients: From basic concepts to the clinical arena. *European Heart Journal, 39*(47), 4150–4158. DOI: 10.1093/eurheartj/ehx568 PMID: 29069341

DeLuca, S., & Bursac, N. (2022). CRISPR library screening in cultured cardiomyocytes. In *Cardiac Tissue Engineering: Methods and Protocols* (pp. 1–13). Springer US. DOI: 10.1007/978-1-0716-2261-2_1

Dixit, S., Kumar, A., Srinivasan, K., Vincent, P. D. R., & Ramu Krishnan, N. (2024). Advancing genome editing with artificial intelligence: Opportunities, challenges, and future directions. *Frontiers in Bioengineering and Biotechnology*, *11*, 1335901. DOI: 10.3389/fbioe.2023.1335901 PMID: 38260726

Duan, L., Ouyang, K., Xu, X., Xu, L., Wen, C., Zhou, X., & Liang, Y. (2021). Nanoparticle delivery of CRISPR/Cas9 for genome editing. *Frontiers in Genetics*, *12*, 673286. DOI: 10.3389/fgene.2021.673286 PMID: 34054927

Dzilic, E., Lahm, H., Dreßen, M., Deutsch, M. A., Lange, R., Wu, S. M., & Doppler, S. A. (2018). Genome editing redefines precision medicine in the cardiovascular field. *Stem Cells International*, *2018*(1), 4136473. DOI: 10.1155/2018/4136473 PMID: 29731778

Fang, T., Cao, X., Ibnat, M., & Chen, G. (2022). Stimuli-responsive nanoformulations for CRISPR-Cas9 genome editing. *Journal of Nanobiotechnology*, *20*(1), 354. DOI: 10.1186/s12951-022-01570-y PMID: 35918694

Fontana, M., Solomon, S. D., Kachadourian, J., Walsh, L., Rocha, R., Lebwohl, D., & Gillmore, J. D. (2024). CRISPR-Cas9 gene editing with nexiguran ziclumeran for ATTR cardiomyopathy. *The New England Journal of Medicine*, *391*(23), 2231–2241. DOI: 10.1056/NEJMoa2412309 PMID: 39555828

Freyermuth, F., Rau, F., Kokunai, Y., Linke, T., Sellier, C., Nakamori, M., & Charlet-Berguerand, N. (2016). Splicing misregulation of SCN5A contributes to cardiac-conduction delay and heart arrhythmia in myotonic dystrophy. *Nature Communications*, *7*(1), 11067. DOI: 10.1038/ncomms11067 PMID: 27063795

Galdos, F. X., Darsha, A. K., Paige, S. L., & Wu, S. M. (2020). Purification of pluripotent stem cell-derived cardiomyocytes using CRISPR/Cas9-mediated integration of fluorescent reporters. In *Cardiac Regeneration: Methods and Protocols* (pp. 223–240). Springer US.

Ganipineni, V. D. P., Gutlapalli, S. D., Danda, S., Garlapati, S. K. P., Fabian, D., Okorie, I., & Okorie, I. J. (2023). Clustered regularly interspaced short palindromic repeats (CRISPR) in cardiovascular disease: A comprehensive clinical review on dilated cardiomyopathy. *Cureus*, *15*(3), e35774. DOI: 10.7759/cureus.35774 PMID: 37025725

Ghahremani, S., Kanwal, A., Pettinato, A., Ladha, F., Legere, N., Thakar, K., & Hinson, J. T. (2024). CRISPR activation reverses haploinsufficiency and functional deficits caused by TTN truncation variants. *Circulation*, *149*(16), 1285–1297. DOI: 10.1161/CIRCULATIONAHA.123.063972 PMID: 38235591

Gong, H., Liu, M., Klomp, J., Merrill, B. J., Rehman, J., & Malik, A. B. (2017). Method for dual viral vector mediated CRISPR-Cas9 gene disruption in primary human endothelial cells. *Scientific Reports*, *7*(1), 42127. DOI: 10.1038/srep42127 PMID: 28198371

Grosch, M., Schraft, L., Chan, A., Küchenhoff, L., Rapti, K., Ferreira, A. M., & Steinmetz, L. M. (2023). Striated muscle-specific base editing enables correction of mutations causing dilated cardiomyopathy. *Nature Communications*, *14*(1), 3714. DOI: 10.1038/s41467-023-39352-1 PMID: 37349314

Gundappa, M., Arumugam, V. A., Hsieh, H. L., Balasubramanian, B., & Shanmugam, V. (2022). Expression of tissue factor and TF-mediated integrin regulation in HTR-8/SVneo trophoblast cells. *Journal of Reproductive Immunology*, *150*, 103473. DOI: 10.1016/j.jri.2022.103473 PMID: 35030354

Gupta, R. M., Schnitzler, G. R., Fang, S., Lee-Kim, V. S., & Barry, A. (2023). Multiomic analysis and CRISPR perturbation screens identify endothelial cell programs and novel therapeutic targets for coronary artery disease. *Arteriosclerosis, Thrombosis, and Vascular Biology*, *43*(5), 600–608. DOI: 10.1161/ATVBAHA.123.318328 PMID: 36994731

Han, J. L., & Entcheva, E. (2023). Gene modulation with CRISPR-based tools in human iPSC-cardiomyocytes. *Stem Cell Reviews and Reports*, *19*(4), 886–905. DOI: 10.1007/s12015-023-10506-4 PMID: 36656467

Han, J. L., Heinson, Y. W., Chua, C. J., Liu, W., & Entcheva, E. (2023). CRISPRi gene modulation and all-optical electrophysiology in post-differentiated human iPSC-cardiomyocytes. *Communications Biology*, *6*(1), 1236. DOI: 10.1038/s42003-023-05627-y PMID: 38062109

Han, S., Xu, Q., Du, Y., Tang, C., Cui, H., Xia, X., & Shang, H. (2024). Single-cell spatial transcriptomics in cardiovascular development, disease, and medicine. *Genes & Diseases*, *11*(6), 101163. DOI: 10.1016/j.gendis.2023.101163 PMID: 39224111

Huang, X., Zhou, G., Wu, W., Duan, Y., Ma, G., Song, J., & Lei, H. (2017). Genome editing abrogates angiogenesis in vivo. *Nature Communications*, *8*(1), 112. DOI: 10.1038/s41467-017-00140-3 PMID: 28740073

Ieki, H., Ito, K., Zhang, S., Koyama, S., Miyazawa, K., Yoshida, H., & Komuro, I. (2024). Machine learning uncovers the contribution of rare genetic variants and enhances risk prediction for coronary artery disease in the japanese population. *Circulation*, *150*(Suppl_1), A4136501–A4136501. DOI: 10.1161/circ.150.suppl_1.4136501

Iyemere, V. P., Proudfoot, D., Weissberg, P. L., & Shanahan, C. M. (2006). Vascular smooth muscle cell phenotypic plasticity and the regulation of vascular calcification. *Journal of Internal Medicine*, *260*(3), 192–210. DOI: 10.1111/j.1365-2796.2006.01692.x PMID: 16918817

Jiang, F. Q., Liu, K., Chen, J. X., Cao, Y., Chen, W. Y., Zhao, W. L., & Qi, X. F. (2022). Mettl3-mediated m6A modification of Fgf16 restricts cardiomyocyte proliferation during heart regeneration. *eLife*, *11*, e77014. DOI: 10.7554/eLife.77014 PMID: 36399125

Kaljanac, M., & Abken, H. (2023). Do treg speed up with CARs? Chimeric antigen receptor treg engineered to induce transplant tolerance. *Transplantation*, *107*(1), 74–85. DOI: 10.1097/TP.0000000000004316 PMID: 36226849

Keepers, B. P., Qian, L., Liu, J., & Shutt, S. (2022). Abstract P1134: Activation of endogenous transcription factors by crispra mediates cardiac reprogramming. *Circulation Research*, *131*(Suppl_1), AP1134–AP1134. DOI: 10.1161/res.131.suppl_1.P1134

Kim, H., Choi, S., Heo, H., Cho, S. H., Lee, Y., Kim, D., & Rhee, S. (2025). Applications of single-cell omics technologies for induced pluripotent stem cell-based cardiovascular research. *International Journal of Stem Cells*, *18*(1), 37–48. DOI: 10.15283/ijsc23183 PMID: 39129179

Kim, M., Hwang, Y., Lim, S., Jang, H. K., & Kim, H. O. (2024). Advances in nanoparticles as non-viral vectors for efficient delivery of CRISPR/Cas9. *Pharmaceutics*, *16*(9), 1197. DOI: 10.3390/pharmaceutics16091197 PMID: 39339233

Klaihmon, P., Lorthongpanich, C., Kheolamai, P., Luanpitpong, S., & Issaragrisil, S. (2022). Distinctive roles of yap and taz in human endothelial progenitor cells growth and functions. *Biomedicines*, *10*(1), 147. DOI: 10.3390/biomedicines10010147 PMID: 35052826

Klasfeld, S. J., Knutson, K. A., Miller, M. R., Fauman, E. B., Berghout, J., Moccia, R., & Kim, H. I. (2025). Common genetic modifiers influence cardiomyopathy susceptibility among the carriers of rare pathogenic variants. *HGG Advances*, *6*(3), 100460. DOI: 10.1016/j.xhgg.2025.100460 PMID: 40411145

Kok, C. L., Koh, Y. Y., Ho, C. K., Thanh, N. T. C., & Teo, T. H. (2024). Artificial intelligence in cardiology. In *2024 IEEE 17th International Symposium on Embedded Multicore/Many-core Systems-on-Chip (MCSoC)* (pp. 462-466). IEEE. DOI: 10.1109/MCSoC64144.2024.00082

Konstantinov, I. E., King, G., & Porrello, E. R. (2022). From genome editing to blastocyst complementation: A new horizon in heart transplantation? *JTCVS Techniques*, *12*, 177–184. DOI: 10.1016/j.xjtc.2022.01.012 PMID: 35403039

Kuchi Bhotla, H., Kaul, T., Balasubramanian, B., Easwaran, M., Arumugam, V. A., Pappusamy, M., Muthupandian, S., & Meyyazhagan, A. (2020). Platelets to surrogate lung inflammation in COVID-19 patients. *Medical Hypotheses*, *143*, 110098. DOI: 10.1016/j.mehy.2020.110098 PMID: 32688185

Lecluze, E., Belhaj, H., Diaz, C., Lo, C. S., Lettre, G., & Lo, K. S. (2024). CRISPRi screening for functional characterization of blood pressure (BP) and coronary artery disease (CAD)-associated genes in endothelial cells. *protocols.io*. https://dx.doi.org/ DOI: 10.17504/protocols.io.5qpvo9ybxv4o/v1

Lee, H., Rho, W. Y., Kim, Y. H., Chang, H., & Jun, B. H. (2025). CRISPR-Cas9 gene therapy: Non-viral delivery and stimuli-responsive nanoformulations. *Molecules (Basel, Switzerland)*, *30*(3), 542. DOI: 10.3390/molecules30030542 PMID: 39942646

Lee, R. G., Mazzola, A. M., Braun, M. C., Platt, C., Vafai, S. B., Kathiresan, S., & Khera, A. V. (2023). Efficacy and safety of an investigational single-course CRISPR base-editing therapy targeting PCSK9 in nonhuman primate and mouse models. *Circulation*, *147*(3), 242–253. DOI: 10.1161/CIRCULATIONAHA.122.062132 PMID: 36314243

Lee, T. H., Liu, P. S., Wang, S. J., Tsai, M. M., Shanmugam, V., & Hsieh, H. L. (2021). Bradykinin, as a reprogramming factor, induces transdifferentiation of brain astrocytes into neuron-like cells. *Biomedicines*, *9*(8), 923. DOI: 10.3390/biomedicines9080923 PMID: 34440126

Lei, J., Jiang, X., Huang, D., Jing, Y., Yang, S., Geng, L., & Qu, J. (2024). Human ESC-derived vascular cells promote vascular regeneration in a HIF-1α dependent manner. *Protein & Cell*, *15*(1), 36–51. DOI: 10.1093/procel/pwad027 PMID: 37158785

Li, Y., Ma, G., Weintraub, N., & Tang, Y. (2015). Enhancing transplanted cardiac progenitor cells homing and survival in ischemic heart by CRISPR-dcas9/sgRNA-based endogenous CXCR4 promoter activation. *Circulation*, *132*(suppl_3), A16620–A16620. DOI: 10.1161/circ.132.suppl_3.16620

Li, Z. H., Wang, J., Xu, J. P., Wang, J., & Yang, X. (2023). Recent advances in CRISPR-based genome editing technology and its applications in cardiovascular research. *Military Medical Research, 10*(1), 12. DOI: 10.1186/s40779-023-00447-x PMID: 36895064

Liang, J., & Wang, Y. G. (2022). In vivo Reprogramming of Mouse Fibroblasts into Cardiac Progenitor Cells for Myocardial Infarction Treatment. *Circulation, 146*(Suppl_1), A13281–A13281. DOI: 10.1161/circ.146.suppl_1.13281

Lin, Y., & Sun, Z. (2022). Klotho deficiency-induced arterial calcification involves osteoblastic transition of VSMCs and activation of BMP signalling. *Journal of Cellular Physiology, 237*(1), 720–729. DOI: 10.1002/jcp.30541 PMID: 34368951

Liu, C., Gu, L., Deng, W., Meng, Q., Li, N., Dai, G., & Fang, H. (2022). N6-methyladenosine RNA methylation in cardiovascular diseases. *Frontiers in Cardiovascular Medicine, 9*, 887838. DOI: 10.3389/fcvm.2022.887838 PMID: 35571209

Liu, C., Shen, M., Manhas, A., Zhao, S., Belbachir, N., Nishiga, M., & Wu, J. (2024). Abstract Or121: Druggable Genome CRISPRi/a screens in iPSC-derived cardiomyocytes identify therapeutic targets for doxorubicin-induced cardiotoxicity. *Circulation Research, 135*(Suppl_1), AOr121–AOr121. DOI: 10.1161/res.135.suppl_1.Or121

Liu, N., & Olson, E. N. (2022). CRISPR modelling and correction of cardiovascular disease. *Circulation Research, 130*(12), 1827–1850. DOI: 10.1161/CIRCRESAHA.122.320496 PMID: 35679361

Liu, Y., & Shanahan, C. M. (2011). Signalling pathways and vascular calcification. *Frontiers in Bioscience, 16*(1302), 14. PMID: 21196233

Lv, W., Shao, Y., Hoshino, A., Arany, Z., Musunuru, K., & McDermott-Roe, C. (2022). CRISPR/Cas9 screens implicate RARA and SPNS1 in doxorubicin cardiotoxicity. bioRxiv, 2022-08.

Marín-García, J., & Akhmedov, A. T. (2015). Epigenetics of the failing heart. *Heart Failure Reviews, 20*(4), 435–459. DOI: 10.1007/s10741-015-9483-x PMID: 25847519

McArthur, K. M., Kay, A. M., Mosier, J. A., Grant, J. N., Stewart, J. A., & Simpson, C. L. (2017). Manipulating the plasticity of smooth muscle cells to regulate vascular calcification. *AIMS Cell Tissue Eng, 1*(2), 165–179. DOI: 10.3934/celltissue.2017.3.165

Medvedev, S. P., & Zakian, S. M. (2015). The CRISPR/Cas9 system-a tool for studying hereditary cardiovascular diseases. *Blood Circulation Pathology and Cardiac Surgery, 19* (S4-2), 113-117.

Meyyazhagan, A., Balasubramanian, B., Easwaran, M., Alagamuthu, K. K., Shanmugam, S., Kuchi Bhotla, H., Pappusamy, M., Arumugam, V. A., Thangaraj, A., Kaul, T., Keshavarao, S., & Cacabelos, R. (2020). Biomarker study of the biological parameter and neurotransmitter levels in autistics. *Molecular and Cellular Biochemistry, 474*(1-2), 277–284. DOI: 10.1007/s11010-020-03851-2 PMID: 32740790

Meyyazhagan, A., Pushparaj, K., Balasubramanian, B., Kuchi Bhotla, H., Pappusamy, M., Arumugam, V. A., Easwaran, M., Pottail, L., Mani, P., Tsibizova, V., & Di Renzo, G. C. (2022). COVID-19 in pregnant women and children: Insights on clinical manifestations, complexities, and pathogenesis. *International Journal of Gynaecology and Obstetrics: the Official Organ of the International Federation of Gynaecology and Obstetrics, 156*(2), 216–224. DOI: 10.1002/ijgo.14007 PMID: 34735717

Miguel-dos-Santos, R., & Cingolani, E. (2023). The hunt for novel AAV capsids with improved cardiac tropism. *Molecular Therapy. Methods & Clinical Development, 31*, 31. DOI: 10.1016/j.omtm.2023.101124 PMID: 37886605

Mohd, Y., Balasubramanian, B., Meyyazhagan, A., Kuchi Bhotla, H., Shanmugam, S.K., Ramesh Kumar, M.K., Pappusamy, M., Alagamuthu, K.K., Keshavarao, S., & Arumugam, V.A. (2021). Extricating the association between the prognostic factors of colorectal cancer. *J Gastrointest Cancer, 52*(3), 022-1028.

Mohd, Y., Kumar, P., Kuchi Bhotla, H., Meyyazhagan, A., Balasubramanian, B., Ramesh Kumar, M. K., Pappusamy, M., Alagamuthu, K. K., Orlacchio, A., Keshavarao, S., Sampathkumar, P., & Arumugam, V. A. (2021). Transmission jeopardy of adenomatosis polyposis coli and methylenetetrahydrofolate reductase in colorectal cancer. *Journal of the Renin-Angiotensin-Aldosterone System : JRAAS, 2021*, 2021. DOI: 10.1155/2021/7010706 PMID: 34956401

Molinaro, C., Scalise, M., Leo, I., Salerno, L., Sabatino, J., Salerno, N., & Marino, F. (2023). Polarizing macrophage functional phenotype to foster cardiac regeneration. *International Journal of Molecular Sciences, 24*(13), 10747. DOI: 10.3390/ijms241310747 PMID: 37445929

Moore, O. M., Aguilar-Sanchez, Y., Lahiri, S. K., Hulsurkar, M. M., Navarro-Garcia, J. A., Word, T. A., & Wehrens, X. H. (2024). Long-term efficacy and safety of cardiac genome editing for catecholaminergic polymorphic ventricular tachycardia. *The Journal of Cardiovascular Aging, 4*(1), 8. DOI: 10.20517/jca.2023.42 PMID: 38464671

Motta, B. M., Pramstaller, P. P., Hicks, A. A., & Rossini, A. (2017). The impact of CRISPR/Cas9 technology on cardiac research: From disease modelling to therapeutic approaches. *Stem Cells International, 2017*(1), 8960236. DOI: 10.1155/2017/8960236 PMID: 29434642

Moya, I. M., & Halder, G. (2019). Hippo-YAP/TAZ signalling in organ regeneration and regenerative medicine. *Nature Reviews. Molecular Cell Biology, 20*(4), 211–226. DOI: 10.1038/s41580-018-0086-y PMID: 30546055

Musunuru, K. (2022). Moving toward genome-editing therapies for cardiovascular diseases. *The Journal of Clinical Investigation, 132*(1), e148555. DOI: 10.1172/JCI148555 PMID: 34981785

Musunuru, K. (2023). CRISPR and cardiovascular diseases. *Cardiovascular Research, 119*(1), 79–93. DOI: 10.1093/cvr/cvac048 PMID: 35388882

Naeem, M., Hoque, M. Z., Ovais, M., Basheer, C., & Ahmad, I. (2021). Stimulus-responsive smart nanoparticles-based CRISPR-cas delivery for therapeutic genome editing. *International Journal of Molecular Sciences, 22*(20), 11300. DOI: 10.3390/ijms222011300 PMID: 34681959

Nie, J., Han, Y., Jin, Z., Hang, W., Shu, H., Wen, Z., & Wang, D. W. (2023). Homology-directed repair of an MYBPC3 gene mutation in a rat model of hypertrophic cardiomyopathy. *Gene Therapy, 30*(6), 520–527. DOI: 10.1038/s41434-023-00384-3 PMID: 36765144

Nishiga, M., Qi, L. S., & Wu, J. C. (2021). CRISPRi/a screening with human iPSCs. In *Pluripotent Stem-Cell Derived Cardiomyocytes* (pp. 261–281). Springer US. DOI: 10.1007/978-1-0716-1484-6_23

Nishiyama, T., Zhang, Y., Mcanally, J., Xu, L., Liu, N., Bassel-Duby, R. S., & Olson, E. N. (2022). Precise genomic correction of RBM20 mutations rescues dilated cardiomyopathy. *Circulation, 146*(Suppl_1), A11164–A11164. DOI: 10.1161/circ.146.suppl_1.11164

Owen, M. C., & Kopecky, B. J. (2024). Targeting macrophages in organ transplantation: A step toward personalized medicine. *Transplantation, 108*(10), 2045–2056. DOI: 10.1097/TP.0000000000004978 PMID: 38467591

Paranitharan, N., Kataria, S., Arumugam, V. A., Hsieh, H., Muthukrishnan, S., & Velayuthaprabhu, S. (2025). Integrin α1 upregulation by tf:fviia complex promotes cervical cancer migration through par2-dependent mek1/2 activation. *Biochemical and Biophysical Research Communications, 742*, 151151. DOI: 10.1016/j.bbrc.2024.151151 PMID: 39657349

Paranitharan, N., Kataria, S., Arumugam, V. A., Hsieh, H., Muthukrishnan, S., & Velayuthaprabhu, S. (2025). Integrin α1 upregulation by tf:fviia complex promotes cervical cancer migration through par2-dependent mek1/2 activation. *Biochemical and Biophysical Research Communications*, *742*, 151151. DOI: 10.1016/j.bbrc.2024.151151 PMID: 39657349

Pollanen, S., & Syan, R. (2024). Gene-based targeting for treatment of hypercholesterolemia and prevention of atherosclerotic cardiovascular disease. *Qapsule: Queen's Undergraduate. Health Science Journal*, *1*(2).

Prasad, K. M., Xu, Y., Yang, Z., Acton, S. T., & French, B. A. (2011). Robust cardiomyocyte-specific gene expression following systemic injection of AAV: In vivo gene delivery follows a Poisson distribution. *Gene Therapy*, *18*(1), 43–52. DOI: 10.1038/gt.2010.105 PMID: 20703310

Pushparaj, K., Kuchi Bhotla, H., Arumugam, V. A., Pappusamy, M., Easwaran, M., Liu, W. C., Issara, U., Rengasamy, K. R. R., Meyyazhagan, A., & Balasubramanian, B. (2022). Mucormycosis (black fungus) ensuing COVID-19 and comorbidity meets - magnifying global pandemic grieve and catastrophe begins. *The Science of the Total Environment*, *805*, 150355. DOI: 10.1016/j.scitotenv.2021.150355 PMID: 34818767

Qi, M., Ma, S., Liu, J., Liu, X., Wei, J., Lu, W. J., & Lan, F. (2024). In vivo base editing of Scn5a rescues type 3 long QT syndrome in mice. *Circulation*, *149*(4), 317–329. DOI: 10.1161/CIRCULATIONAHA.123.065624 PMID: 37965733

Ramya, S., Poornima, P., Jananisri, A., Geofferina, I. P., Bavyataa, V., Divya, M., Priyanga, P., Vadivukarasi, J., Sujitha, S., Elamathi, S., Anand, A. V., & Balamuralikrishnan, B. (2023). Role of hormones and the potential impact of multiple stresses on infertility. *Stresses*, *3*(2), 454–474. DOI: 10.3390/stresses3020033

Risato, G., Brañas Casas, R., Cason, M., Bueno Marinas, M., Pinci, S., De Gaspari, M., & Celeghin, R. (2024). In vivo approaches to understand arrhythmogenic cardiomyopathy: Perspectives on animal models. *Cells*, *13*(15), 1264. DOI: 10.3390/cells13151264 PMID: 39120296

Rocheleau, G., Clarke, S. L., Auguste, G., Hasbani, N. R., Morrison, A. C., Heath, A. S., & Do, R. (2024). Rare variant contribution to the heritability of coronary artery disease. *Nature Communications*, *15*(1), 8741. DOI: 10.1038/s41467-024-52939-6 PMID: 39384761

Rogg, E. M., & Dimmeler, S. (2019). Therapeutic potential of non-coding RNAs for acute and chronic myocardial ischemia. *Current Cardiology*, *8*(03), 223–229.

Roshanravan, N., Tutunchi, H., Najafipour, F., Dastouri, M., Ghaffari, S., & Jebeli, A. (2022). A glance at the application of CRISPR/Cas9 gene-editing technology in cardiovascular diseases. *Journal of Cardiovascular and Thoracic Research*, *14*(2), 77–83. DOI: 10.34172/jcvtr.2022.14 PMID: 35935390

Roth, R., Kim, S., Kim, J., & Rhee, S. (2020). Single-cell and spatial transcriptomics approaches of cardiovascular development and disease. *BMB Reports*, *53*(8), 393–399. DOI: 10.5483/BMBRep.2020.53.8.130 PMID: 32684243

Saeed, A., Ahmad, A., & Husnain, A. (2024). Harnessing AI for advancements in cardiovascular disease management and drug discovery. *International Journal of Science and Research Archive*, *13*(1), 244–249. DOI: 10.30574/ijsra.2024.13.1.1623

Sangeetha, T., Anand, A. V., & Begum, T. N. (2022). Assessment of inter-relationship between anemia and COPD in accordance with altitude. *The Open Respiratory Medicine Journal*, *16*(1), e187430642206270. DOI: 10.2174/18743064-v16-e2206270 PMID: 37273961

Sangeetha, T., Nargis Begum, T., Balamuralikrishnan, B., Arun, M., Rengasamy, K. R. R., Senthilkumar, N., Velayuthaprabhu, S., Saradhadevi, M., Sampathkumar, P., & Vijaya Anand, A. (2022). Influence of SERPINA1 gene polymorphisms on anemia and chronic obstructive pulmonary disease. *Journal of the Renin-Angiotensin-Aldosterone System : JRAAS*, *2022*, 2022. DOI: 10.1155/2022/2238320 PMID: 36320441

Sapp, V., Aguirre, A., Mainkar, G., Ding, J., Adler, E., Liao, R., & Jain, M. (2021). Genome-wide CRISPR/Cas9 screening in human iPS derived cardiomyocytes uncovers novel mediators of doxorubicin cardiotoxicity. *Scientific Reports*, *11*(1), 13866. DOI: 10.1038/s41598-021-92988-1 PMID: 34230586

Schary, Y., Rotem, I., Caller, T., Lewis, N., Shaihov-Teper, O., Brzezinski, R. Y., & Leor, J. (2023). CRISPR-Cas9 editing of TLR4 to improve the outcome of cardiac cell therapy. *Scientific Reports*, *13*(1), 4481. DOI: 10.1038/s41598-023-31286-4 PMID: 36934130

Schmidt, R., Steinhart, Z., Layeghi, M., Freimer, J. W., Nguyen, V. Q., Blaeschke, F., & Marson, A. (2021). CRISPR activation and interference screens in primary human T cells decode cytokine regulation. bioRxiv, 2021-05.

Schoger, E., Lelek, S., Panáková, D., & Zelarayán, L. C. (2022). Tailoring cardiac synthetic transcriptional modulation towards precision medicine. *Frontiers in Cardiovascular Medicine*, *8*, 783072. DOI: 10.3389/fcvm.2021.783072 PMID: 35097003

Schreurs, J., Sacchetto, C., Colpaert, R. M., Vitiello, L., Rampazzo, A., & Calore, M. (2021). Recent advances in CRISPR/Cas9-based genome editing tools for cardiac diseases. *International Journal of Molecular Sciences*, *22*(20), 10985. DOI: 10.3390/ijms222010985 PMID: 34681646

Schweiggert, J., Habeck, G., Hinkel, R., Kehr, D., Cimen, K., Kramp, X., & Busch, M. (2023). Engineered cardiac-specific AAV capsids from a novel next-generation platform (next-cap) with cross-species selection for cardiac-targeted gene therapy. *Circulation*, *148*(Suppl_1), A14481–A14481. DOI: 10.1161/circ.148.suppl_1.14481

Shah, S. H., & Gerszten, R. E. (2024). Sifting for gold in terabytes of data: Illuminating cardiovascular biology in the 'omics age. *Circulation*, *149*(17), 1323–1325. DOI: 10.1161/CIRCULATIONAHA.123.066988 PMID: 38648274

Shan, H., & Fei, T. (2023). CRISPR screening in cardiovascular research. *Frontiers in Cell and Developmental Biology*, *11*, 1175849. DOI: 10.3389/fcell.2023.1175849 PMID: 37123412

Sharma, B., Chauhan, I., Kumar, G., Bhardwaj, K., & Tiwari, R. K. (2024). Nanoparticle carriers: A new era of precise crispr/cas9 gene editing. *MicroRNA (Shariqah, United Arab Emirates)*, *14*(2), 101–111. DOI: 10.2174/012211536631984824102 2092805 PMID: 39484758

Smeir, M., Chumala, P., Katselis, G. S., & Liu, L. (2024). Lymphocyte-Specific protein 1 regulates expression and stability of endothelial nitric oxide synthase. *Biomolecules*, *14*(1), 111. DOI: 10.3390/biom14010111 PMID: 38254711

Song, Y., Zheng, Z., & Lian, J. (2022). Deciphering common long QT syndrome using CRISPR/Cas9 in human-induced pluripotent stem cell-derived cardiomyocytes. *Frontiers in Cardiovascular Medicine*, *9*, 889519. DOI: 10.3389/fcvm.2022.889519 PMID: 35647048

Srivastava, R. A. K. (2023). New opportunities in the management and treatment of refractory hypercholesterolemia using in vivo CRISPR-mediated genome/base editing. *Nutrition, Metabolism, and Cardiovascular Diseases*, *33*(12), 2317–2325. DOI: 10.1016/j.numecd.2023.08.010 PMID: 37805309

Suwito, B. E., Adji, A. S., Wardani, V. A. K., Widjaja, J. S., Angel, S. C. S., & Rahman, F. S. (2022). A review of CRISPR Cas9 for ASCVD: Treatment strategies and could target PSCK9 gene using CRISPR cas9 prevent the patient from atherosclerotic vascular disease? *Bali Medical Journal*, *11*(2), 985–993. DOI: 10.15562/bmj.v11i2.3414

Tada, H., Usui, S., Sakata, K., Takamura, M., & Kawashiri, M. A. (2021). Challenges of precision medicine for atherosclerotic cardiovascular disease based on human genome information. *Journal of Atherosclerosis and Thrombosis*, *28*(4), 305–313. DOI: 10.5551/jat.60087 PMID: 33229815

Tang, R., Long, T., Lui, K. O., Chen, Y., & Huang, Z. P. (2020). A roadmap for fixing the heart: RNA regulatory networks in cardiac disease. *Molecular Therapy. Nucleic Acids*, *20*, 673–686. DOI: 10.1016/j.omtn.2020.04.007 PMID: 32380417

Testa, L. C., Musunuru, K., & Wang, X. (2023). Epigenome editing of the AGT gene as a therapeutic approach for cardiovascular disorders. *Arteriosclerosis, Thrombosis, and Vascular Biology*, *43*(Suppl_1), A430–A430. DOI: 10.1161/atvb.43.suppl_1.430

Tyson, J., Bundy, K., Roach, C., Douglas, H., Ventura, V., Segars, M. F., & Simpson, C. L. (2020). Mechanisms of the osteogenic switch of smooth muscle cells in vascular calcification: WNT signalling, BMPs, mechanotransduction, and EndMT. *Bioengineering (Basel, Switzerland)*, *7*(3), 88. DOI: 10.3390/bioengineering7030088 PMID: 32781528

Velayuthaprabhu, S., & Archunan, G. (2005). Evaluation of anticardiolipin antibodies and antiphosphatidylserine antibodies in women with recurrent abortion. *Indian Journal of Medical Sciences*, *59*(8), 347–352. DOI: 10.4103/0019-5359.16651 PMID: 16129928

Velayuthaprabhu, S., Archunan, G., & Balakrishnan, K. (2007). Placental thrombosis in experimental anticardiolipin antibodies-mediated intrauterine fetal death. *American Journal of Reproductive Immunology (New York, N.Y.)*, *57*(4), 270–276. DOI: 10.1111/j.1600-0897.2007.00474.x PMID: 17362388

Velayuthaprabhu, S., Matsubayashi, H., & Archunan, G. (2016). Beta-2 GPI induced tissue factor and placental apoptosis for the pathophysiology of pregnancy loss in antiphospholipid syndrome. *International Journal of Research in Medical Sciences*, *4*(8), 3109–3113. DOI: 10.18203/2320-6012.ijrms20162267

Velayuthaprabhu, S., Matsubayashi, H., Sugi, T., Nakamura, M., Ohnishi, Y., Ogura, T., & Archunan, G. (2013). Expression of apoptosis in placenta of experimental antiphospholipid syndrome (APS) mouse. *American Journal of Reproductive Immunology (New York, N.Y.)*, *69*(5), 486–494. DOI: 10.1111/aji.12084 PMID: 23398188

Velayuthaprabhu, S., Matsubayashi, H., Sugi, T., Nakamura, M., Ohnishi, Y., Ogura, T., Tomiyama, T., & Archunan, G. (2011). A unique preliminary study on fetal resorption and placental apoptosis in mice with passive immunization of anti-phosphatidylethanolamine antibodies and anti-Factor XII antibodies. *American Journal of Reproductive Immunology (New York, N.Y.)*, *66*(5), 373–384. DOI: 10.1111/j.1600-0897.2011.01008.x PMID: 21623987

Vermersch, E., Jouve, C., & Hulot, J. S. (2020). CRISPR/Cas9 gene-editing strategies in cardiovascular cells. *Cardiovascular Research*, *116*(5), 894–907. DOI: 10.1093/cvr/cvz250 PMID: 31584620

Wang, J., Jonker, T., Cervera I Barea, A., Boender, A. R., Klerk, M., Yang, Y., & Boink, G. J. J. (2024). Adeno-associated virus (AAV) 6 provides efficient gene transfer upon intramyocardial injection in vivo. *Cardiovascular Research*, *120*(Supplement_1), cvae088-073.

Wang, P., Liu, Z., Zhang, X., Li, J., Sun, L., Ju, Z., & Qu, J. (2018). CRISPR/Cas9-mediated gene knockout reveals a guardian role of NF-κB/RelA in maintaining the homeostasis of human vascular cells. *Protein & Cell*, *9*(11), 945–965. DOI: 10.1007/s13238-018-0560-5 PMID: 29968158

Wang, S., Xiang, S., Wu, X., Li, N., & Qi, D. (2016). Regulation of macrophage polarization during atherosclerosis. *Lipid Cardiovasc Res*, *2*, 05-12.

Wang, X., Whittaker, M. N., Jindal, I., Cortez-Alvarado, S. V., Qu, P., & Musunuru, K. (2022). Reversible epigenome editing of PCSK9 as a therapeutic strategy. *Arteriosclerosis, Thrombosis, and Vascular Biology*, *42*(Suppl_1), A125–A125. DOI: 10.1161/atvb.42.suppl_1.125

Westerman, K., Fernández-Sanlés, A., Patil, P., Sebastiani, P., Jacques, P., Starr, J. M., & Ordovás, J. M. (2020). Epigenomic assessment of cardiovascular disease risk and interactions with traditional risk metrics. *Journal of the American Heart Association*, *9*(8), e015299. DOI: 10.1161/JAHA.119.015299 PMID: 32308120

Wilton-Clark, H., & Yokota, T. (2023). CRISPR-Cas9-mediated exon skipping as a cardioprotective strategy in duchenne muscular dystrophy. *Molecular Therapy. Methods & Clinical Development*, *30*, 500–501. DOI: 10.1016/j.omtm.2023.08.013 PMID: 37693945

Wu, J., He, S., Song, Z., Chen, S., Lin, X., Sun, H., & Liu, X. (2023). Macrophage polarization states in atherosclerosis. *Frontiers in Immunology*, *14*, 1185587. DOI: 10.3389/fimmu.2023.1185587 PMID: 37207214

Wu, P. Y., Chandramohan, R., Phan, J. H., Mahle, W. T., Gaynor, J. W., Maher, K. O., & Wang, M. D. (2014). Cardiovascular transcriptomics and epigenomics using next-generation sequencing: Challenges, progress, and opportunities. *Circulation: Cardiovascular Genetics*, *7*(5), 701–710. DOI: 10.1161/CIRCGENETICS.113.000129 PMID: 25518043

Wu, W., Shan, J., Li, Y., Luo, L., Sun, G., Zhou, Y., & Feng, L. (2012). Adoptive transfusion of tolerance dendritic cells prolongs the survival of cardiac allograft: A systematic review of 44 basic studies in mice. *Journal of Evidence-Based Medicine*, *5*(3), 139–153. DOI: 10.1111/j.1756-5391.2012.01191.x PMID: 23672221

Wu, Y., Kong, X. J., Ji, Y. Y., Fan, J., Chen, X. M., Ji, C. C., & Wu, S. H. (2024). Correction of Ito in human induced pluripotent stem Cell-derived cardiomyocyte carrying DPP6 mutation in early repolarization syndrome by CRISPR/Cas9 genome editing. *Experimental Cell Research*, *435*(2), 113929. DOI: 10.1016/j.yexcr.2024.113929 PMID: 38272106

Xiao, R., Zhou, M., Wang, P., Zeng, B., Wu, L., Hu, Z., & Liang, D. (2022). Full-length dystrophin restoration via targeted exon addition in DMD-patient specific iPSCs and cardiomyocytes. *International Journal of Molecular Sciences*, *23*(16), 9176. DOI: 10.3390/ijms23169176 PMID: 36012442

Xie, R., Wang, Y., & Gong, S. (2021). External stimuli-responsive nanoparticles for spatially and temporally controlled delivery of CRISPR-Cas genome editors. *Biomaterials Science*, *9*(18), 6012–6022. DOI: 10.1039/D1BM00558H PMID: 34286726

Xu, S., Shiomi, H., Yamashita, Y., Koyama, S., Horie, T., Baba, O., & Ono, K. (2024). CRISPR-Cas9-guided amplification-free genomic diagnosis for familial hypercholesterolemia using nanopore sequencing. *PLoS One*, *19*(3), e0297231. DOI: 10.1371/journal.pone.0297231 PMID: 38507394

Ye, R., Chen, X., & Li, Y. (2023). CRISPR/Cas systems for therapeutic purposes: Applications in disease prevention, mechanism study, and treatment. *Highlights in Science. Engineering and Technology*, *54*, 1–7.

Zhang, X., Cai, H., Xu, H., Dong, S., & Ma, H. (2023). Critical roles of m6A methylation in cardiovascular diseases. *Frontiers in Cardiovascular Medicine*, *10*, 1187514. DOI: 10.3389/fcvm.2023.1187514 PMID: 37273867

Zhang, Z., Zhang, W., Blakes, R., Sundby, L. J., Shi, Z., Rockey, D. C., & Nam, Y. J. (2022). Fibroblast fate determination during cardiac reprogramming by remodelling of actin filaments. *Stem Cell Reports*, *17*(7), 1604–1619. DOI: 10.1016/j.stemcr.2022.05.004 PMID: 35688153

Zhao, C., Cheng, Y., Huang, P., Wang, C., Wang, W., Wang, M., & Deng, H. (2023). X-ray-guided in situ genetic engineering of macrophages for sustained cancer immunotherapy. *Advanced Materials*, *35*(14), 2208059. DOI: 10.1002/adma.202208059 PMID: 36527738

Zhao, H., Li, Y., He, L., Pu, W., Yu, W., Li, Y., & Zhou, B. (2020). In vivo AAV-CRISPR/Cas9-mediated gene editing ameliorates atherosclerosis in familial hypercholesterolemia. *Circulation*, *141*(1), 67–79. DOI: 10.1161/CIRCULATIONAHA.119.042476 PMID: 31779484

Zhou, W., Wang, C., Chang, J., Huang, Y., Xue, Q., Miao, C., & Wu, P. (2021). RNA methylations in cardiovascular diseases, molecular structure, biological functions and regulatory roles in cardiovascular diseases. *Frontiers in Pharmacology*, *12*, 722728. DOI: 10.3389/fphar.2021.722728 PMID: 34489709

ABBREVIATIONS

AI - Artificial Intelligence
CVD - Cardiovascular Disease
DMD - Duchenne Muscular Dystrophy
PRS - Polygenic Risk Scores
CAD - Coronary Artery Disease
DCM - Dilated Cardiomyopathy
iPSC - Induced Pluripotent Stem Cell
dCas9 - dead Cas9
iPSC-CMs - Induced Pluripotent Stem Cell-Derived Cardiomyocytes
CRISPRi - CRISPR Interference
FH - Familial Hypercholesterolemia
CRISPRa - CRISPR Activation
eNOS - Endothelial Nitric Oxide Synthase
LSP1 - Lymphocyte-Specific Protein 1
VSMC - Vascular Smooth Muscle Cell
miRNAs - microRNAs
lncRNAs - Long Non-Coding RNAs
CPVT - Catecholaminergic Polymorphic Ventricular Tachycardia
m6A - N6-methyladenosine
Mettl3 - methyltransferase-like 3
iCMs - Induced Cardiomyocytes
AAV - Adeno-Associated Virus
cTnT - Cardiac Troponin T
ECs - Endothelial Cells
Tol-DCs - Tolerogenic Dendritic Cells
TTR - Transthyretin
ATTR-CM - Transthyretin Amyloidosis with Cardiomyopathy

Chapter 5
CRISPR-Cas9 Technology for Combating Genetic Disorders

Ajay Sharma
https://orcid.org/0000-0001-6620-4805
Lovely Professional University, India

Pratibha Singh
Lovely Professional University, India

ABSTRACT

The CRISPR-Cas9 technique, which offers an unparalleled degree of precision and efficacy in genome editing, has entirely changed the area of genetic disorders medicine. Since its discovery as a component of the bacterial immune defence system, CRISPR-Cas9 has quickly been modified for use in therapeutic use, most notably in the fight against genetic disease problems. The essential mechanics of CRISPR-Cas9, along with its innovative potential in the treatment of complicated and monogenic genetic illnesses, are discussed in the manuscript. The author has gained insights into effective preclinical and clinical uses, such as the management of sickle cell anaemia, Duchenne muscular dystrophy, and cystic fibrosis, as well as new developments like base editing and prime editing. This chapter also sheds light on important topics such as off-target effects, delivery system constraints, ethical impasses, and regulatory apprehensions. Through a combination of case studies and recent developments, this chapter attempts to give a detailed overview of how CRISPR-Cas9 is changing precision medicine and developing new treatment paradigms for genetic disorders that were previously incurable.

DOI: 10.4018/979-8-3373-4857-5.ch005

INTRODUCTION

CRISPR-Cas9 is a novel genome-editing tool that has transformed the field of genetic therapies and biomedical research within the last ten years. The Cas9 nuclease and the Clustered Regularly Interspaced Short Palindromic Repeats (CRISPR) system allow for precise and programmed DNA sequence change in a variety of animals, including humans (Doudna & Charpentier, 2014). CRISPR-Cas9 provides a more straightforward, effective, and highly customisable method of targeting certain genomic loci than previous gene-editing systems, including zinc-finger nucleases (ZFN) and transcription activator-like effector nucleases (TALENs) (Cox et al., 2015). The CRISPR-Cas9 system works as a molecular scalpel that can identify and cut particular genomic areas, allowing scientists to add, remove, or fix DNA sequences. This skill has great therapeutic potential, especially for treating monogenic disorders, hereditary illnesses caused by single-gene mutations (Barrangou & Doudna, 2016). CRISPR-based treatments have demonstrated notable preclinical and early clinical results in conditions like as Duchenne muscular dystrophy, β-thalassemia, sickle cell disease, and cystic fibrosis (Frangoul et al., 2021).

The global shift towards precision medicine, which seeks to customise therapeutic approaches based on individual genetic profiles, coincides with the emergence of CRISPR-Cas9 (Maeder & Gersbach, 2016). Instead of only treating symptoms in this situation, CRISPR provides a clear path to fixing the underlying genetic flaw. CRISPR-Cas9's versatility enables its use in diagnostic applications, functional genomics research, and even the creation of genetically modified animal models in addition to therapeutic applications (Wang et al., 2020). CRISPR-Cas9 does have certain limits, though, despite its enormous potential. Unintentional genetic alterations in non-specific locations, or off-target consequences, continue to be a major clinical translation risk (Komor et al., 2016). Significant obstacles to the broad therapeutic application of CRISPR include subjects with immunogenicity to bacterially produced Cas9 proteins, ethical concerns about embryo editing, and difficulties with the effective transport of CRISPR machineries to target cells (Anzalone et al., 2019).

These issues have started to be addressed by recent developments. Base editors and prime editors, two more recent CRISPR derivatives, minimise the danger of mutagenesis by allowing specific nucleotide conversions without causing double-strand breaks (Komor et al., 2016; Anzalone et al., 2019). The range of in vivo uses has also increased due to improvements in delivery systems, which include non-viral techniques like lipid nanoparticles and viral vectors like adeno-associated virus (AAV) (Wang et al., 2020). Merging CRISPR with computational methods such as machine learning for off-target prediction and guide RNA optimization represents a new paradigm in precision gene editing. The ethical and social consequences of CRISPR-Cas9 technology must be cautiously examined as it develops. Particularly when it comes to heritable genome alterations, the difference between somatic and germline editing has noteworthy ethical and legal consequences (Cox et al., 2015). To confirm that the advantages of genome editing are realised while reducing the possibility of abuse or injustice, international regulatory outlines and bioethical standards are being developed to strike a balance between novelty and accountability (Barrangou & Doudna, 2016).

The goal of the manuscript is to present a thorough scholarly analysis of CRISPR-Cas9 in relation to the fight against genetic diseases. It will track the system's evolution over time, elucidate its biological processes, investigate its function in monogenic disorders, appraise technology advancements, evaluate therapeutic uses and delivery procedures, and critically analyse ethical and legal matters. This study will conclude with future directions, emphasising how CRISPR is likely to alter medicine by integrating with multi-omic platforms, synthetic biology, and artificial intelligence. Bacterial defence mechanisms that developed as an adaptive immunological response to invasive genetic elements like bacteriophages, plasmids are the source of the CRISPR-Cas9 system. When scientists discovered odd repeated sequences in the Escherichia coli genome in 2002, they came up with the name Clustered Regularly Interspaced Short Palindromic Repeats (CRISPR) (Jansen et al., 2002). But it wasn't until 2005 that comparative genomic investigations showed that the spacers between repetitions matched viral DNA pieces, indicating a role in microbial immunity. Until then, the biological function of these sequences was still unknown (Mojica et al., 2005).

The discovery that CRISPR loci provide Streptococcus thermophilus resistance to bacteriophage infection was made by Barrangou et al. (2007), marking CRISPR as a crucial adaptive immunological mechanism in prokaryotes. In this technique, bacteria use CRISPR arrays to absorb brief viral DNA snippets and incorporate them into their own genomes. Upon subsequent infection, these sequences act as molecular “memory” to identify and target the same viruses. The discovery of the CRISPR-associated (Cas) genes, which encode the helicases and endonucleases that cleave foreign nucleic acids, was a later significant event (Makarova et al., 2011). Cas9 stood out amid them as a key effector protein associated with the Type II CRISPR system. With the help of RNA molecules, the Cas9 protein was able to identify and cut double-stranded DNA at particular locations that were indicated by a brief protospacer adjacent motif (PAM) and the guide RNA sequence. When Jennifer Doudna and Emmanuelle Charpentier successfully reassembled the CRISPR-Cas9 system in vitro in 2012 and showed that it could be trained to cut any desired DNA sequence using a synthetic guide RNA, it marked the beginning of the real revolution (Doudna & Charpentier, 2014). This finding made CRISPR-Cas9 a global genome editing tool rather than just a microbial curiosity. This technique was further modified for usage in mammalian cells by Feng Zhang and associates, confirming its suitability for eukaryotic genome editing (Cong et al., 2013).

The simplicity, affordability, and scalability of CRISPR-Cas9 made it the go-to technique for genome editing in a matter of years. CRISPR's programmability depends just on changing a brief RNA sequence complementary to the DNA target, as opposed to ZFNs and TALENs, which require time-consuming protein engineering for each target (Cox et al., 2015). This adaptability sped up drug development, gene therapy research, and functional genomics investigations by enabling the simultaneous editing of numerous genes (multiplexing). The quick leap from discovery to use is further demonstrated by the evolution of CRISPR throughout time. CRISPR-Cas9 has been used in a number of therapeutic approaches within ten years of its mechanistic elucidation, such as developing gene drives for vector control, engineering immune cells for cancer immunotherapy, and fixing mutations that cause inherited diseases (Barrangou & Doudna, 2016). The system's influence on science and society was further solidified when Doudna and Charpentier were granted the 2020 Nobel Prize in Chemistry.

Three main steps comprise the intricately coordinated process by which the CRISPR-Cas9 system functions: interference, expression, and adaptability (Makarova et al., 2011). Together, these mechanisms allow bacteria to absorb, decode, and apply genetic material from invasive substances to provide protection; scientists have also appropriated these mechanisms to alter the genomes of higher animals. Short foreign DNA pieces called protospacers are removed from the invasive virus or plasmid and incorporated into the host genome's CRISPR array during the adaptation phase. This integration, which acts as a log of previous infections, takes place at the array's leader end. Cas1 and Cas2 proteins work together to create a complex that recognises and integrates additional spacers into the host genome, therefore mediating the acquisition process (Jackson et al., 2017). Through this mechanism, the host's immunological repertoire is increased as the CRISPR locus constantly changes to store new viral information.

The CRISPR array is converted into a lengthy precursor CRISPR RNA (pre-crRNA), which has the repeat-spacer structure, when a bacterium is re-exposed to the same virus. Shorter, mature CRISPR RNAs (crRNAs) are produced from this pre-crRNA, and each one has a unique spacer sequence that is taken from the invader's DNA (Barrangou & Doudna, 2016). The trans-activating CRISPR RNA (tracrRNA) and crRNA combine to create a dual RNA complex in the Type II CRISPR-Cas9 system, which directs Cas9 to its target DNA sequence. These two RNA molecules combine to form a single synthetic molecule called single guide RNA (sgRNA) in the designed genome-editing system. The sgRNA has a scaffold region that interacts with Cas9 and a 20-nucleotide sequence complementary to the target DNA. The CRISPR system was made more easily programmable for use in lab settings by this update (Doudna & Charpentier, 2014). During the interference step, the target DNA sequence is recognised and bound by the Cas9 protein under the guidance of the sgRNA. A PAM sequence, usually "NGG," must be present directly downstream of the target site for this recognition to occur (Jinek et al., 2012). Cas9 causes a double-strand break (DSB) in the DNA when it binds. Non-homologous end joining (NHEJ) and homology-directed repair (HDR), the cell's natural DNA repair processes, are triggered by the break and can be used to accomplish certain genetic changes.

1. Gene disruption is frequently caused by non-homologous end joining (NHEJ), which produces tiny insertions or deletions (indels) at the cleavage site.
2. Conversely, homology-directed repair (HDR) replaces or corrects the target sequence exactly by using an external DNA template (Cox et al., 2015).

Depending on the experimental design, researchers can use these repair mechanisms to carry out precise base repairs, gene knockouts, or knock-ins. CRISPR-Cas9's accuracy and effectiveness have made it essential for both fundamental research and the creation of new treatments. The Streptococcus pyogenes Cas9 (SpCas9) is the most often utilised nuclease; other Cas variations have been created to increase specificity and target range. For example, Cas12a (Cpf1) generates staggered DSBs, which are useful in some applications, by recognising a distinct PAM sequence ("TTTV"). Single-strand nicking and transcriptional regulation are two of the specialised roles of modified Cas9, such as nickase Cas9 (nCas9) and dead Cas9 (dCas9) (Komor et al., 2016). Together, these molecular advancements and mechanistic understandings have made CRISPR-Cas9 a flexible instrument supporting contemporary genome modification. Table 1 shows the timeline for the major discoveries, and Table 2 describes the mechanism and components of CRISPR-Cas9 research.

Table 1. Timeline of Major Discoveries in CRISPR-Cas9 Research

Year	Discoveries	Significance	Reference
1987	Discovery of repetitive DNA sequences in *E. coli*	First observation of CRISPR-like sequences, function unknown.	Ishino et al.
2002	Naming of CRISPR and identification of Cas genes	Defined "Clustered Regularly Interspaced Short Palindromic Repeats."	Jansen et al.
2005	Recognition of CRISPR as part of the bacterial immune system	Linked spacer sequences to viral DNA.	Mojica et al.
2007	Experimental confirmation of antiviral immunity	Demonstrated adaptive immunity in *Streptococcus thermophilus.*	Barrangou et al.

continued on following page

Table 1. Continued

Year	Discoveries	Significance	Reference
2014	Programmable genome editing demonstrated in vitro	Developed sgRNA-Cas9 complex for target DNA cleavage.	Doudna & Charpentier
2013	Application to eukaryotic genome editing	Enabled gene editing in mammalian cells.	Cong et al.
2016	Development of base editors	Precise base substitutions without DSBs.	Komor et al.
2019	Development of prime editing	Precise "search-and-replace" editing without donor templates.	Anzalone et al.
2021	First in vivo human clinical trial success	CRISPR used in patients with transthyretin amyloidosis (NTLA-2001).	Gillmore et al.

Table 2. Mechanism and Components of the CRISPR-Cas9 System

Component / Phase	Function	Key Molecular Players	Outcome
Adaptation	Acquisition of viral DNA sequences (spacers)	Cas1–Cas2 complex	Integration of protospacer into the CRISPR array.
Expression	Transcription and processing of the CRISPR array	crRNA, tracrRNA, RNase III	Formation of mature guide RNA duplex.
Interference	Recognition and cleavage of target DNA	Cas9 endonuclease, sgRNA, PAM	Double-stranded DNA break at the target site.
Repair	Host DNA repair of DSB	NHEJ or HDR pathways	Gene knockout (NHEJ) or precise correction (HDR).

CRISPR-CAS9 IN MONOGENIC DISORDERS

CRISPR-Cas9 is especially useful for correcting monogenic illnesses, which are caused by mutations in a single gene, because genotype and phenotype are directly correlated. The method offers possible treatments for illnesses that were previously incurable by precisely repairing or disrupting faulty alleles. Several noteworthy instances where CRISPR has shown great potential are covered in this section (Table 3).

Figure 1. Image shows the CRISPR-Cas9 for the Monogenic Disorders

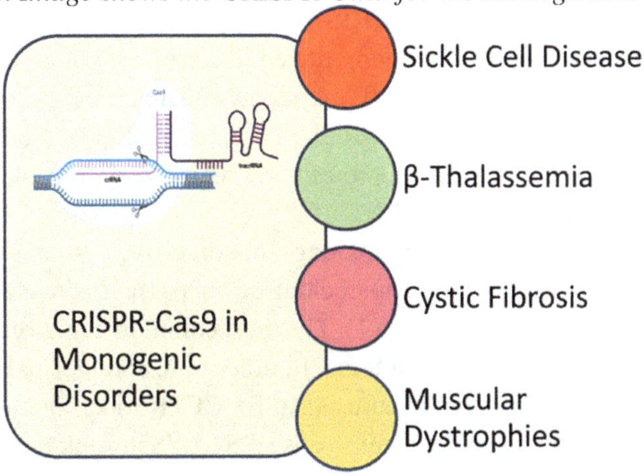

Sickle Cell Disease: A single point mutation (Glu6Val) in the HBB gene, which codes for β-globin, causes sickle cell disease (SCD) by causing aberrant haemoglobin polymerisation and deformation of red blood cells (Figure 1). Conventional therapies, including bone marrow transplantation or hydroxyurea, are either extremely risky or only partially effective. By focusing on and fixing the HBB mutation at the genomic level, CRISPR-Cas9 provides a more straightforward method (Frangoul et al., 2021). Researchers used ex vivo CRISPR editing of patient-derived haematopoietic stem cells (HSCs) in one of the earliest clinical applications. Following myeloablation, these altered cells were reinfused into patients, resulting in a notable improvement in clinical outcomes and persistent production of foetal haemoglobin (HbF) (Park & Bao, 2021). The therapeutic technique entailed reactivating the production of foetal haemoglobin, which compensates for faulty β-globin, by interrupting the BCL11A gene enhancer. This strategy, called CTX001, is being developed in collaboration between Vertex Pharmaceuticals and CRISPR Therapeutics. Early clinical findings showed that for more than a year after therapy, treated patients continued to be transfusion-independent and free of vaso-occlusive crises (Frangoul et al., 2021). The viability of CRISPR-based curative treatments for haematologic diseases has been confirmed by these results.

β-Thalassemia: Another hemoglobinopathy, β-thalassemia, is caused by mutations in the HBB gene that result in inadequate synthesis of β-globin (Figure 1). The adaptation of comparable CRISPR techniques has been made possible by the pathophysiology parallels between SCD and β-thalassemia. To alleviate the anaemic phenotype, CRISPR can revive foetal haemoglobin production by altering the same BCL11A enhancer area (Bueren et al., 2020). Patients with β-thalassemia have also participated in clinical studies with the CTX001 platform, showing improved erythropoiesis and no need for blood transfusions (Frangoul et al., 2021). Crucially, our results demonstrate how CRISPR may use a common therapeutic target to overcome genetic variability among linked illnesses. Ongoing studies are examining the direct repair of particular HBB mutations utilising HDR-based editing using donor DNA templates in addition to BCL11A-based methods. Nonetheless, primary HSCs' HDR efficacy is still low, prominence the necessity for more repair mechanism optimisation (Cox et al., 2015).

Cystic Fibrosis: Mucus builds up in the lungs and digestive system as a result of alterations in the CFTR gene, which impair chloride ion passage across epithelial cells and cause cystic fibrosis (CF) (Figure 1). The most common mutation, ΔF508, causes misfolding and degradation of proteins. In order to restore normal chloride channel function, CRISPR-Cas9 has been used to fix CFTR mutations in airway epithelial cells and induced pluripotent stem cells (iPSCs) (Schwank et al., 2013). Preclinical research has demonstrated that CRISPR-Cas9, in conjunction with HDR templates, may successfully fix CFTR mutations in patient-derived cells (Lee et al., 2020). Furthermore, editing accuracy has been further improved by the use of base editors for single-nucleotide repair without causing DSBs. These results open the door to customised genetic therapy for cystic fibrosis, but practical translation is still difficult because of delivery limitations in lung tissues.

Muscular Dystrophies: Mutations that change the reading frame of the DMD gene, which codes for dystrophin, cause Duchenne muscular dystrophy (DMD), a severe X-linked recessive illness (Figure 1). Muscle degeneration progresses when functioning dystrophin is absent. By deleting faulty exons or fixing point mutations, CRISPR-Cas9 has been used to restore the dystrophin reading frame (Nelson et al., 2019). Systemic administration of AAV-encoded Cas9 and sgRNA improved function in DMD animal models by restoring dystrophin expression in muscle tissues (Amoasii et al., 2018). Comparable outcomes have been documented in canine models, highlighting CRISPR therapy's potential for translation. Contempt the restricted packing capacity of AAV vectors, the transport of long gene segments has been made possible by the use of dual-AAV systems (Wang et al., 2020). Although there are still issues to be resolved, such as immunological reactions to viral vectors and inadequate muscle transduction, human clinical trials for DMD are now in progress. These developments, however, highlight the revolutionary potential of CRISPR-Cas9 in neuromuscular diseases.

Table 3. CRISPR-Cas9 Applications in Monogenic Disorders

Disorder	Target Gene(s)	Therapeutic Strategy	Experimental Model / Trial	Key Outcomes
Sickle Cell Disease (SCD)	*HBB*, *BCL11A* enhancer	Reactivate fetal hemoglobin via enhancer disruption	CTX001 clinical trial	Restored HbF, transfusion independence, symptom remission.
β-Thalassemia	*HBB*, *BCL11A*	Fetal hemoglobin reactivation or direct correction	CTX001 clinical trial	Sustained HbF levels and clinical cure.
Cystic Fibrosis (CF)	*CFTR*	HDR or base editing to correct the ΔF508 mutation	iPSC and airway epithelial models	Restored chloride channel function.
Duchenne Muscular Dystrophy (DMD)	*DMD*	Exon skipping/frame restoration	Mouse and canine models; AAV delivery	Dystrophin expression, muscle recovery.

ADVANCEMENTS IN GENOME EDITING

Genetic engineering has been revolutionised by the standard CRISPR-Cas9 system, although its initial versions were constrained by the production of double-strand breaks (DSBs) and reliance on biological DNA repair processes, which frequently resulted in unwanted mutations. To improve accuracy, effectiveness, and safety, a new generation of genome-editing technologies has emerged. Among them, the groundbreaking developments of base editing, prime editing, and epigenetic modification have expanded the therapeutic potential of CRISPR (Komor et al., 2016; Anzalone et al., 2019).

Base Editing

Base editing is a novel CRISPR-based technique that enables accurate single-nucleotide conversions without the need for donor templates or the introduction of DSBs. The base editor, created by Komor et al. (2016), combines a DNA-modifying enzyme like cytidine deaminase or adenosine deaminase with a catalytically impaired Cas9 (nCas9 or dCas9). Within a predetermined editing window, this fusion complex, which is guided by a single RNA, causes a chemical alteration of certain nucleotides.

There are now two main categories of base editors:

1. **Cytosine Base Editors (CBEs):** Convert cytosine (C)–guanine (G) base pairs into thymine (T)–adenine (A) pairs.
2. **Adenine Base Editors (ABEs):** Convert adenine (A)–thymine (T) base pairs into guanine (G)–cytosine (C) pairs.

Nearly half of all known harmful point mutations, which cause a significant number of hereditary disorders in humans, may be fixed by these methods (Rees & Liu, 2018). For example, in preclinical models, base editing has been used to fix mutations linked to β-thalassemia, progeria, and Tay–Sachs disease, showing promise as a treatment without the cytotoxic side effects of DSBs.

High accuracy, low indel frequency, and independence from the cell cycle phase are the main benefits of base editing. Expanding the customisable sequence window and reducing off-target deamination are still difficult tasks, though. The fidelity and usefulness of base editors in clinical research have significantly increased as a result of ongoing work to create evolving deaminases and better guide RNA design methods (Gaudelli et al., 2017).

Prime Editing: Often called a "search-and-replace" genome-editing technique, prime editing is a further development of CRISPR technology (Anzalone et al., 2019). Using a primary editing guide RNA (pegRNA) that both identifies the target site and encodes the desired genetic change, this technique combines the nCas9 enzyme with a reverse transcriptase enzyme. DNA sequences can be added, removed, or substituted via prime editing without the need for donor templates or DSBs.

This strategy has several noteworthy benefits:

1. It is capable of accurately introducing each of the 12 base-to-base conversions.
2. It broadens the range of curable mutations by enabling the repair of minor insertions and deletions.
3. It reduces undesirable editing results, such as big deletions or indels, which are frequently seen in HDR-dependent CRISPR systems (Anzalone et al., 2019).

Prime editing has effectively fixed the mutations causing Duchenne muscular dystrophy, sickle cell disease, and cystic fibrosis in disease models (Lin et al., 2021). Additionally, it is a particularly appealing platform for in vivo therapeutic editing due to its compatibility with a variety of cell types, including non-dividing cells. The issues with distribution efficiency and the comparatively large size of the prime editor design continue to be barriers to clinical use. To get over these limitations, split-intein system innovations and dual-vector delivery are being investigated (Wang et al., 2020).

Epigenetic Modulation: CRISPR technology has been used to alter gene expression through epigenetic control in addition to altering the DNA sequence. Researchers can either stimulate or repress gene transcription without changing the underlying DNA sequence by fusing transcriptional effectors with catalytically inactive Cas9 (dCas9) (Liu et al., 2016). Studying gene regulatory networks and treating illnesses brought on by abnormal gene expression rather than structural alterations are two areas in which this method excels.

1. To upregulate target genes, CRISPR activation (CRISPRa) uses dCas9 coupled to transcriptional activators like VP64, p300, or SunTag.
2. To silence certain genes, CRISPR interference (CRISPRi) employs dCas9 coupled with repressor domains like KRAB (Gilbert et al., 2014).

In cancer models, epigenetic CRISPR systems have been effectively employed to repress oncogenic drivers and reawaken dormant tumour suppressor genes (Thakore et al., 2015). Furthermore, without permanently changing the genome, CRISPRa/i/i's reversible manipulation of gene expression has therapeutic potential in the treatment of neurological illnesses, metabolic diseases, and immunological regulation. Together, these developments have turned CRISPR-Cas9 from a straightforward nuclease into a versatile genome-engineering tool that can modify, control, and fine-tune the genetic code on several levels. Table 4 works on the comparison of advanced CRISPR technologies.

Table 4. Comparison of Advanced CRISPR Editing Technologies

Editing Type	Key Enzyme / Mechanism	DNA Cleavage	Repair Requirement	Example Applications	Advantages	Limitations
Conventional CRISPR-Cas9	Cas9 nuclease + sgRNA	Double-strand break	NHEJ or HDR	Gene knockout, insertion	Simple, efficient	Off-target mutations, DSB toxicity
Base Editing (CBE/ABE)	nCas9 + Deaminase	No DSB	No donor needed	Point mutation correction	High precision	Limited to transition mutations
Prime Editing	nCas9 + Reverse Transcriptase + pegRNA	Single-strand nick	No donor needed	Small insertions, deletions, and all base changes	Versatile, fewer indels	Large construct size, delivery challenges
Epigenetic Editing	dCas9 + Activator/ Repressor	No cleavage	Not applicable	Gene regulation (CRISPRa/i)	Reversible control	Temporary expression, delivery issues

Delivery Mechanisms

The effective and secure transport of CRISPR-Cas9's constituent parts to target cells and tissues is just as important to the therapeutic use of this technology as accurate editing. Delivery strategies (Table 5) must minimise immunogenicity and off-target dispersion while guaranteeing enough Cas9 and guide RNA expression. In general, there are two types of CRISPR delivery strategies: non-viral systems and viral vector-based systems (Wang et al., 2020).

Viral Vectors: Viral vectors take advantage of viruses' innate capacity to transfer genetic material into host cells. Adeno-associated viruses (AAVs) are the most popular vehicles for delivering CRISPR in vivo because of their long-term gene expression, broad tropism, and low immunogenicity (Figure 2).

Figure 2. The image shows the delivery mechanism of viral vectors

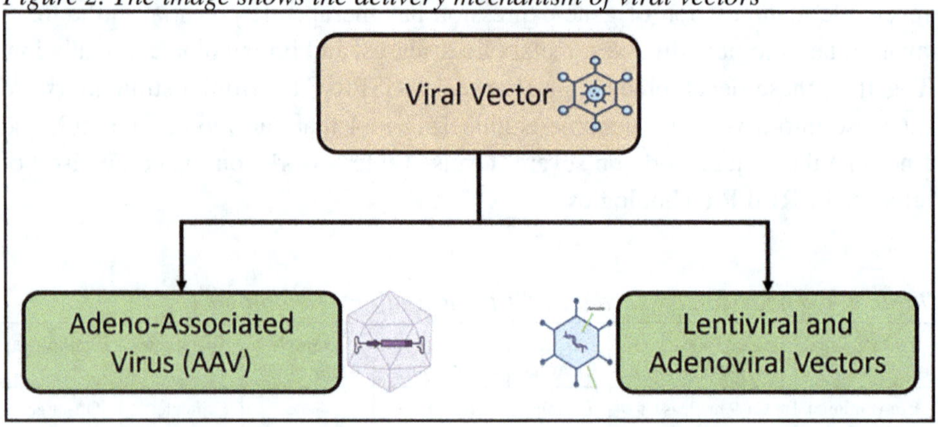

Adeno-Associated Virus (AAV): In animal models, CRISPR components have been effectively delivered via AAV vectors to treat ophthalmic, muscular, and hepatic disorders (Mendell et al., 2021). For instance, AAV-mediated delivery of sgRNA and SpCas9 restored dystrophin expression in several organs in models of Duchenne muscular dystrophy (Amoasii et al., 2018). However, AAV's limited packing capacity (~4.7 kb) makes it impossible for it to transmit big Cas9 variants, which calls for dual-vector or miniaturised Cas9 strategies like Staphylococcus aureus's SaCas9.

Lentiviral and Adenoviral Vectors: Lentiviruses are appropriate for ex vivo usage, such as haematopoietic stem cell modification, because they integrate into the host genome and provide stable expression of CRISPR components in both dividing and non-dividing cells (Maeder & Gersbach, 2016). In contrast, adenoviruses have a greater potential for packaging and temporary expression, but their therapeutic value is limited due to their higher immunogenicity (Cox et al., 2015).

Integration-deficient lentiviral vectors (IDLVs) and self-deleting AAV systems are being researched to enable temporary but effective gene editing without insertional mutagenesis in order to allay safety concerns (Wang et al., 2020).

Non-Viral Systems: Because of their simplicity of large-scale manufacture, decreased immunogenicity, and safety, non-viral delivery methods have drawn more and more interest. These consist of cell-penetrating peptides, polymeric carriers, lipid nanoparticles (LNPs), and electroporation (Figure 3).

Figure 3. Non-Viral delivery mechanism for the CRISPR Cas-9 mechanism and support

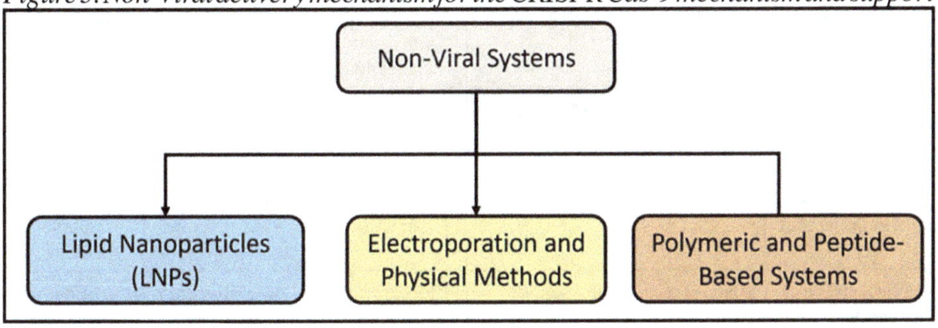

Lipid Nanoparticles (LNPs): Delivering CRISPR components as mRNA or ribonucleoprotein (RNP) complexes has proven very effective when done using LNPs. CRISPR uses have increased as a result of their effectiveness with mRNA vaccine platforms during the COVID-19 pandemic (Liu et al., 2021). For instance, effective gene deletion in hepatocytes was accomplished using LNP-mediated Cas9 mRNA and sgRNA delivery, which may be used to treat metabolic liver illnesses such as transthyretin amyloidosis (ATTR).

Electroporation and Physical Methods: For accurate CRISPR editing of patient-derived stem cells and embryos, ex vivo delivery techniques like as nucleofection and microinjection are frequently employed. These physical techniques are perfect for autologous stem cell-based treatments because they allow for great delivery efficiency without running the danger of viral integration (Park & Bao, 2021).

Polymeric and Peptide-Based Systems: CRISPR RNPs can be encapsulated by cationic polymers and peptides, which can help endosomal escape. These materials show promise for tissue-specific gene editing since they are biodegradable and highly adjustable (Wang et al., 2020). Tissue selectivity and intracellular absorption have been further improved by hybrid delivery methods that combine polymers with lipids or targeting ligands.

Table 5. Delivery Systems for CRISPR Components

Delivery Method	System Type	Delivery Form (DNA/mRNA/RNP)	Advantages	Limitations / Risks	Example Applications
AAV Vectors	Viral	DNA	Stable, tissue-specific expression	Limited cargo size, immune response	In vivo DMD correction, retinal disorders
Lentivirus	Viral	DNA	Integrates into the genome, long-term expression	Insertional mutagenesis	Ex vivo HSC editing for SCD
Adenovirus	Viral	DNA	High capacity, transient expression	Immunogenicity	Liver-targeted editing (preclinical)
Lipid Nanoparticles (LNPs)	Non-viral	mRNA / RNP	Safe, scalable, non-integrating	Limited targeting specificity	NTLA-2001 (ATTR amyloidosis)
Electroporation / Microinjection	Physical	RNP / DNA	High efficiency in vitro	Cell damage, ex vivo use only	iPSC and embryo editing
Polymeric / Peptide Nanocarriers	Non-viral	RNP / mRNA	Biodegradable	Lower efficiency in vivo	Tumor-targeted gene editing

Preclinical and Clinical Applications: CRISPR-Cas9 has advanced very quickly from the bench to the bedside. The method, which targets genetic and haematologic illnesses, has progressed from proof-of-concept investigations to early-phase human clinical trials within ten years of its invention. The two main categories of these applications are ex vivo and in vivo approaches (Frangoul et al., 2021) (Figure 4).

Figure 4. Non-Viral delivery mechanism for the CRISPR Cas-9 mechanism and support

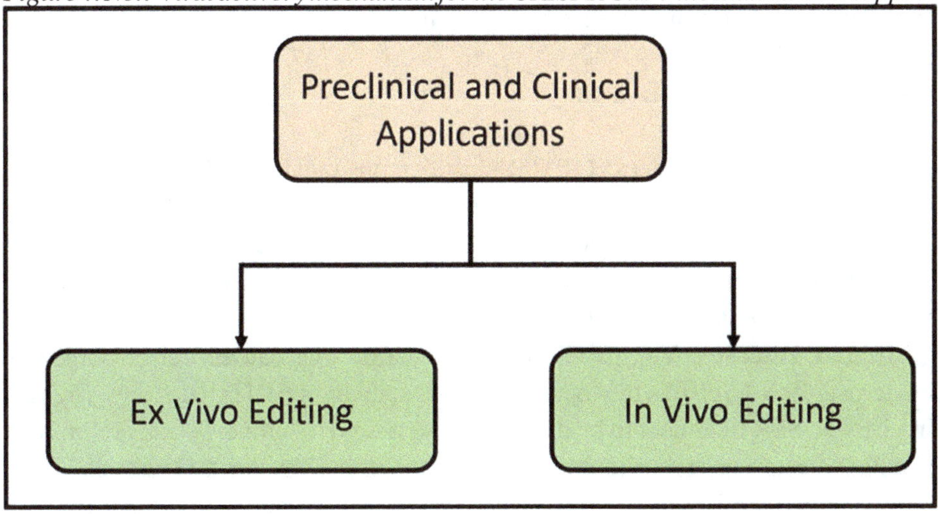

In vivo vs. Ex vivo strategies

Ex Vivo Editing: Ex vivo gene therapy comprises removing the patient's cells, modifying them genetically in a lab, and then reintroducing them into the body. This method offers a regulated setting for genome editing and a comprehensive evaluation of quality before transplantation. Immune cells and haematopoietic stem cells (HSCs) have had the greatest success with it.

With CTX001, the ex vivo method has shown ground-breaking outcomes in the treatment of β-thalassemia and sickle cell disease (Frangoul et al., 2021). CRISPR's function has been extended beyond monogenic illnesses into cancer immunotherapy by means of it to generate CAR-T cells resistant to immune checkpoint suppression or viral reactivation (Eyquem et al., 2017).

In Vivo Editing: Through in vivo editing, CRISPR components are delivered straight into the patient's body, enabling precise tissue modification. To treat transthyretin amyloidosis (Kotit, 2023), for example, Intellia Therapeutics announced the first systemic dose of CRISPR-Cas9 in humans via LNP delivery. A therapeutic milestone for in vivo genome editing was reached when the therapy resulted in a decrease in blood TTR protein levels of more than 85% (Gillmore et al., 2021). In ophthalmologic conditions like Leber congenital amaurosis (EDIT-101), where subretinal administration of AAV-CRISPR components restored photoreceptor function, in vivo techniques have also demonstrated promise (Maeder & Gersbach,

2016). These findings demonstrate the quick transition of CRISPR-based therapies from preclinical evidence to therapeutic reality.

Case Studies and Trials:

Hematologic Disorders: CTX001 has demonstrated long-lasting therapeutic effects, including transfusion independence and symptom relief, in clinical studies for SCD and β-thalassemia (Frangoul et al., 2021). In order to achieve one-step cures without foetal haemoglobin reactivation, other strategies are investigating the direct repair of HBB mutations employing prime editors.

Ocular Diseases: The CEP290 gene mutation that causes Leber congenital amaurosis 10 (LCA10) is the focus of Editas Medicine's EDIT-101 study. The first in vivo CRISPR treatment to be directly administered to human tissue, preliminary results indicate a partial restoration of visual function (Maeder & Gersbach, 2016).

Oncology: Cell-based immunotherapies have also made use of CRISPR. According to Stadtmauer et al. (2020), ex vivo editing of T cells to eliminate TCR and PD-1 genes improves antitumor effectiveness and avoids immunological rejection. These customised CRISPR-engineered immune cells are being investigated for solid tumours, lymphoma, and leukaemia.

Metabolic and Neurological Disorders: In animal models, preclinical research employing CRISPR has fixed mutations that cause spinal muscular atrophy, Huntington's disease, and amyotrophic lateral sclerosis (Yang et al., 2017). Likewise, metabolic abnormalities like phenylketonuria and familial tyrosinemia have been fixed by liver-targeted editing (Wang et al., 2020). These case studies mark the beginning of a new era in molecular medicine by demonstrating CRISPR's unparalleled adaptability in treating a variety of hereditary disease classes.

Ethical, Legal, and Regulatory Challenges

The unmatched ability of CRISPR-Cas9 to change the human genome has raised significant moral, legal, and regulatory issues. Unquestionably, the possibility of curing hereditary illnesses is revolutionary, but it also raises questions about safety, equality, and the ethical confines of genetic intervention (Cox et al., 2015). One of the key topics of discussion in the application of genome-editing technology is how to strike a balance between creativity and accountability (Table 6).

Off-target effects and genomic instability: The possibility of off-target mutations, in which Cas9 causes double-strand breaks at unexpected genomic locations, is one of the key ethical and scientific issues with CRISPR-based treatment (Fu et al., 2013). Unexpected pathogenic consequences or oncogenesis may result from such off-target action that interferes with vital genes or governing components. This presents a thoughtful safety concern in clinical situations, exclusively for germline or systemic in vivo applications.

The creation of high-fidelity Cas9 variants (e.g., eSpCas9, SpCas9-HF1) and computational methods that maximise sgRNA specificity are two attempts to reduce off-target effects (Slaymaker et al., 2016). Preclinical safety assessment is made easier by procedures like GUIDE-seq, Digenome-seq, and CIRCLE-seq, which enable genome-wide identification of unwanted changes (Kim et al., 2020). Despite these developments, long-term monitoring of CRISPR-treated individuals is required for ethical reasons in order to observe genetic stability over cell generations. The mosaicism complicates risk assessment and causes diversity in therapy results when altered and unedited cells coexist inside the same organism (Zuccaro et al., 2020). Maintaining consistent editing in somatic tissues or embryos is still a difficult task that is closely regarded from an ethical standpoint.

Germline editing controversies: One of the most disruptive topics in bioethics is the possibility of human germline editing, which would alter the genome in a way that would be inherited by future generations. Germline editing can permanently change the human gene pool, posing concerns about permission, justice, and human identity, whereas somatic cell editing only affects the treated individual (Lanphier et al., 2015). When Chinese scientist He Jiankui revealed the biological twin girls whose embryos had been altered using CRISPR to delete the CCR5 gene, supposedly imparting resistance to HIV, the issue reached a climax in 2018 (Cyranoski, 2019). The experiment was deemed premature and unethical by the international scientific community because it lacked sufficient safety data, was opaque, and violated medical ethics. The event made clear how urgently strong international governance is needed. Since then, groups including national bioethics councils, the International Society for Stem Cell Research (ISSCR), and the World Health Organisation (WHO) have suggested halting clinical germline editing (WHO, 2021). These frameworks place a strong emphasis on openness, involvement of the public, and diverse separation of enhancement and therapeutic uses. The ethical distinction between treatment and improvement is still hazy, despite the possibility that embryo editing might eradicate serious genetic disorders. "Designer babies," genetic inequity, and the commercialisation of human characteristics are among the concerns raised by opponents (Baltimore et al., 2015). Advocates contend that fair admittance and ethical supervision might stop abuse while facilitating acceptable medical progressions. In the end, most people agree that somatic editing is preferable for the time being, whereas germline treatments are still only used in preclinical or research situations.

Global regulations and policy frameworks: The regulatory environment about CRISPR-Cas9 differs greatly between nations, reflecting differences in ethics, law, and culture. Genome-editing treatments are regulated as biologic pharmaceuticals in the US by the Food and Drug Administration (FDA), which has strict preclinical safety data requirements and requires Investigational New Drug (IND) applications. Funding for human embryo editing research is likewise restricted by the National Institutes of Health (NIH).

On the other hand, genome-edited creatures are categorised as genetically modified organisms (GMOs) by the European Union (EU) in accordance with Directive 2001/18/EC, which requires a thorough ethical and environmental assessment. Nonetheless, a few European countries, including the UK, allow strictly controlled CRISPR research on human embryos for developmental purposes up to 14 days (Dimond & Stephens, 2018).

After the 2018 embryo-editing debacle, China has strengthened its regulatory monitoring despite fast scientific advancements. All human genome-editing research currently requires prior governmental clearance, as required by the Biosecurity Law of 2020 and connected ethical guidelines. A unified governance paradigm is still unattainable on a global scale. To increase openness and foster global collaboration, the WHO's 2021 Advisory Committee on Human Genome Editing suggested creating a worldwide register of genome-editing research (WHO, 2021). However, differences in national laws might open the door for "ethics tourism," in which researchers move to countries with laxer laws.

In conclusion, CRISPR-Cas9 challenges humankind to rethink the ethical and legal bounds of science even as it holds the potential to revolutionise medicine. For CRISPR to be safely and fairly transferred from laboratory innovation to global healthcare, a cooperative, evidence-based, and ethically informed framework will be necessary.

Table 6. Ethical and Regulatory Status of CRISPR Human Genome Editing

Region / Country	Regulatory Authority	Current Status on Germline Editing	Somatic Cell Editing Status	Remarks / Key Policies
United States	FDA / NIH	Prohibited for clinical use	Permitted under IND and FDA oversight	Federal funding ban on embryo editing.
European Union	EMA / National agencies	Strictly prohibited (Directive 2001/18/EC)	Permitted for somatic therapy trials	The EU treats genome editing as GMO.
United Kingdom	HFEA	Research allowed up to 14 days (no implantation)	Permitted	Strong public oversight mechanisms.
China	National Health Commission	Prohibited without prior approval	Permitted with biosecurity regulation	Tightened laws after the 2018 embryo scandal.

continued on following page

Table 6. Continued

Region / Country	Regulatory Authority	Current Status on Germline Editing	Somatic Cell Editing Status	Remarks / Key Policies
Japan	MEXT / MHLW	Prohibited for reproduction	Allowed under clinical ethics review	Focus on somatic safety and transparency.
India	DBT / ICMR	No germline editing permitted	Ex vivo somatic research allowed	Guidelines under the National Biomedical Policy.

FUTURE DIRECTIONS AND INNOVATIONS

The future of CRISPR technology is at the nexus of synthetic biology, data science, and molecular biology as it develops. Advances are broadening the use of CRISPR beyond gene disruption or repair to encompass precise control, disease modelling, and biological engineering at the systems level.

AI in guide RNA design: Machine learning (ML) and artificial intelligence (AI) are now essential techniques for maximising CRISPR specificity and efficiency. Neural networks trained on massive datasets are used by predictive algorithms like DeepCRISPR, CRISPR-Net, and sgRNA-Designer to forecast sgRNA activity and off-target potential (Chuai et al., 2018). These instruments combine thermodynamic factors, chromatin accessibility, and sequence context to produce extremely effective guide RNAs with little danger. Researchers can now predict possible indels or HDR efficiencies by using AI to model repair outcomes after DSBs and simulate Cas protein–DNA interactions (Kim et al., 2020). AI-driven CRISPR design has potential for enhancing patient safety and speeding up therapeutic discovery cycles by combining in silico modelling with wet-lab testing. AI systems may enable personalised gene editing in future therapeutic applications, allowing sgRNA sequences to be customised to each patient's distinct genome and variation profile. With this level of accuracy, personalised genomic therapy would replace generalised editing techniques.

CRISPR and synthetic biology: To generate new cell functionalities, synthetic biology combines biological circuit design with CRISPR-based methods. One such approach is CRISPR-based gene circuits, which employ programmable nucleases to control metabolic processes, biosensors, and intricate gene networks (Nielsen & Keasling, 2016). For example, CRISPR has been integrated into logic gates and synthetic promoters to allow conditional gene expression in response to metabolic or environmental changes. These systems are further used in cellular treatments, environmental cleanup, and biomanufacturing. CRISPR-controlled pathways in engineered bacterial strains can maximise the production of industrial enzymes, biofuels, and medicines. Likewise, CRISPR-based biosensors have been created to quickly identify genetic indicators, environmental pollutants, and infections (Gootenberg et al., 2017). Smart therapies, or cells designed to recognise illness situations and issue therapeutic chemicals on their own, are a promising medical usage of CRISPR-driven synthetic biology. To reduce toxicity, modified T cells can be made to identify tumour antigens and self-regulate the release of cytokines (Nielsen & Keasling, 2016). These hybrid technologies show how CRISPR could be used in next-generation biomedicine as a therapeutic and diagnostic tool.

Potential for polygenic disorders: Many of the most common human ailments, including diabetes, cardiovascular disease, and neuropsychiatric disorders, are polygenic in origin, with difficult interactions among numerous genes and environmental variables, even though current CRISPR applications generally target monogenic diseases. The next area of CRISPR research will focus on treating these conditions (Maeder & Gersbach, 2016). One possible path towards polygenic correction is the creation of multiplexed CRISPR systems, which can edit numerous genomic locations at once. Instead of fixing individual mutations, CRISPRa and CRISPRi technologies may potentially modify gene networks to restore equilibrium across dysregulated pathways (Liu et al., 2016). Systems-level understandings into disease progressions are made possible by merging CRISPR with multi-omics data (genomics, transcriptomics, proteomics, and metabolomics). Researchers may create CRISPR treatments that rewire cellular networks towards healthy phenotypes by finding crucial gene combinations and regulatory hubs. These advances herald a revolutionary age in network medicine, where CRISPR helps as a therapeutic platform and research tool to decode and modify biological complexity, even though the practical realisation of polygenic editing is still a way off.

CONCLUSION

The development of CRISPR-Cas9 technology has expanded the area of genetics and turned theoretical gene therapy into practical medicinal applications. CRISPR is one of the most momentous scientific revolutions of the twenty-first century, having evolved from its modest beginnings as a bacterial immune system to its current position as the foundation of precision medicine.

CRISPR-Cas9 has the rare capacity to fix mutations at their cause in the setting of genetic illnesses, opening the door to likely curative therapies as opposed to symptomatic treatment. Its effectiveness in treating conditions including Duchenne muscular dystrophy, β-thalassemia, sickle cell anaemia, and cystic fibrosis shows that genome editing is feasible and long-lasting in therapeutic circumstances. Its accuracy and reach are still being increased by new techniques, including base editing, prime editing, and epigenetic control. The journey to clinical maturity is still fraught with difficulties, though. Continued research and regulation are needed due to off-target effects, delivery inefficiencies, immunological reactions, and ethical difficulties. The debates over germline editing serve as a reminder that ethical and societal duty must always drive technical development. In the future, merging CRISPR with synthetic biology, AI-driven analytics, and multi-omic data forces led to new developments like programmable cellular systems and personalised gene treatment. These multidisciplinary developments will probably impact a future where genetic diseases may be essentially avoided rather than just treated.

CRISPR-Cas9 is a technique with enormous therapeutic potential, but it also serves as a cue of the great responsibility that comes with having the ability to change humanity's genetic destiny. An assembly confirming that this change proceeds with honesty, equality, and concern is a challenge for equally society and science.

REFERENCES

Amoasii, L., Long, C., Li, H., Mireault, A. A., Shelton, J. M., Sanchez-Ortiz, E., & Olson, E. N. (2018). Single-cut genome editing restores dystrophin expression in a new mouse model of muscular dystrophy. *Science Translational Medicine*, *10*(437), eaap8415. PMID: 29187645

Anzalone, A. V., Randolph, P. B., Davis, J. R., Sousa, A. A., Koblan, L. W., Levy, J. M., Chen, P. J., Wilson, C., Newby, G. A., Raguram, A., & Liu, D. R. (2019). Search-and-replace genome editing without double-strand breaks or donor DNA. *Nature*, *576*(7785), 149–157. DOI: 10.1038/s41586-019-1711-4 PMID: 31634902

Baltimore, D., Berg, P., Botchan, M., Carroll, D., Charo, R. A., Church, G., & Weissman, J. S. (2015). A prudent path forward for genomic engineering and germline gene modification. *Science*, *348*(6230), 36–38. DOI: 10.1126/science.aab1028 PMID: 25791083

Barrangou, R., & Doudna, J. A. (2016). Applications of CRISPR technologies in research and beyond. *Nature Biotechnology*, *34*(9), 933–941. DOI: 10.1038/nbt.3659 PMID: 27606440

Barrangou, R., Fremaux, C., Deveau, H., Richards, M., Boyaval, P., Moineau, S., Romero, D. A., & Horvath, P. (2007). CRISPR provides acquired resistance against viruses in prokaryotes. *Science*, *315*(5819), 1709–1712. DOI: 10.1126/science.1138140 PMID: 17379808

Bueren, J. A., Quintana-Bustamante, O., Almarza, E., Navarro, S., Río, P., Segovia, J. C., & Guenechea, G. (2020). Advances in the gene therapy of monogenic blood cell diseases. *Clinical Genetics*, *97*(1), 89–102. DOI: 10.1111/cge.13593 PMID: 31231794

Chuai, G., Ma, H., Yan, J., Chen, M., Hong, N., Xue, D., Zhou, C., Zhu, C., Chen, K., Duan, B., Gu, F., Qu, S., Huang, D., Wei, J., & Liu, Q. (2018). DeepCRISPR: Optimized CRISPR guide RNA design by deep learning. *Genome Biology*, *19*(1), 80. DOI: 10.1186/s13059-018-1459-4 PMID: 29945655

Cong, L., Ran, F. A., Cox, D., Lin, S., Barretto, R., Habib, N., Hsu, P. D., Wu, X., Jiang, W., Marraffini, L. A., & Zhang, F. (2013). Multiplex genome engineering using CRISPR/Cas systems. *Science*, *339*(6121), 819–823. DOI: 10.1126/science.1231143 PMID: 23287718

Cox, D. B. T., Platt, R. J., & Zhang, F. (2015). Therapeutic genome editing: Prospects and challenges. *Nature Medicine*, *21*(2), 121–131. DOI: 10.1038/nm.3793 PMID: 25654603

Cyranoski, D. (2019). The CRISPR-baby scandal: What's next for human gene-editing. *Nature, 566*(7745), 440–442. DOI: 10.1038/d41586-019-00673-1 PMID: 30809070

Dimond, R., & Stephens, N. (2018). Policy Work and Legitimacy at the Nuffield Council on Bioethics, the Human Fertilisation and Embryology Authority, and the Department of Health. In *Legalising Mitochondrial Donation: Enacting Ethical Futures in UK Biomedical Politics* (pp. 69–88). Springer International Publishing. DOI: 10.1007/978-3-319-74645-6_4

Doudna, J. A., & Charpentier, E. (2014). The new frontier of genome engineering with CRISPR-Cas9. *Science, 346*(6213), 1258096. DOI: 10.1126/science.1258096 PMID: 25430774

Eyquem, J., Mansilla-Soto, J., Giavridis, T., Van Der Stegen, S. J., Hamieh, M., Cunanan, K. M., Odak, A., Gönen, M., & Sadelain, M. (2017). Targeting a CAR to the TRAC locus with CRISPR/Cas9 enhances tumour rejection. *Nature, 543*(7643), 113–117. DOI: 10.1038/nature21405 PMID: 28225754

Frangoul, H., Altshuler, D., Cappellini, M. D., Chen, Y.-S., Domm, J., Eustace, B. K., Foell, J., de la Fuente, J., Grupp, S., Handgretinger, R., Ho, T. W., Kattamis, A., Kernytsky, A., Lekstrom-Himes, J., Li, A. M., Locatelli, F., Mapara, M. Y., de Montalembert, M., Rondelli, D., & Corbacioglu, S. (2021). CRISPR-Cas9 gene editing for sickle cell disease and β-thalassemia. *The New England Journal of Medicine, 384*(3), 252–260. DOI: 10.1056/NEJMoa2031054 PMID: 33283989

Fu, Y., Foden, J. A., Khayter, C., Maeder, M. L., Reyon, D., Joung, J. K., & Sander, J. D. (2013). High-frequency off-target mutagenesis induced by CRISPR-Cas nucleases in human cells. *Nature Biotechnology, 31*(9), 822–826. DOI: 10.1038/nbt.2623 PMID: 23792628

Gaudelli, N. M., Komor, A. C., Rees, H. A., Packer, M. S., Badran, A. H., Bryson, D. I., & Liu, D. R. (2017). Programmable base editing of A• T to G• C in genomic DNA without DNA cleavage. *Nature, 551*(7681), 464–471. DOI: 10.1038/nature24644 PMID: 29160308

Gilbert, L. A., Horlbeck, M. A., Adamson, B., Villalta, J. E., Chen, Y., Whitehead, E. H., Guimaraes, C., Panning, B., Ploegh, H. L., Bassik, M. C., Qi, L. S., Kampmann, M., & Weissman, J. S. (2014). Genome-scale CRISPR-mediated control of gene repression and activation. *Cell, 159*(3), 647–661. DOI: 10.1016/j.cell.2014.09.029 PMID: 25307932

Gillmore, J. D., Gane, E., Taubel, J., Kao, J., Fontana, M., Maitland, M. L., Seitzer, J., O'Connell, D., Walsh, K. R., Wood, K., Phillips, J., Xu, Y., Amaral, A., Boyd, A. P., Cehelsky, J. E., McKee, M. D., Schiermeier, A., Harari, O., Murphy, A., & Lebwohl, D. (2021). CRISPR-Cas9 in vivo gene editing for transthyretin amyloidosis. *The New England Journal of Medicine*, *385*(6), 493–502. DOI: 10.1056/NEJMoa2107454 PMID: 34215024

Gootenberg, J. S., Abudayyeh, O. O., Kellner, M. J., Joung, J., Collins, J. J., & Zhang, F. (2017). Multiplexed and portable nucleic acid detection platform with Cas13, Cas12a, and Csm6. *Science*, *360*(6387), 439–444. DOI: 10.1126/science.aaq0179 PMID: 29449508

Ishino, Y., Shinagawa, H., Makino, K., Amemura, M., & Nakata, A. (1987). Nucleotide sequence of the iap gene, responsible for alkaline phosphatase isozyme conversion in Escherichia coli, and identification of the gene product. *Journal of Bacteriology*, *169*(12), 5429–5433. DOI: 10.1128/jb.169.12.5429-5433.1987 PMID: 3316184

Jansen, R., Embden, J. D. V., Gaastra, W., & Schouls, L. M. (2002). Identification of genes that are associated with DNA repeats in prokaryotes. *Molecular Microbiology*, *43*(6), 1565–1575. DOI: 10.1046/j.1365-2958.2002.02839.x PMID: 11952905

. Jinek, M., Chylinski, K., Fonfara, I., Hauer, M., Doudna, J. A., & Charpentier, E. (2012). A programmable dual-RNA–guided DNA endonuclease in adaptive bacterial immunity. *science*, *337*(6096), 816-821.

Kim, H., Lee, W. J., Oh, Y., Kang, S. H., Hur, J. K., Lee, H., Song, W. J., Lim, K.-S., Park, Y.-H., Song, B.-S., Jin, Y. B., Jun, B.-H., Jung, C., Lee, D.-S., Kim, S.-U., & Lee, S. H. (2020). Enhancement of target specificity of CRISPR–Cas12a by using a chimeric DNA–RNA guide. *Nucleic Acids Research*, *48*(15), 8601–8616. DOI: 10.1093/nar/gkaa605 PMID: 32687187

Komor, A. C., Kim, Y. B., Packer, M. S., Zuris, J. A., & Liu, D. R. (2016). Programmable editing of a target base in genomic DNA without double-stranded DNA cleavage. *Nature*, *533*(7603), 420–424. DOI: 10.1038/nature17946 PMID: 27096365

Kotit, S. (2023). Lessons from the first-in-human in vivo CRISPR/Cas9 editing of the TTR gene by NTLA-2001 trial in patients with transthyretin amyloidosis with cardiomyopathy. *Global Cardiology Science & Practice*, *2023*(1), e202304. DOI: 10.21542/gcsp.2023.4 PMID: 37928601

Lanphier, E., Urnov, F., Haecker, S. E., Werner, M., & Smolenski, J. (2015). Don't edit the human germ line. *Nature*, *519*(7544), 410–411. DOI: 10.1038/519410a PMID: 25810189

Lee, S., Kim, T., Lee, E., Lee, C., Kim, H., Rhee, H., Park, S. Y., Son, H.-J., Yu, S., Park, J. W., Choo, E. J., Park, S., Loeb, M., & Kim, T. H. (2020). Clinical course and molecular viral shedding among asymptomatic and symptomatic patients with SARS-CoV-2 infection in a community treatment center in the Republic of Korea. *JAMA Internal Medicine, 180*(11), 1447–1452. DOI: 10.1001/jamainternmed.2020.3862 PMID: 32780793

Lin, Y., Zhu, Y., Cui, Y., Chen, R., Chen, Z., Li, G., Fan, M., Chen, J., Li, Y., Guo, X., Zheng, X., Chen, L., & Wang, F. (2021). Derepression of specific miRNA-target genes in rice using CRISPR/Cas9. *Journal of Experimental Botany, 72*(20), 7067–7077. DOI: 10.1093/jxb/erab336 PMID: 34283216

Liu, X. S., Wu, H., Ji, X., Stelzer, Y., Wu, X., Czauderna, S., Shu, J., Dadon, D., Young, R. A., & Jaenisch, R. (2016). Editing DNA methylation in the mammalian genome. *Cell, 167*(1), 233–247. DOI: 10.1016/j.cell.2016.08.056 PMID: 27662091

Maeder, M. L., & Gersbach, C. A. (2016). Genome-editing technologies for gene and cell therapy. *Molecular Therapy : the Journal of the American Society of Gene Therapy, 24*(3), 430–446. DOI: 10.1038/mt.2016.10 PMID: 26755333

Makarova, K. S., Haft, D. H., Barrangou, R., Brouns, S. J., Charpentier, E., Horvath, P., Moineau, S., Mojica, F. J. M., Wolf, Y. I., Yakunin, A. F., van der Oost, J., & Koonin, E. V. (2011). Evolution and classification of the CRISPR–Cas systems. *Nature Reviews. Microbiology, 9*(6), 467–477. DOI: 10.1038/nrmicro2577 PMID: 21552286

Mendell, J. R., Al-Zaidy, S. A., Rodino-Klapac, L. R., Goodspeed, K., Gray, S. J., Kay, C. N., Boye, S. L., Boye, S. E., George, L. A., Salabarria, S., Corti, M., Byrne, B. J., & Tremblay, J. P. (2021). Current clinical applications of in vivo gene therapy with AAVs. *Molecular Therapy : the Journal of the American Society of Gene Therapy, 29*(2), 464–488. DOI: 10.1016/j.ymthe.2020.12.007 PMID: 33309881

Mojica, F. J., Díez-Villaseñor, C. S., García-Martínez, J., & Soria, E. (2005). Intervening sequences of regularly spaced prokaryotic repeats derive from foreign genetic elements. *Journal of Molecular Evolution, 60*(2), 174–182. DOI: 10.1007/s00239-004-0046-3 PMID: 15791728

Nelson, C. E., Wu, Y., Gemberling, M. P., Oliver, M. L., Waller, M. A., Bohning, J. D., Robinson-Hamm, J. N., Bulaklak, K., Castellanos Rivera, R. M., Collier, J. H., Asokan, A., & Gersbach, C. A. (2019). Long-term evaluation of AAV-CRISPR genome editing for Duchenne muscular dystrophy. *Nature Medicine, 25*(3), 427–432. DOI: 10.1038/s41591-019-0344-3 PMID: 30778238

Nielsen, A. A., & Keasling, J. D. (2016). Engineering cellular metabolism. *Cell, 164*(6), 1185–1197. DOI: 10.1016/j.cell.2016.02.004 PMID: 26967285

Park, S. H., & Bao, G. (2021). CRISPR/Cas9 gene editing for curing sickle cell disease. *Transfusion and Apheresis Science : Official Journal of the World Apheresis Association : Official Journal of the European Society for Haemapheresis, 60*(1), 103060. DOI: 10.1016/j.transci.2021.103060 PMID: 33455878

Rees, H. A., & Liu, D. R. (2018). Base editing: Precision chemistry on the genome and transcriptome of living cells. *Nature Reviews. Genetics, 19*(12), 770–788. DOI: 10.1038/s41576-018-0059-1 PMID: 30323312

Schwank, G., Koo, B. K., Sasselli, V., Dekkers, J. F., Heo, I., Demircan, T., Sasaki, N., Boymans, S., Cuppen, E., van der Ent, C. K., Nieuwenhuis, E. E. S., Beekman, J. M., & Clevers, H. (2013). Functional repair of CFTR by CRISPR/Cas9 in intestinal stem cell organoids of cystic fibrosis patients. *Cell Stem Cell, 13*(6), 653–658. DOI: 10.1016/j.stem.2013.11.002 PMID: 24315439

Slaymaker, I. M., Gao, L., Zetsche, B., Scott, D. A., Yan, W. X., & Zhang, F. (2016). Rationally engineered Cas9 nucleases with improved specificity. *Science, 351*(6268), 84–88. DOI: 10.1126/science.aad5227 PMID: 26628643

Stadtmauer, E. A., Fraietta, J. A., Davis, M. M., Cohen, A. D., Weber, K. L., Lancaster, E., Mangan, P. A., Kulikovskaya, I., Gupta, M., Chen, F., Tian, L., Gonzalez, V. E., Xu, J., Jung, I., Melenhorst, J. J., Plesa, G., Shea, J., Matlawski, T., Cervini, A., & June, C. H. (2020). CRISPR-engineered T cells in patients with refractory cancer. *Science, 367*(6481), eaba7365. DOI: 10.1126/science.aba7365 PMID: 32029687

Thakore, P. I., D'ippolito, A. M., Song, L., Safi, A., Shivakumar, N. K., Kabadi, A. M., Reddy, T. E., Crawford, G. E., & Gersbach, C. A. (2015). Highly specific epigenome editing by CRISPR-Cas9 repressors for silencing of distal regulatory elements. *Nature Methods, 12*(12), 1143–1149. DOI: 10.1038/nmeth.3630 PMID: 26501517

Wang, D., Zhang, F., & Gao, G. (2020). CRISPR-based therapeutic genome editing: Strategies and in vivo delivery by AAV vectors. *Cell, 181*(1), 136–150. DOI: 10.1016/j.cell.2020.03.023 PMID: 32243786

World Health Organization (WHO). (2021). *Human genome editing: Recommendations*. WHO.

Yang, S., Chang, R., Yang, H., Zhao, T., Hong, Y., Kong, H. E., Sun, X., Qin, Z., Jin, P., Li, S., & Li, X. J. (2017). CRISPR/Cas9-mediated gene editing ameliorates neurotoxicity in mouse model of Huntington's disease. *The Journal of Clinical Investigation*, *127*(7), 2719–2724. DOI: 10.1172/JCI92087 PMID: 28628038

Zuccaro, M. V., Xu, J., Mitchell, C., Marin, D., Zimmerman, R., Rana, B., Weinstein, E., King, R. T., Palmerola, K. L., Smith, M. E., Tsang, S. H., Goland, R., Jasin, M., Lobo, R., Treff, N., & Egli, D. (2020). Allele-specific chromosome removal after Cas9 cleavage in human embryos. *Cell*, *183*(6), 1650–1664.e15. DOI: 10.1016/j.cell.2020.10.025 PMID: 33125898

Chapter 6
CRISPR-Cas9 Technology in Regenerative Medicine and Tissue Engineering

Arthi Gunasekaran
https://orcid.org/0009-0002-7355-6387
Bharathiar University, India

Muthupandi Vanaraj
Bharathiar University, India

Maria Sneha Philip Kumar
RVS College of Nursing, India

Abidharini Jothi Dheivasikamani
Bharathiar University, India

Arthi Boro
Bharathiar University, India

Antony Prabhu Jeyabal Philomenathan
Bharathiar University, India

Alwin Robert Asirvatham
Prince Sultan Military Medical City, Ministry of Defense, Saudi Arabia

Vijaya Anand Arumugam
https://orcid.org/0000-0001-7485-1586
Bharathiar University, India

ABSTRACT

CRISPR/Cas9 technology has emerged as a revolutionary genome editing tool, offering unprecedented precision, efficiency, and versatility in manipulating genetic material. In the fields of regenerative medicine and tissue engineering, CRISPR/Cas9 holds transformative potential for correcting genetic defects, enhancing stem cell therapies, engineering tissue specific cellular functions, and developing disease models. This chapter explores the fundamental mechanisms of CRISPR/Cas9, its integration into regenerative strategies and its application in engineering functional tissues and organs. Key advancements include targeted gene editing in induced pluripotent stem cells (iPSCs), modulation of signaling pathways involved in tissue regeneration and creation of gene-edited scaffolds for organogenesis. The chapter

DOI: 10.4018/979-8-3373-4857-5.ch006

also addresses the current challenges, such as off-target effects, ethical considerations, and regulatory hurdles, while highlighting emerging innovations and future directions for clinical translation. Overall, CRISPR/Cas9 stands at the forefront of next-generation therapeutic strategies, reshaping the landscape of personalized and regenerative medicine.

1. INTRODUCTION TO CRISPR/CAS SYSTEMS

The CRISPR (Clustered Regularly Interspaced Short Palindromic Repeats)/Cas (CRISPR-associated) system was first discovered in Escherichia coli in 1987. However, its role as part of a prokaryotic immune defense was only fully understood in the early 2000s (Barrangou et al., 2007). This defense system allows bacteria to detect and neutralize viral threats by integrating pieces of viral DNA into their genome as "spacers." Later, these segments are transcribed into CRISPR RNAs (crRNAs). Along with Cas proteins, these RNAs guide the destruction of matching viral DNA during reinfection. The groundbreaking study by Jinek et al. (2012) showed that this system could be modified for programmable genome editing in eukaryotic cells, marking the emergence of CRISPR/Cas9 as a revolutionary gene-editing tool.

The main components of CRISPR/Cas9 include a guide RNA (gRNA), the Cas9 nuclease, and a protospacer adjacent motif (PAM). The gRNA, which is usually a combination of crRNA and trans-activating crRNA (tracrRNA), directs Cas9 to a specific DNA sequence. The PAM (usually NGG) helps to enable accurate cleavage (Chen & Doudna, 2017). This creates a double-strand break (DSB) that is repaired through cellular DNA repair pathways. These can be either non-homologous end joining (NHEJ) or homology-directed repair (HDR), allowing for gene disruptions or precise genetic changes. In addition to Cas9, the CRISPR toolbox now features several alternative nucleases with different characteristics. For instance, Cas12a (formerly Cpf1) recognizes thymine-rich PAM sequences (TTTV). It creates staggered cuts and operates using a single crRNA without needing tracrRNA (Shmakov et al., 2017). Furthermore, CasΦ (also called Cas14), obtained from bacteriophages, is a small nuclease that cuts single-stranded DNA without the need for PAM sequences. This is beneficial for gene editing in situations where delivery is limited (Pausch et al., 2020).

Despite its flexibility, the CRISPR/Cas9 system has some limitations. Off-target effects from imperfect gRNA-DNA binding can cause unintentional changes in the genome (Kulcsár et al., 2017). To improve accuracy, modified high-fidelity Cas9 variants like SpCas9-HF1, eSpCas9, and xCas9 have been created. These variants have point mutations that reduce off-target cuts while keeping on-target activity intact. Additionally, new technologies like base editing and prime editing allow for specific nucleotide changes and small sequence insertions without creating DSBs, lowering the risk of genetic damage (Anzalone et al., 2020; Pickar-Oliver & Gersbach, 2019). In summary, CRISPR/Cas systems have evolved from a microbial defense mechanism to a valuable tool for genome engineering. The growth of Cas variants and precision-editing techniques is revolutionizing their use in biomedical research, particularly in regenerative medicine and tissue engineering.

2. PRECISION GENE EDITING TOOLS: BEYOND CAS9

The introduction of CRISPR/Cas9 technology has made great strides in genome engineering. It allows for precise DNA cutting and repair using cellular pathways. However, because it relies on double-stranded DNA breaks (DSBs), needs donor DNA in some cases, and has the potential for off-target effects, researchers have developed more refined genome-editing tools. Two noteworthy advancements are base editing and prime editing. Both techniques enable highly accurate genetic changes without causing DSBs. These innovations offer new opportunities in therapeutic genome editing, particularly in regenerative medicine.

2.1 Base Editors: CBEs and ABEs for Precise Single-Nucleotide Modifications

Base editors are molecular tools made by connecting a deaminase enzyme to a chemically impaired Cas9 protein. This can either be in its nickase (nCas9) or dead (dCas9) form. This setup allows for the conversion of individual DNA bases without breaking the DNA strand or needing a template for homology-directed repair (HDR) (Komor et al., 2016). There are two main types of base editors:

i. Cytosine base editors (CBEs) convert cytosine-guanine (C-G) base pairs into thymine-adenine (T-A) through cytosine deamination.
ii. Adenine base editors (ABEs) change adenine-thymine (A-T) base pairs into guanine-cytosine (G-C) by converting adenine to inosine, which is then read as guanine during DNA replication (Anzalone et al., 2020).

Base editors work well for fixing point mutations, which make up a large portion of known genetic disorders. Unlike traditional CRISPR/Cas9, which depends on HDR an inefficient process in non-dividing or terminally differentiated cells base editors can function at any stage of the cell cycle. This makes them well-suited for treating tissues like the nervous system or skeletal muscle, which generally have low regenerative abilities. In a key study, Komor et al. (2016) showed that CBEs could introduce specific cytosine-to-thymine mutations in mammalian cells without causing DSBs. Since then, improvements in design and enzyme specificity have enhanced the precision and safety of base editors. Anzalone et al. (2020) also demonstrated successful use of base editors in correcting disease-causing mutations in models of sickle cell disease and Duchenne muscular dystrophy (DMD).

2.2 Prime Editing: A Versatile Platform for Complex Genome Edits

While base editors excel at transition mutations, their editing abilities are limited. To overcome this, a more adaptable approach called prime editing has been created. This system combines a Cas9 nickase with a reverse transcriptase enzyme and a specialized RNA molecule known as a prime editing guide RNA (pegRNA). The pegRNA identifies both the genomic target site and the desired sequence change (Anzalone et al., 2019).

Prime editing can perform all 12 base substitutions as well as small insertions and deletions without causing DSBs or needing external DNA templates. Its versatility and accuracy make it one of the safest and most advanced genome-editing methods available. Li et al. (2023) highlighted the effectiveness of prime editing in correcting mutations in several important genes, including HBB, CFTR, and ATP7B. Their results suggest that prime editing could be crucial in developing treatments for a wide range of genetic disorders, especially where maintaining genomic stability is vital for long-term success.

2.3 Applications in Regenerative Models and Disease Correction

Base and prime editing technologies have shown promise in various preclinical models relevant to regenerative medicine, including muscle disorders, skin diseases, and stem cell therapies. Their ability to accurately correct harmful variants in cells taken from patients opens up avenues for personalized treatment strategies.

a) Duchenne Muscular Dystrophy (DMD)

DMD is a progressive muscle-degenerating disorder caused by mutations in the DMD gene. This results in missing functional dystrophin protein. Chemello et al. (2021) used ABEs to fix specific point mutations in cardiomyocytes from induced pluripotent stem cells (iPSCs) of DMD patients. This led to the restoration of dystrophin expression and improved cell performance, demonstrating how base editing can help reverse disease effects. In another study, Anzalone et al. (2019) used prime editing to fix mutations in the DMD gene, restoring full-length dystrophin in human cells. This method, which avoids creating DSBs, offers a safer and potentially long-lasting treatment option for muscle-related genetic conditions.

b) Epidermolysis Bullosa (EB)

Epidermolysis bullosa is a collection of inherited skin disorders that often stem from mutations in the COL7A1 gene. Osborn et al. (2021) used ABE technology on fibroblasts from EB patients to correct harmful nonsense mutations. The base editing system achieved correction rates of up to 45% and successfully restored collagen VII expression, presenting a possible solution for skin repair and regeneration in EB treatment.

c) Human Pluripotent Stem Cells (hPSCs)

Base editors have also been used in hPSCs to model genetic diseases and investigate correction strategies. Yuan et al. (2021) illustrated how ABEs corrected mutations in genes like HBB and G6PD, which relate to β-thalassemia and enzymatic deficiencies, respectively. The edited stem cells continued to self-renew and differentiate into various types, underscoring their potential in regenerative research and future medical applications. Together, these examples highlight the promise of base and prime editing for creating tailored therapies that target specific mutations while minimizing disruption to the genome.

2.4 Comparative Advantages Over Conventional CRISPR/Cas9

Base and prime editors provide several key advantages over traditional CRISPR/Cas9 methods. Both tools remove the need for DSBs, lowering the risks of significant genomic rearrangements, off-target insertions and deletions, and p53 activation. Moreover, they avoid the HDR pathway, allowing effective editing in both dividing and non-dividing cells. This is especially important in regenerative medicine, where many target tissues have limited growth potential.

In regenerative medicine, where ensuring long-term effectiveness, low immune response, and genomic safety are crucial, base and prime editors provide exceptional precision and control. Their capacity to edit genetic sequences with minimal collateral damage positions them as strong candidates for future clinical applications in cell and gene therapies.

3. CRISPR IN INDUCED PLURIPOTENT STEM CELL (IPSC) ENGINEERING

Recent developments in CRISPR genome editing have dramatically advanced the capabilities of induced pluripotent stem cell (iPSC) technologies. The integration of CRISPR into iPSC workflows has enabled precise gene correction, safer and more efficient reprogramming, controlled gene insertion, and lineage-directed differentiation. These applications are transforming the field of regenerative medicine, creating powerful avenues for patient-specific, disease-free, and functional cell-based therapies.

3.1 Reprogramming Somatic Cells via CRISPR Activation (CRISPRa)

Traditional iPSC generation relies on the introduction of exogenous Yamanaka transcription factors (OCT4, SOX2, KLF4, and MYC) into somatic cells. Although widely used, this approach carries risks such as unpredictable expression, low efficiency, and potential insertional mutagenesis. CRISPR activation (CRISPRa) offers a cleaner alternative by using a nuclease-inactive dCas9 fused to transcriptional activators to endogenously activate pluripotency genes. CRISPRa-based reprogramming allows for controlled, integration-free activation of key stemness genes, promoting more consistent and homogeneous iPSC populations. As demonstrated by Sokka et al. (2022), targeting endogenous genes and regulatory elements such as miR-302/367 with CRISPRa enhanced reprogramming fidelity and yielded iPSCs with transcriptomic profiles closely matching standard pluripotent benchmarks.

3.2 Targeted Integration into Safe Harbor Sites

Stable gene expression in iPSCs is essential for regenerative therapies. Random integration, often associated with viral vectors, can disrupt essential genes or lead to silencing. CRISPR-based targeted integration into "safe harbor" loci like AAVS1, ROSA26, or H11 offers a safer and more predictable solution. For instance, Vlassis et al. (2023) utilized Cas9 and MAD7 nucleases to insert transgenes into these loci, maintaining robust expression across differentiation stages. Bharucha et al. (2021) similarly demonstrated successful inducible expression of a prime editing system integrated at AAVS1, enabling precise temporal control over gene editing. These studies underscore the importance of targeted, stable integration in ensuring safety and functional consistency.

3.3 Generation of Disease-Free Autologous iPSCs

Combining CRISPR gene correction with iPSC technology allows the creation of autologous, mutation-free stem cell lines. This approach eliminates the need for immune suppression in transplantation and provides a renewable source of corrected cells. In a study by Ou et al. (2016), patient-derived iPSCs with a mutation in the HBB gene (linked to β-thalassemia) were corrected using CRISPR/Cas9-mediated homologous recombination. The corrected iPSCs differentiated into functional hematopoietic stem cells that expressed normal β-globin and successfully engrafted in model organisms. This exemplifies the clinical potential of CRISPR-corrected iPSCs for personalized gene therapy.

3.4 Enhancing Lineage-Specific Differentiation

Precise control of iPSC differentiation is crucial for generating therapeutically relevant cells. CRISPR has been used to insert lineage-determining transcription factors into safe harbor sites under inducible control, allowing directed differentiation into specific cell types. For example, Davis-Anderson et al. (2023) inserted ISL1 and LHX3 into human iPSCs to drive motor neuron differentiation under doxycycline control. While some off-target effects were observed, the system allowed rapid and efficient neural differentiation. Hendriks et al. (2020) highlighted how CRISPRa and base editing can further refine differentiation protocols by modulating expression at critical developmental checkpoints.

3.5 Therapeutic Integration and Future Outlook

The modular integration of CRISPR technologies across iPSC workflows provides a robust framework for next-generation regenerative medicine. CRISPRa enables non-integrative, high-fidelity reprogramming; safe harbor targeting ensures stable transgene expression; CRISPR-mediated gene correction facilitates personalized, mutation-free therapies; and inducible differentiation systems offer scalable solutions for producing clinically relevant cell types. Together, these approaches enable the creation of genetically precise, patient-compatible, and functionally tailored cell products, accelerating the path to safe and effective stem cell-based interventions.

4. GENOME AND EPIGENOME EDITING FOR CELL REPROGRAMMING5. CRISPR-ENGINEERED ORGANOIDS AND TISSUE CONSTRUCTS

The combination of genome and epigenome editing technologies has notably changed regenerative medicine, especially in reprogramming cell identity. Classical CRISPR/Cas9-based genome editing allows for permanent changes to the DNA sequence. In contrast, epigenome editing provides a groundbreaking alternative by enabling temporary and reversible control of gene activity without altering the DNA itself. This capability is particularly useful in cell conversion scenarios, such as transforming fibroblasts into myocytes or reprogramming astrocytes into neurons. Here, the short-term expression of lineage-specific genes can be enough to trigger the desired cell types.

At the heart of epigenome editing is a modified version of Cas9 called "dead Cas9" (dCas9). This version binds to specific DNA sequences without cutting them. By attaching dCas9 to various epigenetic modifiers, like histone deacetylases, DNA methyltransferases, or demethylases, researchers can accurately target and control gene expression. For example, the dCas9-TET1 fusion can remove methyl groups from CpG sites, leading to gene activation through DNA demethylation (Liu et al., 2016). On the other hand, fusions like dCas9-DNMT3A promote gene silencing by increasing methylation at specific sites (Amabile et al., 2016).

These targeted epigenetic changes have proven especially effective in reprogramming mature cells. A notable example is directing dCas9-TET1 to the promoter of MYOD1, a significant muscle regulator. This induces fibroblasts to take on a muscle-like phenotype by reactivating previously silenced myogenic genes (Zhou et al., 2019). Similarly, using dCas9-HDAC constructs to suppress astrocytic gene programs has shown promise in directly converting astrocytes into functional neurons by reshaping chromatin structure (Heller et al., 2020). These approaches offer a more biologically relevant alternative to transgene-based methods, mimicking natural epigenetic changes during development.

One of the main advantages of epigenome editing is its reversibility. Since the DNA sequence stays the same, the changes in gene expression are not permanent, allowing cells to remain flexible. This quality is particularly important in clinical settings, where control over timing and safety is crucial. Researchers often use inducible expression systems or temporary delivery methods to apply epigenetic changes only when and where they are needed (Morita et al., 2016). This helps reduce potential long-term risks linked to irreversible genome modifications, such as unwanted mutations or tumor formation. Recent developments have also concentrated on multiplexed epigenome editing, which modulates multiple genes simultaneously to manage complex cell fate changes. By introducing multiple guide RNAs that target different gene regulatory elements, researchers can activate or silence groups of genes in a coordinated way. For instance, Kwon et al. (2021) showed that combining dCas9-based gene activation and repression enabled human fibroblasts to be reprogrammed into specific neuron types, effectively mimicking developmental gene expression patterns.

Furthermore, combining CRISPR-mediated epigenome editing with modern omics technologies has led to a deeper understanding of reprogramming pathways. Transcriptomic profiling and assays for chromatin accessibility are now being used alongside editing tools to track dynamic changes during lineage conversion. These methods help identify crucial intermediate states and obstacles, ultimately refining reprogramming protocols (Liu et al., 2022). Beyond the dCas9 system, other CRISPR variants like dCas12a and dCas13 are emerging as flexible tools for epigenetic and transcriptomic regulation. dCas12a targets DNA in a PAM-distinct way, while dCas13 is designed to manipulate RNA without impacting the genome. This opens new possibilities for editing gene expression after transcription, providing a powerful strategy when protecting genomic DNA integrity is essential (Abudayyeh et al., 2017).

5. CRISPR-ENGINEERED ORGANOIDS AND TISSUE CONSTRUCTS

The combination of CRISPR/Cas9 gene editing with organoid and tissue engineering is changing regenerative medicine. This collaboration allows for the creation of patient-specific, functional tissue models and opens new avenues for disease modeling, drug testing, and transplantation. Building on earlier advances with induced pluripotent stem cells (iPSCs) and CRISPR gene editing, this section shows how these technologies move personalized medicine forward. Organoids are three-dimensional, self-organizing structures derived from stem cells that closely resemble the architecture and function of real organs. When combined with CRISPR editing, they become powerful tools for studying development and disease at the genetic level. Before differentiation, CRISPR can introduce or correct mutations, add reporter constructs, or regulate gene expression pathways. This makes these models highly customizable (Yin et al., 2014; Lancaster et al., 2013). A key application is correcting genetic mutations in iPSCs from patients with inherited disorders. For example, in cystic fibrosis, gene editing of the CFTR locus in patient-derived iPSCs has resulted in creating airway organoids with restored chloride transport (Schwank et al., 2013). Similarly, liver organoids from people with alpha-1 antitrypsin deficiency have been genetically modified at the SERPINA1 gene, restoring proper protein folding and secretion (Yusa et al., 2011). These corrected organoids not only function more like healthy tissue but also provide chances for personalized cell therapy.

CRISPR's ability to edit multiple genes at once allows for deeper customization of organoid biology. Modifying signaling pathways like Wnt, Notch, and BMP has led to generating gut and brain organoids with improved physiological features. For instance, brain organoids edited this way have shown layered cortical structures similar to those found in fetal human development. A significant limitation of organoids is their lack of internal blood vessels, which affects growth and scalability. To address this, CRISPR is being used to improve vascularization. By increasing expression of angiogenic genes like VEGFA or decreasing inhibitors such as THBS1, researchers have successfully promoted the formation of vascular networks within organoids. Co-culturing CRISPR-edited endothelial cells with progenitor cells has further enabled making tissues that can be perfused with improved viability (Takebe et al., 2015)(Table 1).

Table 1. Applications of CRISPR in Organoid Systems

Application Area	Target Gene(s)	Organoid Type	Outcome	Reference
Genetic correction	*CFTR*	Intestinal organoids	Recovered ion transport in cystic fibrosis model	Schwank et al., 2013
Genetic correction	*SERPINA1*	Liver organoids	Normalized alpha-1 antitrypsin production	Yusa et al., 2011
Vascularization support	*VEGFA, THBS1*	Liver and gut organoids	Formation of functional capillary-like structures	Takebe et al., 2015
Cardiac tissue function	*HCN4*	Bioengineered heart tissue	Synchronized pacemaker activity and rhythmic contractions	Zhao *et al.*, 2020

Beyond traditional organoid cultures, CRISPR is vital in engineering tissue constructs using 3D bioprinting. Through precise layer-by-layer printing, CRISPR-edited cells can be arranged within scaffolds to form complex tissue structures. For example, cardiac patches made from gene-edited cardiomyocytes expressing pacemaker genes have shown spontaneous electrical activity and rhythmic contraction in vitro. Biodegradable scaffolds, such as those made from polylactic acid (PLA), alginate, or collagen, are typically used to support engineered tissue growth. When filled with CRISPR-modified stem cellsespecially those designed to express osteogenic markers like BMP2 these scaffolds encourage better bone formation and repair, which is promising for orthopedic applications. Additionally, CRISPR technology is enabling detailed lineage tracing and developmental studies within organoid models. By introducing unique DNA barcodes into individual cells, researchers can track cell division and differentiation over time. This technique enhances our understanding of tissue development and helps eliminate undesirable or tumor-prone clones during organoid formation (Raj et al., 2018).

CRISPR-edited organoids are also being used for targeted drug testing. Genetically identical organoid lines that differ only by specific engineered mutations allow researchers to explore how individual gene changes impact drug sensitivity. In colorectal cancer models, for example, CRISPR has been used to create organoids with distinct mutations in genes like KRAS, TP53, and APC, enabling customized drug response analysis (Drost et al., 2015).

6. ENGINEERING SMART BIOMATERIALS WITH CRISPR-RESPONSIVE FUNCTIONS

The combination of CRISPR/Cas systems and biomaterial science has led to a new generation of smart materials that can precisely regulate genes and deliver therapies in tissue engineering. These CRISPR-integrated biomaterials are designed to react to internal biological signals or external environmental changes, like shifts in pH, oxidative state, or specific biomolecules. This capability allows for controlled regeneration over time and space. Three promising strategies (Table 2) in this area include gene-activated scaffolds, 3D-printed constructs with embedded CRISPR circuits, and hybrid biomaterials that merge mechanical structure with genetic control systems.

a) Gene-Activated Scaffolds with CRISPR-Driven Therapeutic Release

Gene-activated scaffolds incorporate CRISPR-based components directly into the material to stimulate gene expression or release therapeutic agents at the injury site. For instance, Liu *et al.* (2018) showed that adding CRISPR/Cas9 systems to biodegradable scaffolds could increase levels of osteogenic markers like BMP2 and VEGF, thereby improving bone repair. Similarly, used non-viral CRISPR activation (CRISPRa) systems embedded in gelatin-based nanofibers to induce VEGF expression locally, which significantly improved blood vessel formation in wound healing models. Further development by Yang *et al.* (2022) showcased electrospun fiber scaffolds with core-sheath architecture containing plasmids encoding BMP2 and VEGF. These constructs allowed for sequential and localized gene expression, closely mimicking the natural healing processes. By avoiding viral vectors and using biomaterial-based gene delivery, these scaffolds minimize potential immune responses while maintaining high therapeutic effectiveness.

b) 3D-Printed Constructs Incorporating CRISPR Circuits

Incorporating CRISPR systems into 3D-printed biomaterials allows for the creation of complex structures with built-in gene regulatory functions. Yin *et al.* (2023) developed hydrogels with embedded CRISPR components that respond to specific triggers, such as light or pH, activating gene expression only under set conditions. This enables precise control over genetic activation in engineered tissues. Another significant advancement came from, who designed hydrogels that contained CRISPR-Cas12a. When this system recognized specific DNA targets, it triggered the release of therapeutic enzymes. This method used the collateral cleavage activity of Cas12a to establish a highly specific and programmable drug delivery system. Integrating CRISPR circuits into 3D-printed materials opens new possibilities for customizable and responsive therapeutic implants.

c) Hybrid Biomaterials for Layered and Localized Regeneration

Hybrid biomaterials merge the mechanical benefits of traditional scaffolds with the biological functions of gene-editing systems. These systems are engineered to release biochemical signals while also triggering targeted gene activation in a controlled way. McMurtrey *et al.* (2016) demonstrated that multi-compartment scaffolds could organize neural progenitor cells spatially and guide their differentiation using biochemical gradients, enhanced by the potential for CRISPR-based modulation. Recent innovations by Patel et al. (2021) involved hydrogel scaffolds embedded with gradients of signaling molecules and CRISPRa components, enabling stepwise activation of pathways for bone and cartilage formation. These materials effectively replicated the complex zones found in tissues like cartilage and bone. Similarly, Zhang et al. (2022) created redox-sensitive nanoparticles within scaffold matrices to activate CRISPR systems in response to low oxygen or oxidative stress, ensuring highly localized gene editing in complex environments.

Table 2. Mechanism of Engineering Smart Biomaterials

Strategy	Mechanism	Application	Key Benefit	References
Gene-activated scaffold	CRISPR-mediated growth factor expression	Bone tissue repair	Localized osteoinductive signaling	Liu et al., 2019
Core–sheath fiber scaffold	Sequential plasmid-based gene activation	Bone regeneration	Spatiotemporal gene expression	Frangoul et al., 2021
Smart hydrogel	Stimulus-responsive CRISPR circuits	On-demand gene regulation	Conditional and precise activation	Yin *et al.*, 2023
Cas12a hydrogel	DNA-sensitive enzyme release	Biosensing/drug delivery	Programmable, target-triggered release	Frangoul et al., 2021

7. DELIVERY SYSTEMS IN REGENERATIVE APPLICATIONS

Efficient and targeted delivery systems are vital for the success of CRISPR-based interventions in regenerative medicine. The selection of a delivery vehicle greatly affects editing efficiency, cellular uptake, and safety profiles. This makes it essential for turning genome editing technologies into therapeutic solutions.

a) **Viral Vectors -** Viral vectors are one of the most common methods for CRISPR delivery because of their high transduction efficiency and stable gene expression. Adeno-associated viruses (AAVs) are preferred for their low immunogenicity and ability to target specific tissues. Lentiviruses allow stable integration into the host genome, ensuring long-term expression in both dividing and non-dividing cells (Smith et al., 2022). However, concerns about insertional mutagenesis and limited packaging capacity necessitate careful vector design and risk assessment (Wang et al., 2020).
b) **Non-Viral Approaches -** Non-viral delivery methods have become popular due to their lower immunogenicity and flexible design. Lipid nanoparticles (LNPs) help encapsulate and protect CRISPR components, allowing efficient cellular uptake through membrane fusion (Patel et al., 2021). Electroporation uses brief electrical pulses to temporarily permeabilize cell membranes, letting nucleic acids and proteins enter directly. This method is especially effective for ex vivo genome editing in stem cells (Kim et al., 2020). Nanoneedle-mediated delivery provides a mechanical and minimally invasive way to transfer CRISPR systems directly into the cytoplasm or nucleus without causing significant cell damage (Li et al., 2022).
c) Biomaterial-Assisted Delivery - Biomaterial-based systems, like CRISPR-loaded hydrogels, offer a localized and sustained release of editing tools at the target

site (Liu et al., 2018). These hydrogels can be designed to respond to environmental signals, allowing for precise control of CRISPR activity over time and space. This approach improves retention at the injury site and reduces systemic exposure, which helps minimize potential off-target effects (Chen et al., 2021).

d) Engineered Exosomes and Extracellular Vesicles

Exosomes and extracellular vesicles (EVs) have shown promise as natural carriers for CRISPR delivery, providing inherent biocompatibility and the ability to cross biological barriers (Zhang et al., 2021). These vesicles can be designed to encapsulate CRISPR-Cas components and showcase targeting ligands, enhancing delivery accuracy to specific cell types. Their ability to avoid immune clearance and allow for repeated dosing further supports their use in regenerative therapies (Table 3).

Table 3. Overview of CRISPR Delivery Platforms in Regenerative Applications

Delivery Method	Advantages	Limitations	Example Application
AAV vectors	High tissue specificity, low immunogenicity	Small cargo capacity (~4.7 kb)	Gene correction in retinal diseases
Lentiviral vectors	Stable integration, long-term expression	Risk of insertional mutagenesis	Engineering iPSCs for organoid development
Lipid nanoparticles	Non-viral, scalable, protects CRISPR components	Potential cytotoxicity, lower efficiency than viral methods	In vivo delivery for liver gene editing
Electroporation	High efficiency in ex vivo applications	Cell damage risk, not suitable for in vivo	Editing hematopoietic stem cells
Nanoneedles	Targeted, localized delivery	Limited scalability	Direct delivery into skin or muscle tissues
CRISPR-loaded hydrogels	Sustained, localized release, customizable	Limited deep-tissue penetration	Regeneration of cartilage or bone

8. IMMUNOENGINEERING AND UNIVERSAL GRAFTS VIA CRISPR9. SYNTHETIC BIOLOGY MEETS REGENERATIVE MEDICINE

CRISPR technology has greatly improved the field of immunoengineering by allowing precise genetic changes to create universal, immune-compatible grafts. One key method is producing hypoimmunogenic stem cells, like human leukocyte antigen (HLA) knockout (KO) lines. These cells can avoid immune detection and reduce the risk of rejection in various patient groups. By disrupting HLA class I and II molecules while keeping some non-classical HLAs, these engineered cells maintain their functions and prevent immune destruction. Besides HLA modification, CRISPR is used to engineer immune checkpoint genes to support graft tolerance. Targeting molecules such as PD-L1, CTLA-4, or CD47 helps control the recipient's immune response. This reduces inflammation and improves graft survival (Xu et al., 2020). This strategy can work with stem cells, differentiated tissues, or bioengineered constructs, allowing for grafts to fit into the host without needing long-term immunosuppression.

Another growing application is CRISPR-based engineering in adoptive immune cell therapies, especially in CAR-macrophage and CAR-T cell platforms. While CAR-T cells are well-known in cancer treatment, their ability to regulate immunity and support tissue repair is getting more attention. Similarly, CAR-macrophages can be designed to target injury sites, enhance blood vessel formation, and manage inflammation. By using CRISPR editing, these immune cells can be refined further for better tissue repair, preventing fibrosis, and achieving targeted regenerative effects.

9. SYNTHETIC BIOLOGY MEETS REGENERATIVE MEDICINE

The convergence of synthetic biology and regenerative medicine is revolutionizing how we control cellular functions and engineer tissues. By combining the programmability of CRISPR-based genome editing with synthetic biology design principles, researchers can construct gene circuits that regulate cell identity, guide tissue formation, and improve therapeutic precision. This integration provides a versatile toolkit for designing living systems that not only replicate developmental cues but also allow external regulation of regenerative processes (Nielsen et al., 2016).

a) CRISPR-Controlled Gene Circuits for Cell-Fate Decisions

CRISPR-regulated gene circuits act as molecular switches that influence whether cells differentiate, reprogram, or remain in a pluripotent state. These circuits use guide RNAs together with Cas proteins to activate or silence transcription factors that drive lineage-specific differentiation (Wroblewska et al., 2015). For instance, CRISPRa can upregulate master regulators involved in cardiac or pancreatic cell formation, while CRISPRi can suppress competing pathways, improving the efficiency of generating desired therapeutic cell types (Nakamura et al., 2021).

In addition, CRISPR circuits can be linked to inducible promoters, which provide temporal control over differentiation. This means that cell-fate decisions can be triggered, halted, or even reversed when needed, giving clinicians and researchers a dynamic tool for tailoring regeneration according to disease-specific or patient-specific needs (Chavez et al., 2015).

b) Logic-Gated Gene Expression for Safety and Specificity

Safety is a central challenge in regenerative medicine, especially when deploying engineered cells in vivo. Synthetic biology addresses this by creating logic-gated CRISPR circuits that function like computational gates such as "AND," "OR," or "NOT." These systems ensure that gene expression is only activated when the correct set of conditions or biomarkers is present (Nissim et al., 2014). Such logic-based control improves therapeutic specificity, reducing off-target risks. For example, regeneration-inducing genes could be turned on only when two biomarkers are detected simultaneously, lowering unintended activation. This approach is highly valuable in immune-related applications, where selective regulation of tolerance-inducing genes may prevent graft rejection without suppressing the entire immune system (Liu et al., 2021). Moreover, safety can be enhanced through "kill-switch" circuits, in which CRISPR programs trigger apoptosis if cells begin proliferating abnormally, providing an additional safeguard against tumorigenesis.

c) Synthetic Morphogen Gradients for Tissue Patterning and Organogenesis

A defining feature of embryonic development is the presence of morphogen gradients that supply positional cues to cells, guiding their differentiation and spatial arrangement. Synthetic biology allows these natural gradients to be recreated in vitro by combining CRISPR-mediated gene control with engineered signaling pathways (Chung et al., 2019).

For example, CRISPR circuits that regulate Wnt or BMP pathways can be programmed to form concentration gradients, directing stem cells to organize into layered tissues such as skin epithelium or intestinal lining (Manfrin et al., 2019). Advanced tools like light-responsive CRISPR systems add another layer of precision, allowing researchers to control morphogen activity in both time and space, leading to the generation of complex tissues such as vascularized organoids and patterned neural constructs (Polstein & Gersbach, 2015).

These engineered gradients are particularly promising for organogenesis, where intricate arrangements of multiple cell types are required. By mimicking the developmental cues observed in natural embryogenesis, CRISPR-enabled synthetic gradients bring engineered tissues closer to fully functional organ substitutes.

10. SPATIOTEMPORAL REGULATION OF CRISPR ACTIVITY

Controlling when and where CRISPR operates is now just as important as choosing what target it edits. New methods are integrating genome and epigenome editors with external cues like light, heat, mechanical forces, and magnetic fields. These techniques are supported by biofabrication technologies that organize signals in space and time. Together, methods such as optogenetic, mechanogenetic, magnetogenetic, and 4D bioprinting allow for precise, reversible, and localized control of CRISPR activity. This level of control is highly relevant in regenerative medicine, where zoning patterns and gradual tissue remodeling are essential (Nielsen et al., 2016).

a) Light-Activated (Optogenetic) CRISPR Systems

Optogenetics offers millisecond-level timing and micrometer-level spatial resolution, making it one of the most precise ways to control CRISPR. A significant design linked dCas9 with blue-light sensitive heterodimerizers (CRY2/CIB1), creating the LACE system. This system activates transcription only when illuminated (Polstein et al., 2015). It allows for reversible CRISPR activation and makes activity adjustable based on light intensity and exposure time. This supports pulse-like programming of developmental regulators or paracrine signals in engineered tissues. Other innovations involve splitting Cas9 into two inactive parts that reassemble under blue light, known as photoactivatable Cas9 (paCas9). This method limits editing to illuminated areas, enabling specific genome editing or lineage tracking within 3D constructs, while leaving nearby regions unaffected (Nihongaki et al., 2015). This precise spatial control is useful in organoid engineering, such as in crypt-villus patterning or cortical layering, where targeted gene editing is crucial for proper tissue organization. Additionally, optical control has expanded to include heat or photothermal actuation, using near-infrared light or ultrasound to create localized heating that activates heat-shock responsive CRISPR circuits. This enables transcriptional regulation in deeper tissues where visible light cannot reach.

b) Mechanogenetic and Magnetogenetic Regulation

Cells can sense and respond to mechanical inputs. This ability has been used to create mechanogenetic CRISPR circuits. For example, synthetic promoters linked to mechanosensors like TRPV4, PIEZO1, or YAP/TAZ pathways have been combined with dCas9 activators and repressors. This creates circuits that respond to mechanical loading (Liu et al., 2022). In cartilage models, these circuits can activate therapeutic pathways under normal strain while switching to protective programs when overloaded, providing a flexible control system for tissue repair. Magnetogenetics supports this approach by using magnetic nanoparticles to deliver thermal or mechanical energy to deep tissue areas that light cannot reach. Alternating magnetic fields can create localized heating to activate heat-shock responsive CRISPR modules, or apply mechanical forces through nanoparticle-receptor complexes to stimulate mechanosensitive pathways (Chen et al., 2024; Zhang et al., 2025). Recently, magnetically responsive actuators, such as engineered MagR/TRPV variants, have shown potential in transforming external magnetic fields into cellular signals that trigger CRISPR responses (Wang et al., 2025). These systems emphasize the potential for wireless, deep-tissue, and reversible CRISPR activation in clinical settings.

c) 4D Bioprinting and Zonal Editing in Layered Constructs

4D bioprinting adds time as a design factor, employing stimuli-responsive bioinks that change their physical or chemical properties after printing. When combined with optogenetic or thermogenetic CRISPR circuits, these biofabricated models allow for zonal gene editing, enabling specific tissue layers to respond to certain cues at defined times (Huang et al., 2025). For example, cartilage and osteochondral tissues require different programs for their superficial and deep zones. The superficial layers benefit from lubricin and collagen II production, while deeper layers (Table 4) need signals for mineralization. Bioprinted constructs with CRISPR logic circuits have facilitated this zonal regulation by activating or repressing transcription factors like SOX9 or RUNX2 at specific times for each layer. Similarly, complex organs like the liver, gut, and brain depend on morphogen signals that are arranged in time and space. Using hydrogels that change stiffness or transparency over time allows for a sequential activation of CRISPR that resembles embryonic development, aiding tissue maturation and integration.

Table 4. Conceptual Framework

Layer	Description	Function/Outcome
Material Layer	Hydrogels engineered with varying optical and thermal properties to control how light, heat, or magnetic stimuli penetrate different zones.	Allows precise regulation of external stimuli penetration into distinct tissue regions.
Circuit Layer	Incorporation of CRISPRa (dCas9-VPR) or CRISPRi (dCas9-KRAB) with logic gates (AND/NOT).	Ensures that gene circuits activate only under defined conditions, improving safety and specificity.
Cue Layer	Application of patterned optical, thermal, or magnetic stimuli on scheduled timelines.	Triggers zone-specific gene activation or repression, enabling stage-specific differentiation.
Outcome	Spatially restricted and time-phased CRISPR activity across tissue constructs.	Promotes anisotropic (uneven) tissue development, enhances maturation, and minimizes off-target remodeling.

11. CRISPR AND MULTI-OMICS-DRIVEN PERSONALIZED REGENERATION

The future of regenerative medicine focuses on personalization. Therapies are designed based on the genetic and molecular profiles of individual patients. When we combine CRISPR with multi-omics technologies, we create new ways to tailor treatments to specific molecular signatures (Wang et al., 2021). By integrating genomics, transcriptomics, epigenomics, proteomics, and metabolomics, researchers can find new therapeutic targets and refine CRISPR applications to improve safety and treatment results (Kobayashi et al., 2023).

a) Integrating single-cell transcriptomics with CRISPR design

Single-cell RNA sequencing (scRNA-seq) offers a clear view for analyzing cellular states and discovering differences among regenerating tissues. When paired with CRISPR perturbation tools, scRNA-seq allows for detailed mapping of gene regulatory networks (Datlinger et al., 2017). This combination enables CRISPR screens with single-cell precision, helping to identify transcription factors that affect lineage specification and cell groups that respond differently to gene editing (Replogle et al., 2020). This integration is especially important in regenerative methods like stem cell therapies, where precise lineage outcomes are crucial for effective repair (Chen et al., 2021).

b) Multi-omics mapping for identifying therapeutic targets

Merging multiple layers of molecular data creates a broad view of tissue regeneration. Epigenomic mapping shows chromatin accessibility, while proteomic and metabolomic data reveal dynamic cellular processes that transcription analysis alone cannot explain (Arora et al., 2022). By combining these datasets with CRISPR-based functional screens, researchers can prioritize therapeutic targets that are both biologically relevant and clinically useful (Zhang et al., 2023). In areas like cartilage and heart repair, multi-omics-guided CRISPR editing is being used to identify causes of pathological remodeling, such as fibrosis and hypertrophy, which leads to personalized treatments (Chaudhuri et al., 2023).

c) AI and machine learning in CRISPR guide RNA design and off-target prediction

AI and machine learning are changing how we apply CRISPR in personalized medicine. These methods analyze large omics datasets to find regulatory hubs and predict effective intervention points (Kim et al., 2021). Machine learning algorithms are especially useful for optimizing guide RNA (gRNA) design by predicting efficiency and reducing off-target effects, thereby enhancing treatment safety (Chuai et al., 2018). When used with patient-specific omics data, AI-driven gRNA design creates personalized editing strategies. This forms a feedback system where CRISPR therapies are tailored to each genetic background (Xu et al., 2022). This approach is especially important in complex tissues like the brain, liver, and heart, where precise editing is needed to restore function without causing adverse remodeling.

12. MITOCHONDRIAL GENOME EDITING IN REGENERATIVE MEDICINE

Mitochondrial genome editing is one of the most complex challenges in genetic engineering. This complexity arises from the unique double-membrane structure of mitochondria and their specific regulatory mechanisms. However, in recent years, the development of specialized tools like DdCBE, mitoZFDs, and TALEDs has started to tackle these challenges. These tools enable accurate and irreversible changes to mitochondrial DNA (mtDNA) in living cells and tissues (Silva-Pinheiro et al., 2022; Cho et al., 2024; Minczuk et al., 2022).

a) Emerging Tools: DdCBE, mitoZFDs, and TALEDs

The DdCBE (DddA-derived cytosine base editor) system, which comes from bacterial cytidine deaminase, allows C→T (or G→A) conversions in mtDNA without causing double-strand breaks. Preclinical studies in mouse cardiac tissue have shown editing efficiencies between 10% and 30%. These results are promising for treating mitochondrial diseases that mainly impact energy-dependent organs like the heart (Cho et al., 2024; Minczuk et al., 2022).

TALEDs (TALE-linked deaminases) mark a significant step forward. They introduce A→G substitutions by combining mitochondrial targeting sequences with TALE arrays and engineered adenine deaminases. A recent variant called sTALEDs has achieved nearly 40% A→G editing efficiency in human cells, with very little off-target activity. This precision makes them especially hopeful for fixing mutations linked to neurodegenerative diseases (Song et al., 2023; Kar et al., 2023).

At the same time, mitoZFDs provide an additional method, expanding the range of mitochondrial editing tools. Together, these new tools signal a new era in which targeted and precise genetic interventions can address mitochondrial disorders.

b) Addressing Heteroplasmy in Degenerative Disorders

One major issue in mitochondrial biology is heteroplasmy, which is the presence of both mutant and normal mtDNA copies within cells. Symptoms of disease often develop once mutant mtDNA exceeds a critical level. Earlier methods, like mitoZFNs and mitoTALENs, worked by selectively breaking down mutant mtDNA to enrich functional wild-type genomes (Minczuk et al., 2022). In contrast, the new base-editing methods such as DdCBE and TALEDs provide a more refined solution. They focus on directly correcting the mutations rather than removing them. This correction keeps the mitochondrial genome intact while reducing the load of harmful variants. Such precise editing holds great potential for mitochondrial myopathies and neurodegenerative disorders like Alzheimer's and Parkinson's disease (Silva-Pinheiro et al., 2022; Kar et al., 2023).

c) Regenerating Energy-Deficient Tissues

Organs with high energy needs, such as the heart, skeletal muscles, and brain, are especially vulnerable to mitochondrial dysfunction. Mutations in mtDNA disrupt oxidative phosphorylation, leading to reduced ATP production and gradual tissue damage. Recent studies using AAV-mediated delivery of DdCBE in mouse cardiac tissue have shown that restoring mitochondria in non-dividing cells is possible, providing a potential solution for cardiac disorders (Minczuk et al., 2022; RegMed-Net, 2022). Additionally, applying DdCBE or TALEDs to patient-derived iPSC neurons and organoids could help restore mitochondrial shape, energy production, and overall function. Fixing mtDNA mutations in these models may improve neuronal resilience, boost axonal repair, and support the recovery of neural networks in energy-deficient brain tissues.

d) Integration with Broader Regenerative Medicine Frameworks

Mitochondrial genome editing should not be viewed in isolation; it can work well alongside other regenerative medicine approaches. For instance, optogenetic and mechanogenetic CRISPRa/i systems (Section 10) may eventually allow researchers to control mitochondrial function with precise timing and location, activating changes only when necessary. Moreover, multi-omics profiling and AI-driven CRISPR design (Section 11) could improve how mitochondrial therapies are personalized. For example, single-cell omics could trace heteroplasmy patterns within affected tissues, enabling specific interventions that use DdCBE or TALEDs on the most impacted cell groups. This integration could lead to more personalized mitochondrial medicine.

13. PRECLINICAL AND CLINICAL APPLICATIONS

The field of CRISPR-based technologies is evolving quickly, particularly in regenerative medicine. Experimental evidence indicates that gene editing can repair, restore, and regenerate damaged tissues. Building on earlier developments like controlling gene expression over time and space, integrating multi-omics data, and editing mitochondrial DNA, CRISPR tools are now used in both preclinical and clinical settings. These efforts include treatments for spinal cord injury, osteoarthritis, and inherited retinal diseases. They also involve using CRISPR-engineered induced pluripotent stem cells (iPSCs) for therapies tailored to individual patients. While these advancements highlight the significant potential of CRISPR, substantial challenges still need to be addressed before it can be used widely in clinical practices (Gillmore et al., 2021; Xu et al., 2023).

a) CRISPR Strategies in Spinal Cord Injury, Osteoarthritis, and Retinal Dystrophy

In spinal cord injury (SCI) research, CRISPR has been investigated as a way to improve neuronal survival and promote regeneration. Preclinical studies suggest that targeting inhibitors of axonal growth, like PTEN or the Nogo receptor, can enhance axonal sprouting and result in measurable improvements in motor function (Chen et al., 2021; Li et al., 2022). In osteoarthritis, CRISPR has been used to silence catabolic enzymes like MMP13 and ADAMTS5, which reduces cartilage damage and slows disease progression in animal models (Mao et al., 2019; Wang et al., 2022). Likewise, inherited retinal dystrophies, including Leber congenital amaurosis, have moved from lab studies to clinical evaluation. Here, directly delivering CRISPR-Cas9 complexes into the subretinal space has shown safety and initial signs of improved vision, demonstrating the potential for in vivo editing in human tissues (Maeder et al., 2019; Gillmore et al., 2021).

b) CRISPR-iPSC-Derived Cell Therapy Platforms

Another innovation combines CRISPR with iPSC technology to create patient-specific cell therapies. By editing harmful mutations in iPSCs and turning them into functional cell typessuch as cardiomyocytes, neurons, or chondrocytes researchers can generate tailored grafts for transplantation (Xu et al., 2023; Wang et al., 2023). For instance, gene-corrected iPSC-derived cardiomyocytes have been found to restore heart function in injury models. Additionally, retinal cells from edited iPSCs have improved photoreceptor survival and visual responses in degenerative conditions

(Liang et al., 2022; Lanza et al., 2022). These processes link molecular correction to functional tissue recovery, paving the way for targeted regenerative therapies.

c)Translational Challenges and Clinical Success Stories

Despite significant progress, transitioning CRISPR therapies from preclinical stages to clinical application is not easy. Efficient and targeted delivery is a major challenge. While adeno-associated viral (AAV) vectors are becoming leading options, their limited carrying capacity and potential to trigger immune responses complicate their use (Mendell et al., 2021). Other concerns include unintended off-target effects, mosaicism, and the unknown long-term safety of edited tissues. Ethical discussions around germline interventions and early human trials also add complexity to clinical translation. Nonetheless, initial clinical trials offer valuable proof-of-concept. The in vivo CRISPR therapy EDIT-101, aimed at Leber congenital amaurosis, has shown safety and promising signs of vision improvement. This represents the first ocular gene editing therapy tested in humans (Maeder et al., 2019). At the same time, ex vivo treatments such as CTX001 for β-thalassemia and sickle cell disease have produced impressive clinical results, with patients achieving functional cures after receiving transplanted edited hematopoietic stem cells (Frangoul et al., 2021). These achievements demonstrate the promise and transformative potential of CRISPR therapies and provide a guide for future applications in degenerative diseases like osteoarthritis, SCI, and heart disorders.

14. CONCLUSION

CRISPR/Cas9 technology has emerged as a transformative tool in regenerative medicine and tissue engineering, offering unprecedented precision and efficiency in genome editing. Its ability to selectively modify genetic sequences enables researchers to correct disease-causing mutations, enhance cell functionality, and guide stem cell differentiation for tissue regeneration. In tissue engineering, CRISPR facilitates the creation of genetically optimized cells and biomaterials that can better integrate, repair, and replace damaged tissues. Both genome and epigenome editing platforms are crucial for advancing cell reprogramming strategies in regenerative medicine. While genome editing offers lasting changes, epigenome editing provides a flexible, adjustable, and safer method by influencing gene function without changing the DNA itself. Together, these technologies are creating a future where precise control of cell fate is not only feasible but also clinically practical, thanks to advancements in delivery systems, multi-gene targeting, and comprehensive molecular profiling.

Despite its remarkable potential, challenges remain such as off-target effects, delivery system limitations, and ethical considerations surrounding germline editing. Continued advancements in CRISPR-based systems, including high-fidelity Cas variants and improved delivery platforms, are expected to overcome these barriers and pave the way for safe clinical translation.

Overall, CRISPR/Cas9 represents a cornerstone technology for the future of regenerative medicine, accelerating the development of personalized and regenerative therapies that can restore tissue function, treat previously incurable conditions, and ultimately redefine the boundaries of modern healthcare.

REFERENCES

Abudayyeh, O. O., Gootenberg, J. S., Essletzbichler, P., Han, S., Joung, J., Belanto, J. J., Verdine, V., Cox, D. B. T., Kellner, M. J., Regev, A., Lander, E. S., Voytas, D. F., Ting, A. Y., & Zhang, F. (2017). RNA targeting with CRISPR–Cas13. *Nature*, *550*(7675), 280–284. DOI: 10.1038/nature24049 PMID: 28976959

Amabile, A., Migliara, A., Capasso, P., Biffi, M., Cittaro, D., Naldini, L., & Lombardo, A. (2016). Inheritable silencing of endogenous genes by hit-and-run targeted epigenetic editing. *Cell*, *167*(1), 219–232.e14. DOI: 10.1016/j.cell.2016.09.006 PMID: 27662090

Anzalone, A. V., Koblan, L. W., & Liu, D. R. (2020). Genome editing with CRISPR–Cas nucleases, base editors, transposases and prime editors. *Nature Biotechnology*, *38*(7), 824–844. DOI: 10.1038/s41587-020-0561-9 PMID: 32572269

Anzalone, A. V., Randolph, P. B., Davis, J. R., Sousa, A. A., Koblan, L. W., Levy, J. M., Chen, P. J., Wilson, C., Newby, G. A., Raguram, A., & Liu, D. R. (2019). Search-and-replace genome editing without double-strand breaks or donor DNA. *Nature*, *576*(7785), 149–157. DOI: 10.1038/s41586-019-1711-4 PMID: 31634902

Arora, R., Sinha, A., & Zhang, Y. (2022). Multi-omics integration in regenerative medicine: From data to therapies. *Trends in Biotechnology*, *40*(11), 1234–1248.

Barrangou, R., Fremaux, C., Deveau, H., Richards, M., Boyaval, P., Moineau, S., Romero, D. A., & Horvath, P. (2007). CRISPR provides acquired resistance against viruses in prokaryotes. *Science*, *315*(5819), 1709–1712. DOI: 10.1126/science.1138140 PMID: 17379808

Bharucha, N., Wong, C., Marti-Gutierrez, N., Nguyen, D., Puri, M., Wu, J., & Wu, J. C. (2021). Generation of AAVS1 integrated doxycycline-inducible prime editor into human iPSCs. *medRxiv*.

Chaudhuri, O., Cooper, J. A., & Mooney, D. J. (2023). Multi-omics and CRISPR in tissue regeneration: Toward precision therapeutics. *Nature Reviews Bioengineering*, *1*(3), 211–225.

Chavez, A., Scheiman, J., Vora, S., Pruitt, B. W., Tuttle, M., Iyer, E. P., & Church, G. M. (2015). Highly efficient Cas9-mediated transcriptional programming. *Nature Methods*, *12*(4), 326–328. DOI: 10.1038/nmeth.3312 PMID: 25730490

Chemello, F., Chai, A. C., Li, H., Rodriguez-Caycedo, C., Sanchez-Ortiz, E., Atmanli, A., & Olson, E. N. (2021). Precise correction of dystrophin mutations in cardiomyocytes derived from Duchenne muscular dystrophy patients via base editing. *Science Advances*, *7*(12), eabg4910. DOI: 10.1126/sciadv.abg4910 PMID: 33931459

Chen, J., Liu, Y., Zhang, Y., & Xu, C. (2021). CRISPR-based modulation of PTEN in spinal cord injury models promotes axonal regeneration. *Nature Communications*, *12*(1), 1–13. PMID: 33397941

Chen, J. S., & Doudna, J. A. (2017). The chemistry of Cas9 and its applications in genome editing. Nature Reviews Chemistry, 1(10), 0078.

Chen, S., Sanjana, N. E., & Shalem, O. (2021). Dissecting neural and stem cell networks using CRISPR and single-cell transcriptomics. *Cell Stem Cell*, *28*(6), 1000–1014.

Cho, S. I., Lim, K., Hong, S., Kim, J. S., & Bae, S. (2024). Engineering TALE-linked deaminases to facilitate precision adenine base editing in mitochondrial DNA. *Cell*, *187*(1), 95–109.e26. DOI: 10.1016/j.cell.2023.11.035 PMID: 38181745

Chuai, G., Ma, H., Yan, J., Chen, M., Hong, N., Xue, D., Zhou, C., Zhu, C., Chen, K., Duan, B., Gu, F., Qu, S., Huang, D., Wei, J., & Liu, Q. (2018). DeepCRISPR: Optimized CRISPR guide RNA design by deep learning. *Genome Biology*, *19*(1), 80. DOI: 10.1186/s13059-018-1459-4 PMID: 29945655

Chung, H. K., Jacobs, C. L., Huo, Y., Yang, J., Krumm, S. A., Plemper, R. K., & Lin, M. Z. (2019). Tunable and reversible drug control of protein production via a self-excising degron. *Nature Chemical Biology*, *15*(5), 520–527. PMID: 26214256

Datlinger, P., Rendeiro, A. F., Schmidl, C., Krausgruber, T., Traxler, P., Klughammer, J., & Bock, C. (2017). Pooled CRISPR screening with single-cell transcriptome readout. *Nature Methods*, *14*(3), 297–301. DOI: 10.1038/nmeth.4177 PMID: 28099430

Davis-Anderson, K., Micheva-Viteva, S., Solomon, E., Hovde, B., Cirigliano, E., Harris, J., Twary, S., & Iyer, R. (2023). CRISPR/Cas9 directed reprogramming of iPSC for accelerated motor neuron differentiation leads to dysregulation of neuronal fate patterning and function. *International Journal of Molecular Sciences*, *24*(22), 16161. DOI: 10.3390/ijms242216161 PMID: 38003351

Drost, J., van Jaarsveld, R. H., Ponsioen, B., Zimberlin, C., van Boxtel, R., Buijs, A., Sachs, N., Overmeer, R. M., Offerhaus, G. J., Begthel, H., Korving, J., van de Wetering, M., Schwank, G., Logtenberg, M., Cuppen, E., Snippert, H. J., Medema, J. P., Kops, G. J. P. L., & Clevers, H. (2015). Sequential cancer mutations in cultured human intestinal stem cells. *Nature*, *521*(7550), 43–47. DOI: 10.1038/nature14415 PMID: 25924068

Frangoul, H., Altshuler, D., Cappellini, M. D., Chen, Y. S., Domm, J., Eustace, B. K., Foell, J., de la Fuente, J., Grupp, S., Handgretinger, R., Ho, T. W., Kattamis, A., Kernytsky, A., Lekstrom-Himes, J., Li, A. M., Locatelli, F., Mapara, M. Y., de Montalembert, M., Rondelli, D., & Corbacioglu, S. (2021). CRISPR–Cas9 gene editing for sickle cell disease and β-thalassemia. *The New England Journal of Medicine*, *384*(3), 252–260. DOI: 10.1056/NEJMoa2031054 PMID: 33283989

Gillmore, J. D., Gane, E., Taubel, J., Kao, J., Fontana, M., Maitland, M. L., & Solomon, S. D. (2021). CRISPR-Cas9 in vivo gene editing for transthyretin amyloidosis. *The New England Journal of Medicine*, *385*(6), 493–502. DOI: 10.1056/NEJMoa2107454 PMID: 34215024

Heller, E. A., Cates, H. M., Peña, C. J., Sun, H., Shao, N., Feng, J., & Nestler, E. J. (2020). Locus-specific epigenetic remodeling controls addiction- and depression-related behaviors. *Nature Neuroscience*, *23*(5), 572–583. PMID: 25347353

Hendriks, D., Hageman, G. J., & Woltjen, K. (2020). Genetic programming of human iPSCs using CRISPR tools: Towards tissue engineering and regenerative medicine. *Cell Stem Cell*.

Huang, J., & Fussenegger, M. (2025). Programming mammalian cell behaviors by physical cues. *Trends in Biotechnology*, *43*(1), 16–42. DOI: 10.1016/j.tibtech.2024.07.014 PMID: 39179464

Jinek, M., Chylinski, K., Fonfara, I., Hauer, M., Doudna, J. A., & Charpentier, E. (2012). A programmable dual-RNA–guided DNA endonuclease in adaptive bacterial immunity. *Science*, *337*(6096), 816–821. DOI: 10.1126/science.1225829 PMID: 22745249

Kar, B., Castillo, S. R., Sabharwal, A., Clark, K. J., & Ekker, S. C. (2023). Mitochondrial base editing: Recent advances toward therapeutic opportunities. *International Journal of Molecular Sciences*, *24*(6), 5798. DOI: 10.3390/ijms24065798 PMID: 36982871

Kim, H. K., Min, S., Song, M., Jung, S., Choi, J. W., & Kim, H. (2021). Deep learning improves prediction of CRISPR–Cpf1 guide RNA activity. *Nature Biotechnology*, *39*(2), 170–178. PMID: 32958957

Kim, S., Kim, D., Cho, S. W., Kim, J., & Kim, J. S. (2020). Highly efficient RNA-guided genome editing in human cells via delivery of purified Cas9 ribonucleoproteins. Genome Research, 30(4), 517–526. Li, Z., Wang, Y., & Xu, X. (2022). Engineering exosomes for targeted CRISPR-Cas9 delivery. *Theranostics*, *12*(1), 131–149.

Kobayashi, T., Sato, M., & Nishikawa, S. (2023). Omics-guided regenerative medicine: From CRISPR editing to patient-specific therapy. *Science Translational Medicine*, *15*(678), eabn1234.

Komor, A. C., Kim, Y. B., Packer, M. S., Zuris, J. A., & Liu, D. R. (2016). Programmable editing of a target base in genomic DNA without double-stranded DNA cleavage. *Nature*, *533*(7603), 420–424. DOI: 10.1038/nature17946 PMID: 27096365

Kulcsár, P. I., Tálas, A., Huszár, K., Ligeti, Z., Tóth, E., & Welker, E. (2017). Crossing enhanced and high-fidelity SpCas9 nucleases to optimize specificity and cleavage. *Genome Biology*, *18*(1), 190. DOI: 10.1186/s13059-017-1318-8 PMID: 28985763

Kwon, D. Y., Zhao, Y. T., Lamonica, J. M., & Zhou, Z. (2021). Locus-specific epigenetic remodeling controls neural differentiation. *Nature Communications*, *12*(1), 2931. PMID: 34006841

Lancaster, M. A., Renner, M., Martin, C. A., Wenzel, D., Bicknell, L. S., Hurles, M. E., Homfray, T., Penninger, J. M., Jackson, A. P., & Knoblich, J. A. (2013). Cerebral organoids model human brain development and microcephaly. *Nature*, *501*(7467), 373–379. DOI: 10.1038/nature12517 PMID: 23995685

Lanza, R., Russell, D. W., & Nagy, A. (2022). Engineering iPSCs with CRISPR for regenerative medicine. *Nature Reviews. Molecular Cell Biology*, *23*(4), 244–261.

Li, C., Zhan, W., Jin, Y., Wang, X., & Jin, Y. (2023). Prime editing: A powerful genome editing tool for human genetic diseases. *Translational Medicine (Sunnyvale, Calif.)*, *21*, 36.

Li, X., Zhang, W., & Chen, Z. (2022). Gene editing strategies for promoting axonal regeneration after spinal cord injury. *Stem Cell Research & Therapy*, *13*(1), 1–12. PMID: 34998430

Liang, P., Xu, Y., Zhang, X., Ding, C., Huang, J., Zhang, Z., & Zhou, C. (2022). CRISPR-engineered iPSC-derived retinal cells for treating degenerative blindness. *Cell Stem Cell*, *29*(3), 420–435.

Liu, X. S., Wu, H., Ji, X., Stelzer, Y., Wu, X., Czauderna, S., Shu, J., Dadon, D., Young, R. A., & Jaenisch, R. (2016). Editing DNA methylation in the mammalian genome. *Cell*, *167*(1), 233–247.e17. DOI: 10.1016/j.cell.2016.08.056 PMID: 27662091

Liu, Y., Lim, J., Teoh, S. H., & Zhang, Y. (2019). Engineering of osteoinductive signals in biodegradable scaffolds for bone tissue engineering using CRISPR. *Biomaterials*, *183*, 232–244.

Liu, Y., Yu, C., Daley, T. P., Wang, F., Cao, W. S., Bhate, S., & Qi, L. S. (2021). CRISPR activation and interference screens in primary human T cells decode cytokine regulation. *Cell*, *184*(2), 403–420.e17.

Liu, Y., Yu, C., Daley, T. P., Wang, F., Cao, W. S., Bhate, S., & Qi, L. S. (2022). CRISPR-based chromatin and epigenetic editing. *Nature Reviews. Methods Primers*, *2*, 6.

Maeder, M. L., Stefanidakis, M., Wilson, C. J., Baral, R., Barrera, L. A., Bounoutas, G. S., & Rebar, E. J. (2019). Development of a CRISPR-Cas9 gene-editing therapy for inherited retinal disease. *Nature Medicine*, *25*(2), 229–233. DOI: 10.1038/s41591-018-0327-9 PMID: 30664785

Manfrin, A., Tabata, Y., Paquet, E. R., Vuaridel, A. R., Rivest, F. R., Naef, F., & Lutolf, M. P. (2019). Engineered signaling centers for the spatially controlled patterning of human pluripotent stem cells. *Nature Methods*, *16*(7), 640–648. DOI: 10.1038/s41592-019-0455-2 PMID: 31249412

Mao, X., Fu, P., Wang, L., & Xu, J. (2019). CRISPR-Cas9-mediated gene editing in osteoarthritis: Targeting MMP13 for cartilage regeneration. *Annals of the Rheumatic Diseases*, *78*(2), 163–171.

Mendell, J. R., Al-Zaidy, S., Rodino-Klapac, L. R., Goodspeed, K., Gray, S. J., Kay, C. N., & Chamberlain, J. R. (2021). Current clinical landscape of AAV-based therapies. *Nature Reviews. Drug Discovery*, *20*(1), 35–55.

Minczuk, M., Silva-Pinheiro, P., Jackson, C. B., & Gammage, P. A. (2022). In vivo mitochondrial base editing via adeno-associated viral delivery to mouse post-mitotic tissue. *Nature Communications*, *13*(1), 750. DOI: 10.1038/s41467-022-28358-w PMID: 35136065

Morita, S., Noguchi, H., Horii, T., Nakabayashi, K., Kimura, M., Okamura, K., Sakai, A., Nakashima, H., Hata, K., Nakashima, K., & Hatada, I. (2016). Targeted DNA demethylation in vivo using dCas9–peptide repeat and scFv–TET1 catalytic domain fusions. *Nature Biotechnology*, *34*(10), 1060–1065. DOI: 10.1038/nbt.3658 PMID: 27571369

Nakamura, M., Srinivasan, P., Chavez, M., Carter, M. A., Dominguez, A. A., La Russa, M., & Qi, L. S. (2021). Anti-CRISPR-mediated control of gene editing and synthetic circuits in eukaryotic cells. *Nature Communications*, *12*(1), 530. PMID: 33483501

Nielsen, A. A. K., Der, B. S., Shin, J., Vaidyanathan, P., Paralanov, V., Strychalski, E. A., & Voigt, C. A. (2016). Genetic circuit design automation. Science, 352(6281), Nissim, L., Perli, S. D., Fridkin, A., Perez-Pinera, P., & Lu, T. K. (2014). Multiplexed and programmable regulation of gene networks with an integrated RNA and CRISPR/Cas toolkit in human cells. *Molecular Cell*, *54*(4), 698–710.

Nihongaki, Y., Kawano, F., Nakajima, T., & Sato, M. (2015). Photoactivatable CRISPR-Cas9 for optogenetic genome editing. *Nature Biotechnology*, *33*(7), 755–760. DOI: 10.1038/nbt.3245 PMID: 26076431

Osborn, M. J., Starker, C. G., McElroy, A. N., Webber, B. R., & Moriarity, B. S. (2021). Base editing-mediated correction of COL7A1 nonsense mutations in dystrophic epidermolysis bullosa. *Molecular Therapy. Nucleic Acids*, *26*, 333–342.

Ou, Z., Niu, X., He, W., Chen, Y., Song, B., Xian, Y., Fan, D., Tang, D., & Sun, X. (2016). The combination of CRISPR/Cas9 and iPSC technologies in the gene therapy of human β-thalassemia in mice. *Scientific Reports*, *6*(1), 32463. DOI: 10.1038/srep32463 PMID: 27581487

Patel, S., Kim, J., Herrera, M., Mukherjee, A., Kabanov, A. V., & Sahay, G. (2021). Development of CRISPR-Cas9 delivery vehicles for in vivo gene editing. *Advanced Drug Delivery Reviews*, *168*, 28–47.

Pausch, P., Al-Shayeb, B., Bisom-Rapp, E., Tsuchida, C. A., Li, Z., Cress, B. F., Knott, G. J., Jacobsen, S. E., Banfield, J. F., & Doudna, J. A. (2020). CRISPR–CasΦ from huge phages is a hypercompact genome editor. *Science*, *369*(6501), 333–337. DOI: 10.1126/science.abb1400 PMID: 32675376

Pickar-Oliver, A., & Gersbach, C. A. (2019). The next generation of CRISPR–Cas technologies and applications. *Nature Reviews. Molecular Cell Biology*, *20*(8), 490–507. DOI: 10.1038/s41580-019-0131-5 PMID: 31147612

Polstein, L. R., & Gersbach, C. A. (2015). A light-inducible CRISPR-Cas9 system for control of endogenous gene activation. *Nature Chemical Biology*, *11*(3), 198–200. DOI: 10.1038/nchembio.1753 PMID: 25664691

Polstein, L. R., & Gersbach, C. A. (2015). A light-inducible CRISPR-Cas9 system for control of endogenous gene activation. *Nature Chemical Biology*, *11*(3), 198–200. DOI: 10.1038/nchembio.1753 PMID: 25664691

Raj, B., Wagner, D. E., McKenna, A., Pandey, S., Klein, A. M., Shendure, J., Gagnon, J. A., & Schier, A. F. (2018). Simultaneous single-cell profiling of lineages and cell types in the vertebrate brain. *Nature Biotechnology*, *36*(5), 442–450. DOI: 10.1038/nbt.4103 PMID: 29608178

RegMedNet. (2022, February 11). In vivo gene editing of mitochondrial DNA shows potential to treat mitochondrial disorders. *Regenerative Medicine Net.*

Replogle, J. M., Norman, T. M., Xu, A., Hussmann, J. A., Chen, J., & Weissman, J. S. (2020). Combinatorial single-cell CRISPR screens by direct guide RNA capture and targeted sequencing. *Nature Biotechnology*, *38*(8), 954–961. DOI: 10.1038/s41587-020-0470-y PMID: 32231336

Schwank, G., Koo, B. K., Sasselli, V., Dekkers, J. F., Heo, I., Demircan, T., Sasaki, N., Boymans, S., Cuppen, E., van der Ent, C. K., Nieuwenhuis, E. E. S., Beekman, J. M., & Clevers, H. (2013). Functional repair of CFTR by CRISPR/Cas9 in intestinal stem cell organoids of cystic fibrosis patients. *Cell Stem Cell*, *13*(6), 653–658. DOI: 10.1016/j.stem.2013.11.002 PMID: 24315439

Shmakov, S., Smargon, A., Scott, D., Cox, D. B. T., Pyzocha, N., Yan, W., Abudayyeh, O. O., Gootenberg, J. S., Makarova, K. S., Wolf, Y. I., Severinov, K., Zhang, F., & Koonin, E. V. (2017). Diversity and evolution of class 2 CRISPR–Cas systems. *Nature Reviews. Microbiology*, *15*(3), 169–182. DOI: 10.1038/nrmicro.2016.184 PMID: 28111461

Silva-Pinheiro, P., Gammage, P. A., & Minczuk, M. (2022). Mitochondrial DNA editing: Recent advances toward therapeutic translation. *Trends in Molecular Medicine*, *28*(6), 514–527.

Smith, C., Abalde-Cela, S., & Abell, C. (2022). Delivery technologies for genome editing. *Nature Reviews. Materials*, *7*(12), 991–1009.

Sokka, J. E., Hendriks, D., Jussila, M., Tuusa, J., Guo, Y., Huhtala, M., & Weltner, J. (2022). CRISPR activation enables high-fidelity reprogramming into human pluripotent stem cells. *Stem Cell Reports*, *17*(3), 620–634. DOI: 10.1016/j.stemcr.2021.12.017 PMID: 35063129

Song, R. J., Han, L., Sun, H. F., & Shen, B. (2023). Advances in mitochondrial DNA base editing technology. *Yi Chuan*, *45*(8), 632–642. PMID: 37609815

Takebe, T., Sekine, K., Enomura, M., Koike, H., Kimura, M., Ogaeri, T., ... & Taniguchi, H. (2015). Vascularized and functional human liver from an iPSC-derived organ bud transplant.

Vlassis, A., Beattie, G. M., Sosa, E., Park, T. S., & Cohen, A. R. (2023). Integration of CRISPR-MAD7 constructs into genomic safe harbor sites enables stable gene expression in iPSC-derived lineages. *Frontiers in Genetics*, *14*, 10682145.

Wang, D., Tai, P. W. L., & Gao, G. (2020). Adeno-associated virus vector as a platform for gene therapy delivery. *Nature Reviews. Drug Discovery*, *19*(5), 325–350. PMID: 30710128

Wang, Y., Wang, F., & Zhang, L. (2021). CRISPR-based regenerative medicine: Target discovery and therapeutic design. *Cell Reports Medicine*, *2*(9), 100389. PMID: 34622226

Wang, Y., Wang, F., & Zhang, L. (2021). CRISPR-based regenerative medicine: Target discovery and therapeutic design. *Cell Reports Medicine*, *2*(9), 100389. PMID: 34622226

Wang, Y., Zhang, H., & Lin, C. (2022). CRISPR-Cas9-based therapeutic approaches in osteoarthritis. *Bone Research*, *10*(1), 1–14. PMID: 34975148

Wang, Z., Chen, Y., & Liu, J. (2023). CRISPR-corrected iPSCs in translational regenerative medicine. *Science Translational Medicine*, *15*(685), eabo6851.

Wroblewska, L., Kitada, T., Endo, K., Siciliano, V., Stillo, B., Saito, H., & Weiss, R. (2015). Mammalian synthetic circuits with RNA binding proteins for RNA-only delivery. *Nature Biotechnology*, *33*(9), 839–841. DOI: 10.1038/nbt.3301 PMID: 26237515

Xu, H., Xiao, T., Chen, C. H., Li, W., Meyer, C. A., Wu, Q., & Liu, X. S. (2022). Machine learning–based personalized CRISPR guide RNA design for therapeutic genome editing. *Nature Communications*, *13*(1), 4556. PMID: 35961959

Xu, W., Wang, J., & Wang, D. (2023). Advances in CRISPR-iPSC-derived cell therapies: Opportunities and challenges. *Nature Biotechnology*, *41*(7), 950–963.

Yin, H., Xue, W., Chen, S., Bogorad, R. L., Benedetti, E., Grompe, M., Koteliansky, V., Sharp, P. A., Jacks, T., & Anderson, D. G. (2014). Genome editing with Cas9 in adult mice corrects a disease mutation and phenotype. *Nature Biotechnology*, *32*(6), 551–553. DOI: 10.1038/nbt.2884 PMID: 24681508

Yuan, J., Ma, Y., Huang, T., Chen, Y., Peng, Y., Li, B., & Li, C. (2021). Efficient base editing for disease modeling in human pluripotent stem cells. *Cell Regeneration (London, England)*, *10*(1), 6. PMID: 33644818

Yusa, K., Rashid, S. T., Strick-Marchand, H., Varela, I., Liu, P. Q., Paschon, D. E., & Lacaud, G. (2011). Targeted gene correction of α1-antitrypsin deficiency in induced pluripotent stem cells. *Nature*, *478*(7369), 391–394. DOI: 10.1038/nature10424 PMID: 21993621

Zhang, X., Li, L., Xu, J., Wang, Z., & Wang, H. (2021). Lipid nanoparticles for genome editing: Progress and prospects. *Advanced Drug Delivery Reviews*, *168*, 137–151.

Zhang, Z., Li, J., & Huang, Y. (2023). Omics-informed CRISPR therapeutics: Precision targeting for regeneration. *Frontiers in Cell and Developmental Biology*, *11*, 1189.

Zhou, Y., Zhu, S., Cai, C., Yuan, P., Li, C., Huang, Y., & Huang, Y. (2019). Induction of muscle progenitor cells from fibroblasts by CRISPR-mediated activation of endogenous MyoD1. *Cell Reports*, *29*(3), 598–610.e4.

Chapter 7
CRISPR-Cas9 for Studying Genetic Effects of Radiation and Heavy Metals

Saloua Mabsor-Zgandaoui
https://orcid.org/0009-0006-4601-9059
Biology and Health Laboratory, Faculty of Sciences, Ibn Tofail University, Kenitra, Morocco

Khawla Rachmoune
https://orcid.org/0009-0001-6483-1060
Laboratory of Water Science, Microbial Biotechnology and Sustainability of Natural Resources, Faculty of Science Semlalia, Cadi Ayyad University, Marrakech, Morocco

Soumia Ed-Day
Biology and Health Laboratory, Faculty of Sciences, Ibn Tofail University, Kenitra, Morocco

ABSTRACT

CRISPR/Cas9 technology is a cutting-edge tool for studying the genetic effects induced by exposure to environmental stressors such as ionizing radiation and heavy metals. These agents are known to cause major molecular perturbations, including DNA damage, increased oxidative stress, enzymatic alterations and inflammatory responses in the nervous system. In this context, CRISPR/Cas9 makes it possible to identify and precisely target the genes involved in the mechanisms of response or vulnerability to these exposures. Thanks to this technology, it is possible to create cellular or animal models that faithfully reproduce the alterations observed in ex-

DOI: 10.4018/979-8-3373-4857-5.ch007

posed organisms, and to explore strategies for correcting or inactivating genes to mitigate toxic effects. The use of CRISPR/Cas9 in this field opens up new prospects for gaining a better understanding of the interactions between the environment and the genome, and ultimately for developing targeted therapeutic approaches to pathologies linked to pollution or radiological exposure.

1. INTRODUCTION

The invention and development of the CRISPR/Cas9 system represents one of the most revolutionary advances in molecular biology, redefining scientists' ability to interrogate, manipulate and rewrite the genome with unprecedented precision. Originally discovered as enigmatic repetitive sequences in prokaryotic genomes by Ishino et al. in 1987, these loci initially had an unknown function and attracted little attention outside the field of microbial genomics. In the following years, however, comparative analyses demonstrated that these clusters consist of short, regularly spaced repeats interspersed with unique spacer sequences derived from exogenous nucleic acids, ultimately leading to the recognition of their role as an adaptive immune system in bacteria and archaea. The conceptual leap occurred in 2002, when Jansen et al. coined the term CRISPR and recognized a conserved cluster of adjacent genes, which they called CAS (CRISPR-associated genes), thus providing a basis for CRISPR-Cas systems as modular defense mechanisms (Ishino et al. 1987).

From an evolutionary and mechanistic standpoint, CRISPR/Cas9, and more specifically the subtype II-A system mediated by Streptococcus pyogenes Cas9, represents a family of programmable nucleases directed by RNAs that target foreign genetic elements for cleavage. This system is defined by three operational phases: adaptation, expression and maturation, and interference. Biochemical and structural studies of the Cas9 endonuclease have revealed its guidance by a dual RNA, a crRNA–tracrRNA, that recognises complementary DNA adjacent to a PAM motif (the protospacer adjacent motif), which triggers double-strand cleavage via its HNH and RuvC dual nuclease domains (Jiang & Doudna, 2017).

Now, its edited form, consisting of a single guide RNA fused to Cas9, has become a universal tool for inducing knockout, knock-in, transcription modulation, and epigenetic editing of target genes. Recent updates to the classification of CRISPR/Cas systems have highlighted their astonishing evolutionary diversification. Makarova et al. (2025) have expanded the classification to two classes, seven types, and forty-six subtypes that distinguish systems with a multi-protein effector complex (class I) from those that mobilize a single-protein effector, such as Cas9, Cas12, or Cas13 (class II). Although class II systems represent a minority of CRISPR loci known in nature, their structural simplicity and programmability have made them the most widely used in biotechnology and medicine.

The advent of CRISPR/Cas9 has opened up new experimental approaches in the fields of genomic biology, radiobiology and environmental toxicology. In radiobiology, CRISPR allows controlled double-strand breaks to be introduced into DNA, enabling the study of DNA repair pathways and the identification of genetic determinants of radiosensitivity and radioresistance (Schmidt et al., 2023; Mladenov et al., 2025). CRISPR enables genetic and epigenetic modelling of heavy metal exposure, functional dissection of detoxification genes, and validation of molecular biomarkers involved in oxidative stress and metal homeostasis in toxicology.

2. FUNDAMENTALS OF THE CRISPR/CAS9 TECHNOLOGY

2.1. CRISPR sequence early identification

In 1987, Ishino et al. reported observing, after the *iap* gene sequence in the *Escherichia coli* genome, a structure consisting of five conserved sequences of 29 nucleotides, separated by unique spacers of 32 nucleotides. A team led by Ishino then highlighted this repeated organization in the *E. coli* genome, comprising five repeated sequences of 29 nucleotides interspersed with unique sequences known as spacers of 32 nucleotides (Ishino et al. 1987). A subsequent study by the same group showed that this sequence has no coding function and is found in other closely related species with no known function (Nakata, Amemura, and Makino 1989). Over the following decade, several teams observed similar structures in bacteria and archaea, giving them different names: Direct Variable Repeat or DVR, Tandem Repeat or TREP, Short Regularly Spaced Repeats or SRSR, Long Cluster of Tandem Repeat or LCTR, Spacers Interspaced Direct Repeats or SPIDR.

In 2002, the acronym CRISPR, standing for Cluster of Regularly Interspaced Short Palindromic Repeats, was proposed by Jansen et al. These unique sequences, unlike conventional repeats, aroused curiosity because they showed no homology in databases and had no common characteristics (Rund Jansen et al. 2002). The repeats, called DRs (direct repeats), contain a palindrome that could have a biological function. The presence of this palindrome and the location of CRISPRs suggest a role in DNA replication or transfer, hypotheses that were later invalidated. The study published in 2002, demonstrates that the repeated structure is transcribed abundantly in *Archaeglobus fulgidus* (Tang et al. 2002). By studying small RNAs involved in regulation, they identified RNAs corresponding to a DR plus a spacer, as well as their multimers, hybridizing at several CRISPR loci (Tang et al. 2002). They suggest that a long precursor is cut into smaller RNAs, a role that will be confirmed in other (Lillestøl et al. 2006). In 2002, Jansen et al. demonstrated the existence of a group of specific genes always located near a CRISPR, and called them *cas* genes for CRISPR-associated. These genes, conserved in several archaea and bacteria, together form the CRISPR-Cas systems. The team also described the leader, a sequence flanking the CRISPR on one side and possessing a TATA motif, which was later identified as the promoter for CRISPR transcription. Analysis of prokaryotic genomes revealed a wide diversity of *cas* genes (Ruud Jansen et al. 2002). Haft et al. (2005) proposed an initial classification into 45 families and several subtypes (Haft et al. 2005).

Studies published in 2005 show that the majority of spacers correspond to viral or plasmid sequences (Pourcel, Salvignol, and Vergnaud 2005; Mojica et al. 2005; Bolotin et al. 2005). Using bioinformatics, 2% of spacers could be linked to a probable origin. Bolotin et al. describe a short Pu-Py-A-A sequence close to the protospacer, aligned at the end of the DR, a sequence later called PAM (Protospacer Adjacent Motif). The study by Pourcel et al. on 108 strains of *Yersinia pestis* shows that CRISPR-Cas acquires new spacers in a polarized manner, making it possible to trace bacterial evolution (Pourcel, Salvignol, and Vergnaud 2005).

2.2. Classification of CRISPR Systems

The evolutionary classification of CRISPR-Cas systems has undergone a considerable revision recently to account for the growing diversity of variants, many of which appear to be rare or have only been discovered recently. According to Makarova et al. (2025), this new classification divides these systems into two classes, seven types and 46 subtypes, compared to six types and 33 subtypes previously. These two classes are defined by the architecture of their effector module: class 1 includes systems in which the effector complex consists of several Cas proteins forming a cr RNA-binding complex, while class 2 is based on a single multidomain protein that performs the entire effector function, such as Cas9, Cas12 or Cas13. Class 1 includes types I, III, IV and the new type VII, while class 2 includes types II, V and VI. Each type is then subdivided into subtypes based on variations in genetic composition, cas gene phylogeny, CRISPR locus structure and functional modalities, such as types III-A to III-I for type III. For example, in class 1, type III is now subdivided into nine subtypes, including the new III-G, III-H, and III-I, each with distinct architectures and mechanisms, such as inactivation of the Palm domain of Cas10 or the absence of an adaptation module (Makarova et al. 2015).

Likewise, in class 2, a new type II subdivision has been added, subtype II-D, which includes small archaeal Cas9 homologues from various archaea of the DPANN superphylum, highlighting the fact that class 2 systems are no longer strictly bacterial. This degree of granularity allows for a better understanding of the diverse evolution of CRISPR-Cas systems; the classification scheme is not limited to architecture, but also incorporates the co-occurrence of functional modules, the phylogeny of characteristic proteins such as Cas1 or Cas10, and connections with mobile genetic elements. It highlights in particular that certain types, such as type V, have developed at least 50 independent events of association between a TnpB-type nuclease and a CRISPR/adaptation locus, further complicating the taxonomic structure and thus requiring a clear distinction between subtypes and variants. The abundance of systems in genomes indicates that the well-characterized types, I to IV, dominate the distribution of prokaryotes, while newly identified variants form a "long tail" of rarity. The fact that the main types now appear to be well described suggests that most of the functional diversity has been detected, but many rare variants remain to be discovered (Makarova et al. 2020) (Figure 1).

Figure 1. CRISPR/Cas System Architecture: From Class I Multi-Effector Complexes to Class II Single Effectors (Choi, Cha, and Kim 2024).

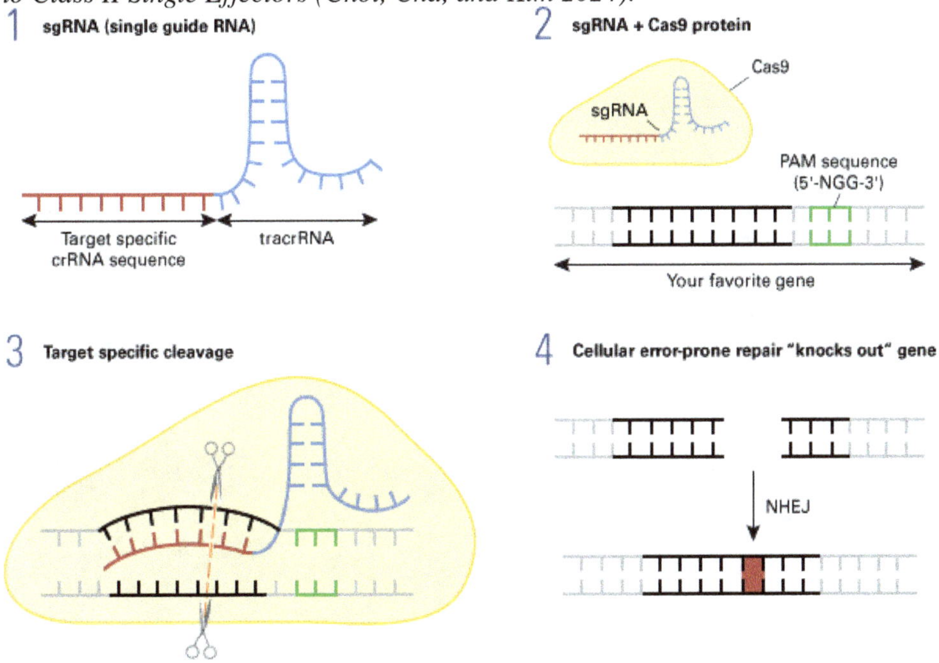

3. MECHANISTIC SEQUENCE OF CRISPR-CAS9 GENOME EDITING

The CRISPR-Cas9 editing process relies on a series of ordered molecular events beginning with the assembly of a ribonucleoprotein complex, followed by genomic target interrogation, double-stranded DNA cleavage, and repair by endogenous cellular pathways. Each step involves precise conformational transitions of the Cas9 endonuclease and strict molecular recognition events, which together ensure editing efficiency and targeting specificity (Mengstie and Wondimu 2021) (Figure 2).

Figure 2. Mechanism of CRISPR/Cas9-Induced Gene Knockout. The sgRNA, composed of a target-specific crRNA fused to a tracrRNA scaffold, associates with the Cas9 endonuclease to form an active ribonucleoprotein complex. Guided by base complementarity, this complex bind and cleaves the target double-stranded DNA. The resulting break is repaired by the error-prone NHEJ pathway, leading to insertions or deletions (INDELs) that disrupt gene function (Takara Bio).

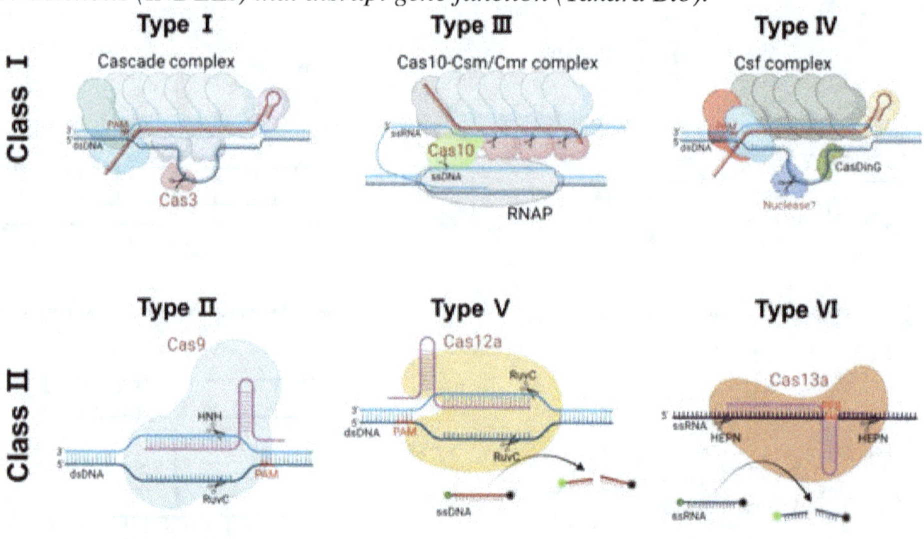

3.1. Mechanistic insight into CRISPR-Cas9: from guide RNA loading to DNA repair outcomes

3.1.1. Assembly of the Cas9-guide RNA ribonucleoprotein complex

The editing cycle starts when Cas9 binds to a guide RNA, producing a highly specific ribonucleoprotein complex. The guide RNA consists of two functional modules, including a scaffold region that allows stable binding to Cas9 and a 20-nucleotide spacer that ensures genomic specificity through complementary base pairing with the target DNA. This entails guide RNA loading, during which Cas9 undergoes a radical conformational change from an inactive bilobed form to an active surveillance conformation. Structural stabilization of the recognition lobe, REC, in particular the REC I and REC II domains, is achieved once the scaffold RNA threads into the amino acid pockets, forming hydrogen bonds and electrostatic interactions. This molecular rearrangement is essential, as Cas9 remains catalytically inert in the absence of RNA. Indeed, it is a search-ready ribonucleoprotein complex that scans genomic DNA in a linear, diffusion-dependent manner (Shao, Xu, and Chen 2016).

□ Target DNA interrogation and PAM detection

In a living cell, the Cas9-ARNg complex does not bind randomly to DNA. It first searches for a short motif, called the protospacer adjacent motif (PAM), which serves as an initial control for target recognition. For the most commonly used variant, Streptococcus pyogenes Cas9, this PAM corresponds to the trinucleotide sequence 5′-NGG-3′. A PAM is recognized by residues in the PAM interaction (PI) domain of Cas9, which inserts itself into the major groove of the DNA and probes the local nucleotide chemistry. Only after PAM binding does Cas9 locally melt the double-stranded DNA structure. This unwinding allows the spacer sequence of the guide RNA to hybridize with its complementary DNA strand. The stability of this RNA-DNA heteroduplex is highest near the PAM region, also known as the primer region. This region is critical: mismatches at this location are rarely tolerated and can prevent editing, while mismatches further away may still allow binding.

□ Double-Strand DNA Cleavage by Cas9 Nuclease Domains

Once the RNA–DNA hybrid is fully established, Cas9 transitions into its catalytic conformation. DNA cleavage is mediated by two distinct nuclease domains housed within Cas9 (Table 1). The two cleavage reactions occur in a coordinated fashion. Structurally, once the HNH domain is correctly positioned at the target strand, the RuvC domain undergoes activation, ensuring simultaneous cleavage and generating a blunt double-strand break located three nucleotides upstream of the PAM. This precise cleavage is the signature event that initiates genome modification (Jiang and Doudna 2017).

Table 1. Functional role of Cas9 nuclease domains (HNH and RuvC) in double-stranded DNA cleavage during CRISPR-Cas9 genome editing.

Nuclease Domain	Function	Cleaved Strand
HNH domain	Catalyzes cleavage at the DNA strand complementary to gRNA	Target strand
RuvC domain	Cuts the non-complementary DNA strand	Opposite strand

□ *DNA Repair Pathways: NHEJ vs. HDR*

After cleavage, the responsibility for editing shifts from CRISPR-Cas9 to the host cell's DNA repair systems. Two main pathways can repair the double-strand break (Table 2). Researchers exploit HDR for precise genome modification, whereas NHEJ is preferred for gene knock-out applications (Yang et al. 2020). However, HDR efficiency depends heavily on cell cycle stage and is limited in non-dividing cells (Liu et al. 2019).

Table 2. Comparison of cellular DNA repair pathways, NHEJ vs. HDR, following CRISPR-Cas9-induced double-strand breaks

Repair Pathway	Characteristics	Effect on Genome
Non-homologous end joining (NHEJ)	Error-prone ligation; does not require a repair template	Insertions/deletions, gene disruption
Homology-directed repair (HDR)	Template-dependent repair; requires donor DNA with homology arms	Precise gene correction or sequence insertion

3.2. Determinants of Cas9 target precision and molecular design

3.2.1. The accuracy and control of off-target activity

Cas9's fidelity in modifying genomic DNA depends mainly on the complementarity between the guide RNA and the target sequence. Immediately adjacent to the PAM, a critical region called the primer sequence plays a decisive role in target recognition. Perfect base pairing in this primer region is a prerequisite for initiating stable binding and DNA cleavage. Conversely, mismatches further away from the PAM are more often tolerated, allowing the Cas9-guide RNA complex to bind and occasionally cleave non-targeted genomic sites. These so-called 'off-target' events can cause insertion/deletion mutations at unwanted loci and compromise the genomic integrity of modified cells, which is a major concern, particularly in therapeutic applications. To minimise these risks, several advanced engineering strategies have been developed, including the use of truncated gRNAs with shortened spacer sequences to improve stringency, high-fidelity Cas9 variants with reduced non-specific interactions, and chemical modification of gRNAs to enhance guide-target pairing accuracy. Together, these improvements significantly reduce off-target activity and improve the safety profile of CRISPR-based interventions (Kanazhevskaya et al. 2025).

3.2.2. Structural organisation and functional dynamics of Cas9

Cas9 is divided into two main lobes that work cooperatively throughout the DNA targeting and cleavage processes. The REC lobe binds to the guide RNA and stabilises it, helping to position the ribonucleoprotein complex in a conformation that allows target surveillance. In contrast, the NUC lobe contains the catalytic domains responsible for DNA cleavage: the HNH domain, which cleaves the DNA strand complementary to the guide RNA, and the RuvC domain, which cleaves the opposite strand. The NUC lobe also contains the PAM interaction domain (PI), which ensures that Cas9 only binds to DNA sequences adjacent to the correct PAM motif. When it binds to a PAM, Cas9 undergoes a series of conformational changes that gradually clamp the DNA, forming an effective clamp. This spatial reorganisation ensures that nuclease activity is not released until RNA-DNA complementarity has been fully verified, thus serving as an internal molecular checkpoint that prevents unintended double-strand breaks (Nishimasu et al. 2014).

4. CRISPR/CAS9 APPLICATIONS IN RADIATION BIOLOGY

4.1. Mechanisms of Radiation-Induced Genetic Damage

Ionizing radiation (IR) exposure induces a variety of cellular damage in living cells, primarily affecting DNA, but also lipids, proteins, and other essential cellular components (Mabsor-Zgandaoui et al. 2025a). This damage results either from direct interactions between radiation and biological molecules, or from indirect reactions mediated by the production of reactive oxygen species (ROS) (Mabsor-Zgandaoui et al. 2025b). Damage observed includes single-strand breaks (SSBs), double-strand breaks (DSBs), base modifications, intra- and inter-strand crosslinks, and DNA-protein adducts (Desouky et al., 2015). These alterations, if not properly repaired, can lead to mutations, chromosomal instability, and, in the long term, to malignant cell transformation.

4.1.1. Double-strand breaks (DSBs): the most critical lesion

Double strand breaks (DSBs) are the most damaging type of lesion and the hardest to fix. Just one DSB that isn't fixed can kill a cell, while fixing it improperly can lead to deletions, inversions, or chromosomal translocations that can mess up the genome. DSBs can be directly induced by radiation or indirectly via EROs generated during irradiation (Hanscom and McVey, 2020). On detection of a DSB, the cell activates a set of molecular signals collectively referred to as DNA Damage Response (DDR). This involves a coordinated cascade of damage detection, signaling, and repair, involving proteins such as ATM (ataxia-telangiectasia mutated), ATR (ATM and Rad3-related), DNA-PKcs, and p53 (Nikfarjam and Singh, 2023). In particular, activation of ATM leads to the phosphorylation of numerous substrates, including H2AX (γ-H2AX), an early and sensitive marker of double-strand breaks (Valente et al., 2022). These changes signal the presence of damage and recruit the appropriate repair proteins.

There are two main repair pathways that contribute to the resolution of DSBs: homologous recombination (HR, *Homologous Recombination*) and the joining of non-homologous ends (NHEJ, *Non-Homologous End Joining*). Considered a reliable pathway, HR uses a homologous sequence as a template and preferentially occurs during the S/G2 phase of the cell cycle. In comparison, NHEJ, which directly joins broken ends without a template, is faster but often prone to errors (Stinson and Loparo, 2021). Choosing between these two pathways depends not only on the cellular context and the type of lesion, but also on the surrounding chromatin and the local epigenetic status.

4.1.2. Oxidative stress and secondary damage

In parallel with DSBs, ionizing radiation generates significant amounts of reactive oxygen species (ROS) such as hydroxyl radicals (•OH), superoxide ($O_2•^-$), and hydrogen peroxide (H_2O_2) (Mabsor-Zgandaoui et al. 2025b). These ROS can persist in the cell and cause prolonged oxidative damage. Purine and pyrimidine bases are particularly vulnerable to these attacks, leading to modifications such as the formation of 8-oxoguanine, a biomarker of oxidative stress (Chatgilialoglu et al., 2021). ROS can also oxidize membrane lipids and proteins, resulting in structural and functional alterations that can disrupt intracellular signaling and compromise cell survival (Mabsor-Zgandaoui el al.2026a).

Recently, studies have shown that chronic oxidative stress induced by radiation amplifies DDR signaling, resulting in persistent alterations in cell cycle regulation, deregulation of repair pathways, and increased susceptibility to secondary mutations (Wang et al.,2025). This means that oxidative damage is not limited to isolated alterations but actively contributes to the spread of genomic instability in irradiated cells and their clonal descendants.

4.2. Response to DNA damage and cellular resistance

The capacity of tumor cells to effectively activate DDR pathways after irradiation is a determining factor in their survival and therapeutic resistance. Studies have demonstrated that certain tumors overexpress genes involved in repair, such as RAD51, BRCA1/2, or DNA-PKcs, which increases their ability to resist radiation (Jurkovicova et al., 2022). In contrast, cells defective in these genes exhibit radiation hypersensitivity.

This finding led to the development of agents capable of interfering with DDR or inhibiting DSB repair, known as radiosensitizers. ATM, ATR, and PARP1 sequestrants are among the most widely studied of these compounds, as they potentiate the effects of radiation by preventing the restoration of damaged DNA (Bright et al., 2022). The synthetically lethal approach takes advantage of specific deficiencies in certain tumor cells, such as BRCA1/2 gene mutations, to induce selective cytotoxicity when a second repair pathway is inhibited. These combined strategies, combining radiotherapy and the targeted inhibition of repair, aim to expand the window of opportunity for treatment by boosting tumor radiosensitivity while minimizing the damage to healthy tissue (Neiger et al., 2021).

4.3. Central role of the NRF2 transcription factor

The NRF2 (Nuclear factor erythroid 2–related factor 2) transcription factor plays an essential role in cellular defense against radiation-induced oxidative stress. Under basal conditions, NRF2 is retained in the cytoplasm by KEAP1, which promotes its degradation. During oxidative stress, NRF2 is released, translocated to the nucleus, and activates the transcription of numerous antioxidant genes, including HO-1 (heme oxygenase-1), NQO1 (NAD(P)H quinone oxidoreductase 1), and enzymes involved in glutathione and NADPH regeneration (Tripathi et al., 2024).

Recent studies have shown that NRF2 activation improves the survival of cells exposed to ionizing radiation by stimulating not only ROS neutralization but also DNA damage repair (Moloudi and Haghdoost, 2025). The persistent activation of NRF2 is, however, ambivalent: while it protects normal tissues and reduces acute radiation syndrome, it can also promote tumor radioresistance. Consequently, the fine modulation of NRF2 controlled activation in healthy tissues, inhibition in tumors represents a promising therapeutic approach to reconcile radiation protection and anticancer efficacy (Occhiuto et al., 2023).

4.4. Target Genes Involved in Radiation Response

A number of genes play key roles in coordinating the cellular response to ionizing radiation. The ATM (ataxia telangiectasia mutated) kinase gene acts as a master regulator of the response to DNA damage: it detects double-strand breaks (DSBs), triggers phosphorylation cascades, and initiates cell cycle arrest and the recruitment of repair mechanisms (Shim, 2025). The TP53 (p53) gene mediates the decision between repair, cell arrest, and apoptosis: depending on the extent of the damage and the cellular context, p53 directs the cell toward survival or death, and TP53 mutations profoundly alter radiosensitivity (Al-Arafat et al., 2024). The RAD51 gene is key in homologous recombination (HR): it enables the correct assembly of DNA ends during DSBs, ensuring genomic stability; its alteration leads to an increased propensity for erroneous repairs. Finally, the NFE2L2 pathway (the Nrf2 transcription factor) regulates the expression of many antioxidant and repair genes: its excitation can confer increased protection to tissues exposed to radiation, but also promote tumor radioresistance.

For instance, in a gastric cancer model, an in silico and in vitro investigation demonstrated that TP53, RAD51, and the BAX gene were over expressed, while ATM was under expressed: after irradiation, reducing the expression of TP53, RAD51, or BAX by gene knockdown significantly decreased the colony formation and cell migration of tumor cells, confirming their role in resistance to radiotherapy (Zhang and He, 2025). Recent research on Nrf2 has also shown that its role is not limited to antioxidant activity: one study showed that Nrf2 cooperates with TOPBP1 to activate the ATR CHK1 pathway in lung cancer cells after irradiation, thereby increasing their DSB repair capacity and radiation resistance (Sun et al., 2024). The data suggested that modulation of these genes (inhibition, activation, mutation) changes the balance between radio-sensitivity and radio-resistance, opening up new avenues for targeted therapeutic strategies.

4.5. Use of CRISPR/Cas9 to Model Radiation-Induced Damage and DNA Repair

The arrival of CRISPR/Cas9 genomic editing technology has totally changed how we study DNA repair mechanisms, in particular when it comes to damage caused by ionizing radiation. This instrument offers the possibility of introducing double-strand breaks (DSBs) in a targeted and precise manner at genome-specific loci, thereby simulating the damage typically caused by irradiation, but with rigorous spatio-temporal control (Schmidt et al., 2023). This enables the mechanisms of DSB detection, signaling, and repair to be dissected under perfectly controlled experimental conditions.

CRISPR/Cas9 is valuable to radiobiology in two ways. On the one hand, it is an experimental model for DNA damage, allowing the generation of controlled numbers of specific DSBs in precise regions of the genome. Such artificially induced fractures closely mimic the damage caused by ionizing radiation, while offering the possibility of monitoring repair in real time using molecular visualization systems. On the other, CRISPR enables the functional manipulation of key genes involved in the DNA damage response (DDR), through inactivation (knockout), activation, or specific modification, in order to assess their individual role in radiosensitivity and repair fidelity (Mladenov et al., 2025).

Using this technology, innovative cell models have been developed that enable the generation of a pre-determined number of double-strand breaks and the study of the dynamics of repair protein recruitment, such as 53BP1, BRCA1, RAD51, and RPA. The models also revealed how repair pathway choices between homologous recombination (HR) and non-homologous end joining (NHEJ) are influenced by chromatin structure, cell cycle phase, and DNA end complexity (Shimada et al.,2025).

In addition, Cas9-derived report systems, such as HR/NHEJ reporter assays or quantitative repair (QR) reporters, have been developed to quantify the choice and efficiency of repair pathways at the cellular level. These tools make it possible to accurately measure how certain mutations, pharmacological treatments, or environmental conditions (hypoxia, oxidative stress) alter the equilibrium between the two principal DSB repair pathways. Using these approaches, several studies have shown that the spatial proximity between different DSBs called clustered DSBs significantly increases the probability of erroneous repairs, thereby generating chromosomal translocations or complex deletions (Tei et al., 2025).

The application of CRISPR/Cas9 technology has also led to the discovery of new regulators of the response to radiation. Genome-wide CRISPR screenings have identified previously unknown genes and pathways involved in radioresistance or radiosensitivity. These studies have highlighted the role of certain proteins involved in redox metabolism, chromatin, and transcriptional control in modulating the cellular response to radiation (Zhou et al.,2023). These findings pave the way for the development of new therapeutic targets designed to improve radiation protection of healthy tissues or, conversely, to enhance radiotoxicity in resistant tumors.

In vitro and in vivo models based on CRISPR/Cas9 have also demonstrated that local chromatin and its state of openness (euchromatin vs. heterochromatin) modulate repair complex exposure to DSBs significantly, affecting repair fidelity and genome stability after irradiation (Xue and Greene, 2021). This confirms that nuclear and chromatin dynamics are key factors in how cells respond to radiation.

4.6. Radiosensitivity, Radioprotective Responses, and Gene-Based Therapeutic Strategies

Recent advances in large-scale CRISPR screening have had a profound impact on our knowledge of the mechanisms of radiosensitivity and radioprotective effects. If applied to tumor cell lines exposed to ionizing radiation, these loss-of-function screens can identify genes whose loss increases sensitivity (radiosensitizers) or, conversely, confers resistance (radioprotectors). The genes belong mainly to key gene pathways such as DNA damage response (DDR), cell cycle regulation, energy metabolism, and oxidative stress management. For instance, selective experiments using CRISPR/Cas9 technology confirmed that inactivation of the ATM gene, a central pivot in DNA damage signaling, leads to a marked increase in radiosensitivity in several tumor types. Also, deleting RAD51, a key part of homologous recombination, messes up how double-strand breaks are fixed and makes radiation deadlier to cells. Conversely, alterations in the TP53 gene produce variable effects depending on the cellular context: complete loss of function can amplify radiosensitivity, while certain gain-of-function mutations promote post-irradiation cell survival. These findings illustrate the complexity of radiation response mechanisms and highlight the major contribution of CRISPR approaches to the identification of new therapeutic targets (Tamaddondoust et al., 2022).

At a parallel level, numerous studies have highlighted the decisive role of endogenous antioxidant pathways in the radiation protection of healthy tissues. In particular, the activation of the transcription factor Nrf2 (Nuclear factor erythroid 2–related factor 2) has proven crucial in cellular defense against oxidative stress induced by ionizing radiation. Nrf2 regulates the transcription of a set of genes encoding detoxification and redox regulation enzymes such as SOD1, HO-1, and GCLM, thereby limiting the formation and deleterious effects of reactive oxygen species (ROS). Such activation contributes to the protection of normal cells while promoting tissue survival (Mabsor-Zgandaoui et al. 2026b). In tumor cells, excessive Nrf2 activation can also induce radioresistance, making it necessary to differentiate between healthy cells and cancer cells (Ngo and Duennwald, 2022). These dual properties offer an interesting therapeutic avenue: activating Nrf2 in healthy tissue to limit radiation toxicity, while selectively inhibiting this pathway in tumors to increase their sensitivity.

From a translational aspect, these advances have paved the way for novel gene therapy approaches aimed at adjusting radiosensitivity based on cell type. For example, targeted inhibition of repair enzymes such as ATM or DNA-PKcs can sensitize tumor cells to lower doses of radiation, while transient overexpression of Nrf2 or inactivation of pro-apoptotic genes such as BAX or PUMA can reduce damage to healthy tissue (Al-Zoubi et al., 2025). In recent years, targeted genetic modulation has been used to enhance the efficacy of radiation therapy. These combined approaches of targeted genetic modulation and radiation therapy represent a major step towards precision medicine, where the radiobiological response can be tailored to the molecular and epigenetic profile of each individual (Viktorsson et al., 2023).

In synthesis, the incorporation of CRISPR technologies with functional radiogenomics approaches offers unprecedented opportunities to understand, predict, and manipulate the molecular determinants of radiosensitivity. These approaches pave the way for personalized, genome-guided radiation therapy, in which tumor vulnerabilities can be harnessed while protecting healthy tissue, thereby optimizing the benefit/risk ratio of modern radiation treatments.

4.7. Emerging Clinical Perspectives in Radiogenomics and Radiation Therapy Enhancement

The radiogenomics field has quickly established itself as a central focus at the interface of genomics and radiation oncology. By combining genomic, transcriptomic, epigenomic, and proteomic data, radiogenomics aims to understand how individual genetic variations influence cellular response to ionizing radiation. This research approach seeks to identify biomarkers that predict radiosensitivity, DNA repair efficiency, and normal tissue toxicity, paving the way for personalized radiation therapy (Guo et al., 2025). Gene polymorphisms in key DNA repair genes, such as ATM, TP53, XRCC1, or BRCA1/2, have been associated with differentially responding to radiation, impacting both treatment efficacy and side effects. In particular, the analysis of gene expression profiles in patients with different responses to radiation therapy has enabled the stratification of patients into groups with similar responses.

Recent progress in multi-omics integration has enhanced the predictive potential of radiogenomic models. The integration of metabolomics and proteomics provides insights into metabolic reprogramming and oxidative stress pathways activated by irradiation. For instance, metabolic biomarkers such as glutathione-related enzymes or lipid peroxidation products are now being investigated as indicators of tissue-specific radiosensitivity (Mabsor-Zgandaoui et al. 2026b). In addition, epigenetic changes, including DNA methylation patterns and histone modifications, appear to be major determinants of the long-term effects of radiation therapy, such as fibrosis, carcinogenesis, or neurocognitive decline. The integration of these different biological layers into predictive algorithms improves the accuracy of individual risk assessment and treatment planning (Maan et al., 2023).

Clinically, these advances are driving a shift toward precision radiation therapy, where the dose, fractionation, and combination with radiosensitizers or radioprotectors can be tailored to each patient's molecular profile. For instance, patients with mutations in DNA repair genes may receive reduced-dose regimens or adjuvant treatments using radioprotective biomaterials to limit damage to healthy tissue (Rachmoune et al. 2026). In contrast, tumors with repair pathway deficiencies can be targeted by synthetic lethality strategies, combining radiotherapy with PARP, ATM, or DNA-PK inhibitors to selectively increase tumor radiosensitivity while sparing normal cells (Shiridokht et al., 2025).

The intersection of CRISPR/Cas9 genome editing and radiogenomic profiling also represents a transformative advance. CRISPR-based screening platforms enable the functional validation of biomarkers identified through omics studies, establishing causal links between genotype and radiation-induced phenotype. In a clinical context, these can be used to model a patient's specific response in organoid systems or to create cell lines that reflect individual radiosensitivity. This synergy between genome editing and multi-omic analyses is set to accelerate translational research, promoting the design of next-generation precision radiotherapy protocols, maximizing tumor control, and minimizing adverse effects (Li et al., 2024).

The integration of radio-genomics into clinical practice heralds a paradigm shift toward data- and biology-driven radiation oncology. The development of AI-assisted predictive models, trained on large-scale clinical and omics datasets, will enable the refinement of personalized treatments and the identification of optimal therapeutic combinations for each patient. With the continuous evolution of technologies and computational tools, the fusion of genomics, biomaterials, and precision radiation therapy is set to redefine cancer care, transforming radiation therapy from a uniform modality into a highly individualized, adaptive, and biologically informed intervention.

4.8. Critical overview of current limitations in genotoxicity and radiotoxicity research

The conventional genotoxicity and radiotoxicity assays, although recognized as the pillars of toxicological research, are progressively being seen as having serious limitations. These approaches, including the comet assay, micronucleus test, and quantification of γH2AX foci, reliably detect general indicators of DNA damage and chromosomal instability, but do not identify the specific genetic loci or molecular determinants involved in these processes. As they typically measure general cellular phenotypes rather than pathway-specific mechanistic outcomes, these conventional assays cannot distinguish the underlying causes of the observed DNA damage. For example, it is impossible to determine whether the damage is due to defects in DSB recognition, chromatin remodeling, or checkpoint signaling itself, not to mention differential responses between cell types depending on chromatin state, timing of replication, or transcriptional activity, which remain masked by traditional assessment criteria (Xue and Greene, 2021; Guo et al., 2025).

4.8.1. Constraints on functional interpretability

Genotoxicity test results often lack functional clarity. The commonly used marker for detecting DSBs, γH2AX, can be induced by various cellular stresses, including replication stress and apoptotic events, and is not specific to canonical DSBs. Therefore, increased γH2AX does not reliably predict whether repair mechanisms are activated, checkpoints are failing, or unrepaired damage is accumulating. Likewise, micronucleus formation highlights chromosomal instability or segregation errors, but does not provide much detail about the cause of the problem: incorrect repair, centrosome dysfunction, or kinetochore instability. It is these ambiguities that make it difficult to establish direct relationships between genes and functions or to identify exploitable regulatory targets from toxicological data sets (Valente et al., 2022).

4.8.2. Variability in methodology and reproducibility

The output of classical assays is often influenced by various methodological factors such as cell culture conditions, number of passages, oxygen levels, and radiation quality, leading to high variability between laboratories and limiting the ability to predict human exposure scenarios. This problem is particularly acute when attempting to model chronic or low-dose exposures relevant to occupational or environmental settings where the sensitivity of tests to subtle changes in DNA repair or adaptive cellular responses is limited. In radiotoxicity research, γH2AX is commonly used as a surrogate for repair kinetics, but current protocols fail to capture the dynamic changes between different repair systems or detect the mutagenic consequences of complex, clustered lesions caused by high-LET radiation. Lack of genomic and epigenetic context another major obstacle to mechanistic toxicology is the lack of genomic, epigenomic, and spatial cellular context provided by conventional genotoxicity methodology (Desouky et al., 2015).

Conventional assays cannot determine how genetic polymorphisms, chromatin structure, transcriptional activity, or three-dimensional genome organization may interact to influence radiosensitivity and responses to damage. Similarly, many genotype-phenotype associations remain correlative rather than experimentally established due to difficulties in identifying robust biomarkers and developing tailored radiation protection strategies. Overall, these various limitations underscore the fact that, while classical genotoxicity tools can quantify overall damage, they cannot reveal the causal molecular mechanisms, site specificity, and individual susceptibility that determine toxicity outcomes. There is an urgent need for advanced, multifaceted approaches capable of determining not only the extent, but also the context and cause of damage, in order to enable a mechanistic understanding that supports precision toxicology.

5. CRISPR/CAS9 APPLICATIONS IN HEAVY METAL TOXICITY STUDIES

5.1. Genetic and epigenetic effects of heavy metals

Environmental contaminants, particularly heavy metals such as lead (Pb), mercury (Hg), copper (Cu), cadmium (Cd), and arsenic (As), are of global concern due to their widespread distribution, accumulation, and persistence for extended periods across different ecological compartments (Adegbola & Adetutu, 2024; Ostrovska et al., 2023). Due to human activities, including agricultural practices, mining, and industrial activities, the concentration of these toxic elements in water, soil, and air has gone beyond natural background values (Ostrovska et al., 2023), which causes serious health problems and affects the balance of ecosystems. The toxicity of heavy metals is associated with genetic and epigenetic alterations leading to genomic instability and disrupted gene function, resulting in various pathological conditions, including cancers (Joneidi et al., 2019; Mishra et al., 2010; Bower et al., 2005).

5.2. Genetic Effects of Heavy Metals

The genotoxicity of heavy metals includes mutations, chromosomal aberrations, and DNA damage, which disrupt normal gene function and potentially contribute to the development of diseases (Joneidi et al., 2019). The mechanisms by which they exert their toxic effects on the genome include oxidative stress (Khan et al., 2024; Joneidi et al., 2019). It is recognized that trace elements such as Cd, Hg, As, Pb, and zinc generate free radicals and reactive oxygen species (ROS) that overwhelm the body's antioxidant defenses such as superoxide dismutase (SOD), catalase (CAT), and glutathione (GSH), resulting in oxidative modifications of genetic material, which cause DNA strand breaks, base modifications, and cross-linking, increasing the likelihood of mutations and chromosomal abnormalities (Khan et al., 2024; Joneidi et al., 2019). Furthermore, certain heavy metals can directly bind DNA, forming adducts or cross-links that interfere with DNA replication and repair processes, leading to mutations or deletions (Joneidi et al., 2019). They can also interfere with gene expression pathways involved in apoptosis and detoxification, further contributing to genetic instability. The accumulation of such DNA damage, if unrepaired, increases the risk of mutations and consequently the development of diseases (Adegbola & Adetutu, 2024).

5.3. Epigenetic effects of heavy metals

Heavy metals not only cause direct genetic damage but also affect gene regulation without altering the DNA sequence through epigenetic mechanisms (Joneidi et al., 2019). Metals, such as cadmium, nickel, arsenic, lead, chromium, and mercury, can interfere with DNA methylation, histone modifications, and non-coding RNA expression, consequently impairing normal cellular functioning and facilitating disease progression, including carcinogenesis and other health disorders (Khan et al., 2024). Studies have indicated that exposure to Cd, As, Pb, and nickel may cause hypomethylation or hypermethylation of specific gene promoters, which can lead to abnormal gene expression, particularly genes involved in cell cycle control, detoxification, and inflammation (Khan et al., 2024; Joneidi et al., 2019; Ryu et al., 2015). Additionally, these metals influence histone modifications, thereby impacting chromatin structure and gene accessibility. Consequently, the epigenetic alterations induced by heavy metal exposure underscore the significant influence of environmental factors in shaping long-lasting and heritable traits, which may increase disease susceptibility across generations (Adegbola & Adetutu, 2024; Joneidi et al., 2019).

5.4. Key genes related to detoxification, metal transport, and cellular defense

Key genes implicated in detoxification, metal transport, and cellular defense form a complex biological network that allows organisms to regulate and alleviate heavy metal toxicity. These include genes encoding metallothioneins, glutathione S-transferases, antioxidant enzymes such as SOD and CAT, and membrane transporters like ATP-binding cassette (ABC) transporters (Table 3).

Table 3. Representative genes involved in metal detoxification, transport and cellular defence.

Functional category	Genes / families	Principal function	References
Metal-binding proteins	Metallothioneins (MT1–MT4)	High-affinity metal chelation; redox buffering	Yang, 2024.
Metal importers	ZIP family (ZIP1, ZIP8, ZIP14)	Uptake of Zn, Fe, Mn (and Cd co-transport)	Hu, 2024; Zhang et al., 2023.
Metal exporters / chaperones	ZnT (SLC30), ATP7A/B, ATOX1	Efflux, intracellular trafficking, chaperoning	Bui & Inaba, 2024.
Phase II detox / thiol metabolism	GCLC / GCLM, GST families (GSTM1, GSTT1)	GSH synthesis; conjugation of xenobiotics	Sha et al., 2025.

continued on following page

Table 3. Continued

Functional category	Genes / families	Principal function	References
Antioxidant enzymes	SOD1/2, CAT, PRDXs	Scavenge ROS produced secondarily to metal stress	Ngo, 2022.
Transcriptional regulators	NFE2L2 (Nrf2), MTF-1	Coordinate inducible expression of detox genes	Ngo, 2022; Yang, 2024.

The detoxification of xenobiotics and the regulation of intracellular metal concentrations depend on a coordinated network of genes responsible for metal binding, transport, and oxidative stress defense. These pathways are evolutionary conserved from bacteria to plants and animals, emphasizing their critical functions in maintaining cellular viability. Three major functional categories of genes can be mentioned: metal-sequestering proteins, membrane metal transporters, and detoxification/antioxidation defense enzymes (Mabsor-Zgandaoui et al. 2026c). Such groups operate in concert with each other: metal transporters modulate cellular exposure, metal-binding proteins buffer surplus ions, and detoxification enzymes stave off cellular injury via ROS (Clemens, 2022; Festa & Thiele, 2023).

5.4.1. Metallothioneins and Metal-Binding Proteins

One of the best-characterized families of metal-binding genes is the metallothioneins. These are small proteins, 6-7 kDa, with a high content of cysteine residues, which show a strong affinity for divalent and monovalent metals due to the presence of thiol-metal coordination complexes (Cruz-Rodríguez et al., 2024). They have the ability to scaffold essential ions like Zn^{2+} and Cu^{2+}, while being able to bind toxic heavy metals like Cd^{2+} and Hg^{2+}. Their transcription is rapidly induced by metal overload, oxidative stress, and cytokine signaling, placing MTs at the crossroads between xenobiotic detoxification and immune response (Bordin et al., 2023).

Recent studies point to MTs as active agents rather than simple metal "sponges." Besides buffering free metal ions, their role in indirectly mitigating oxidative damage includes preventing the Fenton reaction, a metal-mediated mechanism of ROS production (Balzano et al., 2021). Changes in MT expression have recently been associated with cancer and metabolic disorders through transcriptomic studies and identified them as candidate biomarkers of exposure and disease prognosis (Neslund et al., 2022).

5.4.2. Membrane Metal Transporters

Metal transporter genes decide the trafficking of metal ions across cellular compartments and membranes. Their activity dictates intracellular metal exposure and, therefore, influences how strongly stress response pathways are activated.

Recent cryo-EM studies revealed that ZIP transporters operate via a metal shuttling mechanism involving an elevator-like movement across the membrane. This insight, from Zhang et al. (2023), explains how these proteins achieve ion selectivity while preventing uncontrolled influx of toxic ions. The Natural Resistance-Associated Macrophage Protein family, NRAMP, represents transporters of iron and manganese that are conserved from bacteria to mammals. NRAMP1 polymorphisms have been associated with susceptibility to tuberculosis and other infections, underscoring how metal homeostasis can intersect with immune function. Copper regulation involves two groups of transport genes: the CTR family responsible for copper import and P-type ATPases ATP7A/ATP7B responsible for copper export. Dysfunction in ATP7B causes the genetic disorder Wilson's disease due to copper accumulation and neurotoxicity (Weiss et al., 2023). These findings demonstrate that metal transporters represent important gatekeepers: they determine cellular exposure to metals and influence global detoxification capacity.

5.4.3. Detoxification and Antioxidant Defense Genes

Alongside metal sequestration and transport, cells engage an enzymatic detoxification system focused around GSH. GSH is the most abundant intracellular thiol and has been implicated in the neutralization of metals and ROS (Forman, 2023). Synthesis is dependent upon the enzyme complex glutamate–cysteine ligase (GCL) and is induced upon the exposure of cells to xenobiotics.

Glutathione S-transferases (GSTs) conjugate GSH to electrophilic toxins or oxidized biomolecules and enhance their solubility for elimination. Comparative genomics shows the expansion of GST gene families in organisms frequently exposed to metals, including aquatic animals and hyperaccumulator plants (Deng et al., 2024). Functional assays demonstrated that overexpression of GST isoforms improves metal tolerance and decreases lipid peroxidation, confirming that these are indeed determinants of the metal-resistant phenotype (Li et al., 2023).

Other core antioxidant genes include SOD, CAT, and PRDX. These enzymes degrade ROS before they can cause damage to membranes, DNA, or proteins.

5.4.4. Redox-Sensitive Regulatory Pathways: The Nrf2/Keap1 Control System

Nrf2 is a transcription factor that usually mediates the expression of genes involved in response to oxidative and xenobiotic stress. In basal conditions, Nrf2 is retained in the cytoplasm by Keap1 and targeted for degradation. During conditions of oxidative stress or metal exposure, Keap1 is modified through conformational changes, thereby releasing Nrf2 to translocate into the nucleus (Yamamoto & Kensler, 2024).

Once in the nucleus, Nrf2 binds to AREs and drives the expression of:

- Metallothioneins MT1, MT2
- Glutathione synthesis genes (GCL)
- Detoxification enzymes: GST, NQO1
- Heme oxygenase-1 (HMOX1), a cytoprotective enzyme

As indicated in zebrafish and mammalian models, the strong activation of the Nrf2 pathway is related to resistance against heavy metal toxicity and improved survival rates (Zhao et al., 2024).

6. CONCLUSION AND PERSPECTIVES

CRISPR/Cas9 has progressed from a natural microbial immune mechanism to one of the core technologies underpinning modern genetic engineering. Its specificity, modularity and relative simplicity have enabled unprecedented advances in genome editing, functional genomics, radiobiology and environmental toxicology. In addition to facilitating gene correction or disruption, CRISPR is a tool for elucidating biological mechanisms, identifying therapeutic targets and guiding precision medicine. The ongoing refinement of Cas variants and delivery strategies will further accelerate its transition to safe and effective clinical applications.

REFERENCES

Adegbola, P. I., & Adetutu, A. (2024). Genetic and epigenetic modulations in toxicity: The two-sided roles of heavy metals and polycyclic aromatic hydrocarbons from the environment. *Toxicology Reports*, *12*, 502–519. DOI: 10.1016/j.toxrep.2024.04.010 PMID: 38774476

Al-Arafat, T., Mao, A., Katsube, T., & Wang, B. (2024). Exploring the role of p53 in radiosensitivity: A key player in cancer therapy. *Radiation*, *4*(4), 309–324. DOI: 10.3390/radiation4040023

Balzano, L., Tumiatti, V., & Arnesano, F. (2021). Metallothioneins as multitasking mediators of metal homeostasis and redox balance. *Journal of Inorganic Biochemistry*, *225*, 111595. DOI: 10.1016/j.jinorgbio.2021.111595

Bolotin, A., Quinquis, B., Sorokin, A., & Ehrlich, S. D. (2005). Clustered regularly interspaced short palindrome repeats (CRISPRs) have spacers of extrachromosomal origin. *Microbiology*, *151*(8), 2551–2561. DOI: 10.1099/mic.0.28048-0 PMID: 16079334

Bordin, D. L., Silva, T. B., & Pedrosa, F. O. (2023). Regulation and functional diversity of metallothionein isoforms across biological systems. *Environmental Toxicology and Pharmacology*, *104*, 104216. DOI: 10.1016/j.etap.2023.104216

Bower, J. J., Leonard, S. S., & Shi, X. L. (2005). Conference overview: Molecular mechanisms of metal toxicity and carcinogenesis. *Molecular and Cellular Biochemistry*, *279*(1–2), 3–15. https://doi.org/. DOI: 10.1007/s11010-005-8210-7 PMID: 16283510

Bright, S. J., Flint, D. B., Martinus, D. K., Turner, B. X., Manandhar, M., Ben Kacem, M., & Sawakuchi, G. O. (2022). Targeted inhibition of DNA-PKcs, ATM, ATR, PARP, and Rad51 modulate response to X rays and protons. *Radiation Research*, *198*(4), 336–346. DOI: 10.1667/RADE-22-00040.1 PMID: 35939823

Chatgilialoglu, C., Ferreri, C., Krokidis, M. G., Masi, A., & Terzidis, M. A. (2021). On the relevance of hydroxyl radical to purine DNA damage. *Free Radical Research*, *55*(4), 384–404. DOI: 10.1080/10715762.2021.1876855 PMID: 33494618

Choi, W., Cha, S., & Kim, K. (2024). Navigating the CRISPR/Cas landscape for enhanced diagnosis and treatment of Wilson's disease. *Cells*, *13*(14), 1214. Advance online publication. DOI: 10.3390/cells13141214 PMID: 39056796

Clemens, S. (2022). Metal transport and detoxification: Evolutionary insights and molecular mechanisms. *Annual Review of Plant Biology*, *73*, 129–159. DOI: 10.1146/annurev-arplant-102720-031359

Cruz-Rodríguez, L., Guerra-Tschuschke, I., & González-Iglesias, H. (2024). Metallothioneins as biomarkers of metal-induced toxicity and cellular defense. *Toxicology Reports*, *11*, 210–223. DOI: 10.1016/j.toxrep.2024.02.015

Deng, W., Chen, S., & Liu, Y. (2024). Genome-wide analysis of glutathione S-transferases and their roles in heavy metal detoxification. *Ecotoxicology and Environmental Safety*, *282*, 115919. DOI: 10.1016/j.ecoenv.2024.115919

Desouky, O., Ding, N., & Zhou, G. (2015). Targeted and non-targeted effects of ionizing radiation. *Journal of Radiation Research and Applied Sciences*, *8*(2), 247–254. DOI: 10.1016/j.jrras.2015.03.003

Festa, R. A., & Thiele, D. J. (2023). Copper transport in health and disease. *Cell Metabolism*, *35*(4), 567–584. DOI: 10.1016/j.cmet.2023.03.007

Forman, H. J. (2023). Glutathione: The master antioxidant—Mechanisms of synthesis and regulation. *Free Radical Biology & Medicine*, *197*, 87–100. DOI: 10.1016/j.freeradbiomed.2023.01.018

Haft, D. H., Selengut, J., Mongodin, E. F., & Nelson, K. E. (2005). A guild of 45 CRISPR-associated protein families and multiple CRISPR/Cas subtypes exist in prokaryotic genomes. *PLoS Computational Biology, 1*(6), 0474–0483. https://doi.org/DOI: 10.1371/journal.pcbi.0010060

Hanscom, T., & McVey, M. (2020). Regulation of error-prone DNA double-strand break repair and its impact on genome evolution. *Cells*, *9*(7), 1657. DOI: 10.3390/cells9071657 PMID: 32660124

Ishino, Y., Shinagawa, H., Makino, K., Amemura, M., & Nakatura, A. (1987). Nucleotide sequence of the iap gene.... *Journal of Bacteriology*, *169*(12), 5429–5433. DOI: 10.1128/jb.169.12.5429-5433.1987 PMID: 3316184

Jansen, R., Van Embden, J. D. A., Gaastra, W., & Schouls, L. M. (2002). Identification of a novel family of sequence repeats among prokaryotes. *OMICS: A Journal of Integrative Biology*, *6*(1), 23–33. DOI: 10.1089/15362310252780816 PMID: 11883425

Jansen, R., Van Embden, J. D. A., Gaastra, W., & Schouls, L. M. (2002). Identification of genes associated with DNA repeats in prokaryotes. *Molecular Microbiology*, *43*(6), 1565–1575. DOI: 10.1046/j.1365-2958.2002.02839.x PMID: 11952905

Jiang, F., & Doudna, J. A. (2017). CRISPR-Cas9 structures and mechanisms. *Annual Review of Biophysics*, *46*(1), 505–529. Advance online publication. DOI: 10.1146/annurev-biophys-062215-010822 PMID: 28375731

Joneidi, Z., Mortazavi, Y., Memari, F., Roointan, A., Chahardouli, B., & Rostami, S. (2019). The impact of genetic variation on metabolism of heavy metals. *Biomedicine and Pharmacotherapy*, *113*, 108642. DOI: 10.1016/j.biopha.2019.108642 PMID: 30849640

Kanazhevskaya, L. Y., Zhdanova, P. V., Chernonosov, A. A., & Koval, V. V. (2025). Off-target interactions in CRISPR-Cas9 machinery. *Biochemistry and Biophysics Reports*, *43*, 102134. Advance online publication. DOI: 10.1016/j.bbrep.2025.102134 PMID: 40688512

Khan, S. A., Nurulain, S. M., Qureshi, R. N., Zafar, A., Riaz, Z., Shoukat, A., & Shah, S. T. A. (2024). Exposure to heavy metals, antioxidant status, and immune response. *Toxicology Reports*, *11*, 108642. Advance online publication. DOI: 10.1016/j.toxrep.2024.108642

Li, C., Zhang, H., & Wu, X. (2023). Functional characterization of GST gene families. *Plant Physiology and Biochemistry*, *199*, 107952. DOI: 10.1016/j.plaphy.2023.107952

Li, T., Li, S., Kang, Y., Zhou, J., & Yi, M. (2024). Harnessing the evolving CRISPR/Cas9 for precision oncology. *Journal of Translational Medicine*, *22*(1), 749. DOI: 10.1186/s12967-024-05570-4 PMID: 39118151

Lillestøl, R. K., Redder, P., Garrett, R. A., & Brügger, K. (2006). Putative viral defence mechanism in archaeal cells. *Archaea : an International Microbiological Journal*, *2*(1), 59–72. DOI: 10.1155/2006/542818 PMID: 16877322

Liu, M., Rehman, S., Tang, X., Gu, K., Fan, Q., Chen, D., & Ma, W. (2019). Methodologies for improving HDR efficiency. *Frontiers in Genetics*, *9*, 691. Advance online publication. DOI: 10.3389/fgene.2018.00691 PMID: 30687381

Maan, K., Baghel, R., Dhariwal, S., Sharma, A., Bakhshi, R., & Rana, P. (2023). Multi-omics integration reveals radiation-induced altered pathways. *NPJ Systems Biology and Applications*, *9*(1), 42. DOI: 10.1038/s41540-023-00305-5 PMID: 37689794

Mabsor-Zgandaoui, S., Boulmane, H., Rachmoune, K., & Zgandaoui, Y. (2026b). Activation of Enzymes by Radiation: A Frontier for the Discovery of Neuroprotective Drugs. In Tiwari, P., Kadiri, S., & Minocha, N. (Eds.), *Recent Breakthroughs in Neurodegenerative Disorders and Enzyme Therapy* (pp. 107–128). IGI Global Scientific Publishing., DOI: 10.4018/979-8-3373-4627-4.ch005

Mabsor-Zgandaoui, S., Moussaif, A., El Hessni, A., Mesfioui, A., El Mzibri, M., Iddar, A., & Moutaouakkil, A. (2025a). Ionising and Ultraviolet Radiation Induce the Secretion of Melanin in *Tetrahymena Pyriformis. The International Journal of Environmental Studies*, *82*(1), 1–14. DOI: 10.1080/00207233.2024.2412455

Mabsor-Zgandaoui, S., Rachmoune, K., Aftais, I., Elamrani, F. E., Amradi, I., El Housseini, A., Hamdan, Y. A., Zgandaoui, Y., Iddar, A., El Mzibri, M., Moutaouakkil, A., El Hessni, A., & Mesfioui, A. (2026a). Biomedical Polymers: Foundations and Future Innovations in Medicine. In Erişen, D., Khan, F., Chang, C., & Yan, K. (Eds.), *Biomedical Applications and Toxicity of Polymers, Nanoparticles, Biomaterials, and Metal Ions* (pp. 1–30). IGI Global Scientific Publishing., DOI: 10.4018/979-8-3373-0055-9.ch001

Mabsor-Zgandaoui, S., Rachmoune, K., Boulmane, H., Dakir, H., Aftais, I., Ettayri, K., Hamdan, Y. A., Zgandaoui, Y., & Moutaouakkil, A. (2026c). Medical Nanoparticles: Breakthroughs in Diagnostics and Targeted Therapies. In Tiwari, P., Kadiri, S., & Minocha, N. (Eds.), *Recent Breakthroughs in Neurodegenerative Disorders and Enzyme Therapy* (pp. 51–78). IGI Global Scientific Publishing., DOI: 10.4018/979-8-3373-4627-4.ch003

Mabsor-Zgandaoui, S., Rachmoune, K., El Hessni, A., El Mzibri, M., Iddar, A., & Moutaouakkil, A. (2025b). Effect of Hydrogen Peroxide Pre-Treatment on Enhancing the Cellular Resistance of *Tetrahymena Pyriformis* Exposed to Ionizing Radiation. *Toxicological and Environmental Chemistry*, *107*(2), 294–309. DOI: 10.1080/02772248.2025.2460804

Makarova, K. S., Wolf, Y. I., Alkhnbashi, O. S., Costa, F., Shah, S. A., Saunders, S. J., Barrangou, R., Brouns, S. J. J., Charpentier, E., Haft, D. H., Horvath, P., Moineau, S., Mojica, F. J. M., Terns, R. M., Terns, M. P., White, M. F., Yakunin, A. F., Garrett, R. A., van der Oost, J., & Koonin, E. V. (2015). Updated evolutionary classification of CRISPR-Cas systems. *Nature Reviews. Microbiology*, *13*(11), 722–736. DOI: 10.1038/nrmicro3569 PMID: 26411297

Makarova, K. S., Wolf, Y. I., Iranzo, J., Shmakov, S. A., Alkhnbashi, O. S., Brouns, S. J. J., Charpentier, E., Cheng, D., Haft, D. H., Horvath, P., Moineau, S., Mojica, F. J. M., Scott, D., Shah, S. A., Siksnys, V., Terns, M. P., Venclovas, Č., White, M. F., Yakunin, A. F., & Koonin, E. V. (2020). Evolutionary classification of CRISPR–Cas systems. *Nature Reviews. Microbiology*, *18*(2), 67–83. Advance online publication. DOI: 10.1038/s41579-019-0299-x PMID: 31857715

Mengstie, M. A., & Wondimu, B. Z. (2021). Mechanism and applications of CRISPR/Cas9. *Biologics*, *15*, 353–361. Advance online publication. DOI: 10.2147/BTT.S326422

Mishra, S., Dwivedi, S. P., & Singh, R. B. (2010). Epigenetic effect of heavy metals. *The Open Nutraceuticals Journal*, *3*, 188–193. DOI: 10.2174/1876396001003010188

Mladenov, E., Kallies, M., Stuschke, M., Gkika, E., & Iliakis, G. (2025). CRISPR/Cas9-generated DSB clusters mimic complex lesions. *Scientific Reports*, *15*(1), 36480. DOI: 10.1038/s41598-025-22945-9 PMID: 41115973

Mojica, F. J. M., Díez-Villaseñor, C., García-Martínez, J., & Soria, E. (2005). Intervening sequences derive from foreign elements. *Journal of Molecular Evolution*, *60*(2), 174–182. DOI: 10.1007/s00239-004-0046-3 PMID: 15791728

Moloudi, K., & Haghdoost, S. (2025). Ionizing radiation enhances proliferation and differentiation via NRF2. *Antioxidants*, *14*(8), 986. DOI: 10.3390/antiox14080986 PMID: 40867882

Neiger, H. E., Siegler, E. L., & Shi, Y. (2021). Breast cancer predisposition genes and synthetic lethality. *International Journal of Molecular Sciences*, *22*(11), 5614. DOI: 10.3390/ijms22115614 PMID: 34070674

Neslund, J., Carroll, J., & Bryant, R. (2022). Metallothionein expression as prognostic biomarker. *Cancer Biomarkers*, *35*(2), 233–245. DOI: 10.3233/CBM-210536

Ngo, V., & Duennwald, M. L. (2022). Nrf2 and oxidative stress overview. *Antioxidants*, *11*(12), 2345. DOI: 10.3390/antiox11122345 PMID: 36552553

Nikfarjam, S., & Singh, K. K. (2023). DDR signaling — Link between cancer and CVD. *Cancer Medicine*, *12*(4), 4380–4404. DOI: 10.1002/cam4.5274 PMID: 36156462

Nishimasu, H., Ran, F. A., Hsu, P. D., Konermann, S., Shehata, S. I., Dohmae, N., Ishitani, R., Zhang, F., & Nureki, O. (2014). Crystal structure of Cas9. *Cell*, *156*(5), 935–949. DOI: 10.1016/j.cell.2014.02.001 PMID: 24529477

Ostrovska, S. S., Abramov, S. V., Dychko, E. N., Vyselko, A. D., Konovalova, O. S., & Danilchenko, A. K. (2023). Epigenetic effects of heavy metals. *Bulletin of Problems in Biology and Medicine*, *1*(168), 36–43. DOI: 10.29254/2077-4214-2023-1-168-36-43

Pourcel, C., Salvignol, G., & Vergnaud, G. (2005). CRISPR elements in *Yersinia pestis*. *Microbiology*, *151*(3), 653–663. DOI: 10.1099/mic.0.27437-0 PMID: 15758212

Rachmoune, K., Mabsor-Zgandaoui, S., El Housseini, A., Hamdan, Y. A., Bousaid, M., Aknouch, A., Rhazi, M., Gharsallaoui, A., Iddar, A., El Mzibri, M., Moussaif, A., Hassan El Mouden, E., & Moutaouakkil, A. (2026). Radioprotective Effects of Chitosan Polysaccharides and Biomedical Applications. In Erişen, D., Khan, F., Chang, C., & Yan, K. (Eds.), *Biomedical Applications and Toxicity of Polymers, Nanoparticles, Biomaterials, and Metal Ions* (pp. 233–256). IGI Global Scientific Publishing., DOI: 10.4018/979-8-3373-0055-9.ch010

Ryu, H. W., Lee, D. H., & Won, H. R.. (2015). Toxic metals and epigenetic regulation. *Toxicological Research, 31*(1), 1–9. DOI: 10.5487/TR.2015.31.1.001 PMID: 25874027

Schmidt, T. J. N., Berarducci, B., Konstantinidou, S., & Raffa, V. (2023). CRISPR/Cas9 in nanomedicine. *Drug Discovery Today*, *28*(1), 103375. DOI: 10.1016/j.drudis.2022.103375 PMID: 36174966

Schmidt, T. J. N., Berarducci, B., Konstantinidou, S., & Raffa, V. (2023). CRISPR/Cas9 in nanomedicine. *Drug Discovery Today*, *28*, 103375. DOI: 10.1016/j.drudis.2022.103375 PMID: 36174966

Shao, M., Xu, T. R., & Chen, C. S. (2016). Development of CRISPR/Cas9 in animal models. *Zoological Research*. Advance online publication. DOI: 10.13918/j.issn.2095-8137.2016.4.191 PMID: 27469250

Shim, Y. J. (2025). Therapeutic targeting of DNA damage response in TP53/ATM tumors. *Brain Tumor Research and Treatment*, *13*(3), 73–80. DOI: 10.14791/btrt.2025.0017 PMID: 40759474

Shimada, M., Matsumoto, Y., & Otsuka, K. (2025). Live-cell imaging of DNA damage. *The Journal of Biological Chemistry*, *110328*(7), 110328. Advance online publication. DOI: 10.1016/j.jbc.2025.110328 PMID: 40473209

Shiridokht, F., Kehtari, P., Eskandani, M., Farajollahi, A., & Vandghanooni, S. (2025). Harnessing radiosensitizers and nanotechnology. *International Journal of Pharmaceutics: X*, *100419*, 100419. Advance online publication. DOI: 10.1016/j.ijpx.2025.100419 PMID: 41189671

Stinson, B. M., & Loparo, J. J. (2021). DSB repair by NHEJ. *Annual Review of Biochemistry*, *90*, 137–164. DOI: 10.1146/annurev-biochem-080320-110356 PMID: 33556282

Sun, X., Dong, M., Li, J., Sun, Y., Gao, Y., Wang, Y., Du, L., Liu, Y., Ji, K., He, N., Wang, J., Zhang, M., Song, H., Xu, C., & Liu, Q. (2024). NRF2 promotes radiation resistance. *Theranostics*, *14*(2), 681–698. DOI: 10.7150/thno.88899 PMID: 38169561

Takara Bio. (n.d.). *Introduction to the CRISPR–Cas9 system* [Illustration]. https://www.takarabio.com/learning-centers/gene-function/gene-editing/gene-editing-tools-and-information/introduction-to-the-crisprcas9-system

Tamaddondoust, R. N., Wong, A., Chandrashekhar, M., Azzam, E. I., Alain, T., & Wang, Y. (2022). Novel regulators of radiosensitivity. *International Journal of Molecular Sciences*, *23*(15), 8774. DOI: 10.3390/ijms23158774 PMID: 35955908

Tang, T. H., Bachellerie, J. P., Rozhdestvensky, T., Bortolin, M. L., Huber, H., Drungowski, M., Elge, T., Brosius, J., & Hüttenhofer, A. (2002). Identification of 86 candidates for small non-messenger RNAs from the archaeon *Archaeoglobus fulgidus*. *Proceedings of the National Academy of Sciences of the United States of America*, *99*(11), 7536–7541. DOI: 10.1073/pnas.112047299 PMID: 12032318

Tei, C., Hata, S., Mabuchi, A., Okuda, S., Ito, K. K., Genova, M., Fukuyama, M., Yamamoto, S., Chinen, T., Toyoda, A., & Kitagawa, D. (2025). Comparative analysis of multiple DNA double-strand break repair pathways in CRISPR-mediated endogenous tagging. *Communications Biology*, *8*(1), 749. DOI: 10.1038/s42003-025-08187-5 PMID: 40360740

Tripathi, S., Kharkwal, G., Mishra, R., & Singh, G. (2024). NRF2 signaling in metal-induced oxidative stress. *Heliyon*, *10*(18). Advance online publication. DOI: 10.1016/j.heliyon.2024.e37545 PMID: 39309893

Valente, D., Gentileschi, M. P., Guerrisi, A., Bruzzaniti, V., Morrone, A., Soddu, S., & Verdina, A. (2022). Factors to consider for the correct use of γH2AX in the evaluation of DNA double-strand breaks caused by ionizing radiation. *Cancers (Basel)*, *14*(24), 6204. DOI: 10.3390/cancers14246204 PMID: 36551689

Viktorsson, K., Rieckmann, T., Fleischmann, M., Diefenhardt, M., Hehlgans, S., & Rödel, F. (2023). Advances in molecular targeted therapies to increase efficacy of (chemo)radiation therapy. *Strahlentherapie und Onkologie*, *199*(12), 1091–1109. DOI: 10.1007/s00066-023-02064-y PMID: 37041372

Wang, H., Zou, W., & Cao, Y. (2025). Radiation-induced senescence and adaptive response. *Radiation Medicine and Protection*. Advance online publication. DOI: 10.1016/j.radmp.2025.05.002

Weiss, K. H., Stremmel, W., & Gotthardt, D. N. (2023). ATP7B dysfunction in Wilson disease. *Journal of Hepatology*, *78*(4), 789–805. DOI: 10.1016/j.jhep.2022.12.017

Xue, C., & Greene, E. C. (2021). Pathway choices in CRISPR genome editing. *Trends in Genetics*, *37*(7), 639–656. DOI: 10.1016/j.tig.2021.02.008 PMID: 33896583

Yamamoto, M., & Kensler, T. W. (2024). Nrf2 signaling in oxidative and xenobiotic stress: Mechanisms and therapeutic modulation. *Nature Reviews. Molecular Cell Biology*, *25*, 145–161. DOI: 10.1038/s41580-023-00630-7

Yang, H., Ren, S., Yu, S., Pan, H., Li, T., Ge, S., Zhang, J., & Xia, N. (2020). Methods favoring homology-directed repair choice in response to CRISPR/Cas9-induced double-strand breaks. *International Journal of Molecular Sciences*, *21*(18), 6461. DOI: 10.3390/ijms21186461 PMID: 32899704

Zhang, J., & He, P. (2025). In silico and in vitro assessment of TP53, ATM, RAD51, and BAX genes in gastric cancer and their contribution to radiotherapy resistance. *Hereditas*, *162*(1), 1–20. DOI: 10.1186/s41065-025-00496-3 PMID: 40652278

Zhang, Z., Ren, F., & Hohenstein, A. (2023). Structural basis of ZIP family transport. *Science Advances*, *9*(12), eadg8721. Advance online publication. DOI: 10.1126/sciadv.adg8721

Zhao, R., Long, J., & Chen, P. (2024). Nrf2 activation protects against heavy metal–induced oxidative stress in mammalian cells. *Redox Biology*, *68*, 102853. DOI: 10.1016/j.redox.2024.102853

Zhou, Z., Chen, G., Shen, M., Li, J., Liu, K., Liu, M., Shi, S., Yang, D., Chen, W., Chen, S., Yin, Y., Qin, Y., Su, X., Chen, W., & Kang, M. (2023). Genome-scale CRISPR-Cas9 knockout screening in nasopharyngeal carcinoma for radiosensitive and radioresistant genes. *Translational Oncology*, *30*, 101625. DOI: 10.1016/j.tranon.2023.101625 PMID: 36739730

Chapter 8
CRISPR-Cas9 Technology for Medical Diagnostics

Manish Kumar
https://orcid.org/0000-0003-2042-1243
Department of Pharmaceutics, ISF College of Pharmacy, Moga, India

Debnath Das
https://orcid.org/0009-0003-5848-0940
Department of Pharmaceutics, ISF College of Pharmacy, Moga, India

Rupesh K. Gautam
https://orcid.org/0000-0001-5580-5410
Amity Institute of Pharmacy, Amity University, Noida, India

Uma Shanker Maurya
Goel Institute of Pharmacy and Sciences, Lucknow, India

Shruti Srivatava
School of Pharmacy, Chandigarh University, Unnao, India

ABSTRACT

CRISPR-Cas9 is a paradigm shift in the molecular diagnostics field that goes beyond its beginnings as a gene editing technology to become a radically new diagnostic platform. The review discusses the history and uses of CRISPR-based diagnostics in infectious diseases, oncology and genetic screening. CRISPR systems offer explicit, customizable, and molecular recognition features, which can be applied to create efficient, quick, and economical diagnostic systems. The combination of such technologies is one that will bring democratized, universally available molecular diagnostics, and change healthcare in resource-abundant and resource-constrained contexts, and eventually makes CRISPR a foundation of precision diagnostics.

DOI: 10.4018/979-8-3373-4857-5.ch008

14.1 INTRODUCTION

Figure 1. Graphical abstract

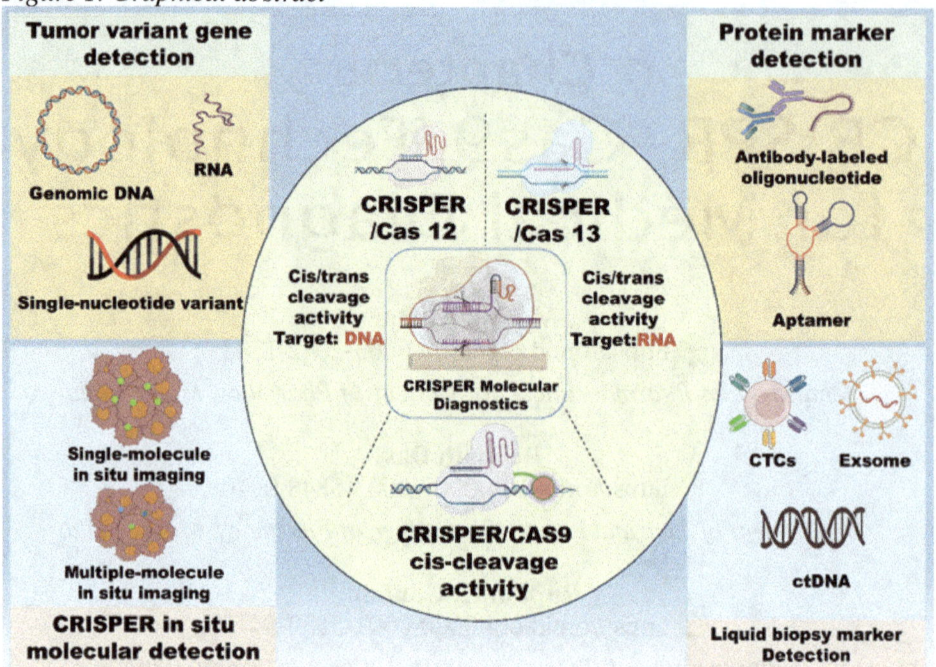

Clustered regularly interspaced short palindromic repeats (CRISPR), initially identified in the Escherichia coli genome in 1987, constitute a component of the adaptive immune system found in prokaryotic organisms, including archaea and bacteria (Mohanraju et al., 2022).Further research into the structure and function of the CRISPR system has led to advancements in gene editing, gene therapy, and diagnostics, building on the original discovery of the system in bacteria as an adaptive immune system (Hossain, 2021). Adapted into an exact gene-editing program, the type II CRISPR system utilizing the Cas9 nuclease, in particular, this programmable system of RNA-guided defence became an appealing system to develop rapid, sensitive, and cost-effective diagnostics. genome modification, the specificity, programmability and molecular recognition properties of CRISPR-Cas9 immediately placed the program as an attractive platform to create rapid, sensitive, cost-effective diagnostics (Zhou & Simonian, 2024).

Diagnostic replacements gene editing takes advantage of the intrinsic capacity of the CRISPR-Cas proteins to find specific nucleic-acid sequences. CRISPR based diagnostics started with the realization that catalytically inactive forms like dCas9 still have a strong affinity to bind DNA but fail to cleave the target (Bhardwaj et al.,

2022).As an example, reconstitution of active luciferase can be achieved by expression of two dCas9 molecules containing complementary fragments of luciferase to bind to two adjacent sequences of a particular gene and producing a measurable light signal that provides evidence of the presence of the pathogen or genetic marker of interest (Dara et al., 2025).Besides the dCas9-based assays, natural cleavage capacity of active Cas9 has been used to develop sequence-enrichment assays of higher specificity. One such innovation is the FLASH (Finding Low Abundance Sequences by Hybridization) platform, which targets specific genomic regions with an endonuclease called Cas9 complexed with custom-designed gRNAs to cut adequately sized fragments of a target genomic region and subsequently subject such fragments to next-generation sequencing (NGS), thus deep profiling the antimicrobial resistance genes in clinical samples with exceptionally low pathogen counts. FLASH-NGS can achieve enrichment by up to five orders of magnitude over background, which is dramatically higher than that of conventional metagenomic sequencing, and is also capable of improving robustness against a single-nucleotide polymorphism and multiplexing capacity, making it an effective diagnosis approach to infectious disease surveillance and antimicrobial resistance surveillance (Quan et al., 2019).

CRISPR-Cas9 has also been used to solve the diagnostic problem in resource-limited environments. Classical molecular techniques like PCR involve thermal cycling, electricity and laboratory facilities and are therefore impractical in the low-resource settings to do point-of-care (PoC) testing. One such recent example the creation of a CRISPR-dCas9-based colorimetric detection of DNA strategy utilizing Recombinase Polymerase Amplification (RPA), dCas9 binding and Rolling Circle Amplification (RCA). The platform allows visible change of colour readings that can be seen by naked eye consistently between temperatures of 20-45°C without the use of any instruments, electricity, and trained technicians. Another example in the Leishmania DNA detection, the method is sensitive to zeptomolar limits and can work with the minimally processed urine or blood samples, as evidenced how CRISPR diagnostics can democratize disease diagnostics in under-served areas (Bengtson et al., 2022).

Although the clinical use of Cas9 in diagnostics is not new, the rest of the CRISPR family, such as Cas12 and Cas13, also have the additional benefit of collateral cleavage, which can be utilized in diagnostic platforms; nevertheless, Cas9 has started to be applied more broadly in diagnostic technologies because of its unique ability to recognize a sequence with high fidelity without collateral cleavage. As an example, CRISPR-based diagnostic solutions have demonstrated sensitive detection of methicillin-resistant Staphylococcus aureus (MRSA) through both Cas9 and dCas9 systems without the need to perform complex sample preparation, thereby highlighting the versatility of the CRISPR-Cas9 systems in diagnostic situations. dCas9-SYBR Green assays, in particular, have allowed to cost-effectively and rapidly detect MRSA DNA through the use of both Cas9 and dCas9 systems as a sensitivity assay, without involving complex sample preparation (Dara et al., 2025).

The CRISPR-Cas9 diagnostics is also becoming widespread in the field of infectious disease Identification in the broader sense. CRISPR has come out as a new generation diagnostic tool that can be used to detect viruses, bacteria, and parasites with a high degree of precision and with a very minimal number of resources. Moreover, due to the capacity of CRISPR systems to discriminate genetic signatures, the diagnostics can identify infection during the early stages when there are limited target nucleic acids, and can be reprogrammed in response to novel pathogenic agents, thus they are useful in epidemic or pandemic preparedness (Wang et al., 2023).

To make CRISPR-Cas9 diagnostics a viable tool in a wide-scale clinical practice, various challenges currently being faced must be addressed. Such problems as off-target binding, CRISPR components delivery into detection systems, and sample preparation workflow optimization remain to be optimized. However, these multidisciplinary approaches to innovating CRISPR, such as signal increase through nanotechnology, machine learning-based gRNA layout, and portably ingrained microfluidic systems, keep on expanding the diagnostic prospects of the technology. CRISPR-Cas9, therefore, is leading the revolution of diagnostics, with unprecedented accuracy, speed, programmability, and accessibility (Zhou et al., 2024).

Conventional diagnostic methods such as PCR and LAMP have a restricted range of use as decentralized or rapid-response testing because such methods incorporate thermal cycling, specialized laboratory infrastructure, and trained personnel. Antigen tests are cheap but not sensitive to allow early identification of infection. CRISPR-based diagnostics is able to solve these limitations by allowing isothermal, sequence-specific nucleic-acid identification by using minimal equipment. Recent advancements, such as the design of guide RNA with the help of AI, microfluidic integration, and engineered Cas12/Cas13/Cas14 effectors, have a significant positive effect on the portability of diagnostic devices, rate, and sensitivity of the analysis. These trends can be incorporated into the background to demonstrate why CRISPR systems can have transformational benefits compared to conventional diagnostics (Du et al., 2025).

In recent years (2019-2025), there is a pronounced change in the type of diagnostic enrichment approaches that rely on the presence of Cas9 to those based on Cas12/Cas13 collateral-cleavage platforms, which can amplify the signal to the point of attomolar sensitivity. Comparative analyses as well reveal shortcomings of traditional PCR diagnostics including slow turnaround, cold-chain, dependency on infrastructure, among others, further contributing to CRISPR use (Sun et al., 2024).

14.1.1 Comparative Overview of CRISPR Effectors

Despite the long-term dominance of Cas9 in the development of CRISPR-based diagnostics, the wider CRISPR family has grown to offer a wider range of molecular tools, especially by Cas12 and Cas13, that can be used to detect nucleic acids. Each effector has its own unique biochemical characteristics that impact on their diagnostic performance, sensitivity, type of target and their capability to be used in point-of-care formats (He et al., 2025).

Cas9 and its derivatives (dCas9, Cas9n) identify targets in the form of dual strands of DNA and act by sequence-specific binding or cleavage, but do not have collateral cleavage activity. This renders Cas9 of special use when specific readout is required without the risk of having non-specific signal amplification.

In contrast, the Cas12 enzymes are directed against a double-stranded DNA target but show strong trans-cleavage activity when they recognize a target i.e., cutting in a non-selective manner any nearby single-stranded DNA reporters. The sensitivity of high measurements is based on this background activity, which enables attomolar detection and fast lateral-flow measurements, which are observed in DETECTR. Cas13 enzymes are extended to RNA such that when targeted RNA binds, the enzyme cleaves neighbouring reporters, which are made of the building blocks of SHERLOCK and other RNA-based diagnostics (Hillary & Ceasar, 2023).

Clinical usefulness of these systems is also different. Cas12 is better in detecting the presence of bacteria in DNA, screening of oncogenic mutations, and assays coupled with isothermal amplification, whereas Cas13 is better at detecting RNA viruses, including SARS-CoV-2, dengue, and Zika.

Table 1. Comparison of Major CRISPR Effectors Used in Diagnostics

Feature	Cas9 / dCas9	Cas13	Cas12
Target type	dsDNA	ssRNA	dsDNA
Collateral cleavage	No	Yes (ssRNA)	Yes (ssDNA)
PAM requirement	Yes (NGG or variants)	None (requires PFS for some orthologs)	Yes (TTTV typical)
Sensitivity potential	Moderate	Very high	Very high
Example platforms	FLASH, dCas9-luciferase assays	SHERLOCK, Cas13 LFA	DETECTR, Cas12 LFA
Typical applications	Sequence enrichment, mutation detection	RNA virus detection, gene expression signatures	Viral DNA, bacterial DNA, ctDNA

Despite its high specificity of sequences and stability of its performance, collateral cleavage does not occur, and signal amplification is therefore restricted, which limits the overall sensitivity of the assay. The high ssDNA collateral activity of Cas12 allows DNA to be readily and highly sensitive detected, but raises the chances of non-specific activation in unoptimal circumstances. Cas13 also offers comparable cleavage of collateral to detect RNA and thus suitable with RNA virus but optimization of the assay should mitigate unwanted cleavage of the reporter. These differences demonstrate the increasing preference of the contemporary diagnostic development toward Cas12/Cas13 systems(Moon & Liu, 2023).

Table 2. Pros and Cons of CRISPR Effectors

Effector	Pros	Cons
Cas9/ dCas9	High specificity; PAM-guided targeting; good for enrichment assays	No collateral cleavage → lower sensitivity; restricted PAM requirements
Cas12	Strong collateral cleavage (high sensitivity); ideal for DNA viruses	Risk of non-specific activation; PAM constraints persist
Cas13	Powerful RNA detection; suitable for viral diagnostics	Reporter background cleavage; sensitive to assay conditions

Various comparative investigations have shown that Cas12 and Cas13 have better sensitivity to analytical cleavage owing to their strong trans-cleavage properties, whereby they can be amplified into a signal without the need to subject the sample to further enzymatic treatment. Cas9 has the disadvantage of not having this property, hence it is not suitable in ultra-low-copy detection assays. The RNA targeting inherent in Cas13 can be specifically used in targeting respiratory viruses but Cas12 allows the detection of mutations in DNA in ultrasensitive conditions. And these mechanistic advantages make the shift of the diagnostic community to Cas12/Cas13 platforms.

14.2 MECHANISMS OF VARIOUS CRISPER TECHNOLOGY

14.2.1 Flash technology

FLASH (Finding Low-Abundance Sequences by Hybridization) reuses the sequence-specific cleavage capability of Cas9 to carry out selective enrichment of the sequences of DNA sequences in advance of sequencing. Its operating principles involve using the programmable endonuclease action of Cas9 to cleave target DNA into fragments that are more likely to be ligated to sequencing adapters to allow deep recovery of clinically relevant genes in cases where they occur by several orders of magnitude more often than host DNA. This process starts with the phosphatase treatment of the starting DNA/cDNA, removing the 5' phosphates on all DNA molecules and eliminating background DNA that would otherwise ligate to adapters when forming a library. This makes sure that the fragments generated by Cas9 are only competent at a later point in the workflow. These newly formed ends alone can be sequenced with adapters and thus transformed into FLASH-cleaved fragments by being transformed into privileged targets to further amplification by downstream amplification. As an example, in respiratory fluid of pneumonia patients, FLASH allowed deep-depth identification of mecA and other chromosomal markers of S. aureus, despite being completely undetected by NGS alone because of low abundance and excessive contamination by human DNA (Quan et al., 2019).

Another significant clinical development of FLASH is FLASH-TB that uses the same mechanism of Cas9 to screen drug-resistant Mycobacterium tuberculosis in sputum. In this system, it has 499 gRNAs that address 52 TB drug-resistance genes. FLASH-TB multiplied the Mtb reads percentage (1.4% with direct WGS) on enriched fragments to 33% of the total reads post-enrichment, and the target depth of FLASH (nearly 2000x deep) to 93.7% of instances of resistance mutations, which is far superior to the performance of direct sequencing methods in low-bacillary-load sputum samples. Since FLASH preserves the precise sequence of enriched fragments, it can simultaneously recover presence/absence data and deep variant data (Tram et al., 2023).

14.2.2 dCas9 binding-based detection

Catalytically inactive Cas9(dCas9) is a highly specific DNA-binding protein, which allows sequence-selective detection of nucleic acid without causing cleavage and is best suited to diagnostic methods based on the production of a signal upon binding, as opposed to turnover. A CRISPR/dCas9 biosensor that combines dCas9 RNP with SMR-based isothermal nucleic acid amplification. An SMR biosensor was used to immobilize sequence-specific primers of targets, while dCas9 RNP was subjected to an isothermal reaction with RPA in a reaction chamber with a single temperature. This allowed for the simultaneous amplification and detection of nucleic acids. Primers are used to expand DNA by binding recombinase enzymes. In order to convert RNA into cDNA, RT-RPA was used to create a primer-recombinase complex. The amplified nucleic acid targets were simultaneously detected by monitoring the wavelength shift. As part of the dCas9 RNP-enhanced nucleic acid amplification procedure on the SMR biosensor, the hybridized primer was used in conjunction with the target templates. The dCas9 RNP, which enables sequence-specific binding to the target nucleic acids (near the waveguide surface at the immobilized forward primer, Supplementary), significantly increased the proportion of each resonant wavelength as a signal enhancer. The SMR biosensors convert the presence of target molecules through alterations in the refractive index that occur near the waveguide surface as a result of binding events. The detection sensitivity is enhanced by increasing the refractive index change, which is achieved when dCas9 attaches to the amplified product on the surface of the sensor. This raises the molecular weight of the sensor surface. Hence, in comparison to SMR alone, the CRISPR/dCas9-mediated biosensor was able to detect the harmful nucleic acids with a high degree of sensitivity within 20 minutes (Koo et al., 2018).

14.2.3 CRISPR Cas9n-drivenFeasibilityanalysis

The practicality of employing the DNA walker in the signal recognition phase was initially examined through the use of PAGE. Mixing the Walker and Protect probes in lane 2 resulted in the production of a W-P double-stranded probe, which in turn led to the generation of a new brand in lane 5. The development of walker support- and protect-target probes, respectively, is seen in lanes 8 and 9, which are similar. The W-P probe and support probe stripes become thinner, and two more bands showed up when lanes 6 and 7 were compared.

This signifies that the engineered DNA walker could proficiently recognize the detection target and then initiate movement. When testing the DNA walker signal output stage, Cas9n's cleavage ability, in the blank sample, we compared Cas9n to the complete target ctDNA detection method to ensure its practicality. Based on the findings, it can be concluded that the presence of Cas9n was necessary to achieve target signal amplification, since only a limited quantity of target ctDNA could be effectively transformed into a large number of DNA fragments. Based on the findings, it can be concluded that the presence of Cas9n was necessary to achieve target signal amplification, since only a limited quantity of target ctDNA could be effectively transformed into a large number of DNA fragments.Upon the addition of Cas9n, a decrease in signal intensity was observed. This change was linked to the effective cleavage of the hp probe, which had been incubated on the electrode by Nb. BbvCI, leading to the release of MB from the electrode (Yu et al., 2023) (Wei et al., 2024).

14.2.4 Collateral Cleavage Mechanisms of Cas12 and Cas13

The main characteristic of Cas12 and Cas13 systems is that they are capable of a collateral (or trans) cleavage activity upon identifying a particular nucleic-acid target. In contrast to Cas9, which only cleaves its target of the intended double-stranded DNA target, Cas12 and Cas13 have a conformational activation allowing non-specific cleavage of the nearby oligonucleotide reporters. This property converts these effectors into high-power signal amplifiers of diagnostics.

Cas12a recognizes and binds its target dsDNA then cleaves adjacent ssDNA reporters with high turnover resulting in the production of visible fluorescence, subsequent production of lateral-flow signals. The latter is the basis of DETECTR which is a quick SARS-CoV-2 test that can provide the results within less than 30 minutes with a sensitivity of more than 95 percent (Prezza et al., 2020) (Singh et al., 2022). The proximity of PAM to Cas12 activity, the complementarity of guide-sequence/target interactions, and the accessibility of targets have been shown to be high priority targets of mechanistic studies showing greater activation of off-targets when PAM adjacent nucleotides assume more flexible conformations (Mahfouz, 2024a).

Cas13 shows collateral cleavage similar to that, however, on ssRNA reporters. This is especially beneficial in the case of RNA virus detection, as was used in the SHERLOCK platform. Cas13a and Cas13b effectors have shown a sensitivity of both attomolar scale when not subjected to thermal cycling with either fluorescent or lateral-flow detection. This type of collateral cleavage allows the approach to be used in amplification-free diagnostic systems, eliminating the equipment requirements and easing the assay protocol.

Cas12 and Cas13, when combined, enhance the diagnostic performance of CRISPR systems because they allow the detection of ultrasensitive targeting with low instrumentation. New design complexities as well, i.e. trade-offs between sensitivity and non-specific activation, are also brought about by these mechanisms, and affect clinical performance.

These mechanistic variations have their consequences that determine, in a direct way, the implementation of CRISPR effectors into diagnostic systems. How Cas9-, Cas12-, and Cas13-based mechanisms are applied into a variety of clinical and point-of-care detection technologies is discussed in the ensuing section (14.3).

Figure 2. Flow chart of the diagnostic target is given bellow

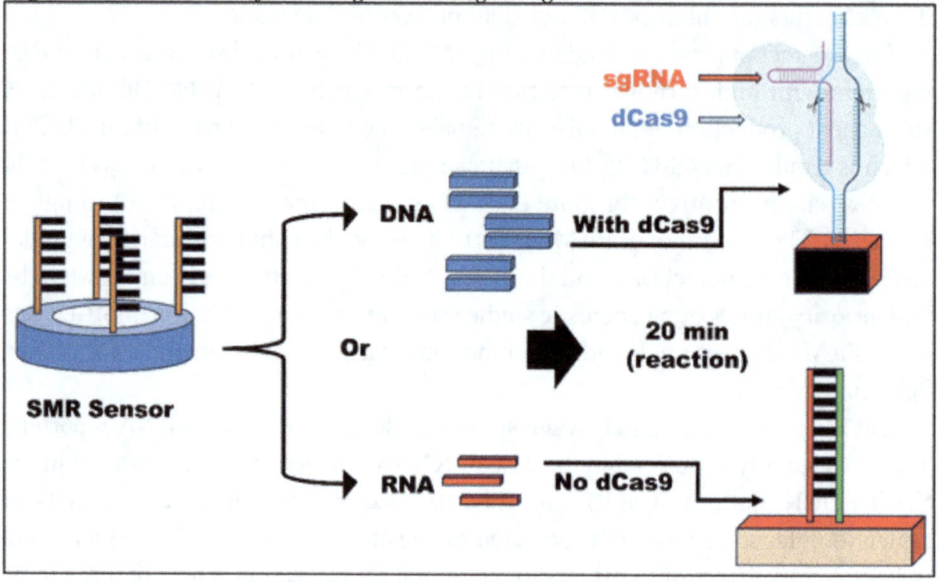

14.3 MAJOR DIAGNOSTIC PLATFORMS

14.3.1 NGS Enrichment Methods

Next-generation sequencing (NGS) enrichment systems form one of the most significant applications of CRISPR-Cas technologies to molecular diagnostics since it increases the recovery of clinically relevant sequences of nucleic acids that would otherwise be lost in the high background level of host DNA. These include FLASH (Finding Low-AbundanceSequencesbyHybridization), which is among the most sophisticated systems, and which uses the programmable double-strand cleavage capacity of Cas9 to identify the target fragments without sequencing them until they are enriched. FLASH is a multistep process, in which the genomic DNA or cDNA purified in a clinical sample undergoes dephosphorylation to block unwanted ligation of the adaptors. These guide RNAs complex with Cas9 to cleave the target DNA into fragments which have a new 5' phosphorylated end which becomes uniquely competent to be ligated with the adaptors. In the library preparation, the ligation of sequencing adapters is only performed on the Cas9-generated ends resulting in exponential amplification of the target sequences during PCR. The scale of enrichment regularly goes up to four to five orders of magnitude, which means that FLASH can detect very low amounts of microbial DNA in complex samples, including respiratory fluid, dried blood spots, or sputum. Your uploaded FLASH paper has provided clinical demonstrations that FLASH 100-100,000-fold enriched AMR genes of Staphylococcus aureus, and the ability to detect the Plasmodium falciparum resistance markers in dried blood spots was completely silent using raw NGS.This platform made it possible to predict multidrug-resistant (MDR) and extensively drug-resistant (XDR) mutations in raw sputum directly, without the need to culture phenotypes, which reflects the paradigm shift of CRISPR-Cas9-based enrichment of NGS in clinical microbiology, as can be seen in the uploaded FLASH-TB file (Malekshoar et al., 2023)(Srivastava et al., 2020a).

Depletion of abundant sequences by hybridization (DASH) uses a cas9 endonuclease to remove unwanted host nucleic acid, such as rRNA transcripts. It depends on the CRISPR/Cas9 system's ability to selectively hybridize to the host DNA in a mixed NGS library and cleave it with the help of host-specific guide RNA molecules (Gu et al., 2016). The DASH approach, first detailed by Gu et al., was shown to be quite successful; after host depletion, pathogenic sequences were observed to be up to four times more abundant. Several studies, whereas not exclusively focused on infectious disease diagnosis, have employed DASH or analogous methodologies, reporting comparatively high levels of host DNA reduction (Prezza et al., 2020). While DASH has been shown to be a cost-effective and robust alternative to direct mNGS, with a reported rRNA depletion efficiency of 81.4% to 88.2% in cerebrospinal fluid samples, it has not been extensively tested for BSI diagnosis using whole blood and/or rapid sequencing library kits/platforms. Useful host depletion strategies like DASH, which are covered in this section, may be evaluated either alone with pure nucleic acid or in combination with pre-DNA extraction procedures.

14.3.2 Electrochemical sensors

Electrochemical biosensors are the most rapidly developing type of CRISPR-based diagnostics with the aim of point-of-care (POC) implementation since it translates molecular recognition events into electrical signals, which do not require intense optical equipment. In contrast to NGS enrichment techniques that rely on post-processing, electrochemical sensors provide real-time operation, high sensitivity, and operation almost without the instrument. CRISPR-Chip, which is based on catalytically inactive dCas9 immobilised to a graphene field-effect transistor (gFET) is the first significant platform in this space. The CRISPR-dCas9/gRNA complex is neuro-engineered onto the graphene surface as a programmable affinity receptor to a particular sequence of DNA. The hybridization and binding of target DNA to the immobilized dCas9 leads to the perturbation of the electrical conductivity of the graphene channel to change the current voltage properties of the transistor (Azizoglu, 2022).

Based on this, E-CRISPR expands the power of detection by relying on the collateral cleavage activity of a few Cas enzymes (usually Cas12a or Cas13a) instead of just dCas9 binding. In the E-CRISPR, reporter oligonucleotides that are redox active molecules are immobilized on the sensor electrode. The binding of Cas12a or Cas13a to its target causes general cleavage of surrounding reporters of ssDNA or ssRNA. This rupture physically removes redox labels of the electrode surface resulting in quantifiable shifts in peak current in voltammetric examination (Weng et al., 2023). CRISPR-powered electrochemical sensors make them ideal candidates to take the place of field-deployable diagnostics in infectious disease outbreaks or low-resource environments because they allow detection of nucleic acids without pre-amplification, their sensitivity is atmolar and can be manufactured in large volumes through the use of screen-printed electrodes. Electrochemical CRISPR diagnostics are still under development towards lab on a chip system that can combine sample preparation, target recognition and electronic readout in closed, disposable cartridges (Dronina et al., 2022).

14.3.3 Optical/colorimetric assays

CRISPR diagnostics based on optical and colorimetric purposes are the most common platforms used in low-resource settings due to the intuitively presented visual outputs, which do not require specific sensors or sequencing technology. The strongest of them are the ones combining Rolling Circle Amplification (RCA) with subsequent lateral flow assays (LFA) to provide high sensitivity and easy to use format. RCA is the method of isothermal amplification of nucleic acid in which a circular DNA template is used, whereby the polymerase is in a position to be extended continuously to produce long repetitive products of ssDNA. In conjunction with CRISPR-based recognition of targets, RCA increases in sensitivity dramatically by connecting sequence-specific detection with exponential signal generation(Qiu & Li, 2021).

In these platforms that merge RCA with CRISPR detection, Cas9 or dCas9 initially binds its target and either directly cleaves a primer to be used in RCA (by Cas9 cleavage) or stabilizes a structure to allow performance of hybridization of circular templates. When RCA is triggered, the amplified products are produced in forms of long ssDNA concatemers, which contain repetitive reporter motifs and hybridize to colorimetric probe in the form of gold nanoparticles, streptavidin-biotin complex or oligonucleotides labeled by dyes. CRISPR-based detection in conjunction with isothermal amplification can be detected using lateral flow strips or handheld fluorometers, and thus point-of-care detection using CRISPR can be achieved with minimal equipment needs as well as signal-resilient performance in the presence of inhibitors in the sample(Khurana et al., 2022).

14.4 CLINICAL APPLICATIONS

14.4.1 Infectious Disease Applications

The new technologies like FLASH-NGS have revolutionized AMR screening since it allows deep enrichment of low-abundance resistance genes directly on patient samples. FLASH has a sensitivity of 98% in the detection of MRSA-associated genes at an extremely low proportion and has been shown to increase on-target sequencing depth by up to five folds.FLASH identified mecA and other AMR determinants not detected by standard metagenomic NGS due to large backgrounds of host DNA has been used in the detection of SARS-CoV-2 with reported sensitivities of 93-95% (Sharma et al., 2021).Viral diagnostics has also used the CRISPR-Cas systems, in particular, Cas12 and Cas13 based systems, including DETECTR and SHERLOCK, have been utilized in SARS-CoV-2 detection with sensitivity. These tests have a high speed (often less than 30-45 minutes), and have a clinical specificity nearly reaching 100 percent, allowing testing at the point of care during crisis stages of the COVID-19 pandemic (Verma et al., 2022). Besides AMR and viral disease diagnostics, CRISPR-based bacterial 16S rRNA gene profiling has demonstrated close to 98 detection rates, enabling the rapid identification of pathogen types in a polymicrobial environment without the need to culture the samples. Besides AMR and viral diagnostics, CRISPR activity has been applied to bacterial 16S rRNA gene profiling, where collateral cleavage activity, which forms the basis of an outbreak-controlling and triage tool, can be used to identify pathogen type in a Clinicians are being enabled to characterize complex infections and host-pathogen interactions much faster and more precisely with this capability to detect and classify bacteria using highly conserved ribosomal markers as evidenced in the review on CRISPR biology in infectious disease (Srivastava et al., 2020b).

Even high sensitivity CRISPR assays can however continue to encounter issues including the batch variability of Cas enzymes, lack of collateral cleavage kinetics, and the requirement of a better multiplexing region where PCR has a competitive edge.

14.4.2 Oncology Applications

CRISPR-based diagnostics have been at the heart of the development of liquid biopsy technologies and especially circulating tumor DNA (ctDNA) diagnostics, where customary sequencing is not sensitive enough to detect an early cancer or track its progression. CRISPR-Cas12/Cas13 (assays) have a sensitivity of 85-95% in ctDNA detection with allele frequencies as low as 0.1, and are more rapid (and cheaper) than qPCR or digital PCR (Singh et al., 2022). CRISPR-based liquid biopsies indicate the value of programmable CRISPR-guided identification of tumor-derived fragments can identify single-nucleotide differences between variants including KRAS G12D, EGFR T790M, and TP53 R248Q to enable clinicians to detect actionable mutations which can guide them to select therapeutic agents and monitor resistance. The summaries of clinical cases in CRISPR oncology studies illustrate situations where CRISPR-enhanced tests were used to establish new mutations in colorectal and lung cancer patients before they started being symptomatic, and thus the patient was treated earlier and survived better. Furthermore, tools of CRISPR enrichment, including the Cas9 cleavage and sequencing, can be used to profile oncogenic hotspots in high depth, and enhance the accuracy of oncology workflows (Zhang et al., 2021).

14.4.3 Genetic Screening Applications

CRISPR diagnostics has also changed the genetic disease screening process, such as prenatal diagnosis, detection of monogenic disorders, and repeat expansion diseases. CRISPR-mediated nucleic acid enrichment has been reported to achieve specificity of 99.9 in prenatal diagnostics, which can be used to identify non-invasively fetal aneuploidies and monogenic variants in fetuses using maternal blood samples (Zhou et al., 2024). simultaneously, CRISPR-based technologies have been found to be highly useful in identifying repeat expansion diseases, such as Huntington's disease and Fragile X syndrome, by identifying expansion-specific sequence motifs or junctions. The reviews presented in your uploaded files emphasize the fact that due to its discriminatory properties, even minor changes in length, CRISPR can be considered a potent alternative to Southern blotting and capillary electrophoresis to intervene earlier and make reproductive decisions based on high-quality molecular evidence (Hillary & Ceasar, 2023).

14.5 CLINICAL CASE STUDIES

On the FLASH clinical series, low-level AMR markers enrichment directly affected treatment choices when dealing with severe pneumonia and when dealing with septic shock. In one instance, FLASH identified a low-abundance mecA allele in a critically ill child patient -allele that was completely missed by conventional NGS and culture -in which case, the anti-MRSA therapy was immediately started and the child was significantly improved and survived. Similar results were obtained in the FLASH-TB study (Tram et al., 2023), where CRISPR-mediated enrichment of TB resistance genes on sputum made it possible to detect multidrug-resistant (MDR) and extensively drug-resistant (XDR) strains within hours, as opposed to weeks of culture reds, which led to timely therapy modification and improved patient stratification outcomes (Zhang et al., 2021).case studies of lung cancer demonstrate that earlier detection of EGFR T790M resistance mutations will lead to clinicians switching to third-generation TKIs before clinical progression. Case studies of genetic diseases also illustrate the influence of CRISPR; an example of this is that CRISPR-based non-invasive prenatal testing identified a pathogenic CAG expansion of HTT at a very early stage of gestation, before counselling and perinatal planning was possible (Singh, 2024).

Table 3. Summarizing Clinical Sensitivity & Specificity Values

Clinical Domain	**Target / Condition**	**CRISPR Platform**	**Specificity**	**Sensitivity**	**Reference**
Oncology (ctDNA)	ctDNA at 0.1% allele frequency	Cas12/Cas13 liquid biopsy	High	85–95%	(Slattery et al., 2025)
Oncology (mutation scanning)	KRAS, EGFR, TP53	Cas12/Cas9 mutation assays	High	80–95%	(Wu & Yang, 2024)
Infectious disease (viral)	SARS-CoV-2 RNA	Cas13 sherlock / cas12 detector	100%	93–95%	(Mahfouz, 2024a)
Infectious disease (AMR)	MRSA AMR genes (*mecA*)	FLASH-NGS	100%	98%	(Tram et al., 2023)
Infectious disease	16S rRNA identification	CRISPR-based ribosomal targeting	High	98%	(Mahfouz, 2024b)
Repeat expansion disorders	Huntington's, Fragile X	Cas9/Cas12 STR detection	Very high	Very high	(Wu & Yang, 2024)
Prenatal genetics	Fetal aneuploidies & monogenic mutations	CRISPR-enhanced NIPT	>99.9% specificity	—	(Wu & Yang, 2024)

Table 4. COMPARATIVE ANALYSIS TABLE OF CRISPR DIAGNOSTIC PLATFORMS

Diagnostic Category	Platform / Technology	Clinical Performance	Key Features	Advantages	Limitations	Reference
POC Viral Detection	Cas12 HPV Lateral Flow	95% sensitivity, 98% specificity, 10 copies/μL LOD	Cleavage of hairpin ssDNA reporters using Cas12	Portable, suitable for rural screening, inexpensive,	Equipment-independent formats exhibit somewhat reduced quantitative resolution.	(Xie et al., 2024)
Off-Target Analysis / gRNA Quality	crispAI (Deep Learning Off-Target Uncertainty Quantifier)	Improved forecasting accuracy compared to the latest ML/ DL models; calculated uncertainty with precision	CNN + bi-LSTM model predicting ZINB-modeled off-target cleavage uncertainty	improves sgRNA design,First model to provide probabilistic uncertainty,	A big training dataset (CHANGE-seq) is necessary, although there has been no direct clinical validation as of yet.	(Ozden & Minary, 2024)
High-Sensitivity Viral RNA Detection	Cas13 SHERLOCK (Zika, Dengue)	Attomolar sensitivity, 100% specificity	collateral cleavage directed by RNA; detection at the atomic level	elimination of extraction steps (HUDSON-SHERLOCK); very sensitive	Strict contamination control is necessary, and there is a multi-step procedure.	(Xie et al., 2024)
POC Infectious Disease Detection	Cas12 DETECTR (SARS-CoV-2)	95% sensitivity, 98% specificity, 10 copies/μL LOD	Cutting strands of single-stranded DNA at the junctions; lateral flow and fluorescence	Short (<30 minutes), requires little equipment, and may be implemented in the field.	No multiplexing; needs to amplify certain targets	(Xie et al., 2024)
Electrochemical CRISPR Biosensors	Cas12a electrochemical sensors (e.g., HPV16 30 pM detection)	Sensitivity of 30 pM, enhanced ECL LOD of 0.48 pM	Electrical detection of ssDNA reporter cleavage tagged with methylene blue	Miniaturizable and very sensitive	More expensive than lateral flow; requires specific electrodes	(Xie et al., 2024)
AI-Integrated CRISPR Diagnostics	CRISPR + ML/ AI for gRNA design & POC optimization	Improving precision while decreasing off-target impacts	AI-powered gRNA creation and automated CRISPR data analysis	Optimizes proof-of-concept processes; boosts the accuracy of predictions	Computing power and model training are necessities.	(Xie et al., 2024)
Bacterial Identification at POC	Cas12 MTB (Tuberculosis)	88.3% sensitivity, 94.6% specificity, 3.13 CFU/mL LOD	Thermolomatic enhancement of Cas12 collateral cleavage	Cheaper and quicker than GeneXpert; suitable for environments with limited resources	Even in the early stages of infection, sensitivity is lower than PCR.	(Xie et al., 2024)

continued on following page

Table 4. Continued

Diagnostic Category	Platform / Technology	Clinical Performance	Key Features	Advantages	Limitations	Reference
Fluorescent / Amplification-Coupled CRISPR Assays	Cas12 fluorescent assays	Identifying cancer-related DNA at concentrations as low as 5 aM	Cas12 cleavage mediated by RCA, EXPAR, or LAMP	Highly sensitive; works with complex biological processes	Equipment required; strips are more convenient for point-of-care testing.	(Xie et al., 2024)
HBV Genotyping POC	CRISPR-HBV (Cas12b + MCDA + Lateral Flow)	10 copies/test sensitivity, 100% specificity	Identification of genotypes B and C; LFB aided by CRISPR	Low-cost, one-hour process; ideal for outlying medical facilities	Only works with two HBV genotypes; further testing is required.	(Xie et al., 2024)
Multiplex Viral Detection	CRISPRD (Multiplex hrHPV Testing)	100% specificity, 10 copies sensitivity	Rapid detection of HPV16 and HPV18 in a single reaction employing AapCas12b, TccCas13a, and HheCas13a	Very promising for use in large-scale screening; fully multiplexed	Still in the early stages; more confirmation is needed	(Xie et al., 2024)
Prenatal Genetic Screening	CRISPR-based prenatal aneuploidy screening	>99.9% specificity (reported across CRISPR-driven NIPT surveys)	Cas-detection of NIPT biomarkers	safer than sequencing-only testing; improves accuracy without invasiveness	Needs a well-managed laboratory setting	(Xie et al., 2024)

14.6 LIMITATIONS AND CHALLENGES

Although the use of CRISPR-based diagnostics has boomed in infectious disease, oncology, and genetic screening, several scientific, technical, and translational issues persist that have limited their complete clinical use. They have direct implications on diagnostic formats about which intracellular CRISPR needs to be deployed, such as CRISPR imaging or intracellular biosensing, limiting their implementation to laboratory systems (Karimi et al., 2025).

Comparatively, PCR-based diagnostics, though viewed as the gold standard of the analytic method, involve thermal cycling, cold-chain reagents, and personnel trained in the laboratory. The strategic value of CRISPR assays, which isothermal and do not need sophisticated machinery to be performed in the field, in the response to outbreaks, is highlighted by the infrastructure requirements (Shuo et al., 2025).

Also notable are the experimentally characterized off-target effects of both Cas9 and Cas12 systems, which pose a threat to diagnostic specificity through the formation of unintended DNA cleavage or binding. The article Off-target effects in

CRISPR-Cas genome editing in human therapeutics describes the fact that CRISPR nucleases are tolerant of mismatches, bulges, and relaxed PAM motifs, leading to background noise and lowering analysis specificity (Kalter et al., 2025).Although diagnostic platforms are often based on collateral cleavage as opposed to genomic editing, the molecular promiscuity here increases background noise and reduces the specificity of analysis. Additionally, PAM-relaxed versions engineered are likely to expand the target range, but can raise off-target activity, which further complicates the practice of diagnostic accuracy. This difficulty is compounded by mechanistic insights of the article PAM-adjacent DNA flexibility tunes CRISPR-Cas12a off-target binding, which demonstrates that Cas12a has the ability to react to unintended DNA when PAM-proximal nucleotides are more flexible or unpaired and hence is more likely to give a positive signal in Cas12a diagnostic assays (Pache et al., 2025).

Amplification bias, primer-dimer formation and contamination are still significant challenges, despite the fact that the integration of Cas12a with enzymatic recombinase amplification minimises equipment requirements and delivers high sensitivity (9 copies/mL), the production of a false positive due to isothermal-amplification artefacts is still a significant issue in implementing CRISPR diagnostics in low-resource or point-of-care (POC) settings. Such limitations restrict the practical performance of assays to be used in clinics distributed around or in the field of testing (Gan et al., 2024).

As an example, in the clinical Cas13a based lateral flow assay of Neisseria gonorrhoeae, even though 100 percent concurrence with NAAT was obtained in the small clinical cohort, the authors note that assay performance will be influenced by environmental factors including temperature unpredictability, freeze-thaw damage of the enzymes, and variable urine content, and as a result, wider validation in diverse populations will be required prior to routine introduction into POC clinics (Allan-Blitz et al., 2025).

Technological limitations also apply to the enzymatic needs of CRISPR such as reliance on PAM sequences and limitations on guide RNA design. The flexibility of CRISPR is also accompanied by the need to have specific recognition of PAM, which limits access to targets in some genomic environments, and by the variability in prediction accuracy based on Cas effectsors and biologic systems (Allen et al., 2025).The design of diagnostic assays is also limited by the accuracy of prediction algorithms to predict accurately depending on Cas effectors and biologic contexts. Some of these problems can be alleviated by including AI, but as the AI-CRISPR review observes, machine learning models will require large training datasets, standardized pipelines and biological validation before they can be trusted to enhance CRISPR performance or minimise off-target effects in practice (Khoshandam et al., 2024).

Lastly, CRISPR diagnostics have ethical, regulatory, and infrastructural challenges. The ethical implications of the possible misuse of the CRISPR systems, data privacy concerns in genetic diagnostics, and disparity in access to the new CRISPR-based technologies are ethical concerns. As depicted by the diagnostic bottlenecks reported in Harnessing CRISPR Effectors for Infectious Disease Diagnostics, the optimal platform should be able to deliver accurate, sensitive, cheap, portable, and broadly deployable testing, but its achievement is subject to the barrier of manufacturing, stability of reagents and variable performance across pathogens and conditions, which continue to be key barriers to achievement of diagnostic reliability and safety (Tripathi et al., 2022).

Another issue is the absence of standardized systems of global manufacturing of CRISPR reagents. Assay performance variability across labs and regions can arise due to the difference in the purity of enzymes used, gRNA stability, the conditions under which the enzymes are lyophilized, and the conditions under which the enzymes are stored. These production differences present serious obstacles to practical implementation that are particularly severe in resource constrained environments (Li et al., 2024).

14.6.1 Regulatory, Commercialization, and Market Adoption Challenges

CRISPR-based diagnostics are not immune to non-technical barriers despite the fast pace of scientific advancement that the field encounters, which influences its conversion into controlled clinical practice. Diagnostics Regulatory Diagnostics Regulatory pathways (FDA Emergency Use Authorization (EUA) or 510(k)/ De Novo approval) should be strong and well-validated, reproducible and contain quality-control information. CRISPR systems such as DETECTR and SHERLOCK showed that it is possible to receive EUA designation in the COVID-19 pandemic, but the granted authorizations were temporary and revealed possible shortcomings in regulatory standards regarding CRISPR systems in the long term (Broughton et al., 2020).

Intellectual-property limitations also pose another threat to commercialization, as the case of patent lawsuits with Cas9 and Cas12 systems. Complexity in licensing raises the cost of development of the diagnostic companies and discourages extensive market coverage. The other commercial barriers are stability of the reagents, lyophilization of the enzymes, and batch-to-batch variability of Cas proteins, which influence stability under real-life situations. Problems with production-related issues like the stability of enzyme activity and the performance of the lateral-flow strips in diverse environmental conditions is still a critical issue (Allan-Blitz et al., 2025)

Further, CRISPR diagnostics must be incorporated into current clinical lab processes. Many point-of-care CRISPR tests are incompatible with any standardized laboratory automation system, making their adoption slow. Subsequently, regulatory compliance, scalability and equitable distribution will play a crucial role in the worldwide adoption of CRISPR-based diagnostics.

The most significant regulatory action was the fact that the DETECTR SARS-CoV-2 assay was approved through the FDA Emergency Use Authorization framework, which implied that CRISPR assays were capable of being clinically administered and of good quality. However, commercialization is further complicated by intellectual-property concerns, namely between the Broad Institute and the University of California that put licensing restrictions on Cas9 and Cas12 systems. Such legal and licensing limitations can cause slower adoption of the market and increase the cost of development (Lafi et al., 2025).

Although EUA approval indicated that it could be done, long term clearance under either FDA 510(k) or De Novo pathways would require large scale multicenter validation, measures of reproducibility, long term stability data and comparability to gold-standard PCR assays. These are some of the requirements that continue to pose significant challenges to commercialization of CRISPR diagnostics (Lau et al., 2025).

14.6.2 Ethical and Societal Implications.

As the CRISPR-based diagnostics enter the growing clinical and public-health domain, the ethical issues gain importance. Despite the speed of these technologies and their ability to be programmed, or detected in rapid ways, there are concerns of privacy of the data, informed consent, and potential abuse of the technologies in their applications. Genetic diagnostics involves sensitive personal information by default and proper protection systems must be established to prevent privacy invasions or genetic discrimination (Lau et al., 2025).

In practice, implementation in the COVID-19 pandemic revealed disparities in the accessibility of diagnostics, with low-resource areas having few opportunities to use CRISPR-based rapid tests, yet having minimal equipment needs. The WHO (2021) guideline on responsible CRISPR governance focuses on equitable access, protection of data privacy, and avoidance of misuse -the values that are necessary to scale CRISPR diagnostics responsibly.

Another major issue is global inequality concerning access to diagnoses. Most low- and middle-income areas cannot even provide cheap point-of-care CRISPR testing because of the limitations of their infrastructure (either no cold-chain or bad enzyme quality) because of the supply chains. Organizational ethical frameworks like the WHO point out that there should be equitable deployment strategies to prevent the expanding of the existing health disparities.

The potential of misuse of CRISPR based assays due to unauthorized genetic surveillance or environmental monitoring is also a biosecurity issue. With the increasing accessibility of CRISPR, it will be important to establish effective governance, openness, and collaboration with the community to curb its abuse. The addition of ethical case studies, including the experience during the COVID-19 pandemic, when rapid diagnostics was an unequal distribution, is an explanation of why social responsibility and accessible CRISPR testing strategies worldwide should be developed (Okesanya et al., 2025).

In the case of the COVID-19 pandemic, CRISPR rapid test kits were supplied to a number of low-income areas without storage facilities. False-negative results were caused by the degradation of enzymes and resulted in delays in treatment and uncontrolled outbreaks. It proves the ethical significance of equitable access to high quality diagnostics and sound supply-chain systems (Subica, 2023).

14.7 FUTURE PERSPECTIVES

The upcoming future of CRISPR-based diagnostics is determined by the combination of microfluidic engineering, artificial intelligence, new Cas effectors, and global health-oriented deployment plans which all focus on making nucleic acid testing more rapid, more affordable, and more accessible. An excellent example of such an improvement can be described by the aerosol-sampling microfluidic LAMP-CRISPR device (ASMD), a device that combines a high-flow aerosol collector (6912 L/min) with an on-chip LAMP-Cas12a detection module, to conduct a SARS-CoV-2 surveillance of large indoor environments in 85 minutes, which also reflects the potential of CRISPR microfluidics to conduct real-time bio surveillance of the environment and decentralized diagnostics. The microfluidic chip reduces reagent usage, has the complete reaction sequence in a closed system, and can be used to automatically enrich, amplify, and detect CRISPR, which is the direction in which diagnostic cartridges of the future may advance to test infectious diseases and perform oncology biomarker assays and genetic screening. With further reduction in fabrication costs and the availability of better materials in which pumps are unnecessary and where the chip in question is operated by capillary forces, microfluidic CRISPR tests are soon going to reach the mass usage in clinics, airports, and community health centers (Zhang et al., 2025).

CRISPR diagnostics are likely to undergo change with the introduction of artificial intelligence (AI). Guide RNA design can be improved by artificial intelligence (AI), assay reading can be improved, and assistance with decisions can be provided.

I have observed the use of intelligence (AI) assist in my work. The production of Artificial intelligence (AI) can eliminate bias as demonstrated by machine-learning based interpretation of Cas13 lateral flow assays. The latest network mediated CRISPR-Cas13 LFA reader had an accuracy of 96.5% in classifying SARS-CoV-2 test strips under light and phone camera conditions. Time (0.2 s) processing with mobile-optimized models was done on the network driven CRISPR-Cas13 LFA reader (Xue et al., 2025). This is a new move that forms a framework, to include AI in at-home CRISPR testing. I believe that the smartphone apps might take over the line quantification, the quality control and the automated reporting to the healthcare system in the nearest future. Another field is the AI guided gRNA design. The off-target risks are reduced by the computational models. The loci are of high-efficiency in the computational models. The optimized computational models are used to optimize multiplex gRNA pools. CRISPR assays that identify pathogens depend on the AI guided gRNA design. The gRNA design of AI is required in CRISPR assays to identify cancer mutation. AI directed gRNA design is necessary, in CRISPR measurements that identify variations (Dou et al., 2025).

The future of CRISPR diagnostics is also being determined by the rapid development of enzyme engineering. Technically, what can be done is already being extended by new versions of Cas12a and Cas13 whose thermal stability is better, their ability to cleave collateral is faster, and PAM is more compatible. The Cas12a-based assay has the capability of genotyping fourteen strains of HPV that are high-risk with a very high analytical sensitivity (as low as two copies per microliter), and flawless clinical specificity. Future emerging enzymes, including Cas14 or Cas7-11, will open diagnostic opportunities to very small DNA/RNA targets, early cancer biomarkers, fetal genetic material and RNA viruses, which are still difficult to detect by standard methods (Hall et al., 2025). The design of Cas enzymes that can perform efficiently at 60-70degC will also enable the consolidation of amplification and detection to be done in actual one-temperature assays and therefore the technology will be more robust when not in the laboratory. It is possible that the most influential frontier is the deployment of global health. CRISPR-based lateral flow assays (LFAs) have the potential to become one of the most democratized types of molecular diagnostics in the sense that they combine the accuracy of PCR with the simplicity of paper strips. Nonetheless, as mentioned in commercial development discussions in amongst your uploaded files, a number of challenges have to be overcome before CRISPR LFAs can be implemented globally: reagent stability, enzyme lyophilization, lot -to-lot consistency and long-lasting packaging that can withstand rural/tropical conditions are also among the main issues that manufacturers have to overcome before CRISPR LFAs can be used on a global scale. Concurrently, the CRISPR-guided enrichment is demonstrating the approach of low-resource adaptation of complex molecular testing. the advances imply that someday CRISPR diagnostics will be thoroughly integrated into social health systems. As microfluidics allows complete automation of testing, AI that ensures reliability and scalability, engineered versions of Cas, which enhance biochemical activity, and global health systems that demand equal distribution, CRISPR-based testing will become the center of prenatal testing or genetic screening on a global scale. CRISPR diagnostics will keep on developing new uses out of laboratory advances to everyday utilization as these two technological pillars continue to mature in parallel and reinforce the healthcare systems and the access to molecular testing in all socioeconomic environments (Kaur et al., 2024).

These novel tendencies directly solve existing drawbacks of diagnosing, such as AI-controlled sequence design lowers off-target activation, microfluidic automation lowers the risk of contamination in the course of sample processing, and engineered variants, such as Cas14, bypassing PAM limitations and enabling the detection of ultrashort DNA fragments (Huang et al., 2024).

These technologies are direct answers to the current diagnostic bottlenecks AI infuses reduced off-target effects, microfluidics infuses reduced contamination and variability of sample handling, engineered Cas variants infuse reduced PAM and temperature restrictions. A combination of all these solutions would result in a roadmap towards standardized globally implementable CRISPR diagnostics.

14.8 CONCLUSION

CRISPR-Cas9 technology has gone beyond its roots of a gene-editing instrument to emerge as an innovative technology in molecular diagnostics. Repurposing the specific nucleic acid recognition capabilities of the system, with both catalytically active (Cas9, Cas12, Cas13) and inactive (dCas9) variants, researchers have designed flexible platforms that are just as sensitive as gold-standard PCR but have the convenience needed to be utilized in point-of-care (POC) systems.These technologies have a significant clinical effect. CRISPR systems have been used in the management of infectious diseases with rapid and highly specific pathogen detection of SARS-CoV-2 to multidrug-resistant tuberculosis. They are defining liquid biopsies anew in the oncology field by making it possible to specifically identify rare circulating tumor DNA (ctDNA) mutations with the help of which cancer can be detected early and its treatment traced. Moreover, CRISPR-mediated screening is offering safer and less invasive methods of prenatal screening and defining of complex repeat expansion diseases in genetic medicine. Although this has been achieved, there are major challenges encountered in terms of clinical usage. Issues like off-target effects, dependence on PAM sequence, and standardized sample preparation protocols should be solved to have similar diagnostic accuracy. Nevertheless, the perspectives are bright in the future. The future with CRISPR technology and artificial intelligence to design the guide RNA to optimize its design, the new microfluidic automation, and new variants of Cas engineered gives a definite direction to the future. With the further development of these innovations, CRISPR diagnostics will be one of the foundations of universal health and will provide speed, precision, and accessibility of molecular testing to everyone in the world.

The chapter is aimed at researchers, clinicians, developers of diagnostic tests, and graduate-level learners interested in an integrated approach of understanding CRISPR diagnostic principles, platforms, regulatory issues, and translational issues.

REFERENCES

Allan-Blitz, L. T., Adams, G., Sanders, G., Shah, P., Ramesh, K., Jarolimova, J., Ard, K. L., Branda, J. A., Klausner, J. D., Sabeti, P. C., & Lemieux, J. E. (2025). Preliminary clinical performance of a Cas13a-based lateral flow assay for detecting Neisseria gonorrhoeae in urine specimens. *MSphere*, *10*(1), e0067724. DOI: 10.1128/msphere.00677-24 PMID: 39688405

Allen, A., Cooper, B. H., Singh, J., Rohs, R., & Qin, P. Z. (2025). PAM-adjacent DNA flexibility tunes CRISPR-Cas12a off-target binding. *Scientific Reports*, *15*(1), 4930. DOI: 10.1038/s41598-025-87565-9 PMID: 39929897

Azizoglu, R. O. (2022). CRISPR-Cas systems in diagnostics: A comprehensive assessment of Cas effectors and biosensors. *Gene and Genome Editing*, *3*, 100019. https://doi.org/https://doi.org/10.1016/j.ggedit.2022.100019. DOI: 10.1016/j.ggedit.2022.100019

Bengtson, M., Bharadwaj, M., Franch, O., van der Torre, J., Meerdink, V., Schallig, H., & Dekker, C. (2022). CRISPR-dCas9 based DNA detection scheme for diagnostics in resource-limited settings. *Nanoscale*, *14*(5), 1885–1895. DOI: 10.1039/D1NR06557B PMID: 35044397

Bhardwaj, P., Kant, R., Behera, S. P., Dwivedi, G. R., & Singh, R. (2022). Next-generation diagnostic with CRISPR/Cas: Beyond nucleic acid detection. *International Journal of Molecular Sciences*, *23*(11), 6052. DOI: 10.3390/ijms23116052 PMID: 35682737

Broughton, J. P., Deng, X., Yu, G., Fasching, C. L., Servellita, V., Singh, J., Miao, X., Streithorst, J. A., Granados, A., Sotomayor-Gonzalez, A., Zorn, K., Gopez, A., Hsu, E., Gu, W., Miller, S., Pan, C.-Y., Guevara, H., Wadford, D. A., Chen, J. S., & Chiu, C. Y. (2020). CRISPR-Cas12-based detection of SARS-CoV-2. *Nature Biotechnology*, *38*(7), 870–874. DOI: 10.1038/s41587-020-0513-4 PMID: 32300245

Dara, M., Shafieipour, N., Saffar, M., Dianatpour, M., Tabei, S.-M.-B., & Dastgheib, S.-A. (2025). Diagnosis of Infectious Diseases by CRISPR/Cas System. *OBM Genetics*, *9*(2), 1–39. DOI: 10.21926/obm.genet.2502289

Dou, H. Y., Huang, T. S., Wu, H. C., Hsu, C. H., Chen, F. J., & Liao, Y. C. (2025). Targeted sputum sequencing for rapid and broad drug resistance of Mycobacterium tuberculosis. *Infection*, *53*(4), 1413–1423. DOI: 10.1007/s15010-024-02463-y PMID: 39821740

Dronina, J., Samukaite-Bubniene, U., & Ramanavicius, A. (2022). Towards application of CRISPR-Cas12a in the design of modern viral DNA detection tools [Review]. *Journal of Nanobiotechnology*, *20*(1), 41. DOI: 10.1186/s12951-022-01246-7 PMID: 35062978

Du, K., Zeng, Q., Jiang, M., Hu, Z., Zhou, M., & Xia, K. (2025). CRISPR/Cas12a-Based Biosensing: Advances in Mechanisms and Applications for Nucleic Acid Detection. *Biosensors (Basel)*, *15*(6), 360. DOI: 10.3390/bios15060360 PMID: 40558442

Gan, T., Yu, J., Deng, Z., & He, J. (2024). ERA-CRISPR/Cas12a system: A rapid, highly sensitive and specific assay for Mycobacterium tuberculosis. *Frontiers in Cellular and Infection Microbiology*, *14*, 1454076. DOI: 10.3389/fcimb.2024.1454076 PMID: 39233906

Gu, W., Crawford, E. D., O'Donovan, B. D., Wilson, M. R., Chow, E. D., Retallack, H., & DeRisi, J. L. (2016). Depletion of Abundant Sequences by Hybridization (DASH): Using Cas9 to remove unwanted high-abundance species in sequencing libraries and molecular counting applications. *Genome Biology*, *17*(1), 41. DOI: 10.1186/s13059-016-0904-5 PMID: 26944702

Hall, T., Gulati, S., Sang, R., Jia, Z., Mckinnirey, F., Vesey, G., Goldys, E., & Deng, F. (2025). CRISPR biosensing with lateral flow assays for point of care Diagnostics: Overcoming commercial development challenges. *Trends in Analytical Chemistry*, *118275*, 118275. https://doi.org/https://doi.org/10.1016/j.trac.2025.118275. DOI: 10.1016/j.trac.2025.118275

He, X., Liang, J., Zhang, J., Fang, W., Liu, J., Zhang, M., Wang, L., & Song, C. (2025). CRISPR/Cas13a triggered-DNA walker amplified SERS sensor for ultrasensitive detection of cancer-related exosomal miRNA. *Biosensors & Bioelectronics*, *289*, 117924. DOI: 10.1016/j.bios.2025.117924 PMID: 40886435

Hillary, V. E., & Ceasar, S. A. (2023). A Review on the Mechanism and Applications of CRISPR/Cas9/Cas12/Cas13/Cas14 Proteins Utilized for Genome Engineering. *Molecular Biotechnology*, *65*(3), 311–325. DOI: 10.1007/s12033-022-00567-0 PMID: 36163606

Hossain, M. A. (2021). CRISPR-Cas9: A fascinating journey from bacterial immune system to human gene editing. *Progress in Molecular Biology and Translational Science*, *178*, 63–83. DOI: 10.1016/bs.pmbts.2021.01.001 PMID: 33685600

Huang, B., Guo, L., Yin, H., Wu, Y., Zeng, Z., Xu, S., Lou, Y., Ai, Z., Zhang, W., Kan, X., Yu, Q., Du, S., Li, C., Wu, L., Huang, X., Wang, S., & Wang, X. (2024). Deep learning enhancing guide RNA design for CRISPR/Cas12a-based diagnostics. *iMeta*, *3*(4), e214. DOI: 10.1002/imt2.214 PMID: 39135699

Kalter, N., Fuster-Garcia, C., Silva, A., Ronco-Diaz, V., Roncelli, S., Turchiano, G., Gorodkin, J., Cathomen, T., Benabdellah, K., Lee, C., & Hendel, A. (2025). Off-target effects in CRISPR-Cas genome editing for human therapeutics: Progress and challenges. *Molecular Therapy. Nucleic Acids*, *36*(3), 102636. DOI: 10.1016/j.omtn.2025.102636 PMID: 40777742

Karimi, M. A., Paryan, M., Behrouzian Fard, G., Sadeghian, H., Zarrinfar, H., & Hosseini Bafghi, M. (2025). Challenges and Opportunities in the Application of CRISPR-Cas9: A Review on Genomic Editing and Therapeutic Potentials. *Medical Principles and Practice : International Journal of the Kuwait University, Health Science Centre*, •••, 1–17. DOI: 10.1159/000547334 PMID: 40675140

Kaur, R., Gupta, S., Chauhan, A., Mishra, V., Sharma, M. K., & Singh, J. (2024). Harnessing the power of clustered regularly interspaced short palindromic repeats (CRISPR) based microfluidics for next-generation molecular diagnostics. *Molecular Biology Reports*, *51*(1), 896. DOI: 10.1007/s11033-024-09840-8 PMID: 39115550

Khoshandam, M., Soltaninejad, H., Hosseinkhani, S., Saltanatpour, Z., Goudarzi, M. T. H., & Hamidieh, A. A. (2024). CRISPR/Cas and artificial intelligence to improve precision medicine: Future perspectives and potential limitations. *Human Gene*, *42*, 201356. DOI: 10.1016/j.humgen.2024.201356

Khurana, A., Sayed, N., Singh, V., Khurana, I., Allawadhi, P., Rawat, P. S., Navik, U., Pasumarthi, S. K., Bharani, K. K., & Weiskirchen, R. (2022). A comprehensive overview of CRISPR/Cas 9 technology and application thereof in drug discovery. *Journal of Cellular Biochemistry*, *123*(10), 1674–1698. https://doi.org/https://doi.org/10.1002/jcb.30329. DOI: 10.1002/jcb.30329 PMID: 36128934

Koo, B., Kim, D. E., Kweon, J., Jin, C. E., Kim, S. H., Kim, Y., & Shin, Y. (2018). CRISPR/dCas9-mediated biosensor for detection of tick-borne diseases. *Sensors and Actuators. B, Chemical*, *273*, 316–321. DOI: 10.1016/j.snb.2018.06.069 PMID: 32288252

Lafi, Z., Ata, T., & Asha, S. (2025). CRISPR in clinical diagnostics: Bridging the gap between research and practice. *Bioanalysis*, *17*(4), 281–290. DOI: 10.1080/17576180.2025.2459520 PMID: 39902531

Lau, C. H., Li, X., Liang, Q. L., Guo, R., Ren, X., Xu, Z., Zhu, Y., Lai, Y., Liu, G., Huang, Y., Wu, W., Zhu, H., Chen, J., & Zhang, X. (2025). CRISPR technology for diagnosis and treatment of human brucellosis. *Journal of Microbiological Methods*, *240*, 107339. DOI: 10.1016/j.mimet.2025.107339 PMID: 41274609

Li, X., Wang, H., Qi, X., Ji, Y., Li, F., Chen, X., Li, K., & Li, L. (2024). PCR Independent Strategy-Based Biosensors for RNA Detection. *Biosensors (Basel)*, *14*(4), 200. DOI: 10.3390/bios14040200 PMID: 38667193

Mahfouz, M. (2024). Revolutionizing Point-of-Care Diagnostics via CRISPR Systems. *ACS Synthetic Biology*, *13*(2), 411–412. DOI: 10.1021/acssynbio.3c00763 PMID: 38197893

Mahfouz, M. (2024). *Revolutionizing point-of-care diagnostics via CRISPR systems.* ACS Publications. DOI: 10.1021/acssynbio.3c00763

Malekshoar, M., Azimi, S. A., Kaki, A., Mousazadeh, L., Motaei, J., & Vatankhah, M. (2023). CRISPR-Cas9 targeted enrichment and next-generation sequencing for mutation detection. *The Journal of Molecular Diagnostics : JMD*, *25*(5), 249–262. DOI: 10.1016/j.jmoldx.2023.01.010 PMID: 36841425

Mohanraju, P., Saha, C., van Baarlen, P., Louwen, R., Staals, R. H. J., & van der Oost, J. (2022). Alternative functions of CRISPR-Cas systems in the evolutionary arms race. *Nature Reviews. Microbiology*, *20*(6), 351–364. DOI: 10.1038/s41579-021-00663-z PMID: 34992260

Moon, J., & Liu, C. (2023). Asymmetric CRISPR enabling cascade signal amplification for nucleic acid detection by competitive crRNA. *Nature Communications*, *14*(1), 7504. DOI: 10.1038/s41467-023-43389-7 PMID: 37980404

Okesanya, O. J., Ahmed, M. M., Ogaya, J. B., Amisu, B. O., Ukoaka, B. M., Adigun, O. A., Manirambona, E., Adebusuyi, O., Othman, Z. K., Oluwakemi, O. G., Ayando, O. D., Tan, M. I. R. S., Idris, N. B., Kayode, H. H., Oso, T. A., Ahmed, M., Kouwenhoven, M. B. N., Ibrahim, A. M., & Lucero-Prisno, D. E.III. (2025). Reinvigorating AMR resilience: Leveraging CRISPR-Cas technology potentials to combat the 2024 WHO bacterial priority pathogens for enhanced global health security-a systematic review. *Tropical Medicine and Health*, *53*(1), 43. DOI: 10.1186/s41182-025-00728-2 PMID: 40176174

Ozden, F., & Minary, P. (2024). Learning to quantify uncertainty in off-target activity for CRISPR guide RNAs. *Nucleic Acids Research*, *52*(18), e87. DOI: 10.1093/nar/gkae759 PMID: 39275984

Pache, M. M., Rutuja, R. P., Nikam, S. V., Rode, R. B., & Darekar, A. B. (2025). CRISPR-Cas9 in Pharmaceutical Research: Applications, Challenges, Ethical Considerations and Future Directions. *Asian Journal of Pharmacy and Technology.*, *15*(3), 296–4.

Prezza, G., Heckel, T., Dietrich, S., Homberger, C., Westermann, A. J., & Vogel, J. (2020). Improved bacterial RNA-seq by Cas9-based depletion of ribosomal RNA reads. *RNA (New York, N.Y.)*, *26*(8), 1069–1078. DOI: 10.1261/rna.075945.120 PMID: 32345633

Qiu, M., & Li, P. (2021). CRISPR/Cas-based diagnostics and gene therapy. *BIO Integration.*, *2*(3), 121. DOI: 10.15212/bioi-2020-0048

Quan, J., Langelier, C., Kuchta, A., Batson, J., Teyssier, N., Lyden, A., Caldera, S., McGeever, A., Dimitrov, B., King, R., Wilheim, J., Murphy, M., Ares, L. P., Travisano, K. A., Sit, R., Amato, R., Mumbengegwi, D. R., Smith, J. L., Bennett, A., & Crawford, E. D. (2019). FLASH: A next-generation CRISPR diagnostic for multiplexed detection of antimicrobial resistance sequences. *Nucleic Acids Research*, *47*(14), e83. DOI: 10.1093/nar/gkz418 PMID: 31114866

Sharma, G., Sharma, A. R., Bhattacharya, M., Lee, S. S., & Chakraborty, C. (2021). CRISPR-Cas9: A Preclinical and Clinical Perspective for the Treatment of Human Diseases. *Molecular Therapy : the Journal of the American Society of Gene Therapy*, *29*(2), 571–586. DOI: 10.1016/j.ymthe.2020.09.028 PMID: 33238136

Shuo, Y., Feng, X., Zhao, C., Zhang, J., Wang, Y., Wang, J., Ding, Y., Shi, X., Li, J., Song, X., & Wang, J. (2025). A dual enhancement strategy for ultrasensitive detection of norovirus based on catalytic hairpin assembly-assisted CRISPR/Cas12a sensing system. *Journal of Dairy Science*, *108*(5), 4726–4733. DOI: 10.3168/jds.2024-26150 PMID: 40139379

Singh, D. (2024). Revolutionizing Lung Cancer Treatment: Innovative CRISPR-Cas9 Delivery Strategies. *AAPS PharmSciTech*, *25*(5), 129. DOI: 10.1208/s12249-024-02834-6 PMID: 38844700

Singh, M., Bindal, G., Misra, C. S., & Rath, D. (2022). The era of Cas12 and Cas13 CRISPR-based disease diagnosis. *Critical Reviews in Microbiology*, *48*(6), 714–729. DOI: 10.1080/1040841X.2021.2025041 PMID: 35164636

Slattery, J. R., Naung, N. Y., Kalinna, B. H., & Pal, M. (2025). CRISPR-Powered Liquid Biopsies in Cancer Diagnostics. *Cells*, *14*(19), 1539. DOI: 10.3390/cells14191539 PMID: 41090767

Srivastava, S., Upadhyay, D. J., & Srivastava, A. (2020). Next-generation molecular diagnostics development by CRISPR/Cas tool: Rapid detection and surveillance of viral disease outbreaks. *Frontiers in Molecular Biosciences*, *7*, 582499. DOI: 10.3389/fmolb.2020.582499 PMID: 33425987

Srivastava, S., Upadhyay, D. J., & Srivastava, A. (2020). Next-Generation Molecular Diagnostics Development by CRISPR/Cas Tool: Rapid Detection and Surveillance of Viral Disease Outbreaks. *Frontiers in Molecular Biosciences*, *7*, 582499. DOI: 10.3389/fmolb.2020.582499 PMID: 33425987

Subica, A. M. (2023). CRISPR in Public Health: The Health Equity Implications and Role of Community in Gene-Editing Research and Applications. *American Journal of Public Health*, *113*(8), 874–882. DOI: 10.2105/AJPH.2023.307315 PMID: 37200601

Sun, Y., Wen, T., Zhang, P., Wang, M., & Xu, Y. (2024). Recent Advances in the CRISPR/Cas-Based Nucleic Acid Biosensor for Food Analysis: A Review. *Foods*, *13*(20), 3222. DOI: 10.3390/foods13203222 PMID: 39456285

Tram, T. T. B., Ha, V. T. N., Trieu, L. P. T., Ashton, P. M., Crawford, E. D., Thu, D. D. A., Quang, N. L., Thwaites, G. E., Walker, T. M., Anscombe, C., & Thuong, N. T. T. (2023). FLASH-TB: An Application of Next-Generation CRISPR to Detect Drug Resistant Tuberculosis from Direct Sputum. *Journal of Clinical Microbiology*, *61*(4), e0163422. DOI: 10.1128/jcm.01634-22 PMID: 37010411

Tripathi, S., Khatri, P., Fatima, Z., Pandey, R. P., & Hameed, S. (2022). A Landscape of CRISPR/Cas Technique for Emerging Viral Disease Diagnostics and Therapeutics: Progress and Prospects. *Pathogens (Basel, Switzerland)*, *12*(1), 56. DOI: 10.3390/pathogens12010056 PMID: 36678404

Verma, M. K., Roychowdhury, S., Sahu, B. D., Mishra, A., & Sethi, K. K. (2022). CRISPR-based point-of-care diagnostics incorporating Cas9, Cas12, and Cas13 enzymes advanced for SARS-CoV-2 detection. *Journal of Biochemical and Molecular Toxicology*, *36*(8), e23113. DOI: 10.1002/jbt.23113 PMID: 35642647

Wang, M., Wang, H., Li, K., Li, X., Wang, X., & Wang, Z. (2023). Review of CRISPR/Cas Systems on Detection of Nucleotide Sequences. *Foods*, *12*(3), 477. DOI: 10.3390/foods12030477 PMID: 36766007

Wei, Y., Fu, Y., Li, C., Chen, S., Xie, L., & Chen, M. (2024). Ultrasensitive detection of circulating tumor DNA using a CRISPR/Cas9 nickase-driven 3D DNA walker based on a COF-AuNPs sensing platform. *Microchimica Acta*, *191*(11), 671. DOI: 10.1007/s00604-024-06749-8 PMID: 39404875

Weng, Z., You, Z., Yang, J., Mohammad, N., Lin, M., Wei, Q., Gao, X., & Zhang, Y. (2023). CRISPR-Cas biochemistry and CRISPR-based molecular diagnostics. *Angewandte Chemie International Edition, 62*(17), e202214987. DOI: 10.1002/anie.202214987 PMID: 36710268

Wu, H.-p., & Yang, Z.-l. Advances in the applications of CRISPR/Cas system for tumor molecular diagnostics. LabMed Discovery. 2024;1(1):100011. https://doi.org/https://doi.org/10.1016/j.lmd.2024.100011

Xie, S., Yue, Y., & Yang, F. (2024). Recent Advances in CRISPR/Cas System-Based Biosensors for the Detection of Foodborne Pathogenic Microorganisms. *Micromachines, 15*(11), 1329. DOI: 10.3390/mi15111329 PMID: 39597141

Xue, M., Gonzalez, D. H., Osikpa, E., Gao, X., & Lillehoj, P. B. (2025). Rapid and automated interpretation of CRISPR-Cas13-based lateral flow assay test results using machine learning. *Sensors & Diagnostics, 4*(2), 171–181. DOI: 10.1039/D4SD00314D PMID: 39817182

Yu, L., Peng, Y., Sheng, M., Wang, Q., Huang, J., & Yang, X. (2023). Sensitive and Amplification-Free Electrochemiluminescence Biosensor for HPV-16 Detection Based on CRISPR/Cas12a and DNA Tetrahedron Nanostructures. *ACS Sensors, 8*(7), 2852–2858. DOI: 10.1021/acssensors.3c00806 PMID: 37402133

Zhang, B., Li, J., & Yu, W. (2025). Integration of CRISPR/dCas9-Based methylation editing with guide positioning sequencing identifies dynamic changes of mrDEGs in breast cancer progression. *Cellular and Molecular Life Sciences : CMLS, 82*(1), 46. DOI: 10.1007/s00018-024-05562-z PMID: 39833630

Zhang, H., Qin, C., An, C., Zheng, X., Wen, S., Chen, W., Liu, X., Lv, Z., Yang, P., Xu, W., Gao, W., & Wu, Y. (2021). Application of the CRISPR/Cas9-based gene editing technique in basic research, diagnosis, and therapy of cancer. *Molecular Cancer, 20*(1), 126. DOI: 10.1186/s12943-021-01431-6 PMID: 34598686

Zhou, H., Ye, P., Xiong, W., Duan, X., Jing, S., He, Y., Zeng, Z., Wei, Y., & Ye, Q. (2024). Genome-scale CRISPR-Cas9 screening in stem cells: Theories, applications and challenges. *Stem Cell Research & Therapy, 15*(1), 218. DOI: 10.1186/s13287-024-03831-z PMID: 39026343

Zhou, J., Li, Z., Seun Olajide, J., & Wang, G. (2024). CRISPR/Cas-based nucleic acid detection strategies: Trends and challenges. *Heliyon, 10*(4), e26179. DOI: 10.1016/j.heliyon.2024.e26179 PMID: 38390187

Zhou, L., & Simonian, A. L. (2024). CRISPR/Cas Technology: The Unique Synthetic Biology Genome-Editing Tool Shifting the Paradigm in Viral Diagnostics, Defense, and Therapeutics. *Annual Review of Biomedical Engineering*, *26*(1), 247–272. DOI: 10.1146/annurev-bioeng-081723-013033 PMID: 38346278

Chapter 9

CRISPR–Cas9 Technology for Epigenome Editing in Precision Medicine:

Opportunities and Challenges

M. M. Nidhilangelo
https://orcid.org/0000-0003-1497-9865
Pondicherry University, India

Ranjith Kumavath
https://orcid.org/0000-0002-0751-5491
Pondicherry University, India

ABSTRACT

CRISPR–Cas9, originally derived from a bacterial adaptive immune system, has rapidly evolved into a transformative platform for genome and epigenome engineering with broad applications in biology and precision medicine. Guided by single-guide RNAs, the Cas9 nuclease introduces targeted DNA modifications through cellular repair pathways. Meanwhile, engineered variants, such as high-fidelity nucleases, nickases, and nuclease-deficient dCas9, extend their functionality to precise gene regulation, epigenetic modulation, RNA editing, and live-cell genomic imaging. dCas9-based effectors enable locus-specific control of transcription by directing DNA methylation, histone modifications, or chromatin remodelling, offering reversible and non-permanent alternatives to conventional epigenetic drugs. In vivo models, including somatically engineered mouse models and xenografts, enable physiologically relevant studies of tumorigenesis and therapeutic response, while base and prime editors offer safer, DNA break–free approaches for the precision correction of pathogenic variants. Clinically, CRISPR is driving innovation in adoptive immunotherapies, regenerative medicine, and targeted delivery systems,

DOI: 10.4018/979-8-3373-4857-5.ch009

with nanoparticle- and ligand-modified carriers improving tissue-specific targeting and reducing systemic toxicity. Despite these advances, major challenges remain, including off-target activity, inefficient and tissue-restricted delivery, instability of epigenetic edits, and ethical concerns surrounding germline modification and long-term safety. Ongoing innovations in CRISPR engineering, delivery technologies, and regulatory oversight are addressing these barriers. Collectively, CRISPR-based platforms represent a cornerstone of next-generation personalised therapeutics, offering programmable, scalable, and clinically relevant strategies to understand and treat human disease.

1. ABBREVIATIONS

- CRISPR – Clustered Regularly Interspaced Short Palindromic Repeats
- Cas9 – CRISPR-associated protein 9
- gRNA / sgRNA – guide RNA / single-guide RNA
- dCas9 – dead Cas9 (catalytically inactive Cas9)
- PAM – Protospacer Adjacent Motif
- NHEJ – Non-Homologous End Joining
- HDR – Homology-Directed Repair
- DSB – Double-Strand Break
- AAV – Adeno-Associated Virus
- iPSCs – Induced Pluripotent Stem Cells
- DNMT3A / DNMT3L – DNA (cytosine-5)-methyltransferase 3A / 3L
- TET1 – Ten–Eleven Translocation methylcytosine dioxygenase 1
- KRAB – Krüppel-associated box
- VP64 – Viral Protein 64
- LOF – Loss-of-Function
- GOF – Gain-of-Function

2. GLOSSARY

- CRISPR-Cas9 – A bacterial-derived gene-editing system that uses a guide RNA and Cas9 nuclease to target specific DNA sequences. Adapted for epigenome editing with catalytically inactive Cas9 (dCas9).
- Cas9 – DNA-cutting enzyme used in CRISPR to introduce double-strand breaks at targeted genomic loci.

- dCas9 (Dead/Catalytically Inactive Cas9) – A Cas9 variant that binds DNA without cutting it, used to recruit epigenetic modifiers for gene regulation or imaging.
- Guide RNA (gRNA / sgRNA) – RNA sequence that directs Cas9 or dCas9 to a specific DNA site; synthetic single-guide RNA (sgRNA) combines crRNA and tracrRNA functions.
- Epigenome Editing – Modulation of gene expression without changing DNA sequence via DNA methylation, histone modifications, or chromatin remodelling.
- CRISPRi / CRISPRa – dCas9-based repression (CRISPR interference) or activation (CRISPR activation) of gene expression through fused epigenetic effector domains.
- Epigenetic Modifiers – Proteins fused to dCas9 to modify chromatin.
- KRAB – transcriptional repressor
- DNMT3A / DNMT3B – DNA methyltransferases
- TET1 – demethylase
- p300 – histone acetyltransferase
- Chromatin Accessibility – Degree to which DNA is exposed and accessible for binding; affects efficiency of epigenome editing and CRISPR-based imaging.
- High-Fidelity Cas9 Variants – Engineered Cas9 proteins (SpCas9-HF1, eSpCas9, HypaCas9) designed to reduce off-target binding while maintaining on-target activity.
- Off-Target Effects – Unintended binding or editing of non-target genomic sites by CRISPR systems, which may alter gene expression.
- Off-Target Epigenetic Changes – Unintended alterations in DNA methylation, histone modifications, or chromatin state at non-target loci caused by epigenome editors.
- Transient vs. Durable Edits – Transient edits are temporary; durable edits persist through cell divisions. Epigenome edits may be reversed by cellular chromatin remodelling.
- CRISPR-ChAP-MS – Chromatin affinity purification combined with mass spectrometry to study protein-DNA interactions at targeted loci; efficiency depends on chromatin state.
- Signal Amplification in Imaging – Techniques to enhance visualisation of loci in live cells, e.g., SunTag, RNA aptamer systems, SplitGFP, Casilio system.
- Adeno-Associated Virus (AAV) – Efficient but limited packaging capacity; often requires split-dCas9.

- Lentiviral Vectors – Allow large constructs and persistent expression; risk of insertional mutagenesis.
- Non-Viral Systems – Lipid nanoparticles (LNPs), polymeric/gold nanoparticles, electroporation; safer but limited in vivo efficiency.
- Tissue Tropism & Barriers – Many vectors accumulate in the liver/spleen; brain delivery requires BBB-crossing strategies.
- Immune Response to CRISPR – Pre-existing antibodies to Cas proteins or viral vectors can reduce efficacy and cause toxicity.
- iPSC-Derived Therapies – Epigenome editing of induced pluripotent stem cells requires monitoring for tumorigenicity and genomic stability.
- Regulatory and Ethical Considerations – Oversight of germline risk, dual-use potential, equitable access, and safety under frameworks like ISSCR, NIH, and FDA guidance.
- Enhancers and Promoters – DNA regulatory regions targeted by dCas9–epigenetic effectors to modulate gene expression.
- Telomeres – Repetitive DNA regions often used for CRISPR imaging; require fewer sgRNAs compared to non-repetitive loci.
- Multiplex Editing / Gene Modulation – Simultaneous targeting of multiple loci or genes using dCas9-based systems.

3. INTRODUCTION

CRISPR-Cas9, or Clustered Regularly Interspaced Short Palindromic Repeats and CRISPR-associated protein 9, is a powerful gene-editing technology that allows scientists to precisely modify DNA sequences within living cells. It originates from a bacterial immune system that defends against viral infections. In biotechnology, CRISPR-Cas9 has been adapted to cut specific DNA sequences using the Cas9 enzyme, which acts like molecular scissors. A guide RNA (gRNA) directs Cas9 to the targeted site in the genome, enabling the modification, deletion, or replacement of genes to study their function or correct genetic mutations (Aljabali et al., 2024). CRISPR-Cas9 technology is emerging as a central tool in precision medicine, particularly in oncology (**Figure 1**). It enables high-throughput functional screening, including genome-wide loss-of-function (LOF) and gain-of-function (GOF) studies, to identify tumour-driver genes, synthetic lethal targets, and drug resistance pathways. These screens help uncover therapeutic targets such as metabolic enzymes and chromatin regulators, and identify biomarkers associated with treatment sensitivity or resistance. Moreover, CRISPR-based approaches can identify genes and pathways involved in resistance to chemotherapies, targeted therapies, and immunotherapies, thereby contributing to the development of more effective, personalized cancer treatments (Xing et al., 2020). The emergence of epigenome editing was enabled by the development of catalytically inactive Cas9 (dCas9), which retains guide-RNA–directed DNA binding but lacks nuclease activity. This converted CRISPR from a DNA-cutting tool into a programmable platform for gene regulation. By recruiting transcriptional regulators or chromatin modifiers, dCas9 systems enable reversible control of gene expression without altering the DNA sequence. These technologies underpin a wide range of applications across cancer biology, regenerative medicine, neurology, immunology, and other fields (Seem et al., 2024). CRISPR-based epigenome editing can therefore modulate transcription through changes in DNA methylation, histone modifications, chromatin structure, non-coding RNA interactions, and 3D genome organisation. Its high specificity and flexibility arise from the fact that only the guide RNA needs to be changed to target different loci, allowing for multiplexed and precise gene targeting with minimised off-target effects, especially with advanced systems like CRISPRoff and CRISPRon (Chen et al., 2023). This approach has been used to study enhancer–gene interactions, transcriptional regulation, and cell differentiation. In cancer therapeutics, dCas9-effector fusions have been employed to reactivate tumour suppressor genes, silence oncogenes, and reverse aberrant epigenetic silencing (Rajanathadurai et al., 2024). In regenerative medicine, CRISPR-based tools enable precise control of stem cell differentiation and tissue repair by activating genes required for the generation of induced pluripotent stem cells, facilitating direct cellular reprogramming and organoid development, and correcting epigenetic defects, such as those seen in Fragile X syndrome (Fadul et al., 2023). In neurology, CRISPR and CRISPRi/a systems facilitate the modeling and reversal of disorders such as Alzheimer's, Parkinson's, Huntington's disease, and ALS by targeting genes involved in neuronal survival, synaptic plasticity, and repeat expansion disorders (Dräger et al., 2022). In immunology, CRISPR is utilized to engineer T cells, B cells, NK cells, and macrophages, thereby enhancing CAR-T persistence and facilitating the modulation of autoreactive pathways in conditions such as lupus, rheumatoid arthritis, and Type I diabetes (Stadtmauer et al., 2020; Tao et al., 2024). Infectious disease research benefits from Cas13-based antiviral strategies, gene screens for host dependency factors, and CRISPR diagnostics such as SHERLOCK and DETECTR (Abbott et al., 2020). Beyond human health, CRISPR plays a major role in agriculture by generating high-yield, stress-tolerant, and disease-resistant crops, as well as editing traits in livestock to enhance disease resistance, productivity, and suitability for xenotransplantation (Rasheed et al., 2021). Environmental applications include the use of gene drives to control vector populations and the development of engineered microbes for bioremediation of pollutants. In synthetic biology, CRISPR facilitates metabolic engineering for the production of biofuels, pharmaceuticals, industrial enzymes, and advanced biomaterials through multiplex and epigenome editing (Joshi & Mishra, 2022; Rafeeq et al., 2023). Together, these diverse applications highlight CRISPR's broad impact across biology, medicine, agriculture, and environmental science. Altogether, CRISPR-Cas9 and its epigenome-editing derivatives have transformed our ability to interrogate gene function and reprogram cellular states, positioning them as cornerstone technologies for advancing precision medicine and developing next-generation therapeutics.

Figure 1. **Overview of CRISPR-based genome and epigenome editing strategies** - *CRISPR-Cas9 generates double-strand breaks repaired by NHEJ or HR. Base editors (BEs) introduce A-to-G or C-to-T substitutions using Cas9 nickase deaminase fusions. Prime editors (PE) combine Cas9 nickase with reverse transcriptase to mediate precise insertions, deletions, or substitutions. CRISPRi and CRISPRa employ regulated transcription. Epigenome editors enable the targeted demethylation or methylation of DNA at specific loci.*

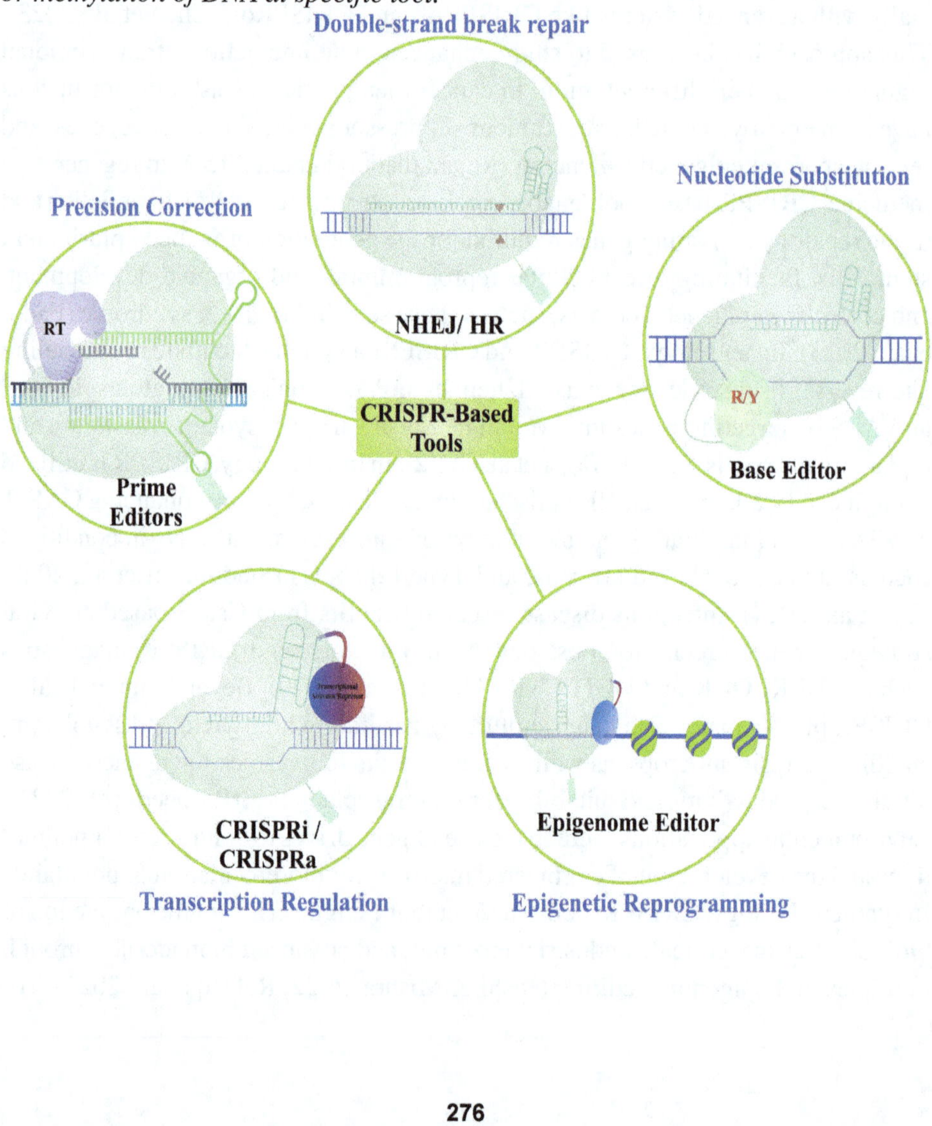

4. MECHANISMS AND EVOLUTION OF THE CRISPR-CAS9 SYSTEM

The classical CRISPR-Cas9 system originates from a prokaryotic adaptive immune mechanism that defends against viruses by using RNA-guided nucleases to cleave foreign DNA (**Figure 2**). In genome engineering, this system has been simplified into two core components: The Cas9 endonuclease and a single-guide RNA (sgRNA), which fuses crRNA and tracrRNA into a programmable RNA scaffold (Jinek et al., 2012). The sgRNA directs Cas9 to a complementary genomic locus proximal to a protospacer adjacent motif (PAM), typically 5′-NGG-3′ for *Streptococcus pyogenes* Cas9 (SpCas9). Cas9 triggers a double-strand break (DSB) three base pairs upstream of the PAM via its two nuclease domains, RuvC and HNH. Subsequent to cleavage, cells repair the DSB through either non-homologous end joining (NHEJ), which tends to produce insertions or deletions that interfere with gene function, or homology-directed repair (HDR), which enables accurate sequence replacement via a donor template (Zhang et al., 2021). A critical milestone in CRISPR evolution was the innovation of catalytically inactive Cas9, or dead Cas9 (dCas9), generated through the introduction of mutations D10A (RuvC) and H840A (HNH), which upholds DNA-binding ability but lacks nuclease activity, facilitating locus-specific modulation without instating breaks (Sapozhnikov & Szyf, 2021; Schubert et al., 2021). When combined with epigenetic modifiers such as KRAB, DNMT3A, TET1, p300, or histone demethylases, dCas9 supports the programmable modulation of chromatin states, encompassing DNA methylation, histone acetylation, transcriptional activation or repression, and chromatin remodeling. These tools have been widely implemented in functional genomics and in coordinating transcriptional programs across various mechanisms, such as cellular reprogramming and lineage conversion (Nuñez et al., 2021; Zhu, 2022).

CRISPR systems are widely classified into two classes. Class 1 systems utilize multi-protein effector complexes, including Types I, III, and IV, such as Cas3, Csm, and Cmr. Class 2 system relies on a single large effector Cas protein to conduct interference, with key examples being Cas9 (Type II), Cas12 (Type V), and Cas13 (Type VI) (Pickar-Oliver & Gersbach, 2019). Among these, Cas9 is the most extensively applied nuclease, with numerous natural and engineered variants. Common examples include SpCas9, SaCas9, NmeCas9, StCas9, eSpCas9, SpCas9-HF1, xCas9, Cas9 nickase, and dead Cas9 (**Table 1**). SpCas9, derived from *Streptococcus pyogenes*, forms a complex with crRNA and tracrRNA (or a synthetic sgRNA) to target specific DNA sequences adjacent to an NGG PAM. While highly efficient, wild-type SpCas9 can tolerate mismatches between the sgRNA and target DNA, causing unintended off-target cleavage due to excessive, non-specific DNA strand partition (Wang & Li, 2021). To address this, an augmented specificity Cas9 was engineered, like

- (eSpCas9), formulated by Slaymaker et al. (2016) through targeted amino acid substitutions (K848A, K1003A, R1060A) in positively charged residues of the non-target strand groove. These adjustments minimize DNA unwinding and enhance dependence on precise sgRNA–DNA pairing, improving specificity while maintaining NGG PAM recognition and compatibility with base editors, prime editors, and transcriptional regulators.
- SpCas9-HF1, documented by Kleinstiver et al. (2016), integrates four-point mutations (N497A, R661A, Q695A, Q926A) that dampen non-specific DNA backbone associations, creating an increasingly reliance on optimising guide complementarity. It sustains high on-target efficiency (>85% of wild-type) and NGG PAM identification. Subsequent variants (HF2–HF4) with additional mutations further mitigate the residual off-target effects. SpCas9-HF1 also reveals improved efficacy in homology-directed repair (HDR), particularly when consolidated with cell-cycle–regulated delivery systems (Allemailem et al., 2023; Matsumoto et al., 2024).
- xCas9, created by Hu et al. (2018) using phage-assisted sustained evolution, discerns a greater diversity of PAMs (NG, GAA, GAT) through mutations in its PAM-interfacing domain, curbing guanine reliance and facilitating access to formerly unreachable genomic zones. It sustains high specificity and is congruent with genome, base, and prime editing (Kim et al., 2020).

Additional Cas9 orthologs widen the conveyance and targeting possibilities.

- StCas9, from *Streptococcus thermophilus*, acknowledges the PAM sequence NNAGAAW and encompasses of a REC lobe (guide RNA–DNA recognition), NUC lobe (RuvC and HNH nuclease domains), and PAM-interfacing domain. It initiates DSBs that are repaired by NHEJ or HDR, and offers distinct targeting options in contrast to SpCas9 (Kostyushev et al., 2019).
- SaCas9, from *Staphylococcus aureus*, is ~300 amino acids smaller than SpCas9, creating it well-suited for adeno-associated virus (AAV) delivery. It detects a 5′-NNGRRT-3′ PAM and displays the equivalent function domains as SpCas9. Its compressed size and discrete PAM exactness expand its applicability in in vivo genome editing (Yang et al., 2023).
- NmeCas9, from *Neisseria meningitidis*, is a smaller ortholog (~1,368 amino acids) with a PAM requirement of NNNNGATT. It has a fundamentally high specificity and is notably beneficial for AAV-mediated deployment in precise genome editing purposes (Ibraheim et al., 2018).

- Cas9 nickase is created through inactivating one nuclease domain D10A mutation in RuvC (cuts non-target strand) or H840A in HNH (cuts target strand) to initiate single-strand breaks. Applications comprise paired nickases for highly selective DSBs, HDR stimulation with minimal NHEJ activation, base editing when fused to deaminases, and prime editing when coupled with reverse transcriptase. Because single-strand breaks are less mutagenic than DSBs, nickases reduce off-target mutations and indels (Lee et al., 2023).

Together, these mechanistic insights and engineered variants have transformed CRISPR–Cas9 from a bacterial immune system into a versatile toolkit for precise genome and epigenome engineering across biological research, biotechnology, and clinical applications.

Figure 2. **A chronology of significant turning points in the field of CRISPR discovery and technological advancement -** *From the initial finding of CRISPR sequences to the identification of additional Cas effector families, blue dots stand for significant biological discoveries. Green dots indicate significant advancements in technology and medicine, such as programmable CRISPR-Cas9 editing, the creation of precision editors and epigenome-editing tools, CRISPR-based diagnostic systems, and clinical successes that led to the approval of the first CRISPR therapy.*

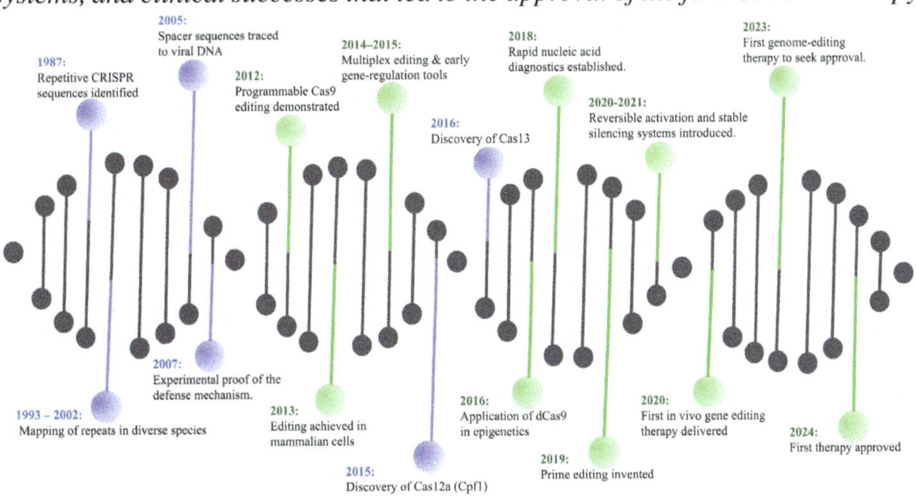

Table 1. Different Variants of Cas9 and Their Properties for CRISPR-based Genome Editing

Cas9 Variant	Source Organism	PAM Sequence	Key modifications	Advantages	References cited
SpCas9	*Streptococcus pyogenes*	NGG	Wild-type Cas9 forms a complex with crRNA+tracrRNA or sgRNA.	Efficient genome editing.	Wang & Li, 2021
eSpCas9	Engineered from SpCas9		K848A, K1003A, and R1060A mutations reduce non-target strand unwinding.	Higher specificity; compatible with base/prime editing and transcriptional regulation	Slaymaker et al., 2016
SpCas9-HF1			N497A, R661A, Q695A, Q926A mutations weaken non-specific DNA contacts; HF2–HF4 variants further improve.	High fidelity; >85% on-target efficiency; improved HDR	Kleinstiver et al., 2016
Cas9 Nickase			D10A (RuvC) cuts non-target strand; H840A (HNH) cuts target strand	Paired nickases for precise DSBs; HDR stimulation; base/prime editing; reduced off-target effects.	Lee et al., 2023
Dead Cas9 (dCas9)			Catalytically inactive (both nuclease domains mutated)	DNA binding without cleavage; used for CRISPRi, CRISPRa, epigenetic modulation	Seem et al., 2024
xCas9		NG, GAA, GAT	Mutations in the PAM-interacting domain broaden PAM recognition and reduce guanine dependence	Expanded targeting range; high specificity; compatible with base/prime editing	Hu et al., 2018
StCas9	*Streptococcus thermophilus*	NNAGAAW	Distinct REC, NUC, and PAM-interacting domains	Unique PAM targeting; DSB induction for NHEJ or HDR	Kostyushev et al., 2019
SaCas9	*Staphylococcus aureus*	NNGRRT	Smaller size for AAV delivery; same core domains as SpCas9	Compact; suited for in vivo editing; distinct PAM expands target sites	Yang et al., 2023
NmeCas9	*Neisseria meningitidis*	NNNNGATT	Compact ortholog with inherently high specificity	AAV-compatible; precise genome editing	Ibraheim et al., 2018

5. EPIGENETIC CONTROL OF HEALTH AND DISEASE

Epigenetic modifications are heritable yet reversible changes that regulate gene expression without altering the DNA sequence. These include DNA methylation, histone modifications, chromatin remodelling, and non-coding RNAs (ncRNAs), all of which play crucial roles in health and disease. DNA methyltransferases (DNMTs) add methyl groups to cytosine residues at CpG sites, thereby repressing transcription by blocking transcription factor binding or recruiting repressive proteins, which maintains genomic stability and regulates tissue-specific gene expression (Akram et al., 2024). DNA methylation primarily occurs at the 5th carbon of cytosine (5mC) within CpG dinucleotides (Stevens et al., 2022). It regulates gene expression by inhibiting the access of transcription factors, recruiting repressive complexes, and collaborating with histone modifiers to shape the chromatin structure. Dynamic methylation and demethylation cycles during development establish lineage-specific gene expression patterns, which are essential for cellular identity (Chang et al., 2023). Environmental factors, such as maternal intake of methyl-donor nutrients like betaine and folate, can influence methylation of imprinted genes (e.g., *Igf2/H19*), impacting brain development, metabolism, and other traits (Socha et al., 2024; von Holdt et al., 2023). DNA methylation also affects neurodevelopment and behaviour, with aberrations linked to imprinting syndromes, cancer, metabolic, and neuropsychiatric disorders (Rooney & Sadikovic, 2022; Alshammari et al., 2022; Dhar et al., 2021). In head and neck squamous cell carcinoma (HNSCC), methylation plays a critical role in modulating immune checkpoint genes, including *CD28*, *CTLA-4*, *ICOS*, *CD80*, and *CD86*. While promoter methylation typically suppresses gene expression, methylation within gene bodies can have the opposite effect, depending on specific genomic contexts. Notably, methylation of *CD28* influences nearby genes *CTLA-4* and *ICOS* due to shared regulatory elements, with these epigenetic patterns correlating with tumour features, immune cell infiltration, tumour mutational burden, and interferon-γ signatures. Such profiles hold promise as biomarkers predicting responses to immune checkpoint blockade therapies, especially anti-CTLA-4 and potentially anti-ICOS treatments. Moreover, *CD80* methylation is associated with a poor prognosis, and HPV infection can further modulate the methylation of immune-related genes, thereby affecting treatment outcomes (de Vos et al., 2020).

Histone modifications, including acetylation, methylation, phosphorylation, ubiquitination, and Sumoylation, regulate chromatin architecture and influence the cell cycle, DNA repair, differentiation, and stem cell pluripotency. Aberrant H3K27 trimethylation driven by EZH2 overexpression promotes cancer progression, while reduced histone acetylation contributes to neurodegenerative diseases (Hou et al., 2022). In B-cell lymphoma, histone deacetylase 3 (HDAC3) represses *PD-L1* expression by maintaining compact chromatin at its promoter. HDAC3 inhibition increases histone acetylation, recruits BRD4, and elevates *PD-L1* transcription, facilitating immune evasion. Combining HDAC3 inhibitors with PD-L1 blockade enhances tumour regression synergistically in preclinical models (Deng et al., 2019). ATP-dependent chromatin remodelers, such as the SWI/SNF, ISWI, CHD, and INO80 complexes, reposition nucleosomes to regulate DNA accessibility, thereby supporting transcription, development, differentiation, and DNA repair. Mutations in SWI/SNF components, such as SMARCA4 (also known as BRG1), are implicated in various cancers and autoimmune diseases (Feoktistov et al., 2022). In colorectal cancer (CRC), BRG1 remodels chromatin at the *CD44* promoter, facilitating recruitment of the transcription factor RUNX2. Loss of BRG1 disrupts this process, reducing *CD44* expression and impairing epithelial-mesenchymal transition (EMT), invasiveness, and cancer stemness (Yan et al., 2020). Regulatory non-coding RNAs, such as microRNAs (miRNAs), long non-coding RNAs (lncRNAs), and small interfering RNAs (siRNAs), fine-tune gene expression at both transcriptional and post-transcriptional levels, thereby impacting immune responses, development, and metabolism. Dysregulated miRNAs, such as oncomiRs, promote tumorigenesis, while altered long non-coding RNAs (lncRNAs) contribute to metastasis and neurological disorders (Loganathan & Doss, 2023). In non-small cell lung cancer (NSCLC), overexpression of *CD73* enhances tumour proliferation, EMT, immune evasion, and therapy resistance. The tumour-suppressive miRNA miR-30a-5p targets the 3′-UTR of *CD73* mRNA, suppressing its expression; however, miR-30a-5p is often downregulated in NSCLC, allowing elevated *CD73* levels and tumour progression (Chen et al., 2020).

6. ADVANCES IN CRISPR-MEDIATED EPIGENOME EDITING

CRISPR-dCas9-based epigenome editing enables precise manipulation of epigenetic marks, including DNA methylation, histone modifications, and chromatin remodelling, at defined genomic loci without altering the underlying DNA sequence. The dCas9 retains DNA-binding capability via a guide RNA but lacks nuclease activity, allowing it to serve as a programmable scaffold for epigenetic modifier domains to regulate transcription reversibly.

- **Transcriptional Activation:** One approach, histone acetylation-based stimulation, uses dCas9 merged to the histone acetyltransferase P300, which catalyses acetylation of histone H3 at lysine 27 (H3K27ac) and lysine 9 (H3K9ac), marks related to active transcription. In colorectal cancer (CRC), Wang et al. (2024) tailored sgRNAs aiming at the promoter of ZNF334, a putative tumour suppressor. Enlisting the dCas9-P300 complex at this locus prompted site-specific histone acetylation, mobilized ZNF334 transcription, and resulted in reduced proliferation and increased apoptosis in CRC cells. CRISPR activation (CRISPRa) tactics utilize dCas9 in conjunction with transcriptional activators such as VP64, p65, or Rta to recruit the transcriptional machinery. The multiplexed dCas9–SAM system has been used to concurrently upregulate the antigen-presenting costimulatory molecules CD70, CD80, and CD86 in tumour cells, thereby facilitating tumour immunogenicity (Alves et al., 2021).

- **Transcriptional Repression:** CRISPR interference (CRISPRi) leverages dCas9 conjugated to repressor domains like KRAB to inhibit transcription initiation or elongation. KRAB enlists KAP1, which in turn lures the histone methyltransferase SETDB1 to deposit repressive H3K9me3 marks, aggregating chromatin into a heterochromatic state that is inaccessible to transcription factors. This method has been used to silence immune checkpoint proteins, such as CD47, thereby boosting macrophage-mediated phagocytosis, and PD-L1, thereby restoring T cell cytotoxicity, without incorporating permanent DNA changes (Stuart et al., 2021; Hayes et al., 2022).
- **DNA Methylation Editing:** Directed DNA methylation can be accomplished by coupling dCas9 to DNA methyltransferases, such as DNMT3A, to incorporate methyl groups on cytosine residues within CpG sites at gene promoters. This adjustment restricts transcription factor binding and tightens chromatin, competently downregulating transcription. For example, promoters of EpCAM, CD71, and CXCR4 have been methylated through dCas9 DNMT3A fusion assemblies to elicit steady transcriptional silencing (Carroll & Giacca, 2023).
- **Chromatin Remodelling:** dCas9 can be combined with ATP-dependent chromatin remodelers, featuring BRG1 (SMARCA4, SWI/SNF family), CHD1 (an ATPase that supports open chromatin), or ISWI family motors playing a role in nucleosome spacing to directly reposition, evict, or slide nucleosomes at promoters, enhancers, or other modulatory elements. Such localised remodelling can either aid transcription-factor recruitment and gene activation or increase nucleosome compaction to repress transcription (Donovan et al., 2019).
- **RNA-Level Editing:** CRISPR-mediated RNA editing promotes temporary transcript adjustment, rectification of pathogenic mutations, and tuning of noncoding RNAs. The ADAR enzyme, or its deaminase domain fused to Cas9 or Cas13b, catalyses adenosine-to-inosine (A→I) conversions in double-stranded RNA. This strategy, implemented as REPAIR or RESCUE systems, provides temporary modulation of mRNA transcripts and has been used to study genes and noncoding RNAs, as well as to minimize harmful

transcripts in diseases such as ALS and Huntington's disease (Marina et al., 2020).

This flexibility makes CRISPR–dCas9–based epigenome editing a potent platform for exploring gene function, cellular differentiation, and disease mechanisms, as well as for developing therapeutic strategies that modulate gene expression without altering the underlying DNA sequence (**Figure 3**).

7. THE ROLE OF CRISPR-CAS9 IN PRECISION MEDICINE

CRISPR–Cas9 technologies have evolved into a central to precision medicine by allowing programmable control of gene expression and epigenetic states. The dCas9-based tools are applied to therapeutically relevant genes in diverse disease contexts **(Table 2)**. For example, epigenetic activation systems have been leveraged to elevate tumour suppressor genes such as IL33 and BRCA1, reestablishing their expression in cancer models (Sapozhnikov & Szyf, 2021). Similarly, CRISPR-based transcriptional activators have been shown to amplify the expression of regulatory genes, such as PTEN, revealing strong therapeutic potential in contexts where gene dosage is pivotal (Umeh-Garcia et al., 2022). Conversely, CRISPR-mediated repression has been leveraged to repress disease-associated genes, including immune regulators such as FOXP3 in breast cancer models and PTEN in neuronal mechanisms (Moses et al., 2020). Programmable DNA methylation tools have enabled the stable silencing of clinically relevant genes, including IL6ST, EpCAM, CXCR4, BACH1, TFRC, SMARCA2, Id2, and Id4, highlighting their utility for extended transcriptional control in cancer and immune disorders. As well, targeted histone deacetylation has been employed to repress transcriptional activity at loci such as FOS and DPP4, facilitating its use for pathway modulation (Cai et al., 2023). Beyond DNA-targeted approaches, evolving RNA-editing platforms enable the short-term correction of pathogenic mutations or modulation of non-coding RNA pathways without requiring permanent genome refinements, offering a complementary layer of specificity control (Marina et al., 2020). Together, these CRISPR-based regulatory systems provide a multipurpose therapeutic toolkit that aids in disease modeling, functional genomics, and the advancement of next-generation targeted interventions.

Figure 3. **CRISPR–dCas9 epigenome editing tools** - *dCas9 fused to effector domains enables locus-specific epigenetic modulation without DNA cleavage. (i) dCas9–TET1 induces DNA demethylation and gene activation. (ii) dCas9–DNMT3A mediates DNA methylation and repression. (iii) dCas9–p300 acetylates histones to promote transcription. (iv) dCas9–KRAB recruits repressors to silence gene expression.*

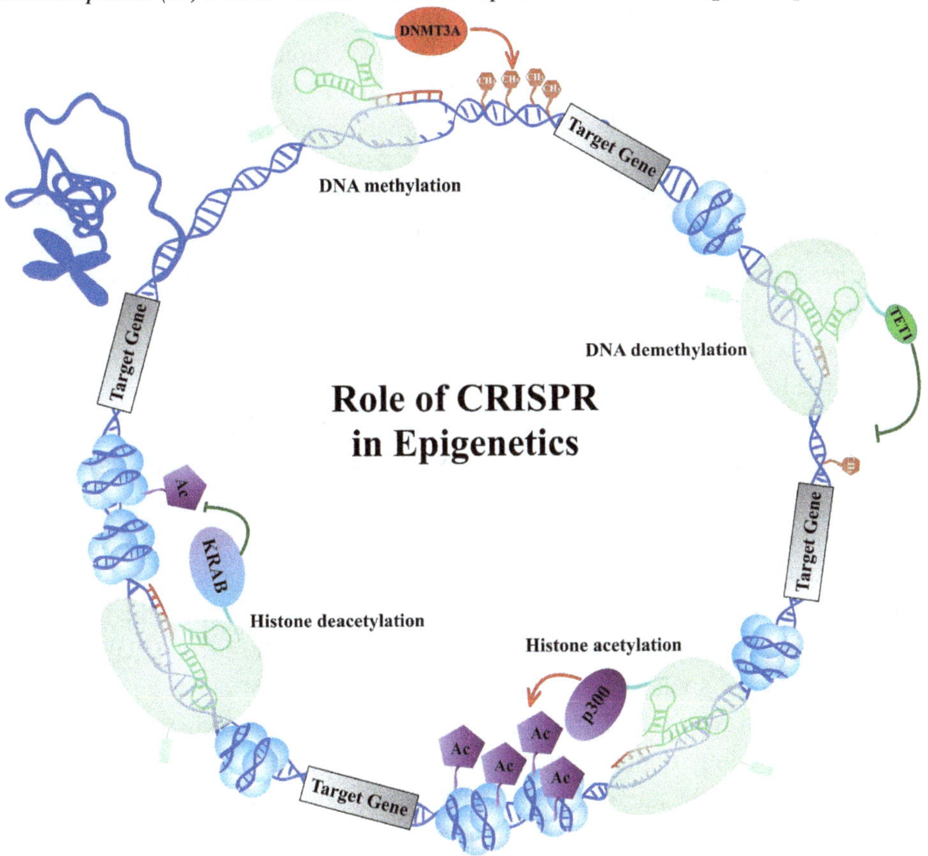

Catalytically inactive Cas9 (dCas9) has emerged as a powerful tool for live-cell genomic imaging due to its ability to bind specific DNA sequences without inducing double-strand breaks. By fusing dCas9 to fluorescent proteins (FPs) and guiding it with single-guide RNAs (sgRNAs), researchers can label genomic loci without genome editing. Early studies demonstrated the successful visualization of telomeres and non-repetitive loci, such as MUC4, as well as regulatory elements like enhancers and promoters, although single-locus signals are often weak. To enhance detection, signal amplification strategies have been developed, including the SunTag system, RNA aptamer extensions (e.g., MS2), SplitGFP, the ArrayG nanobody enhancer, and the Casilio system, which enable multi-color imaging and visualization of chromatin interactions in live cells. These approaches allow the study of spatial relationships between loci, such as promoter super-enhancer distances, and dynamic changes following protein depletion, with photobleaching-resistant strategies supporting long-term imaging (van Staalduinen et al., 2023). In functional genomics, CRISPR has revolutionised the identification of therapeutic targets and the mapping of synthetic lethal interactions. Genome-wide and combinatorial screens, particularly those integrating gRNA–UMI (unique molecular identifier) strategies, enable clone tracing and control for tumour heterogeneity, improving reliability in patient-derived organoids (PDOs) and xenografts (PDXs) (Das et al., 2022). Loss-of-function approaches, such as CRISPR knockout (CRISPRko) and CRISPR interference (CRISPRi), consistently identify genes associated with drug resistance. In contrast, CRISPR activation (CRISPRa) screens reveal genes whose overexpression drives resistance or enhances sensitivity. The parallel application of loss- and gain-of-function screens enables the cross-validation of hits and the comprehensive mapping of gene networks that modulate drug response (Sage et al., 2018). CRISPR-Cas9 also enables the precise engineering of immune cells, particularly CAR-T cells, to enhance anti-tumor efficacy and safety. Knockouts of inhibitory molecules, such as PD-1, LAG-3, and TIM-3, boost T and NK cell cytotoxicity, while targeted knock-ins allow for the controlled expression of therapeutic genes or cytokine modulators, minimizing excessive inflammatory responses. Multiplex editing enhances CAR-T expansion and persistence, and next-generation tools, such as base and prime editors, enable precise, double-strand break-free modifications for fine-tuning cytokine pathways (Dimitri et al., 2022). Moreover, CRISPR-Cas9 has transformed personalised disease modelling using patient-derived induced pluripotent stem cells (iPSCs) (**Table 2**). By introducing or correcting patient-specific mutations, including single-nucleotide variants, researchers can create accurate disease models for mechanistic studies and therapeutic testing. The integration of CRISPR with iPSCs provides high-throughput capabilities, isogenic controls, and minimized off-target effects, overcoming the limitations of traditional cell lines and animal models while accounting for genetic background variability (Sen & Thummer, 2022).

Table 2. CRISPR-Based Gene Editing Tools and Their Mechanisms in Precision Medicine.

CRISPR Tool	Application Area	Mechanism	References
dCas9-TET1 (Catalytic domain)	DNA demethylation	Site-specific DNA demethylation at promoter CpG sites, leading to transcriptional activation.	Sapozhnikov & Szyf, 2021
dCas9-VPR	Transcriptional activation	Recruitment of strong transcriptional activators to enhance gene expression.	Umeh-Garcia et al., 2022
dCas9–KRAB	Transcriptional silencing	Recruitment of KAP1 and chromatin modifiers (HDACs, SETDB1, HP1) to repress transcription.	Moses et al., 2020
dCas9–DNMT3A	Targeted DNA Methylation	Cytosine methylation at CpG sites blocks TF binding or recruiting methyl-binding proteins (e.g., MeCP2).	Park et al., 2022
dCas9–HDAC	Histone Deacetylation	Removal of acetyl groups from histones, compacting chromatin and reducing transcription.	Cai et al., 2023
Cas9/Cas13b–ADAR	RNA Editing	Adenosine-to-inosine (A→I) editing in dsRNA.	Marina et al., 2020
dCas9–FP	Live-cell Genomic Imaging	Fluorescent labelling of genomic loci.	van Staalduinen et al., 2023
CRISPRko, CRISPRi, CRISPRa	Functional Genomics (Screening)	Genome-wide loss- and gain-of-function screens; gRNA–UMI integration for clone tracing.	Das et al., 2022; Sage et al., 2018
CRISPR knockout/ knock-in, base editing, prime editing	Immunotherapy (CAR-T/NK engineering)	Knockout of inhibitory receptors; targeted cytokine gene insertion; multiplex editing for persistence.	Dimitri et al., 2022
CRISPR editing in patient-derived iPSCs	Personalised Disease Modelling	Introduction or correction of patient-specific mutations; creation of isogenic controls.	Sen & Thummer, 2022

8. EMERGING OPPORTUNITIES IN THE CLINICAL TRANSLATION OF CRISPR-CAS9 MEDIATED EPIGENOME EDITING

CRISPR technology is widely used to model cancer and identify genetic vulnerabilities, such as loss of tumour suppressor genes, in organoids or 3D cultures. These studies can be performed in vivo by transplanting CRISPR-engineered organoids or delivering CRISPR components directly into somatic tissues, enabling disease modelling in physiologically relevant environments. Somatically Engineered Mouse Models (SEMMs) are generated by introducing CRISPR components into blastocysts or adult tissues, allowing targeted mutations in organs such as the lung, liver, and pancreas. This approach supports HDR-mediated knock-in of specific variants, such as distinct KRAS mutations, to study their tissue-specific tumorigenic and therapeutic responses. In vivo modeling of chromosomal translocations, such as EML4-ALK fusions in lung cancer, has also been achieved through the delivery of Cas9 with dual sgRNAs (Sakari et al., 2020; Lima & Maddalo, 2021). The use of multiple gRNAs enables simultaneous targeting of various genes to investigate complex phenotypes, though excessive double-strand breaks can cause toxicity. To address this, dCas9-based epigenome and transcriptome editing offers a safer alternative, enabling multiplexed gene modulation without DNA cleavage. CRISPR-based base editing has also been applied in zebrafish to model precise single-nucleotide changes, supporting applications in xenograft studies and rapid drug-response testing (Ravichandran & Maddalo, 2023). Beyond disease modelling, CRISPR plays a key role in targeted drug delivery. Engineered nonviral carriers, such as nanoparticles, can be surface-modified with ligands or antibodies to recognise molecular markers on specific cell types. These systems deliver CRISPR/Cas components directly to diseased tissues, enabling precise gene editing, silencing, or activation while sparing healthy cells. Such targeted delivery increases treatment specificity, reduces systemic toxicity, and advances personalised medicine (Chen et al., 2020). In epigenetic therapy, CRISPR's precision enables targeted modification of genes or regulatory elements to reverse disease-associated chromatin changes. Its adaptability across diverse organisms and cell types, including human cell lines, supports patient-specific interventions. For instance, patient-derived iPSCs can be edited to correct pathogenic mutations prior to therapeutic transplantation (Lučanský et al., 2023).

8.1. Emerging CRISPR System Beyond Cas9

Although Cas9-based platforms remain the core of ongoing epigenome editing technologies, a range of next-generation CRISPR systems has broadened the toolkit for gene and epigenetic alterations.

- Cas12a (Cpf1) grants a separate PAM prerequisite (TTTV), staggered DNA cuts, and a compact coding size that optimises viral delivery. Its crRNA handling competence enables multiplex epigenome editing using a single transcript, offering a benefit for the synchronized adjustment of gene networks (Bandyopadhyay et al., 2020).
- Cas13 aims at RNA rather than DNA, supporting reversible transient transcript-level treatments devoid of reconfiguring the genome. Cas13-based tools have been optimised for RNA knockdown, RNA base editing, and tran-

scriptomic screens, delivering therapeutic potential in scenarios where transient or non-heritable regulation is favoured (Zhan et al., 2023).

- Precision-editing tools such as base editors (mediating C-to-T or A-to-G conversions) and prime editors (capable of template insertions, deletions, and substitutions) deliver high-accuracy genome regulation with diminished double-strand breaks. While not traditionally employed for epigenome editing, these platforms demonstrate how CRISPR systems are advancing toward greater precision, safety, and programmability.

Embedding these emerging technologies expands the range of CRISPR implementations and highlights future prospects for synergistic or complementary frameworks in precision medicine.

6.2 Computational and AI-driven Methods for CRISPR

Computational methods and machine-learning (ML) models are rapidly enhancing sgRNA design, off-target prediction, and editing proficiency forecasting capabilities, which are essential for safe and effective epigenome editing. Modern tools unite trait engineering from sequence context, chromatin approachability, and epigenomic state with deep-learning architectures to forecast guide success and accuracy more precisely than prior rule-based frameworks. Representative breakthroughs include CRISPRon for refined on-target proficiency inference, which reduces the experimental probe range for productive guides and facilitates subject allocation for authentication (Xiang et al., 2021). In addition to sgRNA design, AI frameworks are being used to forecast adjustment outcomes (including base-editor and prime-editor windows), model reconstitution outcomes, and embed multi-omics traits that impact editing in a given cell type. Current hybrid tools that fuse expression/prediction layers, such as EXPosition or graph-based models, illustrate how computational pipelines now extend from guide prioritization to anticipated phenotypic consequences. Embedding these estimative pipelines into preclinical protocols expedites candidate selection, minimises off-target risk, and sustains regulatory dossiers by offering in silico evidence of specificity and capability.

- **AI-Driven sgRNA Design:** Unlike preceding rule-based or sequence-motif techniques, AI models unite multiple layers of details like the sgRNA sequence, PAM context, thermodynamic stability, chromatin reachability, nucleotide composition, and sometimes 3D genome attributes to estimate how well a guide will function in a targeted cell type or with a specific nuclease variant. Tools such as DeepCRISPR, CRISPRon, and DeepHF utilize convolutional or recurrent neural networks to identify minor sequence features correlated with maximum activity or minimized off-target cutting (Chuai et al., 2018; Xiang et al., 2021). These models can rank extensive numbers of candidate guides, suggest the most optimised sgRNAs, and mark sequences likely to trigger inadvertent revisions. Notably, novel AI systems incorporate transfer learning and editor-specific training, for example, for high-fidelity Cas9 or dCas9-based epigenetic inducers, thereby making predictions more accurate across tissues and editing platforms. Overall, AI-driven sgRNA design enhances robustness in CRISPR experiments, reduces the risk of off-target events, and accelerates the progression of therapeutic epigenome editing by facilitating significantly streamlined, context-specific guide prioritization, a core requirement for clinical translation.

- **Off-target prediction with ML:** Modern off-target predictors utilise convolutional neural networks (CNNs), recurrent neural networks (RNNs), and hybrid architectures to infer sequence patterns and contextual attributes that drive unexpected fragmentation. These models take as input the candidate lead and prospective off-target loci, often incorporating mismatch position/identity, PAM context, spacer–target thermodynamics, and, when accessible, chromatin properties such as DNA accessibility and nucleosome occupancy. Example tool families include CNN-based predictors and hybrid pipelines that incorporate a sequence model with chromatin weighting. For example, CNN-Crispr-style models and succeeding deep models. They offer genome-wide risk maps and rank off-target sites by likelihood, promoting prioritisation and filtering. However, numerous models trained on cell lines have minimized accuracy in primary cells or other species; chromatin inputs refine accuracy but require matched epigenomic data. It is proposed to use at least two independent ML off-target predictors, cross-check high-score sites with exploratory assays (GUIDE-seq, DISCOVER-seq, SITE-seq), and incorporate chromatin data from the targeted cell type where possible.
- **Editing-outcome forecasting:** Prediction of editing results outperforms a binary cut/no-cut model to forecast the spectrum of edits that occur after a pause or base/prime editing event. Specialised models, such as inDelphi and FORECasT, predict indel distributions for nuclease cuts, while base-editing predictors like BE-Hive and DeepBE variants calculate the editing window and the probability of C-to-T or A-to-G adjustments covering sequence contexts. For prime editing, recent tools like PrimeDesign and ML refinements anticipate the proficiency of template combination and likely by-products. These models embed local sequence motifs, microhomology signals, repair pathway predispositions, and intermittently cell-line–specific repair predispositions. They enable researchers to opt for guides forecasted to result in the targeted allelic impact, for example, a preferred frameshift or precise base

conversion, thereby diminishing unproductive trials. Despite this, the estimations are probabilistic and can vary with cell cycle state, DNA competence, and delivery layout. It is proposed to use outcome predictors to shortlist and guide/edit designs, authenticate predicted outcomes based on data through amplicon sequencing, and iterate model tuning with pilot data for the target cell type.

- **Integrated pipelines:** Integrated design pipelines integrate on-target scoring, off-target risk examination, and outcome estimating into a single prioritisation framework and, at times, layer in downstream functional prediction. For example, a foreseen expression variation after promoter or enhancer modification. Examples of this class integrate tools that pair guide design with chromatin-aware grading, or platforms that estimate not only editing capability but also likely phenotypic impact by linking editing targets to expression models or regulatory network maps, such as EXPosition-style integrated systems. These pipelines automate the manual stitching of tools, generate ranked candidate sets alongside multi-criteria scoring, and can output prioritized experiments prepared for authentication. Even so, integrated tools may obscure hypotheses, such as how scores are unified, and can disseminate bias from their training datasets. The superior approach examines component scores, tunes weightings to the experimental framework, and regularly performs orthogonal functional validation, such as transcript evaluation, phenotypic assays, and preceding large-scale deployment.

Applying AI-driven design and integrated pipelines offers numerous concrete practical benefits for translational CRISPR work.

- ✓ They remarkably minimise the experimental exploration space in place of testing dozens of candidate guides per locus. Researchers typically need only a few top-ranked guides to achieve success, saving time and resources.
- ✓ Optimised screening elevates "first-pass" hit incidences, hastening timelines from target selection to functional confirmation.
- ✓ Computational documentation, like rankings, estimated off-target maps, and outcome patterns, reinforces supervisory and security dossiers by enabling in-silico threat appraisals that can co-occur with pilot off-target assays.
- ✓ Iterative model tuning, using small pilot datasets from the intended cell type or delivery format, improves predictive accuracy and supports reproducibility across labs.

Figure 4. Opportunities and problems of evolving CRISPR technology - An overview of the main advantages of CRISPR in therapeutics and contemporary medicine, together with an overview of the major problems associated with delivery.

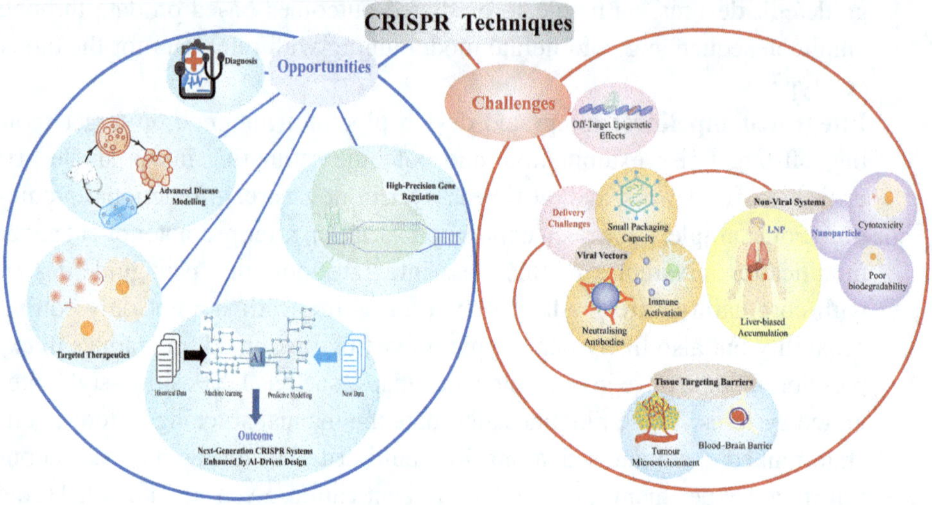

9. CHALLENGES AND LIMITATIONS OF CRISPR-BASED EPIGENOME EDITING

CRISPR-based epigenome editing faces multiple technical, biological, and safety limitations. Off-target effects remain a significant concern, as dCas9–epigenetic effector complexes can be recruited to unintended genomic sites, altering the regulation of unrelated genes. Low-specificity sgRNAs in CRISPRi/a screens have been shown to cause phenotypic changes by affecting off-target loci near essential genes (Tycko et al., 2019). To address this, structure-guided Cas9 variants such as SpCas9-HF1, eSpCas9, and HypaCas9 have been engineered to reduce off-target binding while maintaining high on-target activity (Allemailem et al., 2023). In CRISPR-based imaging, repetitive loci, such as telomeres, are more easily visualized with single sgRNAs, whereas non-repetitive sequences require multiple guides for adequate signal. While live-cell tracking of telomere dynamics and single-gene movement through mitosis demonstrates feasibility, resolution and temporal stability remain limiting factors. Multicolour dCas9 imaging enables the simultaneous visualisation of multiple loci, but requires precise spectral separation and calibration. Moreover, chromatin accessibility affects imaging efficiency, as techniques like CRISPR-ChAP-MS are less effective in repressive chromatin regions (Laufer & Singh, 2015).

Productive in vivo administration of large dCas9 effector conjugates is an additional bottleneck. Viral vectors remain the most widely used delivery vehicles for CRISPR-based epigenome editing, but each system has significant limitations. Adeno-associated viruses (AAVs) exhibit strong tissue tropism and enhanced transduction efficiency, but their minimal packaging capacity (4.5–5 kb) is insufficient for substantial dCas9–effector fusions. Split-dCas9 systems have been developed, but they significantly decrease efficiency and hinder manufacturing (Moreno et al., 2018; Yang et al., 2024). AAV re-dosing is insufficient due to the presence of neutralizing antibodies, and immune activation remains a barrier to systemic delivery. Lentiviral vectors can enable expanded constructs and integrate into the genome for persistent expression; however, this introduces the liability of insertional mutagenesis, persistent effector operation, and long-term safety concerns, remarkably for epigenome editors aimed to function non-permanently. Non-viral frameworks, such as lipid nanoparticles (LNPs), polymeric nanoparticles, and gold nanoparticles, as well as physical delivery approaches (e.g., electroporation), enable less hazardous choices but still encounter significant constraints. LNPs are highly productive for mRNA delivery; however, encapsulating high-volume dCas9–protein complexes or multi-component epigenetic editors remains suboptimal, frequently causing protein denaturation or degradation during the formulation process. Unmodified ionizable LNPs primarily aggregate in the liver, hindering delivery to tissues such as muscle, lung, pancreas, and brain (Kazemian et al., 2022). Endosomal escape also remains a rate-limiting step, with <2% of internalised cargo penetrating the cytosol. Polymeric nanoparticles and gold nanocarriers can transfer nucleic acids or proteins; however, they are hindered by issues associated with cytotoxicity, poor biodegradability, and limited in vivo targeting. Electroporation is exceptionally capable ex vivo, as demonstrated in the clinical application of CAR-T cell construction, but is unsuitable for whole-organism administration due to tissue impairment and limited permeation depth (Abou-el-Enein et al., 2021; Kazemian et al., 2022). Overall, non-viral systems are encouraging for ex vivo editing, but in vivo delivery of high-volume epigenetic editors remains a significant bottleneck. A key hindrance to therapeutic epigenome editing is the complication of delivering CRISPR systems to precise organs and cell populations. Most systemically delivered nanoparticles and viral vectors accumulate in the liver, spleen, and lungs, with limited penetration into other tissues. Focusing the brain is particularly challenging due to the blood–brain barrier, necessitating intrathecal administration, receptor-mediated transcytosis, or the use of engineered capsids. Pancreas, muscle, and heart tissues manifest similarly poor absorption (Wu et al., 2023). Solid tumors pose additional barriers, including a tightly structured extracellular matrix, aberrant vasculature, hypoxia, and heterogeneous receptor expression, all of which impede the entry of nanoparticles and viral vectors. Immune cells, including macrophages and dendritic cells, promptly sequester circulating vectors, thereby curtailing their bioavailability at the intended tissues. Beyond this, cell-specific expression of membrane receptors reduces targeted editing in heterogeneous tissues, necessitating the formulation of ligand-decorated LNPs, engineered capsids, or cell-type–specific promoters. Together, these targeting constraints remain among the most notable complications to non-harmful and high-performing CRISPR-based epigenome engineering in vivo.

Epigenetic modifications induced by CRISPR are often unstable, especially outside CpG-dense or transcriptionally inactive regions. Cellular chromatin remodelling pathways can actively reverse artificial changes, and edits may be diluted during cell division. Achieving durable effects without altering DNA sequence may require more robust multi-domain dCas9 fusions or repeated dosing, though outcomes remain context-dependent (Whittaker et al., 2023).

10. ETHICAL AND SAFETY CONSIDERATIONS

CRISPR-based epigenome editing poses significant ethical and safety challenges that must be addressed before its widespread clinical implementation. The risk of intentional or inadvertent germline adjustment remains a foundational ethical margin. Therapeutic actions that might influence gametes or embryos could generate heritable modifications and therefore demand strict restriction, oversight, and transparent international governance consistent with ISSCR and NIH guidance (Lovell-Badge et al., 2021).

A key risk factor is the immune system's response to CRISPR components. Multiple individuals carry pre-existing antibodies that counter bacterial Cas proteins, and viral vectors can rapidly elicit pronounced humoral and cellular immune responses, potentially impeding dosing and intensifying the toxicity threat. Tackling these challenges will require integrating immune-mapping into patient selection, administering short-term immunosuppression as needed, formulating diminished immunogenic nucleic acid variants, and engineering next-generation capsids with minimized immune visibility. Tumorigenicity and genomic stability are notable uncertainties for cell-based therapeutic pipelines. iPSC-derived products and *ex vivo* edited cellular therapies require thorough screening for insertional mutagenesis, chromosomal abnormalities, and epigenetic variability that could promote malignant transformation (Wiley et al., 2025). Equity, access, and dual-use questions necessitate ethical deliberation that extends beyond laboratory-level safety. Accessibility to high-cost gene and epigenome therapies poses a threat to exacerbate health disparities; in contrast, the potential for non-therapeutic enhancement or military misuse warrants well-defined regulatory limits. Authors and funders should embrace unbiased trial designs, collaborate with stakeholders, and publish risk–benefit and cost-effectiveness assessments (Lee & Sawai, 2024). Finally, ongoing regulatory frameworks, such as ISSCR statements, NIH gene-editing policies, and FDA guidance on human gene therapy products, offer starting points but should be adapted to address the specificities of epigenome editing, including transient versus durable edits, reversibility, and off-target epigenetic changes.

11. CONCLUSION

The evolution of CRISPR-Cas9 from a bacterial defence mechanism to a versatile genome engineering platform has revolutionised modern biology and precision medicine. Advances in Cas9 derivatives, including high-fidelity nucleases, base and prime editors, and dCas9-based effectors, have extended its functionality beyond precise DNA cleavage to encompass transcriptional regulation, epigenome editing, RNA modulation, and live-cell genomic imaging. These innovations have enabled applications ranging from reactivating silenced tumour suppressors and repressing oncogenes to engineering immune cells, modelling patient-specific diseases, and mapping therapeutic resistance networks. In particular, CRISPR–dCas9-mediated epigenome editing offers unprecedented precision compared to conventional epigenetic drugs, allowing targeted and reversible modulation of DNA methylation, histone modifications, and chromatin structure without altering the underlying genome. This approach holds great promise in oncology, regenerative medicine, and the treatment of epigenetic disorders, while also providing powerful tools for dissecting disease mechanisms. Despite these advances, several challenges remain for clinical translation. Off-target activity, limited efficiency of in vivo delivery systems, instability of epigenetic modifications, and unresolved ethical and safety concerns continue to hinder the widespread adoption of therapeutic approaches. Addressing these issues through engineered Cas9 variants, optimised delivery strategies, and robust regulatory oversight will be essential to ensure both efficacy and safety. Collectively, CRISPR technologies represent a transformative step in precision medicine, offering patient-tailored solutions that integrate molecular specificity with therapeutic versatility. As ongoing innovations converge with clinical translation, CRISPR is poised to become a cornerstone of next-generation, personalised therapeutics.

ACKNOWLEDGMENTS

The authors acknowledge the Department of Biotechnology, School of Life Sciences, Pondicherry University, for assistance. RK thanks the ANRF, Government of India (ANRF/IRG/2024/000974/LS), for financial support.

COMPETING INTERESTS

The authors declare there were no potential conflicts of interest

12. REFERENCES

Abbott, T. R., Dhamdhere, G., Liu, Y., Lin, X., Goudy, L., Zeng, L., Chemparathy, A., Chmura, S., Heaton, N. S., Debs, R., Pande, T., Endy, D., La Russa, M. F., Lewis, D. B., & Qi, L. S. (2020). Development of CRISPR as an antiviral strategy to combat SARS-CoV-2 and influenza. *Cell*, *181*(4), 865–876. DOI: 10.1016/j.cell.2020.04.020 PMID: 32353252

Abou-el-Enein, M., Elsallab, M., Feldman, S. A., Fesnak, A. D., Heslop, H. E., Marks, P., Till, B. G., Bauer, G., & Savoldo, B. (2021). Scalable manufacturing of CAR T cells for cancer immunotherapy. *Blood Cancer Discovery*, *2*(5), 408–422. DOI: 10.1158/2643-3230.BCD-21-0084 PMID: 34568831

Akram, F., Tanveer, R., Andleeb, S., Shah, F. I., Ahmad, T., Shehzadi, S., Akhtar, A. M., & Syed, G. (2024). Deciphering the epigenetic symphony of cancer: Insights and epigenetic therapies implications. *Technology in Cancer Research & Treatment*, *23*, 15330338241250317. DOI: 10.1177/15330338241250317 PMID: 38780251

Aljabali, A. A., El-Tanani, M., & Tambuwala, M. M. (2024). Principles of CRISPR-Cas9 technology: Advancements in genome editing and emerging trends in drug delivery. *Journal of Drug Delivery Science and Technology*, *92*, 105338. DOI: 10.1016/j.jddst.2024.105338

Allemailem, K. S., Almatroodi, S. A., Almatroudi, A., Alrumaihi, F., Al Abdulmonem, W., Al-Megrin, W. A. I., Aljamaan, A. N., Rahmani, A. H., & Khan, A. A. (2023). Recent advances in genome-editing technology with CRISPR/Cas9 variants and stimuli-responsive targeting approaches within tumor cells: A future perspective of cancer management. *International Journal of Molecular Sciences*, *24*(8), 7052. DOI: 10.3390/ijms24087052 PMID: 37108214

Alshammari, E., Zhang, Y., Sobota, J., & Yang, Z. (2022). Aberrant DNA methylation of tumor suppressor genes and oncogenes as cancer biomarkers. *Genomic and Epigenomic Biomarkers of Toxicology and Disease: Clinical and Therapeutic Actions*, 251-271.

Alves, E., Taifour, S., Dolcetti, R., Chee, J., Nowak, A. K., Gaudieri, S., & Blancafort, P. (2021). Reprogramming the anti-tumor immune response via CRISPR genetic and epigenetic editing. *Molecular Therapy. Methods & Clinical Development, 21*, 592–606. DOI: 10.1016/j.omtm.2021.04.009 PMID: 34095343

Bandyopadhyay, A., Kancharla, N., Javalkote, V. S., Dasgupta, S., & Brutnell, T. P. (2020). CRISPR-Cas12a (Cpf1): A versatile tool in the plant genome editing tool box for agricultural advancement. *Frontiers in Plant Science, 11*, 584151. DOI: 10.3389/fpls.2020.584151 PMID: 33214794

Cai, R., Lv, R., Shi, X. E., Yang, G., & Jin, J. (2023). CRISPR/dCas9 tools: Epigenetic mechanism and application in gene transcriptional regulation. *International Journal of Molecular Sciences, 24*(19), 14865. DOI: 10.3390/ijms241914865 PMID: 37834313

Carroll, M. S., & Giacca, M. (2023). CRISPR activation and interference as investigative tools in the cardiovascular system. *The International Journal of Biochemistry & Cell Biology, 155*, 106348. DOI: 10.1016/j.biocel.2022.106348 PMID: 36563996

Chang, W., Zhao, Y., Rayêe, D., Xie, Q., Suzuki, M., Zheng, D., & Cvekl, A. (2023). Dynamic changes in whole genome DNA methylation, chromatin and gene expression during mouse lens differentiation. *Epigenetics & Chromatin, 16*(1), 4. DOI: 10.1186/s13072-023-00478-7 PMID: 36698218

Chen, C., Tang, J., Xu, S., Zhang, W., & Jiang, H. (2020). Research Article miR-30a-5p Inhibits Proliferation and Migration of Lung Squamous Cell Carcinoma Cells by Targeting FOXD1.

Chen, F., Alphonse, M., & Liu, Q. (2020). Strategies for nonviral nanoparticle-based delivery of CRISPR/Cas9 therapeutics. *Wiley Interdisciplinary Reviews. Nanomedicine and Nanobiotechnology, 12*(3), e1609. DOI: 10.1002/wnan.1609 PMID: 31797562

Chen, X., Torres, J. P. M., Li, Y., Lugones, L. G., & Wösten, H. A. (2023). Inheritable CRISPR based epigenetic modification in a fungus. *Microbiological Research, 272*, 127397. DOI: 10.1016/j.micres.2023.127397 PMID: 37141850

Chuai, G., Ma, H., Yan, J., Chen, M., Hong, N., Xue, D., Zhou, C., Zhu, C., Chen, K., Duan, B., Gu, F., Qu, S., Huang, D., Wei, J., & Liu, Q. (2018). DeepCRISPR: Optimized CRISPR guide RNA design by deep learning. *Genome Biology*, *19*(1), 80. DOI: 10.1186/s13059-018-1459-4 PMID: 29945655

Das, S., Bano, S., Kapse, P., & Kundu, G. C. (2022). CRISPR based therapeutics: A new paradigm in cancer precision medicine. *Molecular Cancer*, *21*(1), 85. DOI: 10.1186/s12943-022-01552-6 PMID: 35337340

de Vos, L., Grünwald, I., Bawden, E. G., Dietrich, J., Scheckenbach, K., Wiek, C., Zarbl, R., Bootz, F., Landsberg, J., & Dietrich, D. (2020). The landscape of CD28, CD80, CD86, CTLA4, and ICOS DNA methylation in head and neck squamous cell carcinomas. *Epigenetics*, *15*(11), 1195–1212. DOI: 10.1080/15592294.2020.1754675 PMID: 32281488

Deng, S., Hu, Q., Zhang, H., Yang, F., Peng, C., & Huang, C. (2019). HDAC3 inhibition upregulates PD-L1 expression in B-cell lymphomas and augments the efficacy of anti–PD-L1 therapy. *Molecular Cancer Therapeutics*, *18*(5), 900–908. DOI: 10.1158/1535-7163.MCT-18-1068 PMID: 30824609

Dhar, G. A., Saha, S., Mitra, P., & Nag Chaudhuri, R. (2021). DNA methylation and regulation of gene expression: Guardian of our health. *Nucleus (Austin, Tex.)*, *64*(3), 259–270. PMID: 34421129

Dimitri, A., Herbst, F., & Fraietta, J. A. (2022). Engineering the next-generation of CAR T-cells with CRISPR-Cas9 gene editing. *Molecular Cancer*, *21*(1), 78. DOI: 10.1186/s12943-022-01559-z PMID: 35303871

Donovan, D. A., Crandall, J. G., Banks, O. G., Jensvold, Z. D., Truong, V., Dinwiddie, D., McKnight, L. E., & McKnight, J. N. (2019). Engineered chromatin remodeling proteins for precise nucleosome positioning. *Cell Reports*, *29*(8), 2520–2535. DOI: 10.1016/j.celrep.2019.10.046 PMID: 31747617

Dräger, N. M., Sattler, S. M., Huang, C. T. L., Teter, O. M., Leng, K., Hashemi, S. H., Hong, J., Aviles, G., Clelland, C. D., Zhan, L., Udeochu, J. C., Kodama, L., Singleton, A. B., Nalls, M. A., Ichida, J., Ward, M. E., Faghri, F., Gan, L., & Kampmann, M. (2022). A CRISPRi/a platform in human iPSC-derived microglia uncovers regulators of disease states. *Nature Neuroscience*, *25*(9), 1149–1162. DOI: 10.1038/s41593-022-01131-4 PMID: 35953545

Fadul, S. M., Arshad, A., & Mehmood, R. (2023). CRISPR-based epigenome editing: Mechanisms and applications. *Epigenomics*, *15*(21), 1137–1155. DOI: 10.2217/epi-2023-0281 PMID: 37990877

Feoktistov, A. V., Georgieva, S. G., & Soshnikova, N. V. (2022). Role of the SWI/SNF chromatin remodeling complex in regulation of inflammation gene expression. *Molecular Biology*, *56*(2), 182–195. DOI: 10.1134/S0026893322020054 PMID: 35403618

Hayes, B. H., Zhu, H., Andrechak, J. C., & Discher, D. E. (2022). Titrating CD47 by mismatch CRISPRi reveals incomplete repression can eliminate IgG-opsonized tumors but CD47 heterogeneity limits induction of anti-tumor IgG. bioRxiv, 2022-09.

Hou, C., Xiao, L., Ren, X., Cheng, L., Guo, B., Zhang, M., & Yan, N. (2022). EZH2-mediated H3K27me3 is a predictive biomarker and therapeutic target in uveal melanoma. *Frontiers in Genetics*, *13*, 1013475. DOI: 10.3389/fgene.2022.1013475 PMID: 36276954

Hu, J. H., Miller, S. M., Geurts, M. H., Tang, W., Chen, L., Sun, N., Zeina, C. M., Gao, X., Rees, H. A., Lin, Z., & Liu, D. R. (2018). Evolved Cas9 variants with broad PAM compatibility and high DNA specificity. *Nature*, *556*(7699), 57–63. DOI: 10.1038/nature26155 PMID: 29512652

Ibraheim, R., Song, C. Q., Mir, A., Amrani, N., Xue, W., & Sontheimer, E. J. (2018). All-in-one adeno-associated virus delivery and genome editing by Neisseria meningitidis Cas9 in vivo. *Genome Biology*, *19*(1), 137. DOI: 10.1186/s13059-018-1515-0 PMID: 30231914

Jinek, M., Chylinski, K., Fonfara, I., Hauer, M., Doudna, J. A., & Charpentier, E. (2012). A programmable dual-RNA–guided DNA endonuclease in adaptive bacterial immunity. *science, 337*(6096), 816-821.

Joshi, S., & Mishra, S. (2022). Recent advances in biofuel production through metabolic engineering. *Bioresource Technology*, *352*, 127037. DOI: 10.1016/j.biortech.2022.127037 PMID: 35318143

Kazemian, P., Yu, S. Y., Thomson, S. B., Birkenshaw, A., Leavitt, B. R., & Ross, C. J. (2022). Lipid-nanoparticle-based delivery of CRISPR/Cas9 genome-editing components. *Molecular Pharmaceutics*, *19*(6), 1669–1686. DOI: 10.1021/acs.molpharmaceut.1c00916 PMID: 35594500

Kim, H. K., Lee, S., Kim, Y., Park, J., Min, S., Choi, J. W., Huang, T. P., Yoon, S., Liu, D. R., & Kim, H. H. (2020). High-throughput analysis of the activities of xCas9, SpCas9-NG and SpCas9 at matched and mismatched target sequences in human cells. *Nature Biomedical Engineering*, *4*(1), 111–124. DOI: 10.1038/s41551-019-0505-1 PMID: 31937939

Kleinstiver, B. P., Pattanayak, V., Prew, M. S., Tsai, S. Q., Nguyen, N. T., Zheng, Z., & Joung, J. K. (2016). High-fidelity CRISPR–Cas9 nucleases with no detectable genome-wide off-target effects. *Nature*, *529*(7587), 490–495. DOI: 10.1038/nature16526 PMID: 26735016

Kostyushev, D., Brezgin, S., Kostyusheva, A., Zarifyan, D., Goptar, I., & Chulanov, V. (2019). Orthologous CRISPR/Cas9 systems for specific and efficient degradation of covalently closed circular DNA of hepatitis B virus. *Cellular and Molecular Life Sciences : CMLS*, *76*(9), 1779–1794. DOI: 10.1007/s00018-019-03021-8 PMID: 30673820

Laufer, B. I., & Singh, S. M. (2015). Strategies for precision modulation of gene expression by epigenome editing: An overview. *Epigenetics & Chromatin*, *8*(1), 34. DOI: 10.1186/s13072-015-0023-7 PMID: 26388942

Lee, J., Lim, K., Kim, A., Mok, Y. G., Chung, E., Cho, S. I., Lee, J. M., & Kim, J. S. (2023). Prime editing with genuine Cas9 nickases minimizes unwanted indels. *Nature Communications*, *14*(1), 1786. DOI: 10.1038/s41467-023-37507-8 PMID: 36997524

Lee, T. L., & Sawai, T. (2024). Navigating equity in global access to genome therapy expanding access to potentially transformative therapies and benefiting those in need requires global policy changes. *Frontiers in Genetics*, *15*, 1381172. DOI: 10.3389/fgene.2024.1381172 PMID: 38638119

Lima, A., & Maddalo, D. (2021). SEMMs: Somatically engineered mouse models. A new tool for in vivo disease modeling for basic and translational research. *Frontiers in Oncology*, *11*, 667189. DOI: 10.3389/fonc.2021.667189 PMID: 33968774

Loganathan, T., & Doss, C. G. P. (2023). Non-coding RNAs in human health and disease: Potential function as biomarkers and therapeutic targets. *Functional & Integrative Genomics*, *23*(1), 33. DOI: 10.1007/s10142-022-00947-4 PMID: 36625940

Lovell-Badge, R., Anthony, E., Barker, R. A., Bubela, T., Brivanlou, A. H., Carpenter, M., Charo, R. A., Clark, A., Clayton, E., Cong, Y., Daley, G. Q., Fu, J., Fujita, M., Greenfield, A., Goldman, S. A., Hill, L., Hyun, I., Isasi, R., Kahn, J., & Zhai, X. (2021). ISSCR guidelines for stem cell research and clinical translation: The 2021 update. *Stem Cell Reports*, *16*(6), 1398–1408. DOI: 10.1016/j.stemcr.2021.05.012 PMID: 34048692

Lučanský, V., Holubeková, V., Kolková, Z., Halašová, E., Samec, M., & Golubnitschaja, O. (2023). Multi-faceted CRISPR/Cas technological innovation aspects in the framework of 3P medicine. *The EPMA Journal*, *14*(2), 201–217. DOI: 10.1007/s13167-023-00324-6 PMID: 37275547

Marina, R. J., Brannan, K. W., Dong, K. D., Yee, B. A., & Yeo, G. W. (2020). Evaluation of engineered CRISPR-Cas-mediated systems for site-specific RNA editing. *Cell Reports*, *33*(5), 108350. DOI: 10.1016/j.celrep.2020.108350 PMID: 33147453

Matsumoto, D., Matsugi, E., Kishi, K., Inoue, Y., Nigorikawa, K., & Nomura, W. (2024). SpCas9-HF1 enhances accuracy of cell cycle-dependent genome editing by increasing HDR efficiency, and by reducing off-target effects and indel rates. *Molecular Therapy. Nucleic Acids*, *35*(1), 102124. DOI: 10.1016/j.omtn.2024.102124 PMID: 38328481

Moreno, A. M., Fu, X., Zhu, J., Katrekar, D., Shih, Y. R. V., Marlett, J., Cabotaje, J., Tat, J., Naughton, J., Lisowski, L., Varghese, S., Zhang, K., & Mali, P. (2018). In situ gene therapy via AAV-CRISPR-Cas9-mediated targeted gene regulation. *Molecular Therapy : the Journal of the American Society of Gene Therapy*, *26*(7), 1818–1827. DOI: 10.1016/j.ymthe.2018.04.017 PMID: 29754775

Moses, C., Hodgetts, S. I., Nugent, F., Ben-Ary, G., Park, K. K., Blancafort, P., & Harvey, A. R. (2020). Transcriptional repression of PTEN in neural cells using CRISPR/dCas9 epigenetic editing. *Scientific Reports*, *10*(1), 11393. DOI: 10.1038/s41598-020-68257-y PMID: 32647121

Nuñez, J. K., Chen, J., Pommier, G. C., Cogan, J. Z., Replogle, J. M., Adriaens, C., Ramadoss, G. N., Shi, Q., Hung, K. L., Samelson, A. J., Pogson, A. N., Kim, J. Y. S., Chung, A., Leonetti, M. D., Chang, H. Y., Kampmann, M., Bernstein, B. E., Hovestadt, V., Gilbert, L. A., & Weissman, J. S. (2021). Genome-wide programmable transcriptional memory by CRISPR-based epigenome editing. *Cell*, *184*(9), 2503–2519. DOI: 10.1016/j.cell.2021.03.025 PMID: 33838111

Park, H., Shin, J., Kim, Y., Saito, T., Saido, T. C., & Kim, J. (2022). CRISPR/dCas9-Dnmt3a-mediated targeted DNA methylation of APP rescues brain pathology in a mouse model of Alzheimer's disease. *Translational Neurodegeneration*, *11*(1), 41. DOI: 10.1186/s40035-022-00314-0 PMID: 36109806

Pickar-Oliver, A., & Gersbach, C. A. (2019). The next generation of CRISPR–Cas technologies and applications. *Nature Reviews. Molecular Cell Biology*, *20*(8), 490–507. DOI: 10.1038/s41580-019-0131-5 PMID: 31147612

Rafeeq, H., Afsheen, N., Rafique, S., Arshad, A., Intisar, M., Hussain, A., Bilal, M., & Iqbal, H. M. (2023). Genetically engineered microorganisms for environmental remediation. *Chemosphere*, *310*, 136751. DOI: 10.1016/j.chemosphere.2022.136751 PMID: 36209847

Rajanathadurai, J., Perumal, E., & Sindya, J. (2024). Advances in targeting cancer epigenetics using CRISPR-dCas9 technology: A comprehensive review and future prospects. *Functional & Integrative Genomics*, *24*(5), 164. DOI: 10.1007/s10142-024-01455-3 PMID: 39292321

Rasheed, A., Gill, R. A., Hassan, M. U., Mahmood, A., Qari, S., Zaman, Q. U., Ilyas, M., Aamer, M., Batool, M., Li, H., & Wu, Z. (2021). A critical review: Recent advancements in the use of CRISPR/Cas9 technology to enhance crops and alleviate global food crises. *Current Issues in Molecular Biology*, *43*(3), 1950–1976. DOI: 10.3390/cimb43030135 PMID: 34889892

Ravichandran, M., & Maddalo, D. (2023). Applications of CRISPR-Cas9 for advancing precision medicine in oncology: From target discovery to disease modeling. *Frontiers in Genetics*, *14*, 1273994. DOI: 10.3389/fgene.2023.1273994 PMID: 37908590

Rooney, K., & Sadikovic, B. (2022). DNA methylation episignatures in neurodevelopmental disorders associated with large structural copy number variants: Clinical implications. *International Journal of Molecular Sciences*, *23*(14), 7862. DOI: 10.3390/ijms23147862 PMID: 35887210

Sage, C. L., Lawo, S., Panckier, P., Scales, T. M., Rahman, S. A., Little, A. S., McCarthy, N. J., Moore, J. D., & Cross, B. C. (2018). Dual-direction CRISPR transcriptional regulation screening uncovers gene networks driving drug resistance. *Cancer Research*, *78*(13, Supplement), 409–409. DOI: 10.1158/1538-7445.AM2018-409

Sakari, M., Suzuki, T., Yano, S., & Tsukahara, T. (2020). Effective RNA knockdown using CRISPR-Cas13a and molecular targeting of the EML4-ALK transcript in H3122 lung cancer cells. *International Journal of Molecular Sciences*, *21*(23), 8904. DOI: 10.3390/ijms21238904 PMID: 33255340

Sapozhnikov, D. M., & Szyf, M. (2021). Unraveling the functional role of DNA demethylation at specific promoters by targeted steric blockage of DNA methyltransferase with CRISPR/dCas9. *Nature Communications*, *12*(1), 5711. DOI: 10.1038/s41467-021-25991-9 PMID: 34588447

Schubert, M. S., Thommandru, B., Woodley, J., Turk, R., Yan, S., Kurgan, G., McNeill, M. S., & Rettig, G. R. (2021). Optimized design parameters for CRISPR Cas9 and Cas12a homology-directed repair. *Scientific Reports*, *11*(1), 19482. DOI: 10.1038/s41598-021-98965-y PMID: 34593942

Seem, K., Kaur, S., Kumar, S., & Mohapatra, T. (2024). Epigenome editing for targeted DNA (de) methylation: A new perspective in modulating gene expression. *Critical Reviews in Biochemistry and Molecular Biology*, *59*(1-2), 69–98. DOI: 10.1080/10409238.2024.2320659 PMID: 38440883

Sen, T., & Thummer, R. P. (2022). CRISPR and iPSCs: Recent developments and future perspectives in neurodegenerative disease modelling, research, and therapeutics. *Neurotoxicity Research*, *40*(5), 1597–1623. DOI: 10.1007/s12640-022-00564-w PMID: 36044181

Slaymaker, I. M., Gao, L., Zetsche, B., Scott, D. A., Yan, W. X., & Zhang, F. (2016). Rationally engineered Cas9 nucleases with improved specificity. *Science*, *351*(6268), 84–88. DOI: 10.1126/science.aad5227 PMID: 26628643

Socha, M. W., Flis, W., & Wartęga, M. (2024). Epigenetic genome modifications during pregnancy: The impact of essential nutritional supplements on DNA methylation. *Nutrients*, *16*(5), 678. DOI: 10.3390/nu16050678 PMID: 38474806

Stadtmauer, E. A., Fraietta, J. A., Davis, M. M., Cohen, A. D., Weber, K. L., Lancaster, E., Mangan, P. A., Kulikovskaya, I., Gupta, M., Chen, F., Tian, L., Gonzalez, V. E., Xu, J., Jung, I., Melenhorst, J. J., Plesa, G., Shea, J., Matlawski, T., Cervini, A., & June, C. H. (2020). CRISPR-engineered T cells in patients with refractory cancer. *Science*, *367*(6481), eaba7365. DOI: 10.1126/science.aba7365 PMID: 32029687

Stevens, A. J., de Jong, L., & Kennedy, M. A. (2022). The dynamic regulation of G-quadruplex DNA structures by cytosine methylation. *International Journal of Molecular Sciences*, *23*(5), 2407. DOI: 10.3390/ijms23052407 PMID: 35269551

Stuart, W. D., Fink-Baldauf, I. M., Tomoshige, K., Guo, M., & Maeda, Y. (2021). CRISPRi-mediated functional analysis of NKX2-1-binding sites in the lung. *Communications Biology*, *4*(1), 568. DOI: 10.1038/s42003-021-02083-4 PMID: 33980985

Tao, R., Han, X., Bai, X., Yu, J., Ma, Y., Chen, W., Zhang, D., & Li, Z. (2024). Revolutionizing cancer treatment: Enhancing CAR-T cell therapy with CRISPR/Cas9 gene editing technology. *Frontiers in Immunology*, *15*, 1354825. DOI: 10.3389/fimmu.2024.1354825 PMID: 38449862

Tycko, J., Wainberg, M., Marinov, G. K., Ursu, O., Hess, G. T., Ego, B. K., Aradhana, , Li, A., Truong, A., Trevino, A. E., Spees, K., Yao, D., Kaplow, I. M., Greenside, P. G., Morgens, D. W., Phanstiel, D. H., Snyder, M. P., Bintu, L., Greenleaf, W. J., & Bassik, M. C. (2019). Mitigation of off-target toxicity in CRISPR-Cas9 screens for essential non-coding elements. *Nature Communications*, *10*(1), 4063. DOI: 10.1038/s41467-019-11955-7 PMID: 31492858

Umeh-Garcia, M., O'Geen, H., Simion, C., Gephart, M. H., Segal, D. J., & Sweeney, C. A. (2022). Aberrant promoter methylation contributes to LRIG1 silencing in basal/triple-negative breast cancer. *British Journal of Cancer*, *127*(3), 436–448. DOI: 10.1038/s41416-022-01812-8 PMID: 35440669

van Staalduinen, J., van Staveren, T., Grosveld, F., & Wendt, K. S. (2023). Live-cell imaging of chromatin contacts opens a new window into chromatin dynamics. *Epigenetics & Chromatin*, *16*(1), 27. DOI: 10.1186/s13072-023-00503-9 PMID: 37349773

von Holdt, B. M., Kartzinel, R. Y., van Oers, K., Verhoeven, K. J., & Ouyang, J. Q. (2023). Changes in the rearing environment cause reorganization of molecular networks associated with DNA methylation. *Journal of Animal Ecology*, *92*(3), 648–664. DOI: 10.1111/1365-2656.13878 PMID: 36567635

Wang, G., & Li, J. (2021). Review, analysis, and optimization of the CRISPR Streptococcus pyogenes Cas9 system. *Medicine in Drug Discovery*, *9*, 100080. DOI: 10.1016/j.medidd.2021.100080

Wang, Q., Ma, C., Mao, H., & Wang, J. (2024). Epigenome editing by a CRISPR-Cas9-based acetyltransferase activates the ZNF334 gene to inhibit the growth of colorectal cancer. *International Journal of Biological Macromolecules*, *277*, 134580. DOI: 10.1016/j.ijbiomac.2024.134580 PMID: 39122070

Whittaker, M. N., Testa, L. C., Quigley, A., Jindal, I., Cortez-Alvarado, S. V., Qu, P., Yang, Y., Alameh, M.-G., Musunuru, K., & Wang, X. (2023). Epigenome editing durability varies widely across cardiovascular disease target genes. *Arteriosclerosis, Thrombosis, and Vascular Biology*, *43*(10), 2075–2077. DOI: 10.1161/ATVBAHA.123.319748 PMID: 37589141

Wiley, L., Cheek, M., LaFar, E., Ma, X., Sekowski, J., Tanguturi, N., & Iltis, A. (2025, June). The ethics of human embryo editing via CRISPR-Cas9 technology: A systematic review of ethical arguments, reasons, and concerns. [). Dordrecht: Springer Netherlands.]. *HEC Forum*, *37*(2), 267–303. DOI: 10.1007/s10730-024-09538-1 PMID: 39302534

Wu, D., Chen, Q., Chen, X., Han, F., Chen, Z., & Wang, Y. (2023). The blood–brain barrier: Structure, regulation and drug delivery. *Signal Transduction and Targeted Therapy*, *8*(1), 217. DOI: 10.1038/s41392-023-01481-w PMID: 37231000

Xiang, X., Corsi, G. I., Anthon, C., Qu, K., Pan, X., Liang, X., Han, P., Dong, Z., Liu, L., Zhong, J., Ma, T., Wang, J., Zhang, X., Jiang, H., Xu, F., Liu, X., Xu, X., Wang, J., Yang, H., & Luo, Y. (2021). Enhancing CRISPR-Cas9 gRNA efficiency prediction by data integration and deep learning. *Nature Communications*, *12*(1), 3238. DOI: 10.1038/s41467-021-23576-0 PMID: 34050182

Xing, H., & Meng, L. H. (2020). CRISPR-Cas9: A powerful tool towards precision medicine in cancer treatment. *Acta Pharmacologica Sinica*, *41*(5), 583–587. DOI: 10.1038/s41401-019-0322-9 PMID: 31792341

Yan, X., Han, D., Chen, Z., Han, C., Dong, W., Han, L., Zou, L., Zhang, J., Liu, Y., & Chai, J. (2020). RUNX2 interacts with BRG1 to target CD44 for promoting invasion and migration of colorectal cancer cells. *Cancer Cell International*, *20*(1), 505. DOI: 10.1186/s12935-020-01544-w PMID: 33071648

Yang, X., Bui, T. A., Mei, H., Aksoy, Y. A., Deng, F., Hutvagner, G., & Deng, W. (2024). Exploring the potential and challenges of CRISPR delivery and therapeutics for genetic disease treatment. *Advanced Functional Materials*, *34*(38), 2402630. DOI: 10.1002/adfm.202402630

Yang, Z. X., Fu, Y. W., Zhao, J. J., Zhang, F., Li, S. A., Zhao, M., Wen, W., Zhang, L., Cheng, T., Zhang, J.-P., & Zhang, X. B. (2023). Superior fidelity and distinct editing outcomes of SaCas9 compared with SpCas9 in genome editing. *Genomics, Proteomics & Bioinformatics*, *21*(6), 1206–1220. DOI: 10.1016/j.gpb.2022.12.003 PMID: 36549468

Zhan, Y., Cao, C., Li, A., Mei, H., & Liu, Y. (2023). Enhanced RNA knockdown efficiency with engineered fusion guide RNAs that function with both CRISPR-CasRx and hammerhead ribozyme. *Genome Biology*, *24*(1), 9. DOI: 10.1186/s13059-023-02852-w PMID: 36650600

Zhang, H., Qin, C., An, C., Zheng, X., Wen, S., Chen, W., Liu, X., Lv, Z., Yang, P., Xu, W., Gao, W., & Wu, Y. (2021). Application of the CRISPR/Cas9-based gene editing technique in basic research, diagnosis, and therapy of cancer. *Molecular Cancer*, *20*(1), 126. DOI: 10.1186/s12943-021-01431-6 PMID: 34598686

Zhu, Y. (2022). Advances in CRISPR/Cas9. *BioMed Research International*, *2022*(1), 9978571. DOI: 10.1155/2022/9978571 PMID: 36193328

Chapter 10
Advances in CRISPR-Cas9-Based Mitochondrial Gene Editing and Delivery Systems for Precision Medicine:
Targeting the Powerhouse

Sawani Rodrigo
https://orcid.org/0009-0000-9962-1832
International Institute of Health Sciences, Sri Lanka

Hettige Kaveesha Sandani Dinudya
https://orcid.org/0009-0007-4858-2651
International Institute of Health Sciences, Sri Lanka

Anne Sithuvili Hettiarachchige Done
https://orcid.org/0009-0009-5595-8455
International Institute of Health Sciences, Sri Lanka

Aresha Manamperi
University of Kelaniya, Sri Lanka

Y. I. Nilmini Silva Gunawardene
https://orcid.org/0000-0002-2528-4236
University of Kelaniya, Sri Lanka

DOI: 10.4018/979-8-3373-4857-5.ch010

ABSTRACT

The powerhouse of cells, mitochondria, play a crucial role in homeostasis, energy metabolism, apoptosis and stress response. Mitochondria possess a separate set of DNA (mtDNA), which is associated with a wide range of metabolic, degenerative and age-related disorders. The CRISPR-Cas 9 system revolutionised medicine with its discoveries in the nucleus; however, its application in mitochondria remains constrained due to the challenges of delivery, due to the double membrane and repair mechanisms, since mitochondria lack the conventional repair mechanisms, unlike the nucleus. These limitations directed CRISPR-independent and CRISPR-inspired strategies aiming at overcoming the hurdles. This chapter discusses the obstacles in achieving precise mitochondrial genome editing in order to shed light on new therapeutic possibilities.

INTRODUCTION

Imagine stepping into a sprawling metropolis, where every building, street, and park has a purpose. In this city, energy flows like electricity through power lines, information is transmitted through networks of sensors, and the citizens follow complex rules to maintain order. This is not a city on Earth; it is one within a cell. At the heart of this microscopic city lie mitochondria, the organelles often referred to as the cell's powerhouses. Yet, as in any well-planned urban centre, power generation is only one aspect of their role. Mitochondria are also command centres, capable of sensing stress, orchestrating responses to fluctuating resources, and determining the fate of the cell. Understanding this organisation is key to grasping how mitochondria contribute to health and, when hijacked, disease.

As the central regulators of energy metabolism and apoptosis, mitochondria play essential roles in cellular function. Mutations in mitochondrial DNA (mtDNA) contribute to a wide range of degenerative and metabolic diseases. The CRISPR/Cas9 system has revolutionised nuclear genome editing and offers numerous therapeutic possibilities. However, despite its feat in the nucleus, mitochondrial editing remains difficult due to hurdles in delivering guide RNA and repairing the resulting DNA breaks. Therefore, recent studies are shifting toward CRISPR-inspired, but CRISPR-independent approaches.

This chapter aims to study the potential of emerging genome-editing technologies in mitochondrial genetic engineering by highlighting the advances, challenges and future directions.

MITOCHONDRIAL GENETICS AND DISEASE

Structure, Inheritance and Heteroplasmy of mtDNA

The mitochondrial network serves as the central orchestrator of cellular energy, a semi-autonomous organelle found in the cytoplasm, covered in two membranes, where the inner membrane is folded, creating cristae to increase the surface area (Figure 1). Inside is a compact genome, called mitochondrial DNA (mtDNA), a small, circular, double-stranded molecule located within the matrix of the mitochondrion, which is primarily responsible for cellular energy production. MtDNA is 16,569 base pairs long, encoding only 37 genes which code for 13 proteins essential for oxidative phosphorylation (OXPHOS) for ATP generation, 22tRNAs for translating mitochondrial mRNA and 2rRNAs forming the core of the mitochondrial ribosomes. These genes are densely packed, with only a few non-coding regions, and one strand is guanine-rich, while the other is cytosine-rich. The maternal inheritance due to the elimination of paternal mtRNA from the father's sperm during fertilisation has profound implications for genetics and evolution. For instance, mtDNA is used as a powerful tool in tracing direct maternal lineages back through time, by mutating faster than nuclear DNA by acting as a "molecular clock". In addition, a mother carrying pathogenic mtDNA mutations will pass them on to all her children, while her daughters pass the same on to their offspring.

Heteroplasmy is a critical concept in mitochondrial diseases; it refers to when an individual cell contains a mixture of both wild-type (normal/healthy) and mutant mtDNA molecules. A typical cell can tolerate a certain percentage of mutant DNA called mutant load. Whereas homoplasmy is where all mtDNA within a cell is identical with either wild-type or mutant form. This can be due to cell division (mitosis), when this gene is randomly distributed to daughter cells, leading to a bottleneck effect, where the genetic drift will significantly change the level of heteroplasmy both between generations and in tissues of the same individual.

The clinical manifestations of mitochondrial disorders depend largely on the degree of mitochondrial dysfunction. Because heteroplasmy causes variable mutant loads, symptoms can differ among family members. Tissues with high energy requirements are particularly vulnerable and develop disease once their threshold of mutant genomes is exceeded.

Figure 1. Structure of mitochondria: Mitochondria are a double membrane organelle, whereas the inner membrane is folded into cristae, and electron transport chain (ETC) complexes are located in the inner mitochondrial membrane (created in Biorender.com).

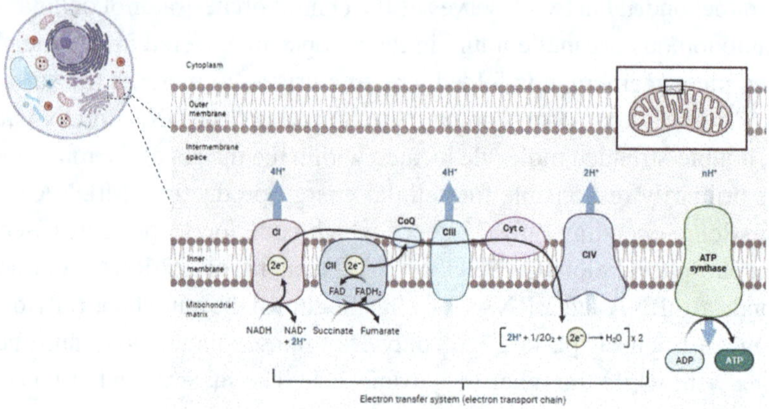

Mutation Rate

mtDNA has long fascinated researchers because of its remarkably high mutation rate compared to nuclear DNA, a feature that shapes both the physiology and pathology of human cells. It is estimated to mutate roughly twenty times faster than the nuclear genome, a contrast driven by several properties unique to mitochondria. As the primary producers of reactive oxygen species (ROS) generated during ETC in OXPHOS, mitochondria constantly generate mutagenic molecules capable of damaging bases and inducing strand breaks. Because the mitochondrial genome is situated in close physical proximity to these ROS, it is endlessly exposed to oxidative stress that increases the mutation rate (Habbane et al., 2021).

Unlike nuclear DNA, which has the protection of histones and extravagant repair systems, mtDNA is more vulnerable because it is not packaged in histone proteins and relies on a less efficient DNA repair machinery. Nuclear DNA damage can be adjusted through multiple pathways, including nucleotide excision repair, mismatch repair, and homologous recombination (HR), while mitochondrial DNA repair is mostly limited to base excision repair and a few secondary mechanisms. These limitations leave mtDNA highly vulnerable to damage, and its continuous replication, even in non-dividing cells like neurons and muscle fibres, further increases age-related copying errors.

The consequences of these weaknesses are noteworthy. Mutations in mtDNA can interrupt the genes encoding proteins for the OXPHOS complexes, compromising the cell's ability to generate ATP proficiently and its ability to increase the production of ROS. This creates a destructive, vicious cycle, which has relevance to cancer. Mutations in mtDNA are progressively identified in various tumours, commonly affected genes include MT-ND1 and MT-ND4 of Complex I and MT-ATP6 of Complex V, where the electron flow is interrupted and shifts cells toward glycolysis. This metabolic reprogramming aligns with the Warburg effect, where cancer cells favorably metabolize through glycolysis even when oxygen is present, allowing rapid proliferation and survival under hypoxic conditions typical of the tumour microenvironment (Welch et al., 2022).

Heteroplasmy further contributes to tumour growth by generating mitochondrial diversity within a single tumour. Subpopulations with high mutant loads may flourish under severe hypoxia, while those with more functional mitochondria thrive in better-oxygenated regions, together forming a highly resilient tumour ecosystem. Additionally, as mtDNA is maternally inherited, certain pathogenic variants can confer inherited cancer risk.

Maternal Inheritance

Uniparental transmission of mtDNA results from active processes during fertilisation that ensure only those from the oocyte persist in the zygote. This feature begins a matrilineal line of inheritance for the mitochondrial genome, with significant implications for disease transmission and population genetics (W. Lee et al., 2023). A clinically vital feature of this pattern is the mitochondrial genetic bottleneck. A woman carrying even a relatively low proportion of mutated mtDNA may have children whose cells range from nearly free of mutations to fully homoplasmic for a pathogenic variant. This randomness clarifies why siblings who inherit the same mutation often experience very different disease severities, and why predicting inheritance patterns remains so challenging (Zaidi et al., 2019).

Figure 2. Mitochondrial heteroplasmy. Coexistence of wild-type (green) and mutant (red) mtDNA within a cell. The proportion of mutant mtDNA influences mitochondrial function and disease manifestation (created in Biorender.com)

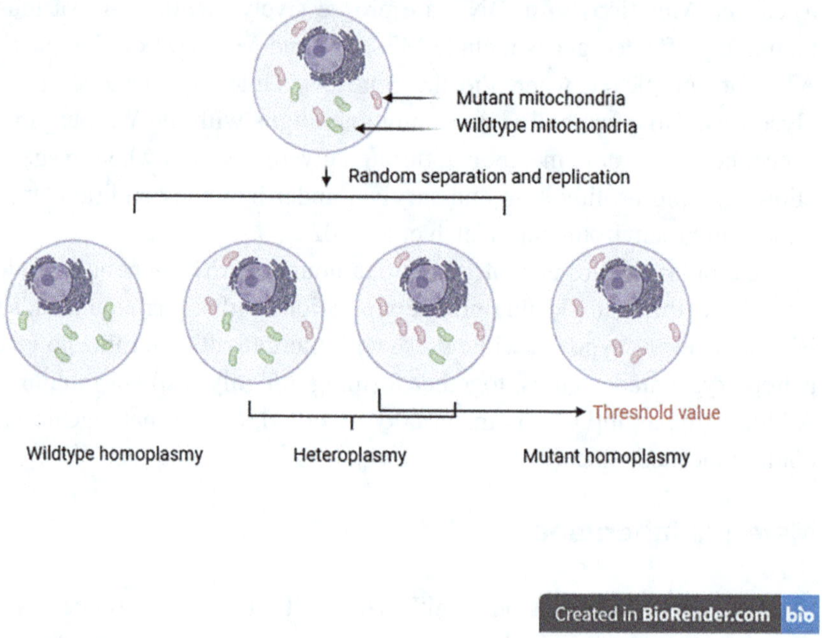

Maternal inheritance has key implications in cancer biology as well. While many mtDNA mutations related to cancer happen somatically within tumour tissue, some are maternally transmitted. These inherited variants can predispose individuals to cancer by lessening mitochondrial efficiency, increasing ROS production, or weakening metabolic flexibility. For instance, a mild defect in OXPHOS may remain clinically silent yet prime cells for oncogenic transformation under metabolic or environmental stress. The mitochondrial bottleneck further adds to this unpredictability, as siblings can inherit markedly different proportions of a mutation.

From an evolutionary standpoint, the combination of maternal transmission and the bottleneck helps maintain mitochondrial diversity while simultaneously permitting harmful variants to persist in populations. Clinically, this persistence poses challenges because maternally inherited mtDNA mutations can predispose multiple generations to both classical mitochondrial disorders and malignancies. Thus, mtDNA transmission links reproductive biology with disease risk, emphasising that cancer susceptibility is shaped not only by the nuclear genome but also by the maternal mitochondrial genome.

OVERVIEW OF THE CRISPR/CAS9 GENOME EDITING SYSTEM AS A GENE EDITING TOOL IN MITOCHONDRIA.

Clustered Regularly Interspaced Short Palindromic Repeats, in short CRISPR, "the genetic scissors," has emerged as a transformative tool in genetic engineering, enabling precise, efficient, and programmable genome editing across diverse organisms. Once the DNA is cut, the natural repair system of the cell will either introduce the respective changes or stitch it back together. Thus, CRISPR is named as a powerful tool in changing genes in living organisms and found to be crucial in agriculture, medicine, etc. CRISPR was derived from a bacterial adaptive immune system in around 50% of bacteria and 90% of archaea and was used in defence from invading bacteriophages. CRISPR-associated (cas) genes encode proteins such as nucleases, helicases, and polymerases, which participate in spacer acquisition and in destroying invading DNA. The component comprises two parts: the Cas9 endonuclease, which is responsible for inducing double-strand break (DSBs) at the target site of the DNA, and a single-guide RNA (sgRNA), which directs the Cas9 protein to the specific genomic location/ loci. Cas 9 protein is versatile due to a common Protospacer adjacent motif (PAM) sequence, which allows targeting of virtually any genomic region with complementary sgRNA design. The DSBs are repaired by endogenous pathways, non-homologous end joining (NHEJ) or homology-directed repair (HDR), enabling targeted gene disruption or precise sequence correction. The simplicity of guide RNA design, combined with high editing efficiency, has made CRISPR/Cas9 a foundational tool for modelling disease, correcting pathogenic variants, and developing next-generation therapeutics.

Figure 3. DNA repair mechanism, (1) NHEJ and (2) HDR followed by CRISPR/ Cas 9 gene editing. CRISPR/Cas9 introduces site-specific double-strand breaks in the target DNA. Cellular repair mechanisms correct the error via (1) NHEJ by directly ligating the DNA and (2) HR by repairing using the donor DNA template (created in Biorender.com)

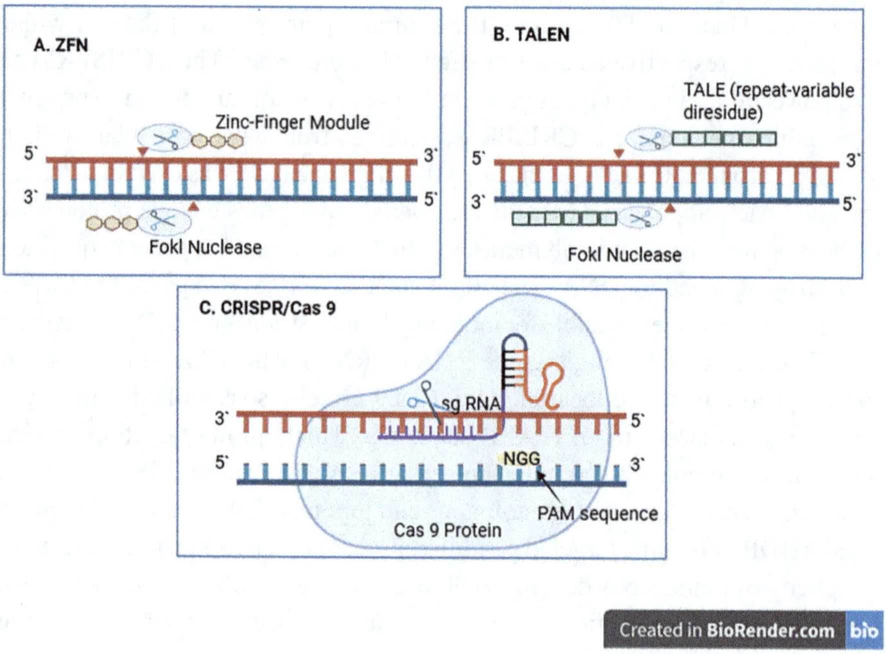

Pathogenic Mutations

Pathogenic mtDNA mutations can give rise to upsetting clinical syndromes, particularly in energy-dependent tissues such as the brain, heart, and skeletal muscle (Wen et al., 2025). One well-known example is the tRNA^Leu(UUR) mutation, responsible for most cases of MELAS (Mitochondrial Encephalomyopathy, Lactic Acidosis, and Stroke-like Episodes). This defect harshly disrupts mitochondrial protein synthesis and OXPHOS. Intriguingly, the same mutation can also lead to MIDD (Maternally Inherited Diabetes and Deafness), demonstrating the broad phenotypic effects of a single mtDNA variant.

Another extensively studied mutation affects Complex I and results in Leber Hereditary Optic Neuropathy (LHON), a condition characterised by acute, permanent vision loss in young adults. Similarly, the m.8344A>G variant in MT-TK underlies MERRF (Myoclonic Epilepsy with Ragged-Red Fibres), marked by seizures, muscle weakness, and ragged-red fibres on biopsy. Mutations in MT-ATP6, such as m.8993T>G/C, further demonstrate the worth of heteroplasmy: moderate levels may cause NARP (Neuropathy, Ataxia, and Retinitis Pigmentosa), while higher levels can lead to far severe Leigh syndrome.

In MELAS, neurological symptoms classically appear when heteroplasmy exceeds this threshold in neurons, while LHON-related vision loss emerges when about 70% of the optic nerve mtDNA carries the mutation. Lower heteroplasmy levels may predispose individuals to slightly weaker metabolic dysfunctions such as diabetes, whereas near-homoplasmic states often result in fatal outcomes like Leigh syndrome or perinatal death (Figure 2) (Nissanka & Moraes, 2020). Richhariya *et al* have shown that when neurons face chronic mitochondrial dysfunction, they undergo a metabolic shift similar to the Warburg effect observed in cancer (Richhariya et al., 2025). Using cell-specific CRISPR knockouts of mitochondrial fission (Drp1) and fusion (Opa1, Marf) genes in Drosophila neurons, the study revealed that loss of mitochondrial fusion, particularly Opa1, triggers the strongest effects. Fusion-deficient neurons activate a cancer-like metabolic rewiring, marked by a major increase in glycolysis and a strong upregulation of lactate dehydrogenase (LDH). This shift compensates for reduced ATP production caused by defective mitochondria and acts as a neuroprotective response. When LDH is removed together with Opa1, ATP levels collapse, and severe neurodegeneration occurs, confirming that this Warburg-like adaptation is essential for neuronal survival. The transcription factor ATF4 drives this protective glycolytic program, linking mitochondrial stress to metabolic reprogramming. Overall, the study highlights that cells, whether cancer cells or stressed neurons, can adopt similar survival strategies when mitochondrial function fails. This strengthens the emerging connection between mitochondrial dysfunction, metabolic adaptation, and disease biology, bridging cancer metabolism with neurodegeneration.

These principles extend straight into cancer biology. Mutations in mtDNA genes such as MT-ND4 and MT-ATP6 disrupt OXPHOS, driving cells toward glycolysis and increasing ROS levels, which in turn endorse genomic instability and tumorigenesis.

Mitochondrial Mutations as Cancer Biomarkers

Beyond their roles in disease pathogenesis, mitochondrial mutations hold great promise as biomarkers for cancer detection, prognosis, and monitoring. Numerous features of mtDNA make it chiefly attractive for biomarker development. Its high copy number within cells means that even minor tumours can produce measurable amounts of mutant mtDNA into circulation (Ho & Ho, 2025). Additionally, certain mutations occur in hotspots, such as MT-ND1, MT-ND4, and MT-ATP6, which are regularly associated with specific cancer types. These recurrent patterns allow researchers to link specific mutations with distinct malignancies, refining diagnostic specificity (Shelton et al., 2024). Mendelian Randomisation (MR) leverages genetic variants as instrumental variables to deduce the correlation between mitochondrial dysfunction and six different cancer types (Tang et al., 2025). mtDNA copy number (mtCN) and mutations have tissue-specific effects in cancer: increased mtDNA raises colorectal cancer and glioblastoma risk but lowers lung cancer risk, while impaired Complex I links to pancreatic cancer. Mutations like MT-ND5 in pancreatic tumours boost glycolysis yet rely on OXPHOS under nutrient stress. Lung adenocarcinoma cells can acquire healthy mitochondria from astrocytes, increasing chemoresistance. ROS have dual roles, low levels promote survival via HIF-1α/NF-κB, while high ROS causes apoptosis. mtDNA-mutant cancers show elevated ROS but compensate with antioxidants, creating vulnerabilities. ROS-modulating agents like metformin can selectively kill these cells, reducing tumours by ~70%, highlighting mitochondria and ROS as key, context-dependent therapeutic targets (Wallace, 2012). Given the strong variation in mtCN across tissues and its association with resistance to redox stress, we should think of mtCN not just as a passive readout but as a potential contributor to cancer cell survival under stress. CRISPR screens across cell lines could shed more light on how modulating mtCN impacts therapy response.

Circulating mtDNA can be identified in body fluids such as blood, urine, and saliva, presenting minimally invasive options for cancer screening. For example, urine samples from patients with bladder cancer every so often contain mutations in MT-ND1 and MT-ND4, permitting early detection. In breast and colorectal cancers, mitochondrial variants have been linked with poor prognosis and resistance to treatment, stressing their potential as prognostic biomarkers. Because mtDNA occurs in far greater abundance than nuclear circulating tumour DNA, it suggests higher sensitivity in liquid biopsy approaches.

Variations in the ratio of mutant to wild-type mtDNA over time can reflect tumour progression, rise of resistant subclones, or response to therapy. Observing heteroplasmy in circulating mtDNA could hence provide real-time insights into tumour dynamics, contributing a noteworthy advantage over traditional biopsies that habitually fail to capture the genetic heterogeneity of tumours (Pérez-Amado et al., 2021).

Maternal inheritance further improves the value of mtDNA as a biomarker. Certain inherited variants have been associated with increased cancer risk, and recognising these variants within families can aid in stratifying individuals for tailored surveillance. This is chiefly valuable when nuclear DNA mutations in well-known cancer susceptibility genes, such as BRCA1 or BRCA2, are absent, but familial cancer risk remains unsolved.

Nonetheless, drawbacks remain. Distinguishing pathogenic mutations from age-related variants in normal tissues is tough, and calculating low levels of heteroplasmy with accuracy necessitates extremely sensitive sequencing technologies. In spite of these challenges, developments in next-generation sequencing and bioinformatics are progressively improving the uncovering and interpretation of mitochondrial variants. Mitochondria-targeted restriction endonucleases (mtREs) have been shown to reduce mitochondrial heteroplasmy, but their application is limited because only a few pathogenic mutations create unique restriction sites suitable for targeting. On the other hand, yeast tRNA (tRK1) contains unique structural elements such as the F-arm or the D-hairpin, which can be recognised by mitochondrial RNA import machinery. As mammalian cells do not have an innate system for RNA import into mitochondria, by fusing these motifs into synthetic RNAs such as guide RNAs of CRISPR, the whole component can be transported to mitochondria for editing, repair or functional rescue. Research also showed that RNAs based on tRK1 could selectively inhibit replication of mutant mtDNA once imported into mitochondria. In vivo evidence suggests that preKARS2 (cytosolic precursor of human mitochondrial lysyl-tRNA synthetase) plays a direct role in mitochondrial RNA import within human cells, leading to a new opening of mitochondrial import methods (Yin et al., 2022).

Mitochondria and Cancer

Mitochondria's evolutionary origins shape their role in cancer. Once free-living bacteria, they became key regulators of energy metabolism, redox balance, and apoptosis. In healthy cells, they generate ATP and help remove damaged cells, but in cancer, these functions are rewired to support survival, growth, metabolic adaptation, and immune evasion. This shift highlights how mitochondrial dysfunction drives tumour progression and reveals potential therapeutic vulnerabilities (Lee et al., 2025). Likewise, oxidative stress within cancer cells drives metabolic reprogramming, boosting the Warburg effect, in which cells preferentially rely on glycolysis even in the presence of oxygen (Tufail et al., 2024).

One of the initial indicators of mitochondrial participation in cancer is evident in their genetic material. Mutations in mtDNA tend to accumulate rapidly, particularly in regions that code for components of the ETC. These alterations impair OXPHOS, resulting in changes to energy metabolism and giving rise to the phenomenon historically termed the Warburg effect (Wu et al., 2025). Earlier perspectives proposed that mitochondria were nonessential in cancer, yet modern research has demonstrated that mitochondria remain crucial, adjusting to metabolic and environmental pressures. Instead of vanishing, they reconfigure their functions, enabling tumour cells to satisfy elevated biosynthetic and energy requirements, withstand stress, and sustain survival under adverse conditions. Mutated mtDNA contributes not only to disrupted metabolism but also to tumour heterogeneity, affecting proliferation rates, metastatic capabilities, and responsiveness to therapy (Carinci et al., 2021).

Mitochondrial ROS play a dual role in cancer. At moderate levels, they act as signalling molecules that promote proliferation, angiogenesis, and stress adaptation. At high levels, they cause DNA and cellular damage, leading to apoptosis or senescence. Cancer cells maintain a delicate balance by raising ROS enough to drive growth while boosting antioxidant systems, such as glutathione, SOD, and thioredoxin, to prevent cell death. This controlled redox state helps tumours survive but also creates a therapeutic vulnerability: treatments that disrupt this balance by increasing ROS or weakening antioxidant defences can selectively target cancer cells while sparing healthy tissue (Khan et al., 2025).

Mitochondria also serve as central regulators of apoptosis, the programmed cell death mechanism that removes damaged or potentially harmful cells. Under normal physiological conditions, intrinsic apoptosis is governed by the balance between pro- and anti-apoptotic Bcl-2 family proteins, with mitochondria functioning as critical decision-making centres. Permeabilisation of the mitochondrial outer membrane permits the release of cytochrome c, assembly of the apoptosome, and activation of executioner caspases, orchestrating an orderly dismantling of the cell. In cancer, however, these pathways are subverted to promote survival. Overexpression of anti-apoptotic proteins, such as Bcl-2 and Bcl-xL, blocks cytochrome c release, while mutations in tumour suppressors like p53 impair damage detection, enabling cells to bypass checkpoints and accumulate genetic alterations. Even under chemotherapy or radiotherapy, many tumour cells evade death, with their mitochondria acting as fortified strongholds. Paradoxically, this reliance on mitochondrial control creates therapeutic opportunities: BH3 mimetics, which emulate pro-apoptotic signals, can overcome this blockade, compelling tumour cells to confront their intrinsic vulnerabilities (Barbi et al., 2025).

Mitochondria in cancer cells are highly dynamic, constantly undergoing fusion, fission, and mitophagy. Fusion helps stabilise energy output and dilute damage, while fission produces smaller mitochondria that support mobility, adaptation, and preparation for mitophagy. Mitophagy then removes dysfunctional organelles to prevent excess ROS and maintain homeostasis. In cancer, these processes are often altered to enhance survival, metastasis, and resistance to stress. Targeting mitochondrial dynamics, such as disrupting fusion, fission, or mitophagy, can disturb energy balance and sensitise tumour cells to treatment. Emerging CRISPR-based tools aimed at modulating these pathways offer new opportunities to study and therapeutically manipulate mitochondrial behaviour in cancer (Changaei et al., 2025).

Cancer stem cells (CSCs) are highly dependent on mitochondrial OXPHOS for energy, survival, and tumour regeneration. Their mitochondria support redox balance, stress resistance, and metabolic flexibility, aided by regulators such as PGC-1α, which boosts biogenesis and antioxidant defences (Chen et al., 2025). Mitochondrial fusion, fission, and mitophagy are tightly controlled to maintain organelle quality.

This reliance creates a therapeutic vulnerability: disrupting OXPHOS, inhibiting mitophagy, or targeting mitochondrial regulators can selectively eliminate CSCs and reduce recurrence (Lv et al., 2025).

Mitochondria shape both the tumour microenvironment and immune responses. When damaged mitochondria release mtDNA, the cGAS–STING pathway activates type I interferons and recruits immune cells (Riley & Tait, 2020). While this normally enhances defence, tumours exploit chronic low-level signalling to promote immune tolerance. Within the tumour, T cells and NK cells face nutrient depletion and oxidative stress, leading to mitochondrial dysfunction, reduced ATP, and impaired cytotoxicity. Excess ROS further drives T-cell exhaustion, while cancer cells maintain their own redox balance. By manipulating mitochondrial–immune interactions, tumours evade surveillance and resist immunotherapy. Therapeutically, restoring mitochondrial fitness in immune cells or selectively inducing mitochondrial stress in tumour cells offers promising strategies to strengthen anti-tumour immunity (Chen et al., 2025). Although CRISPR has been combined with STING activation (e.g., to knock out PD-L1), the technology has not yet been widely used to directly target mtDNA to exploit this immune axis (Lu et al., 2022).

Mitochondrial metabolism also plays a key role in determining the polarisation of tumour-associated macrophages (TAMs) and the functional capacity of dendritic cells. TAMs with altered mitochondrial activity acquire immunosuppressive phenotypes, promoting angiogenesis, tissue remodelling, and tumour progression. Similarly, dysfunctional dendritic cells are unable to efficiently present antigens, thereby impairing T cell activation. Interventions targeting mitochondrial pathways in these immune cell populations can reprogram the tumour microenvironment from an immunosuppressive to an immunostimulatory state, enhancing the effectiveness of immunotherapies such as checkpoint inhibitors or CAR T cell treatments (Carinci et al., 2021).

Metastasis is another context in which mitochondria play a crucial role. The high energy requirements of cellular migration, invasion, and colonisation of distant tissues depend on properly distributed and functional mitochondria. Fission produces small, highly mobile mitochondria at the leading edge of migrating cells, while mitophagy eliminates damaged organelles, preventing localised ROS-induced apoptosis. Cancer stem cells (CSCs) exploit these mitochondrial dynamics to initiate secondary tumours, highlighting that targeting mitochondrial function is essential not only for eliminating primary tumours but also for restricting metastatic dissemination (Tufail et al., 2025).

Mitochondria's dual role in cancer, as drivers of tumour growth and points of vulnerability, has shaped several therapeutic strategies. OXPHOS inhibitors such as metformin reduce ATP and increase ROS in respiration-dependent tumours, while BH3 mimetics restore apoptosis by blocking anti-apoptotic proteins. Disrupting mitochondrial dynamics or mitophagy destabilises energy distribution and sensitises tumour cells to therapy. Nanoparticle delivery and mitochondria-targeted compounds improve precision and limit off-target damage. When combined with standard treatments, immunotherapy, or ROS-modulating agents, these approaches exploit cancer's reliance on mitochondrial function while strengthening anti-tumour immunity (Barbi et al., 2025). A recent study used CRISPR-mediated knockout of key mitochondrial-dynamics genes (Opa1 and Drp1), showing that KRas-mutant lung adenocarcinoma relies specifically on Opa1-regulated mitochondrial fusion to maintain cristae structure, ETC function, and NAD^+ regeneration, processes essential for rapid tumour proliferation. These CRISPR experiments reveal Opa1 as a selective metabolic dependency and potential therapeutic target, validated through both genetic deletion and pharmacological inhibition. Importantly, insights from CRISPR-based mitochondrial genome editing, such as the ability to induce mtDNA breaks and disrupt ETC subunits, suggest future therapeutic possibilities where CRISPR tools could be engineered to mimic this ETC collapse in cancer cells, positioning CRISPR as a tool that identifies mitochondrial weaknesses in tumours, but also as a potential precision strategy to exploit the mitochondrial dependencies that drive cancer growth (Sessions et al., 2022).

Emerging technologies are enabling increasingly precise mitochondrial interventions. CRISPR-based mitochondrial editing, gene therapies targeting metabolic regulators, and synthetic lethality strategies now exploit tumour-specific vulnerabilities. Advances in imaging further allow real-time tracking of mitochondrial dynamics, ROS, and metabolic flux, supporting adaptive, personalised treatment. By profiling mitochondrial function in tumours and immune cells, clinicians can tailor therapies that disrupt malignant resilience while preserving normal physiology (Lee et al., 2025).

In cancer, mitochondria regulate survival, proliferation, metabolism, and immune signalling, but these essential roles also create therapeutic vulnerabilities, disrupting OXPHOS, mitophagy, or redox balance can selectively target tumour cells. Emerging mitochondrial-targeted therapeutics increasingly exploit the triphenylphosphonium (TPP^+) cation as a delivery motif to selectively accumulate drugs within the mitochondrial matrix. Because TPP^+ is a lipophilic cation, it readily traverses mitochondrial membranes. It concentrates in response to the organelle's negative membrane potential, making it a powerful tool for directing anticancer agents to the ETC. A key example is MitoTAM, a tamoxifen derivative conjugated to TPP^+, which simultaneously inhibits Complex I and uncouples OXPHOS to eliminate senescent or metabolically vulnerable cancer cells selectively. Comparisons with structurally similar OXPHOS inhibitors such as MS-L6, which lack the TPP^+ tag and therefore fail to accumulate in mitochondria, underscore the essential role of TPP^+ in mitochondrial targeting. This strategy highlights a growing therapeutic paradigm: designing mitochondria-penetrating drugs that exploit mitochondrial bioenergetics to achieve selective anticancer effects (Al Assi et al., 2024; Du et al., 2025; Sivagnanam et al., 2024).

UNMET NEEDS IN CURRENT THERAPIES

Despite key progress in mitochondrial genetics, there are still no curative treatments for mtDNA-related disorders or for cancers induced by mitochondrial dysfunction. Current management is mainly supportive. Patients with conventional mitochondrial syndromes may receive antioxidants, vitamins, or metabolic cofactors to reduce oxidative stress or boost residual mitochondrial activity, but these measures do not fix the causal mutations (Ball et al., 2025).

A key obstacle is the hurdle of directly modifying the mitochondrial genome. Unlike nuclear DNA, mtDNA is resistant to precise editing or selective removal. Without tools that unfailingly shift heteroplasmy toward healthy genomes, clinicians cannot correct or eliminate pathogenic variants. In cancer, this constraint is intensified by the metabolic plasticity of tumour cells, which quickly reorganise energy pathways when required.

Mitochondrial delivery remains another daunting challenge. The double membrane of mitochondria limits the entry of therapeutic agents, regardless of whether they are small molecules, peptides, or genetic tools. While some advancements have been made in developing mitochondria-targeting carriers, such as lipophilic cations, peptides, and nanoparticles, their clinical use remains imperfect due to issues of specificity, safety, and efficiency.

Diagnostics also face noteworthy gaps. Reliable biomarkers to monitor mtDNA mutation load and heteroplasmy dynamics are lacking, and current techniques every so often necessitate invasive sampling or fail to sense subtle but clinically significant changes. In cancer, heterogeneity within tumours further muddies interpretation, as different regions of the same tumour may harbour distinct mtDNA variants (Gadicherla et al., 2024).

Collectively, these challenges stress the need for innovative strategies. Effective therapies for mitochondrial disorders will eventually need to remove or correct mutant genomes and restore mitochondrial function. In oncology, interventions must exploit mitochondrial vulnerabilities without initiating compensatory pathways that enable cancer cell survival. Across both fields, progress will depend on improved biomarkers, advanced delivery systems, and deeper insight into mitochondrial dynamics.

Emerging genetic engineering tools

The challenges mentioned above have driven the advancement of new generations of genetic engineering tools aimed at treating mitochondrial diseases and cancer. Latest efforts focus on fixing pathogenic mtDNA mutations through two major strategies. The first involves selectively removing mutant genomes to shift heteroplasmy toward the healthy population. Tools such as mitoZFNs, mitoTALENs, and ARCUS meganucleases have shown the ability to lower mutant load and restore mitochondrial function in disease models.

The second strategy directly amends the mitochondrial genome. This is possible with guide-RNA–independent base editors. DddA-derived cytosine base editors (DdCBEs), which fuse programmable TALE proteins to a split toxin, allow targeted C-to-T conversions and have been confirmed in cell culture and animal systems. More recently, TALED adenine base editors have developed this capability to A-to-G editing, with engineering improvements markedly lowering off-target activity.

Of course, editing tools are only useful if they can reach the mitochondria. That's where delivery systems come in. Lipid-based carriers such as the MITO-Porter have been designed to cross the mitochondrial membranes and deliver proteins, nucleic acids, or small molecules. Excitingly, recent work has even demonstrated the successful delivery of Cas9 protein complexes into isolated mitochondria, proving that direct protein targeting is possible. Other approaches use cationic molecules like triphenylphosphonium (TPP^+) or mitochondrial-targeting peptides to selectively transport therapeutic cargo. Together, these advances are opening the door to treatments for inherited mitochondrial diseases as well as new ways to target cancer cells that rely heavily on OXPHOS.

Recent research studies provide remarkable evidence that mitochondrial base editing is not only feasible but also therapeutically powerful in disease prevention and treatment. mitoZFNs and mitoTALENs are more labour-intensive and costly compared to CRISPR, where only the guide RNA (sgRNA) needs to be changed. CRISPR is one of the newest interventions in gene editing, which has shown remarkable results in nuclear gene editing.

To understand and appreciate the powerful tool CRISPR, it is helpful to compare it with earlier tools such as zinc-finger nucleases (ZFNs) and transcription-factor effector nucleases (TALENs) (Figure 4).

Table 1. Comparison between gene editing tools ZFN, TALENS and CRISPR

Feature	CRISPR/Cas	ZFN (Zinc Finger Nucleases)	TALENs (Transcription Activator-Like-Effector Nucleases)
Targeting Mechanism	RNA-guided: a short guide RNA (~20 nt) directs a Cas nuclease to complementary DNA (or RNA for Cas13). Simple programmable base-pairing.	Protein-based: engineered arrays of zinc-finger DNA-binding domains (each finger ≈3 bp) fused to FokI nuclease. Requires designing protein modules for each new target.	Protein-based: modular TALE repeats (each repeat recognises 1 bp via RVD code) fused to FokI nuclease. Modular code makes sequence targeting predictable but still requires protein assembly.
Ease of Design	Very easy: design and order guide RNAs; little protein engineering. Enables rapid prototyping.	Difficult: bespoke protein engineering expertise or commercial services needed; design iterations common.	Moderate: protein construction required but TALE repeat code is straightforward to assemble (more user-friendly than ZFNs).
Length of New Design Required	sgRNA ≈20 nt (plus PAM requirement adjacent to target). PAM constraint varies by Cas variant.	The recognition site is usually composed of several zinc-finger modules (typical target ~18–36 bp across a pair), each module ≈3 bp.	Target recognition commonly ~30–40 bp (two TALEN arms flanking spacer); each TALE repeat recognises 1 bp.
Multiplexing	High: multiple sgRNAs expressed simultaneously (multiplex screens, combinatorial edits). Widely used for pooled and multiplex editing.	Low: each additional target requires engineering a new zinc-finger protein. Not practical for high-order multiplexing.	Moderate: possible but requires constructing multiple TALEN pairs — more laborious than CRISPR.
Cost & Time Efficiency	Low cost & rapid (days to weeks): cheap oligos for gRNAs and plasmids/IVT or RNPs; many ready-to-use reagents.	High cost & slow: custom protein design, cloning, and validation increase time and expense	Moderate cost/time: protein assembly required, but faster than ZFNs thanks to TALE modularity.
Scalability	Highly scalable (functional genomics, pooled CRISPR screens, therapeutic pipelines).	Poor scalability for large projects due to the custom protein needs.	Moderate scalability; used successfully in many labs and in plants/animals, but less scalable than CRISPR for large libraries.

continued on following page

Table 1. Continued

Feature	CRISPR/Cas	ZFN (Zinc Finger Nucleases)	TALENs (Transcription Activator-Like-Effector Nucleases)
Precision	High on-target efficiency but variable off-target rates depending on guide, context, and Cas variant; many detection and high-fidelity variants developed. Comparative studies show CRISPR can have higher off-target signals than optimised ZFNs/TALENs in some assays, but improvements (HiFi Cas, base editors, prime editors) have reduced off-target concerns.	High specificity when well-designed; off-targets can be low with obligate heterodimeric FokI and careful zinc-finger selection, but design complexity makes consistent high specificity harder to achieve in practice.	High specificity; TALENs often show lower off-target activity than early CRISPR systems in some comparative studies, but are larger and harder to deliver.
Applications	Gene knockout, knock-in (HDR), base editing, prime editing, transcriptional regulation, diagnostics (Cas12/Cas13), pooled screens, therapeutic development. Widely adopted across biology and medicine.	Clinical and preclinical gene therapy (ex vivo and some in vivo trials), early gene-editing demonstrations, and targeted therapies where bespoke high specificity is required.	Agriculture (plant genome edits), animal models, functional genomics, and some therapeutic development, where high specificity and reduced off-target activity are priorities.

Sources: Bi et al., 2022; Kumar et al., 2025; Yin et al., 2022

Figure 4. Comparison between A. ZFN, B. TALEN and C. CRISPR/Cas 9 gene editing techniques. A. ZFNs and B. TALENs use engineered DNA-binding proteins coupled with the FokI nuclease to generate double-strand breaks (DSBs). ZFNs recognise triplet base pairs via zinc finger motifs, while TALENs recognise single bases through TALE repeats, where target recognition in both is protein-DNA. In contrast, C. CRISPR/Cas9 employs a guide RNA (gRNA) to direct the Cas9 nuclease to a complementary DNA target adjacent to a PAM sequence, enabling RNA-guided cleavage (RNA-DNA target) (Created by Biorender.com).

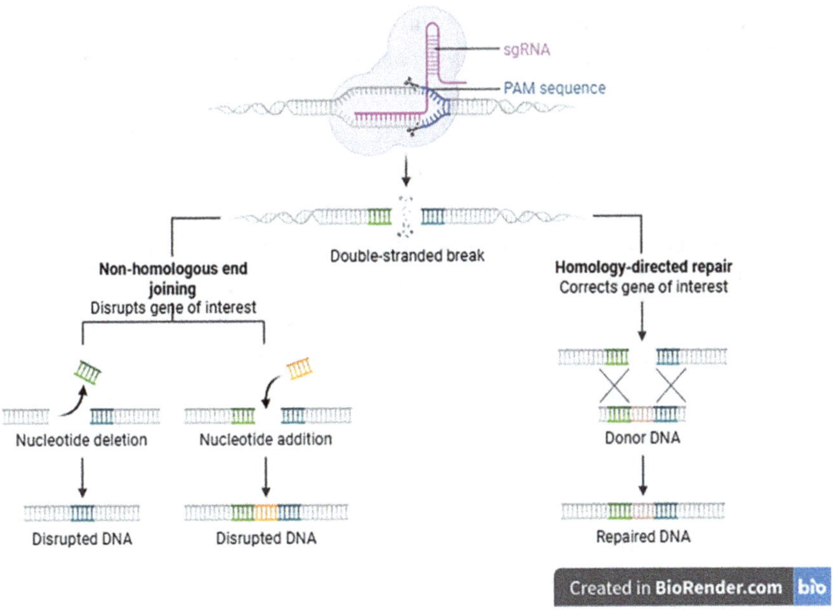

FUNDAMENTAL BARRIERS OF CRISPR IN MITOCHONDRIAL GENE EDITING

Despite the success in the nucleus, CRISPR-Cas9 in mitochondria is hindered by several incompatibilities. Cas9, one of the commonest endonucleases, which acts as molecular scissors to cut the DNA, and the cell's own repair mechanism will either disrupt the gene or rewrite with the complementary strand's information, acting as the fastest, cheapest and most reliable system in gene editing. The CRISPR-Cas 9 system usually works with a guide RNA carrying 20 bases. Thus, it could bind to complementary DNA sequences in the genome, causing off-target cuts, leading to accidental cuts in the genome. To eliminate this occurrence, a new modification to Cas9 was done, which required 2 guides to be targeted before cutting. Moreover, the CRISPR-Cas 9 system is proven to be very effective in locating and analysing functional genes in mitochondria as well as creating precise models of mitochondrial diseases for study purposes. In addition, to confirm the disease-causing mutations, isogenic controls can be created using this technology, which are cell lines that are genetically identical but differ only at the mutation site to facilitate a clear background for functional studies. In 2015, the first trial of mitochondrial genome editing with CRISPR-Cas9 was reported. Researchers created mitoCas9, which targeted MT-COX1 and MT-COX3 genes along with the sgRNAs and mitochondrial-targeting sequences (MTS). Human Embryonic Kidney cells (HEK-293T) are usually used in CRISPR/Cas9 experiments, which originally came from kidney tissues of human embryos in the 1970s. Another debate about mitochondrial CRISPR editing is the rapid degradation due to the lack of classical repair pathways, such as NHEJ. In the nucleus, CRISPR-Cas9-edited DNA is repaired by mechanisms such as NHEJ, which directly joins the broken DNA ends without the homologous template's guide and HR, which uses the sister chromatid or the homologous sequence (Figure 3). In the nucleus, CRISPR mostly targets NHEJ due to its quick nature. On the contrary, mitochondria don't have NHEJ machinery; thus, double-strand breaks (DSBs) are degraded as a quality control system. Since mitochondria contain many copies of DNA, this mechanism is proven safe, but it causes a major barrier for CRISPR editing in mitochondria. When the edited/ cut DNA is degraded instead of repaired, mtCN can be reduced, leading to inefficient copies for energy production. Research advancement for the mentioned challenges with CRISPR independent strategies, such as using RAD51, a crucial protein in the HR DNA repair pathway. Treating cells with RAD51 agonist RS-1 enhance exogenous single-stranded DNA oligonucleotides (ssODN)-mediated knock-in efficiency in crude mtDNA by fivefold and further elevated in purified mtDNA (Bi et al., 2022).

Even though the continuous approaches suggest the potential therapeutic applications, reproducibility, efficiency and proof for mtDNA repair accuracy remain unresolved, and CRISPR-Cas9 is valuable in generating disease models, performing large-scale genetic screens, and exploring mitochondrial biology. Although its use as a therapy for human mitochondrial disorders is still distant, advances in delivery systems and editing precision may eventually bridge this gap.

STRATEGIES TO OVERCOME GRNA IMPORT & REPAIR CHALLENGES.

In contrast to Cas9, AsCas12a nuclease, which is a type of V CRISPR system, creates staggered DNA cuts that are smaller in size, thus requiring smaller RNAs, supporting it as a promising candidate in mitochondrial gene editing. Su9 MTS, a mitochondrial targeting signal in the mito-AsCas12a system, ensures that the Cas12a nuclease actually reaches the mitochondria, overcoming one of the main barriers to mitochondrial genome editing. As a result, the mito-AsCas12a system induced precise mtDNA deletions. Moreover, expression of the mito-AsCas12a doesn't disrupt the normal function of mitochondria. Although the efficiency of mtDNA remains very low, combination techniques provide promising results (Nikitchina et al., 2025). Even though Su9 MTSs are most effective, *Lachnospiraceae bacterium (Lb)* Cas12a with MTSs nucleases also localise better when fused. Precise targeting is essential to obtain heteroplasmy purification, and compared to Cas9, Cas12a contains a smaller guide RNA and staggered cuts to enhance efficiency (Antón et al., 2020).

Even though not studied well enough, another strategy is to modify the guide RNA with stem-loop motifs such as H1 RNA stem loop from nuclear RNase P complex, F arm or D-hairpin structures from yeast cytosolic tRNALys, and the 7-2 RNA, which is a small non-coding RNA that is linked with signal recognition particle (SRP). These can be included in sgRNA or crRNA (CRISPR-RNA) scaffolds to facilitate their entrance into the mitochondria (Yin et al., 2022).

DELIVERY SYSTEMS (PROTEIN-BASED, NANOCARRIERS, VIRAL, AND BASE EDITORS)

The dual membrane of mitochondria leads to major hurdles in delivering. The outer membrane is relatively permeable to small molecules, whereas the inner membrane is a formidable barrier. This structure helps maintain the essential proton gradient for ATP production, and when delivering the CRISPR system, it must cross both these barriers, and efficient delivery is important in achieving therapeutic relevance, especially considering the fact that mitochondria contain many copies of mtDNA and the absence of classical DNA repair pathways. There are several delivery strategies, such as protein-based delivery, nanocarriers and lipid-based systems, viral vectors and base editors coupled with DNA-binding modules that are studied by scientists.

Firstly, proteins of short positively charged peptide sequences, such as nucleases, are fused with MTSs, which are short peptides that signal import into the mitochondrial matrix. Nucleases such as Vas9 or Cas12a are used. Recognition of the MTS at the N-terminus of the precursor proteins, followed by unfolding and transportation of the proteins from the intermembrane space into the matrix, is done by Translocase of the outer membrane (TOM) and translocase of inner membrane (TIM), respectively (Figure 5). For example, Su9 is the most potent and commonly used MTS, which is an ATP synthase subunit 9 and Lb-Cas12a from the *Lachnospiraceae bacterium* Cas12a. Delivering pre-assembled Cas protein with the gRNA allows reducing off-target effects and immunogenicity due to by bypassing the DNA transcription, thus reducing the triggering of the cellular immune responses to the foreign DNA, but challenges such as transporting the large RNP complex through the mitochondrial import machinery exist. According to a recent finding of Taylor and the team, the PNPase (polynucleotide phosphorylase) protein, which is present in the space of the mitochondrial membrane, helps import RNAs through KH and S1 binding domains. When the study clustered IncRNAs, they found that low GC content, increased unpaired nucleotides and specific K-mer frequencies would favour the PNPase interaction, suggesting IncRNA-like structure preference by PNPase (Taylor et al., 2024).

Figure 5. Mitochondrial delivery of CRISPR components via MTS-fused RNP complex. The pre-assembled CRISPR RNP (Cas protein + guide RNA) is fused with a mitochondrial targeting sequence (MTS) at the N-terminus. The MTS directs the complex to the mitochondrion, where it is recognised by the TOM (Translocase of the Outer Membrane) and TIM (Translocase of the Inner Membrane) complexes, allowing import into the mitochondrial matrix. Once inside, the RNP complex targets and edits circular mitochondrial DNA (mtDNA). The MTS may be cleaved off after import, leaving the RNP active in the matrix (created in Biorender.com).

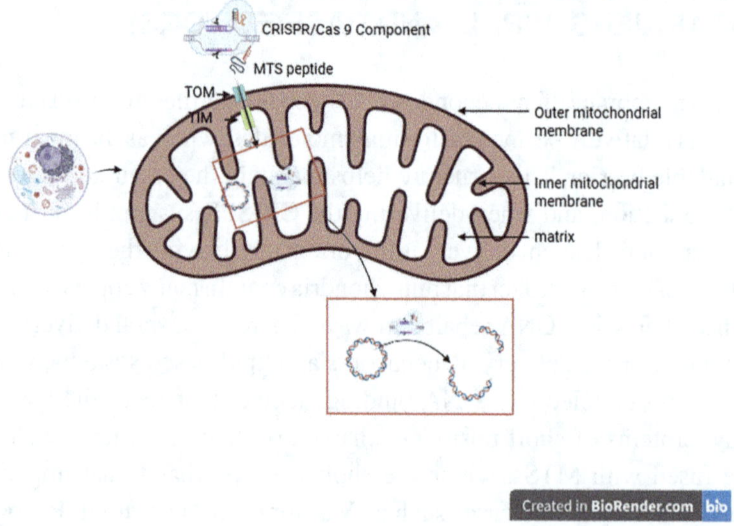

Secondly, nanocarriers and lipid-based systems are able to trick the mitochondrial membranes. MITO-Porter mechanisms are a highly versatile method which allows the delivery of proteins, RNA or DNA. Where a sophisticated liposome with a dual-layer lipid structure. The outer layer is fused with the cell membrane to deliver the cargo into the cytoplasm. The inner layer with the modified mitochondrial-fusogenic lipids and positive charges then fuses with the membrane of mitochondria by releasing the components into the matrix. RNP-MITO-porter, modified with STR-R8, a positively charged peptide found to be effective in cellular uptake and mitochondrial targeting. (Norota et al., 2025) Cas9–sgRNA complexes are efficiently encapsulated into ~70 nm particles using a microfluidic system, and the carrier enables robust mtDNA cleavage. In isolated mitochondria, it achieved up to 75% targeted mtDNA cutting, which was not possible with standard transfection reagents. In live cells, it produced rapid cleavage (~20% within 1.5 hours), and in disease-model fibroblasts, it selectively reduced mutant mtDNA levels by ~35%. Although fluorescence imaging confirms the carrier localisation, the specificity needs to be verified by conducting experiments using wild-type fibroblasts that lack the target mutation, to distinguish on-target cleavage from non-specific degradation or cytotoxicity. However, a key limitation in mitochondrial genome editing is the low efficiency of delivering editing tools into mitochondria. In the plasmid-based mito-Cas9 system, knock-in efficiency is extremely low (0.03–0.23%), likely due to poor mitochondrial transport. Similarly, the RNP-MITO-Porter system achieves only ~0.1% delivery efficiency. Across the field, improving mitochondrial and tissue-specific delivery remains a major challenge for advancing precise mitochondrial therapeutics toward clinical use. Even when efficient delivery is achieved, the transient activity and degradation of the RNP complex mean that the initial cleavage achieved (e.g., ≈20% at 1.5 hours in HeLa cells) decreases rapidly over time (to ≈5% at 6 hours), reflecting the degradation of the cleaved DNA. Thus, degradation after DSBs also poses a major hurdle. Even though TPP+ provide efficient drug delivery to mitochondria, considering the size and complexity of CRISPR components, TPP+ can't be used to deliver guide RNA into mitochondria.

Thirdly, viral vectors can be used, such as AAV9 (Adeno-Associated Virus serotype 9). According to Zhou and colleagues, a novel AAV9-based delivery system with cardiac/muscle tropism improved tissue-specific targeting, reducing off-target effects in liver/spleen by 78% compared to conventional AAVs ($p = 0.01$) (Zhou et al., 2024). As AAV is not naturally made to deliver genes into mitochondria but to the nucleus, a strategy of delivering AAV to the nucleus and consequently the cell produces the MTS-fused CRISPR proteins, which can be imported to the mitochondria. This is an indirect route which may be inefficient, because mitochondrial targeting with viral vectors is still in experimental stages.

Lastly, Base editors coupled with DBA-binding modules can be the most promising strategy. Owing to the fact that mtDNA lacks the conventional repair mechanisms, cutting mtDNA would lead to degradation and potentially cause catastrophic loss of function. In this mode, fusion proteins like base editors are used, which consist of a DNA-binding module and a deaminase enzyme. The module provides specificity, especially for alternative binders like TALEs and ZFNs. The enzyme is able to convert bases. For instance, Cytidine Deaminase converts C to A; an Adenine Deaminase converts A to C. During the process, TALE or ZFN would guide the whole component to the specific mutant site on the mtDNA, followed by chemical deamination without breaking the DNA strand. Since there are no DSBs, the mechanism reduces the degradation, and the correction of the specific mutation increases precision and efficiency.

CRISPR-INSPIRED TOOLS FOR MITOCHONDRIAL GENOME EDITING

CRISPR/Cas9 has revolutionised genome editing in nuclear DNA, and researchers have tried to extend it to mitochondria. However, the usage of CRISPR in mtDNA is in its early stages yet, and the double membrane of mitochondria acts as a barrier and mitochondria-targeted nucleases can target specific mtDNA and can knock down the mutated genes in heteroplasmic cells. Cytosine base editors (CBEs) and adenine base editors (ABEs) are designed to convert faulty nucleotides C → T and A → G, respectively by offering precise repairing of disease-causing point mutations. When coupling this technique with programmable DNA-binding modules. For instance, Cas9, TALE proteins, or Zinc-finger proteins can direct these editors to the exact site in the genome. Transcription Activator-Like Effectors (TALE) are coming from plant-infecting bacteria like *Xanthomonas*, which attach very specifically to the plant's DNA to turn certain genes on or off like a switch. The natural DNA recognition ability of TALEs can be reused in the lab to bind them to the gene of interest, like customizable DNA GPS systems. Additionally, these can guide enzymes such as Cas9 and deaminases to the exact place in the genome where the changes are needed. For instance, research identified a new bacterial toxin-derived cytidine deaminase from *Pseudomonas syringae*, called SsdAtox. Which they engineered into a new cytosine base editor (SsCBE) by fusing SsdAtox with nickase Cas9 (nCas9) and TALEs. This combination, unlike traditional base editors, makes single-base changes without cutting the DNA. This enzyme can precisely change one DNA letter, cytosine (C), into another, thymine (T), by first turning C into uracil (U), which the cell later reads as T. SsCBE works as well as older editors like BE4max but with fewer unintended changes in DNA or RNA. SsCBE can be delivered through multiple methods, including viral vectors, ribonucleoproteins (RNPs), and extracellular vesicles, and is capable of editing both nuclear and mitochondrial DNA. This discovery opens up new areas for mitochondrial gene editing, which was previously impossible. Interestingly, when paired with TALE proteins, SsCBE effectively edits mitochondrial DNA without causing off-target changes, expanding its potential applications. Overall, this work demonstrates that SsCBE is a compact, precise, and versatile tool for genome editing, offering promising possibilities for both research and therapeutic interventions, including correcting mitochondrial mutations linked to human diseases, which is a breakthrough in medicine (Kweon et al., 2025). TALE has been used in other organelles, such as chloroplasts, where a custom TALE DNA-binding array, split DddAtox (a bacterial cytidine deaminase), and TadA-8e (an adenine deaminase) allows targeting editing inside organelles. Another research conducted by a group of Chinese scientists demonstrates that targeting specific genes like rrn16, psbA, and psaA, high editing efficiencies (up to 99%) can be achieved, inducing traits such as herbicide resistance or albinism. Importantly, these edits were stably inherited in the next generation, and some edited plants were transgene-free, which is significant for crop breeding and regulatory purposes. Like chloroplasts, mitochondria have double membranes and internal cristae, making genome editing challenging. The TALED approach shows that combining organelle-targeted DNA-binding proteins with base editors with or without CRISPR can overcome these barriers, especially in heteroplasmic cells (Tang & Du, 2025).

CRISPR Interference (CRISPRi) can be used in mitochondrial gene editing studies to reversibly silence mitochondrial genes by "turning off" a mitochondrial regulator temporarily, which doesn't create any actual breaks in the DNA, thus safer than the traditional CRISPR-Cas 9 to study dose-dependent effects by reducing the gene expression partially to target the challenges of gRNA import challenges, and off-target effects. For instance, mitochondrial dysfunctional diseases such as dominant optic atrophy can be treated with OPA1 gene expression can be repressed using a doxycycline-inducible dCas0-KRAB system via CRISPRi-mediated inhibition of DRP1. Meanwhile, OPA1 knockdown results in mitochondrial fragmentation, reduced OXPHOS, and impaired cell viability. This methodology of mimicking and studying the pathogenic mitochondrial states offers more promising results. Moreover, nuclear-encoded mitochondrial genes are more tractable than mitochondrial genes themselves (Jie Cheng et al., 2023).

Researchers at the Whitehead Institute identified mitochondrial carrier homolog 2 (MTCH2), a protein located in the outer mitochondrial membrane, which facilitates the insertion of tail-anchored proteins that play a crucial role in communication, signalling and apoptosis. While standard CRISPR/Cas systems face major hurdles in mitochondria, careful MTS selection (like Su9) and transient expression of smaller nucleases like Cas12a show promise. mitoCRISPR remains experimental, but linking CRISPR with TALE-based base editors could open new possibilities for precise mtDNA manipulation. Together, these CRISPR-based and CRISPR-inspired strategies could overcome the current limitations by addressing the unmet needs of shifting heteroplasmy, safe and efficient delivery and non-invasive monitoring of therapeutic outcomes. According to recent studies, CRISPR-Cas9 editing in mitochondria faces barriers such as poor sgRNA import, rapid degradation of mtDNA after double-strand breaks, and low knock-in efficiency. Literature suggests several strategies to overcome these issues, including adding mitochondrial targeting sequences to Cas9, using longer homologous arms for ssODN delivery, activating RAD51 to boost HR, and improving mitochondrial RNA/DNA transport. These approaches remain experimental, but they provide possible pathways for enhancing mtDNA editing. Moreover, while both primary editing studies represent breakthroughs, future research must address detailed specificity validation in order to confirm the effective use of CRISPR/Cas 9 in mitochondrial gene editing.

ETHICAL AND SAFETY CONSIDERATIONS

Mitochondrial genome editing raises significant ethical and safety concerns, particularly because mtDNA is maternally inherited, meaning germline changes could affect future generations. Each technology has specific risks: nucleases like mito-TALENs or ZFNs can cause excessive mtDNA loss or cleave nuclear mitochondrial sequences, while base editors may produce unintended "bystander" edits or RNA alterations. Newer TALEDs have improved precision and reduced RNA off-targets by over 99%. Risk–benefit considerations vary: in severe mitochondrial diseases, shifting heteroplasmy may be life-saving, but in cancer therapy, editing could harm healthy tissues through excessive ROS or toxicity. Preclinical studies highlight the need for precise delivery, organ-specific targeting, and long-term safety evaluation. For CRISPR/Cas9, mitochondrial delivery remains challenging, so CRISPR-free approaches, such as protein-only nucleases or base editors, are currently favoured for clinical development.

FUTURE DIRECTIONS

Looking ahead, the field is rapidly advancing toward next-generation mitochondrial editors with narrower activity windows, higher strand selectivity, and safeguards against off-target effects, enabling more precise targeting of mutations. Delivery technologies, including nanoparticles and hybrid lipid–peptide systems like advanced MITO-Porters, are being refined to carry large proteins or RNP complexes safely, avoiding DNA integration and minimising immune responses. The therapeutic scope is expanding beyond rare inherited disorders; in oncology, editing mitochondrial genes could reveal vulnerabilities in OXPHOS-dependent cancer stem-like cells. Combining these approaches with metabolic inhibitors or immunotherapies may enhance efficacy. Long-term, integrating editing tools with diagnostics such as mtDNA heteroplasmy mapping and off-target tracking will ensure precise, personalised interventions. Clinical translation will require careful attention to heteroplasmy, off-target effects, delivery, and patient stratification based on mtDNA copy number, mutation burden, and stress-response capacity.

REFERENCES

Al Assi, A., Posty, S., Lamarche, F., Chebel, A., Guitton, J., Cottet-Rousselle, C., Prudent, R., Lafanechère, L., Giraud, S., Dallemagne, P., Suzanne, P., Verney, A., Genestier, L., Castets, M., Fontaine, E., Billaud, M., & Cordier-Bussat, M. (2024). A novel inhibitor of the mitochondrial respiratory complex I with uncoupling properties exerts potent antitumor activity. *Cell Death & Disease*, *15*(5), 311. Advance online publication. DOI: 10.1038/s41419-024-06668-9 PMID: 38697987

Antón, Z., Mullally, G., Ford, H. C., van der Kamp, M. W., Szczelkun, M. D., & Lane, J. D. (2020). Mitochondrial import, health and mtDNA copy number variability seen when using type II and type V CRISPR effectors. *Journal of Cell Science*, *133*(18), jcs248468. Advance online publication. DOI: 10.1242/jcs.248468 PMID: 32843580

Ball, M., van Bergen, N. J., Compton, A. G., Thorburn, D. R., Rahman, S., & Christodoulou, J. (2025). Therapies for Mitochondrial Disease: Past, Present, and Future. *Journal of Inherited Metabolic Disease*, *48*(4), e70065. DOI: 10.1002/jimd.70065 PMID: 40714961

Barbi, V., De Martino, S., Aiello, A., Gottardi Zamperla, M., Negri, S., Cis, L., Pecci, V., Nanni, S., Farsetti, A., Martelli, F., Gaetano, C., & Atlante, S. (2025). Non-coding RNAs as novel biomarkers and therapeutic targets in breast cancer. *Oncology Reviews*, *19*, 1621144. DOI: 10.3389/or.2025.1621144 PMID: 40951838

Bi, R., Li, Y., Xu, M., Zheng, Q., Zhang, D.-F., Li, X., Ma, G., Xiang, B., Zhu, X., Zhao, H., Huang, X., Zheng, P., & Yao, Y.-G. (2022). Direct evidence of CRISPR-Cas9-mediated mitochondrial genome editing. *Innovation (Cambridge (Mass.))*, *3*(6), 100329. DOI: 10.1016/j.xinn.2022.100329 PMID: 36275864

Carinci, M., Vezzani, B., Patergnani, S., Ludewig, P., Lessmann, K., Magnus, T., Casetta, I., Pugliatti, M., Pinton, P., & Giorgi, C. (2021). Different Roles of Mitochondria in Cell Death and Inflammation: Focusing on Mitochondrial Quality Control in Ischemic Stroke and Reperfusion. *Biomedicines*, *9*(2), 169. DOI: 10.3390/biomedicines9020169 PMID: 33572080

Changaei, M., Azimzadeh Tabrizi, Z., Karimi, M., Kashfi, S. A., Koochaki Chahardeh, T., Hashemi, S. M., & Soudi, S. (2025). From powerhouse to modulator: Regulating immune system responses through intracellular mitochondrial transfer. *Cell Communication and Signaling : CCS*, *23*(1), 232. DOI: 10.1186/s12964-025-02237-5 PMID: 40394666

Chen, X., Zhang, J., Gao, F., Liu, N., Du, H., Li, J., Li, Z., & Chen, R. (2025). Exercise therapy: An effective approach to mitigate the risk of cancer metastasis. *World Journal of Surgical Oncology*, *23*(1), 192. DOI: 10.1186/s12957-025-03846-7 PMID: 40380237

Cheng, Jie, Duan, Yukan, & Cynthia, A Berlinicke, & Donald J Zack. (2023). CRISPR interference knockdown of Dynamin-Related Protein 1 (DRP1) restores mitochondrial homeostasis disrupted by inhibition of Optic Atrophy 1 (OPA1) expression. *Investigative Ophthalmology & Visual Science*, •••, 64.

Du, H., Xu, T., Yu, S., Wu, S., & Zhang, J. (2025). Mitochondrial metabolism and cancer therapeutic innovation. In *Signal Transduction and Targeted Therapy* (Vol. 10, Issue 1). Springer Nature. https://doi.org/DOI: 10.1038/s41392-025-02311-x

Gadicherla, R., Rai, N., Othayoth, R., & Kamma, S. (2024). Characterization of mitochondrial DNA mutations in colorectal cancer progression by in silico approach and use as potential biomarkers for diagnosis and prognosis. *The Egyptian Journal of Medical Human Genetics*, *25*(1), 126. DOI: 10.1186/s43042-024-00599-y

Habbane, M., Montoya, J., Rhouda, T., Sbaoui, Y., Radallah, D., & Emperador, S. (2021). Human Mitochondrial DNA: Particularities and Diseases. *Biomedicines*, *9*(10), 1364. DOI: 10.3390/biomedicines9101364 PMID: 34680481

Ho, M. K., & Ho, J. W. K. (2025). Mitochondrial DNA common deletion in cancer: Emerging evidence and methodological challenges. *Computational and Structural Biotechnology Journal*, *27*, 3929–3940. DOI: 10.1016/j.csbj.2025.09.010

Khan, S. S., Rangraze, I. R., Wali, A. F., Jhancy, M., Attia, R. A., Elshamly, H. A. H., Adam, S., & Elbeshbeishy, R. A. M. (2025). Repurposing Metformin in Precision Oncology: Mechanistic Insights, Biomarker-Guided Strategies, and Translational Imperatives. *Medicina*, *61*(9), 1577. DOI: 10.3390/medicina61091577 PMID: 41010968

Kumar, R., Bargoti, T., Sengar, S., Singh, D., & Nain, V. (2025). A Decade of Genome Editing: Comparative Review of ZFN, TALEN, and CRISPR/Cas9. International Journal of Innovative Science and Research Technology, 3708–3717. https://doi.org/DOI: 10.38124/ijisrt/25apr2221

Kweon, J., Park, S., Jeon, M. Y., Lim, K., Jang, G., Jang, A.-H., Lee, M., Seok, C., Lee, C., Park, S., Ahn, J., Jang, J., Kim, N., Sung, Y. H., Kim, D., & Kim, Y. (2025). High-efficiency base editing for nuclear and mitochondrial DNA with an optimized DYW-like deaminase. *Molecular Therapy : the Journal of the American Society of Gene Therapy*, *33*(11), 5611–5623. Advance online publication. DOI: 10.1016/j.ymthe.2025.08.007 PMID: 40781772

Lee, E., Ozigbo, A. A., Varon, J., Halma, M., Laezzo, M., Ang, S. P., & Iglesias, J. (2025). Mitochondrial Reactive Oxygen Species: A Unifying Mechanism in Long COVID and Spike Protein-Associated Injury: A Narrative Review. *Biomolecules*, *15*(9), 1339. DOI: 10.3390/biom15091339 PMID: 41008646

Lee, H., Kim, B., Park, J., Park, S., Yoo, G., Yum, S., Kang, W., Lee, J.-M., Youn, H., & Youn, B. (2025). Cancer stem cells: Landscape, challenges and emerging therapeutic innovations. *Signal Transduction and Targeted Therapy*, *10*(1), 248. DOI: 10.1038/s41392-025-02360-2 PMID: 40759634

Lee, W., Zamudio-Ochoa, A., Buchel, G., Podlesniy, P., Gutierrez, N. M., Puigròs, M., Calderon, A., Tang, H.-Y., Li, L., Mikhalchenko, A., Koski, A., Trullas, R., Mitalipov, S., & Temiakov, D. (2023). Molecular Basis for Maternal Inheritance of Human Mitochondrial DNA. *Nature Genetics*, *55*(10), 1632–1639. DOI: 10.1038/s41588-023-01505-9 PMID: 37723262

Lu, Q., Chen, R., Du, S., Chen, C., Pan, Y., Luan, X., Yang, J., Zeng, F., He, B., Han, X., & Song, Y. (2022). Activation of the cGAS-STING pathway combined with CRISPR-Cas9 gene editing triggering long-term immunotherapy. *Biomaterials*, *291*, 121871. DOI: 10.1016/j.biomaterials.2022.121871 PMID: 36323073

Lv, Y., Shu, F., Luo, D., Jia, R., Huang, Y., & Xu, C. (2025). Targeting tumor angiogenesis and metabolism: A new perspective in pediatric thoracic tumor therapy. *Frontiers in Cell and Developmental Biology*, *13*, 1558403. DOI: 10.3389/fcell.2025.1558403 PMID: 40213391

Nikitchina, N., Heckel, A.-M., Shebanov, N., Mazunin, I., Tarassov, I., & Entelis, N. (2025). Targeted deletions in human mitochondrial DNA engineered by Type V CRISPR–Cas12a system. *NAR Molecular Medicine*, *2*(2), ugaf021. Advance online publication. DOI: 10.1093/narmme/ugaf021 PMID: 41255706

Nissanka, N., & Moraes, C. T. (2020). Mitochondrial DNA heteroplasmy in disease and targeted nuclease-based therapeutic approaches. *EMBO Reports*, *21*(3), e49612. DOI: 10.15252/embr.201949612 PMID: 32073748

Norota, K., Ishizuka, S., Hirose, M., Sato, Y., Maeki, M., Tokeshi, M., Ibrahim, S. M., Harashima, H., & Yamada, Y. (2025). Lipid nanoparticle delivery of the CRISPR/Cas9 system directly into the mitochondria of cells carrying m.7778G>T mutation in MtDNA (mt-Atp8). *Scientific Reports*, *15*(1), 18717. Advance online publication. DOI: 10.1038/s41598-025-03671-8 PMID: 40537532

Pérez-Amado, C. J., Bazan-Cordoba, A., Hidalgo-Miranda, A., & Jiménez-Morales, S. (2021). Mitochondrial Heteroplasmy Shifting as a Potential Biomarker of Cancer Progression. *International Journal of Molecular Sciences*, *22*(14), 7369. DOI: 10.3390/ijms22147369 PMID: 34298989

Richhariya, S., Shin, D., Schlichting, M., & Rosbash, M. (2025). Metabolic rewiring prevents neurodegeneration caused by chronic mitochondrial dysfunction. *Current Biology*, *35*(22), 5443–5459.e5. DOI: 10.1016/j.cub.2025.09.063 PMID: 41151583

Riley, J. S., & Tait, S. W. (2020). Mitochondrial DNA in inflammation and immunity. *EMBO Reports*, *21*(4), e49799. DOI: 10.15252/embr.201949799 PMID: 32202065

Sessions, D. T., Kim, K. B., Kashatus, J. A., Churchill, N., Park, K. S., Mayo, M. W., Sesaki, H., & Kashatus, D. F. (2022). Opa1 and Drp1 reciprocally regulate cristae morphology, ETC function, and NAD+ regeneration in KRas-mutant lung adenocarcinoma. *Cell Reports*, *41*(11), 111818. Advance online publication. DOI: 10.1016/j.celrep.2022.111818 PMID: 36516772

Shelton, S. D., House, S., Martins Nascentes Melo, L., Ramesh, V., Chen, Z., Wei, T., Wang, X., Llamas, C. B., Venigalla, S. S. K., Menezes, C. J., Allies, G., Krystkiewicz, J., Rösler, J., Meckelmann, S. W., Zhao, P., Rambow, F., Schadendorf, D., Zhao, Z., Gill, J. G., & Mishra, P. (2024). Pathogenic mitochondrial DNA mutations inhibit melanoma metastasis. *Science Advances*, *10*(44), eadk8801. Advance online publication. DOI: 10.1126/sciadv.adk8801 PMID: 39485847

Sivagnanam, S., Das, K., Pan, I., Stewart, A., Barik, A., Maity, B., & Das, P. (2024). Engineered triphenylphosphonium-based, mitochondrial-targeted liposomal drug delivery system facilitates cancer cell killing actions of chemotherapeutics. *RSC Chemical Biology*, *5*(3), 236–248. DOI: 10.1039/D3CB00219E PMID: 38456034

Tang, J., & Du, K. (2025). Mitochondrial base editing: From principle, optimization to application. *Cell & Bioscience*, *15*(1), 9. DOI: 10.1186/s13578-025-01351-8 PMID: 39856740

Tang, J., Zhang, J., Yang, R., Chen, H., Yu, X., Peng, W., & Zeng, P. (2025). The causal relationships between mitochondria and six types of cancer: A Mendelian randomization study. *BMC Cancer*, *25*(1), 794. DOI: 10.1186/s12885-025-14201-0 PMID: 40295943

Taylor, A. D., Hathaway, Q. A., Kunovac, A., Pinti, M. V., Newman, M. S., Cook, C. C., Cramer, E. R., Starcovic, S. A., Winters, M. T., Westemeier-Rice, E. S., Fink, G. K., Durr, A. J., Rizwan, S., Shepherd, D. L., Robart, A. R., Martinez, I., & Hollander, J. M. (2024). Mitochondrial sequencing identifies long noncoding RNA features that promote binding to PNPase. *American Journal of Physiology. Cell Physiology, 327*(2), C221–C236. DOI: 10.1152/ajpcell.00648.2023 PMID: 38826135

Tufail, M., Jiang, C.-H., & Li, N. (2024). Altered metabolism in cancer: Insights into energy pathways and therapeutic targets. *Molecular Cancer, 23*(1), 203. DOI: 10.1186/s12943-024-02119-3 PMID: 39294640

Tufail, M., Jiang, C.-H., & Li, N. (2025). Wnt signaling in cancer: From biomarkers to targeted therapies and clinical translation. *Molecular Cancer, 24*(1), 107. DOI: 10.1186/s12943-025-02306-w PMID: 40170063

Wallace, D. C. (2012). Mitochondria and cancer. *Nature Reviews. Cancer, 12*(10), 685–698. DOI: 10.1038/nrc3365 PMID: 23001348

Welch, D. R., Foster, C., & Rigoutsos, I. (2022). Roles of mitochondrial genetics in cancer metastasis. *Trends in Cancer, 8*(12), 1002–1018. DOI: 10.1016/j.trecan.2022.07.004 PMID: 35915015

Wen, H., Deng, H., Li, B., Chen, J., Zhu, J., Zhang, X., Yoshida, S., & Zhou, Y. (2025). Mitochondrial diseases: From molecular mechanisms to therapeutic advances. *Signal Transduction and Targeted Therapy, 10*(1), 9. DOI: 10.1038/s41392-024-02044-3 PMID: 39788934

Wu, Y., Chen, Z., Chen, X., Hu, Y., & Ma, Y. (2025). Recent advances in the investigation of the regulatory network underlying reactive nitrogen species-mediated tumorigenesis: Molecular mechanisms and targeted therapeutic strategies. *Redox Report : Communications in Free Radical Research, 30*(1), 2564593. DOI: 10.1080/13510002.2025.2564593 PMID: 41036604

Yin, T., Luo, J., Huang, D., & Li, H. (2022). Current Progress of Mitochondrial Genome Editing by CRISPR. *Frontiers in Physiology, 13*, 883459. Advance online publication. DOI: 10.3389/fphys.2022.883459 PMID: 35586709

Zaidi, A. A., Wilton, P. R., Su, M. S.-W., Paul, I. M., Arbeithuber, B., Anthony, K., Nekrutenko, A., Nielsen, R., & Makova, K. D. (2019). Bottleneck and selection in the germline and maternal age influence transmission of mitochondrial DNA in human pedigrees. *Proceedings of the National Academy of Sciences of the United States of America, 116*(50), 25172–25178. DOI: 10.1073/pnas.1906331116 PMID: 31757848

Zhou, Q., Cao, T., Li, F., Zhang, M., Li, X., Zhao, H., & Zhou, Y. (2024). Mitochondria: A new intervention target for tumor invasion and metastasis. *Molecular Medicine (Cambridge, Mass.)*, *30*(1), 129. DOI: 10.1186/s10020-024-00899-4 PMID: 39179991

Chapter 11
Nano-Delivery for CRISPR-Cas9 Technology in Precision Medicine

Suresh Waghamare
https://orcid.org/0009-0001-8909-2209
NIMS Institute of Pharmacy, NIMS University, Jaipur, India

Deepak Uttamrao Bhusare
Digambarrao Bindu ACS College, India

Annasaheb Kudhekar
Nandkumar Shinde College of Pharmacy, India

Jahed Sayyad
NIMS Institute of Pharmacy, NIMS University, Jaipur, India

Amol Darwade
S.N.D. College of Pharmacy, India

Poonam Kodag
Rashtriya College of Pharmacy, India

Pooja Paulbudhe
Oyster Institute of Pharmacy, India

Vikas Ashok Deshmane
Oyster Institute of Pharmacy, India

ABSTRACT

The integration of CRISPR/Cas9 genome editing and nanotechnology is transforming precision medicine through the targeted, efficient, and safe delivery of gene-editing agents. With its great promise in the treatment of genetic disorders, the clinical application of CRISPR/Cas9 is hampered by delivery issues such as degradation, immunogenicity, and off-target effects. This chapter discusses the evolution and use of nano-delivery systems like lipid nanoparticles, polymeric carriers, dendrimers, gold nanoparticles, and exosomes to deliver CRISPR/Cas9 elements across biological barriers in a highly specific manner. Focus is given to nanoformulation strategies that maximize cellular uptake, endosomal escape, and

DOI: 10.4018/979-8-3373-4857-5.ch011

nuclear localization with minimal systemic toxicity. Case studies of preclinical and clinical progress in nano-delivered CRISPR therapies for cancer, genetic diseases, and viral diseases are also presented. The chapter is wrapped up with perspectives on regulatory affairs and future directions on combining nanocarriers with genome editing for personalized medicine.

1. INTRODUCTION

1.1. Overview of CRISPR/Cas9 and its significance in genome editing.

CRISPR/Cas9 (Clustered Regularly Interspaced Short Palindromic Repeats/CRISPR-associated protein 9) is an innovative genome editing tool. Originally identified as part of the immune system of bacteria and archaea, it is now widely used by scientists to detect and eliminate viral infections by capturing viral DNA fragments and integrating them into their genomes. This natural defense has been transformed into a highly versatile and programmable gene editing system capable of precise modifications across almost all organisms (T. Li et al., 2023). The CRISPR/Cas9 system functions through a straightforward yet effective mechanism. A synthetic single-guide RNA (sgRNA), designed to match a specific DNA sequence, guides the Cas9 nuclease to the target location within the genome. Upon arrival, Cas9 introduces a double-stranded break (DSB) in the DNA. The cell then activates its repair processes, primarily utilizing two pathways: non-homologous end joining (NHEJ), which often results in insertions or deletions (indels) that can interfere with gene function; and homology-directed repair (HDR), allowing accurate insertion or correction of DNA sequences with the help of a homologous template (Redman et al., 2016).

CRISPR/Cas9 is notable for its accuracy, simplicity, and adaptability. Unlike earlier genome-editing tools like zinc finger nucleases (ZFNs) and transcription activator-like effector nucleases (TALENs), which require complex protein engineering for each target, CRISPR/Cas9 uses a flexible RNA molecule, making it much easier and cheaper to use. This accessibility has democratized gene editing technology and sped up progress in various scientific fields (Redman et al., 2016). CRISPR/Cas9 has a broad and influential range of applications. In biomedical research, it is frequently employed to generate gene knockouts, study disease mechanisms, and create animal models of human conditions. Clinically, CRISPR-based treatments are under development for several genetic disorders, such as sickle cell anemia, β-thalassemia, and hereditary blindness. The agriculture industry also leverages CRISPR to engineer crops with enhanced resistance to pests, greater environmental resilience, and better nutritional content. Furthermore, in biotechnology, CRISPR is used to improve microbes for the production of biofuels, enzymes, and medicines (Ansori et al., 2023). Although CRISPR/Cas9 offers great promise, it also presents various ethical, safety, and regulatory challenges. Concerns like off-target effects, spontaneous mutations, and germline editing implications continue to be studied and debated. Regulatory bodies worldwide are actively creating guidelines to ensure responsible use, especially in clinical and agricultural settings (Chehelgerdi et al., 2024).

The integration of nanotechnology with CRISPR/Cas9 technology has revolutionized precision medicine by addressing key limitations in gene editing delivery, such as poor cellular uptake, nuclease degradation, and off-target effects. Traditional viral vectors, while effective, often provoke immune responses and have limited cargo capacity, prompting the shift toward non-viral nano-delivery systems. These nanomaterials encapsulate CRISPR components—ranging from plasmid DNA and mRNA to ribonucleoprotein (RNP) complexes enabling targeted, efficient, and safer genome modifications. Recent reviews highlight how nanosystems leverage physicochemical properties such as size, charge, and surface functionalization to enhance endosomal escape, nuclear localization, and tissue-specific homing, paving the way for therapeutic applications in genetic disorders, cancer, and infectious diseases.(Duan, Ouyang, Xu, et al., 2021; T. Li et al., 2023a; Rauf et al., 2025) Diverse classes of nanomaterials have emerged as frontrunners in CRISPR/Cas9 delivery, each offering unique advantages tailored to precision medicine needs. Lipid nanoparticles (LNPs), such as ionizable LNPs (iLNPs) and selective organ-targeting (SORT) variants, excel in systemic delivery due to their biocompatibility and ability to achieve over 80% editing efficiency in vivo, particularly in the liver for conditions like transthyretin amyloidosis (NTLA-2001 trials). Polymer-based carriers, including polyethyleneimine (PEI), chitosan, and poly(β-amino esters) (PBAE), facilitate proton sponge-mediated endosomal escape and pH-responsive release, as seen in multistage delivery nanoparticles (MDNPs) for tumor targeting. Inorganic options like gold nanoparticles (AuNPs) and metal-organic frameworks (MOFs) provide stability and multifunctionality, such as photothermal synergy for cancer therapy, while extracellular vesicles (EVs) like exosomes mimic natural cellular communication to evade immune clearance and deliver RNPs with minimal immunogenicity. These systems not only protect CRISPR cargos from degradation but also enable stimuli-responsive behaviors—e.g., ROS or hypoxia-triggered disassembly—enhancing precision in diseased microenvironments.(T. Li et al., 2023a; Rauf et al., 2025)

In precision medicine applications, nano-delivered CRISPR/Cas9 has shown transformative potential across therapeutic domains. In oncology, LNPs and AuNP hybrids have disrupted oncogenes like PLK1 in hepatoma models, suppressing tumor growth by up to 77%, while exosome-mediated editing targets KRAS mutations in pancreatic cancer for immunotherapy enhancement. For genetic diseases, CRISPR-Gold nanoparticles reversed fragile X syndrome phenotypes in mouse brains via mGluR5 knockout, and EV-based systems corrected dystrophin in muscular dystrophy models with over 90% efficiency in induced pluripotent stem cells. Cardiovascular and metabolic applications include LNP knockdown of Fbn1 for aortic disorders and exosome activation of HNF4α to attenuate liver fibrosis, demonstrating broad translational promise evidenced by ongoing trials like CTX110 for hematological malignancies.(Duan, Ouyang, Xu, et al., 2021)

Despite these advances, challenges persist in applying nano-CRISPR systems to clinical precision medicine. Key hurdles include ensuring consistent particle size for large-scale production, overcoming endosomal entrapment and nuclear barriers in non-dividing cells, and minimizing residual immunogenicity or off-target effects, especially with larger Cas9 payloads. Systemic delivery typically targets the liver, requiring advanced targeting ligands like iRGD peptides or multi-responsive systems (e.g., pH/ROS dual-locking) for extrahepatic areas like the brain or tumors. Biosafety concerns, such as long-term clearance and potential toxicity of inorganic components, emphasize the need for biodegradable hybrid materials. Future research will likely focus on high-fidelity Cas variants, AI-optimized formulations, and combination therapies—such as CRISPR paired with photothermal ablation—to enhance efficacy and decrease risks. These developments will solidify nano-delivery as a crucial part of personalized genomics by 2030.(T. Li et al., 2023a)

1.2. The Role of Precision Medicine in the Era of Personalized Therapies.

Precision medicine is a cutting-edge healthcare strategy that focuses on customizing medical treatments based on a patient's genetics, lifestyle, and environment. Unlike the conventional "one-size-fits-all" approach, it aims to provide more precise diagnoses, targeted therapies, and better treatment results by considering each person's unique biological features. This approach is especially important in the age of targeted therapies, where the goal is to tailor medical interventions to individual needs and characteristics, ultimately improving therapeutic effectiveness and reducing adverse side effects. (Marques et al., 2024)

Precision medicine integrates genomics, proteomics, metabolomics, and advanced bioinformatics to identify the molecular roots of diseases. By detecting specific genetic mutations, biomarkers, or molecular signatures associated with a condition, clinicians can select targeted therapies more likely to benefit individual patients. For instance, in cancer treatment, this approach has resulted in therapies that focus on oncogenic mutations like EGFR, HER2, and BRAF, thereby enhancing prognosis and survival chances. In pharmacogenomics, understanding genetic variations in drug-metabolizing enzymes allows for personalized drug selection and dosing, which can decrease toxicity and improve treatment outcomes. (Molla & Bitew, 2024)

Advances in high-throughput sequencing technologies and data-driven clinical decision support have quickly sped up the adoption of precision medicine in specialty clinics. For rare gene disorders, whole-genome and exome sequencing are becoming crucial for accurate diagnosis and treatment planning. In neurodegenerative and autoimmune diseases, it has helped identify subgroups that respond better to specific biologics or immunotherapies through molecular-level patient stratification. Precision medicine has also advanced personalized immunotherapies like CAR-T cell therapy, which involves genetically engineering a patient's own immune cells to target cancer. (Saparov & Zech, 2025)

1.3. Efficient and Safe Delivery Systems

The effectiveness of existing therapeutics, especially in precision medicine and gene editing, heavily relies on the development of efficient and safe delivery systems. As therapeutic molecules become more complex—ranging from small molecules and proteins to nucleic acids and CRISPR reagents the challenge is not only their design but also their targeted and controlled delivery to specific tissues or cells. An effective delivery system must protect therapeutic agents from degradation, assist in transport across biological barriers, enhance their bioavailability, and reduce systemic toxicity and off-target effects. (T. Li et al., 2023b)

Conventional drug delivery methods are often plagued by poor solubility, high clearance, non-specific distribution, and limited cellular uptake. All these drawbacks can reduce therapeutic efficacy as well as increase the likelihood of side effects. For example, chemotherapy includes non-selective drug delivery of small molecule cytotoxic agents that can produce excessive toxicity to normal tissues and result in severe adverse effects. Gene therapy and RNA drugs also include overwhelming obstacles in the sense that nucleic acids are unstable, even in in vivo environments, and delivery of these through cell membranes is difficult without activating the immune system. (Elayaperumal et al., 2025)

To overcome these deficiencies, various novel drug delivery systems have been engineered. Nanocarriers such as liposomes, polymeric nanoparticles, dendrimers, micelles, and lipid nanoparticles have also been studied intensively over the past few decades due to their ability to deliver therapeutic agents, protect them from enzymatic degradation, and deliver them selectively to the target site for the disease. These nanocarriers can be engineered to deliver controlled release, ligand-initiated activation, and active targeting via cell-surface receptor-specific ligands. For instance, lipid nanoparticles have revolutionized mRNA vaccine delivery, as evidenced during the COVID-19 pandemic, establishing proof of concept for future nucleic acid treatments. (Sur et al., 2019)

Besides, the safety profile of delivery systems is a critical area, especially in clinical translation. Materials for delivery systems must be non-immunogenic, biodegradable, and biocompatible. Long-term biodistribution, metabolism, and material toxicity must be carefully assessed to ensure patient safety. There is a growing regulatory requirement to provide detailed preclinical and clinical data to assess the pharmacokinetics, immunogenicity, and toxicity of novel delivery systems. (Geszke-Moritz & Moritz, 2024)

1.4. Scope of the chapter.

This chapter offers a detailed overview of the link between CRISPR/Cas9 gene-editing technology and the emerging field of precision medicine, highlighting the critical role of next-generation drug delivery systems. It reviews the history and mechanism of CRISPR/Cas9, its importance in targeted genome editing, and its therapeutic uses. The chapter explains how precision medicine is revolutionizing healthcare through personalized treatments tailored to individual genetic and molecular profiles, enhancing effectiveness and minimizing side effects. (Mengstie & Wondimu, 2021) It also underscores the urgent need for efficient, safe delivery methods to translate these innovations from research to clinical practice. Special attention is given to the design and use of nanocarriers, lipid nanoparticles, and other biomaterial-based platforms that enable targeted delivery of CRISPR components and gene therapies. Issues such as stability, targeting accuracy, immunogenicity, and regulatory factors are covered. Overall, the chapter provides a foundational framework for researchers, clinicians, and biotechnologists to understand the integration of genome editing, personalized medicine, and drug delivery technologies. It sets the stage for future sections on design strategies, recent advancements, clinical applications, safety evaluations, and the outlook for CRISPR-based therapies in personalized medicine.

2. CHALLENGES IN CRISPR/CAS9 DELIVERY

2.1. Stability and degradation of CRISPR components

The therapeutic utility of CRISPR/Cas9 systems largely rests on the stability and degradation kinetic behavior of their constituent parts, particularly the Cas9 protein and the single-guide RNA (sgRNA). The parts must remain structurally stable and functionally sound for a considerable amount of time to be transported to the target site, bind with the genome, and induce the desired genetic edit. Under physiological conditions, they are susceptible to degradation by endogenous enzymes, which can negatively influence their editing efficiency and specificity. (Aljabali et al., 2024)

The Streptococcus pyogenes Cas9 protein is a large and immunogenic protein, which will be cleared or neutralized by the host immune system within a few hours if it is not given with protection. Intravascular proteolytic enzymes or the intracellular compartments may destroy Cas9 before it enters the nucleus, reducing its gene-editing function. Moreover, immune recognition of bacterial Cas proteins raises concerns about immunotoxicity and repeated dosing during treatment. (Mengstie & Wondimu, 2021)

Another significant challenge is stability of the sgRNA. As a short, single-stranded RNA molecule, the sgRNA is inherently unstable and very susceptible to degradation by nucleases such as RNases present in extracellular fluids and intracellularly. Abrupt degradation of the sgRNA can result in reduced availability of guide RNA to direct Cas9 to the point of action, resulting in reduced editing efficacy and increased risk of off-target effects due to transient or partial targeting.(Wang & Han, 2024)

To these issues, various strategies have been devised in an effort to enhance the bioavailability and stability of CRISPR elements. Modifications of sgRNAs, such as 2'-O-methylation, phosphorothioate bonds, and locked nucleic acids (LNAs), have been shown to enhance nuclease resistance and prolong functional lifespan. Similarly, Cas9 protein can be stabilized by engineering with stabilizing domains, administered as a part of ribonucleoprotein (RNP) complexes, or packaged within protective delivery vehicles such as lipid nanoparticles (LNPs), polymeric carriers, or exosomes. These systems not only shield CRISPR components from enzymatic degradation but also facilitate cellular uptake and nuclear delivery.(Wang & Han, 2024)

Understandings of the degradation mechanisms and stability profiles of CRISPR components will be key to the design of effective delivery strategies and optimal therapeutic impact. Preclinical studies will need to address not only the editing efficiency but also the pharmacokinetics, biodistribution, and degradation kinetics of CRISPR tools in vivo. These results will be the key to the safe and reproducible application of genome-editing tools in the clinic. (Du et al., 2023) Briefly speaking, the stability of CRISPR components is a pillar to the success of gene-editing therapeutics. Challenging degradation through chemical modification, protein engineering, and enhanced delivery systems can significantly enhance the precision, safety, and consistency of CRISPR-based treatments.

2.2. Off-target effects and immune responses.

Despite the revolutionary potential of CRISPR/Cas9 technology in genome editing, there are serious challenges related to off-target effects and immune response, which are key hurdles for its clinical translation. Off-target effects refer to unintended cutting of DNA at genomic sites beyond the target site sequence, which could lead to unwanted mutations, genomic instability, or disruption of essential genes. These indeterminate alterations result from partial complementarity between the single-guide RNA (sgRNA) and DNA strands with partial homology, leading to insertion-deletion mutations (indels) or chromosomal rearrangements that may produce deleterious effects. (Kolanu, 2024)

Minimization of off-target effects is necessary to render CRISPR-based therapies safe and useful. Multiple strategies have been designed to increase targeting specificity, for example, generating high-fidelity Cas9 enzymes with reduced levels of nonspecific binding, using truncated or chemically altered sgRNAs, and introducing paired nickases that cause single-strand breaks requiring two proximal target sites for double-stranded cleavage. In addition, computational tools with sophisticated bioinformatics and genome-wide off-target screening platforms, such as GUIDE-seq and Digenome-seq, enable the identification and quantification of off-target sites, thereby guiding more precise editing protocols. (Asmamaw Mengstie et al., 2024)

Another significant hurdle is the immune response elicited by CRISPR components. The Cas9 protein, produced of bacteria such as Streptococcus pyogenes or Staphylococcus aureus, is not native to the human immune system. This recognition can provoke both innate and adaptive immune responses, including the formation of neutralizing antibodies and the activation of cytotoxic T cells. This immunity not only reduces gene-editing efficiency by prematurely eliminating CRISPR components but can also provoke inflammation or harmful immunological consequences, limiting repeated administration and chronic treatment. (Mehta & Merkel, 2020)

Immunomodulation strategies entail the use of Cas9 orthologs from less commonly employed bacterial species with lower immunogenicity, short-term delivery methods such as ribonucleoprotein (RNP) complexes minimizing exposure duration, and immunosuppressive protocols during treatment. Packaging CRISPR reagents within biocompatible stealthy delivery systems such as lipid nanoparticles or polymeric carriers can further shield them from immune detection. Active research tries to design hypoimmunogenic Cas9 variants by protein engineering and gain clearer insight into immune interactions to improve the safety profile of CRISPR medicines.(S. Zhang et al., 2021)

2.3. Obstacles to intracellular and nuclear targeting.

Successful genome editing using CRISPR/Cas9 requires delivery of its components not only into target cells but also into the nucleus, where genomic DNA is located. Despite this, several biological barriers prevent efficient intracellular and nuclear delivery of CRISPR/Cas9 agents, thereby affecting their therapeutic efficacy. The cellular membrane is the first obligatory barrier, a selectively permeable lipid bilayer that prevents the passage of large biomolecules, such as the Cas9 protein and sgRNA. Passive diffusion is not possible due to the size and polarity of such macromolecules, and delivery must occur via active transport systems or endocytic routes.(Seijas et al., 2025)

Once within the cell, CRISPR reagents must escape the endosomal-lysosomal route. After endocytosis, drug molecules are trapped within endosomes and then trafficked to lysosomes, where hydrolytic enzymes and an acidic pH degrade them, thereby diminishing their bioavailability. Successful endosomal escape is therefore crucial to deliver CRISPR reagents into the cytoplasm before destruction. Several methods, including the utilization of pH-sensitive carriers, fusogenic lipids, and membrane-disruptive peptides, have been developed to enable endosomal escape. (Bareford & Swaan, 2007)

The final and likely most challenging barrier is the nuclear envelope, a double lipid bilayer membrane that regulates macromolecular traffic between the nucleus and cytoplasm. CRISPR machinery must diffuse through nuclear pore complexes (NPCs) to access the genome and induce edits. Small molecules (<40 kDa) may diffuse passively through NPCs, whereas larger proteins, such as Cas9 (~160 kDa), require active transport mechanisms. Nuclear localization signals (NLS) designed into the Cas9 protein or delivery vehicles can facilitate importin recognition and ensuing nuclear translocation. Yet, nuclear uptake efficiency may be cell type- and physiology-dependent, influencing editing efficacy.(Van Der Zanden et al., 2023)

Other intracellular delivery-restrictive factors include cytosolic nucleases that degrade unprotected nucleic acids, intracellular trafficking barriers, and immune recognition risk within the cellular environment. Such multilayered barriers underscore the need for advanced delivery systems that can successfully traverse cellular compartments and achieve functional gene editing. Several strategies nanocarriers, viral vectors, and cell-penetrating peptides have been explored to bypass these challenges and enhance both cytoplasmic and nuclear delivery of CRISPR components (see Figure 01 below for an overview of cell delivery barriers to CRISPR).(Zhou et al., 2004)

Figure 1. Barriers to CRISPR Delivery in Cells

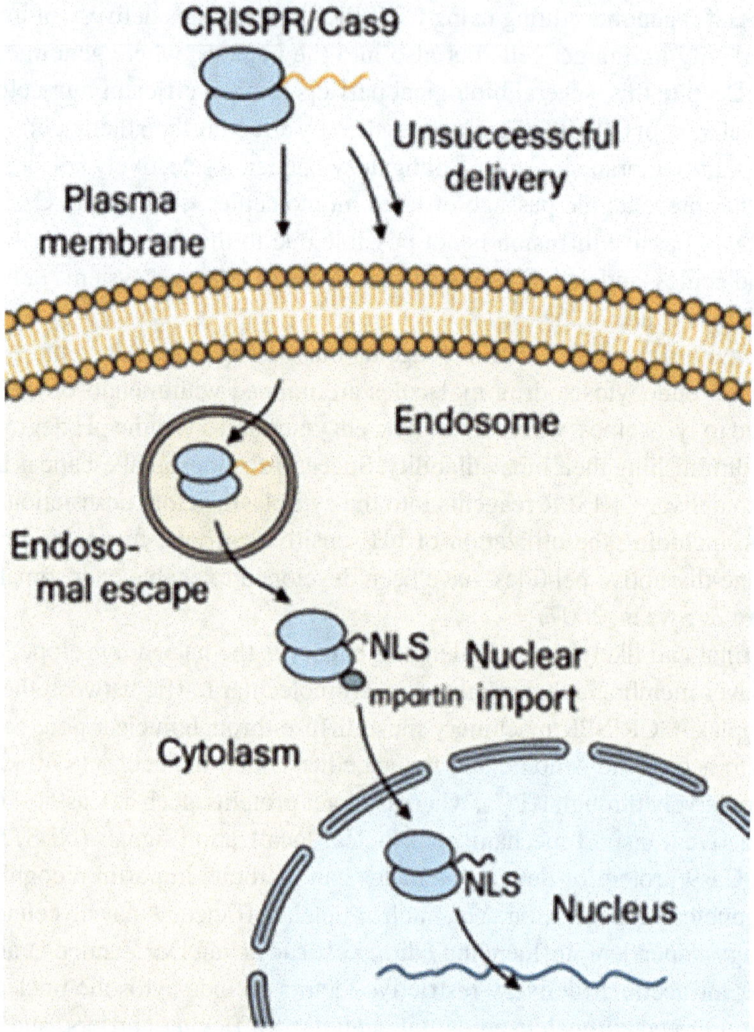

3. NANOTECHNOLOGY-ENABLED DELIVERY PLATFORMS

3.1 Lipid-Based Nanocarriers

3.1.1. Lipid nanoparticles (LNPs) for mRNA and RNP delivery

Lipid nanoparticles (LNPs) have emerged as the most effective and promising delivery systems for nucleic acid-based therapies, such as messenger RNA (mRNA) and CRISPR-associated ribonucleoprotein (RNP) complexes. Their ability to encapsulate, stabilize, and deliver fragile biomolecules across biological barriers has revolutionized the field of gene therapy and genome editing. LNPs typically consist of ionizable lipids, phospholipids, cholesterol, and PEG-lipids that together form a stable and biocompatible vesicular structure capable of delivering nucleic acids into target cells.(Jung et al., 2022)

For mRNA transduction of mRNA, LNPs provide a shielded environment that protects mRNA molecules from nuclease-mediated extracellular degradation, enhances cellular uptake through endocytosis, and facilitates endosomal escape, allowing the mRNA to reach the cytoplasm where it can be translated into therapeutic proteins. The effectiveness of LNPs in delivering mRNA was demonstrated globally throughout the COVID-19 pandemic, as LNP-delivered mRNA vaccines exhibited exceptional safety, immunogenicity, and the ability to be produced on a rapid scale. This has sparked interest in using LNPs to deliver a broad range of mRNA therapies, including protein replacement therapy, cancer immunotherapy, and infectious disease vaccines.(Wei et al., 2024)

Similarly, LNPs were engineered to deliver CRISPR/Cas9 ribonucleoprotein complexes, namely, Cas9 pre-complexed with a guide RNA. Delivery of CRISPR as an RNP has several advantages, such as transient activity that reduces off-target effects and immune stimulation. Entrapment of RNPs into LNPs protects the macromolecules from degradation and enhances their intracellular delivery and nuclear targeting. The ionizable lipids in LNPs play critical roles by having a positive charge at acidic pH of the endosomes, disrupting the endosomal membrane, and enabling release of the RNPs into the cytoplasm and subsequently into the nucleus where genome editing occurs.(S. Zhang et al., 2021)

The structure of LNPs can be tailored to precisely regulate particle size, surface charge, and lipid composition, thereby influencing biodistribution, cellular uptake, and immune compatibility. PEGylation of the LNP surface optimizes circulation times by reducing opsonization and reticuloendothelial system clearance. Targeting ligands can be incorporated to facilitate tissue-specific delivery, additionally improving therapeutic efficacy and safety.(Mendonça et al., 2023)

3.1.2. Clinical advances

The translational use of CRISPR/Cas9 gene-editing technology in the clinic has accelerated significantly over the last few years, with multiple ongoing and completed clinical trials targeting various genetic diseases, cancers, and viral disorders. These kinds of trials show both the therapeutic potential and challenge of CRISPR-based treatments in human patients (T. Li et al., 2023b)

One of the most important early successes has been in the treatment of β-thalassemia and sickle cell disease (SCD). They are both monogenic blood disorders caused by mutations in the β-globin gene. Clinical trials such as those conducted by CRISPR Therapeutics and Vertex Pharmaceuticals (e.g., the CTX001 trial) have used ex vivo CRISPR editing of hematopoietic stem cells (HSCs) to reactivate fetal hemoglobin production by disrupting the enhancer of the BCL11A gene. Edited patients have shown impressive increases in fetal hemoglobin, mitigating disease symptoms and reducing transfusion burdens, with long-term engraftment of the edited cells and acceptable safety profiles. These experiments are a breakthrough in gene editing therapies, moving towards possible cures for these genetic disorders. (Almotiri et al., 2025)

In inherited blindness, Editas Medicine's EDIT-101 therapy is at the forefront of the first in vivo CRISPR treatment for Leber congenital amaurosis type 10 (LCA10), a retinal dystrophy caused by a CEP290 gene mutation. The therapy infuses CRISPR elements using an adeno-associated virus (AAV) vector directly into the eye to correct the genetic defect. Preliminary clinical trials have yielded encouraging safety and early signs of visual gain, demonstrating the promise of localized CRISPR editing in delicate tissues.(Maeder et al., 2019)

Formulations of LNP delivery have also enabled in vivo editing applications beyond ex vivo cell therapies. Intellia Therapeutics, for example, has reported promising outcomes in its ATTR trial of a systemic condition caused by abnormal deposition of transthyretin protein. Using intravenous delivery of guide RNA and CRISPR/Cas9 mRNA in LNPs, the therapy specifically edits the liver cells that produce the faulty protein. The treated patients had excellent reductions in circulating transthyretin, with acceptable side effects, a much improvement from non-viral, systemic CRISPR delivery. (T. Zhang et al., 2024)

CRISPR has also been applied to explore cancer immunotherapy. Clinical trials of CRISPR-edited T cells are ongoing, in which the native T cell receptors or immune checkpoint genes are being knocked out to enhance anti-tumour activities. For instance, the University of Pennsylvania and other institutions have conducted first-in-human trials with CRISPR-enhanced T cells engineered to target cancer types such as non-small cell lung cancer, multiple myeloma, and sarcoma. These engineered T cells were stable in patients and showed potential anticancer activity without eliciting severe immune responses, opening the door to next-generation immunotherapies with higher specificity and longer duration. (Vimal et al., 2024)

In addition to these, CRISPR clinical trials have been extended to diseases like hereditary transthyretin amyloidosis, hereditary angioedema, and HIV infection, both to correct direct viral genome editing and to boost immune responses. Furthermore, base editing and prime editing technologies, which are less invasive and more precise gene-editing methods, are advancing into preclinical and early clinical trials.

3.2 Polymeric Nanoparticles

3.2.1 PLGA, PEG-PLA, chitosan-based carriers

Polymeric vectors, such as poly(lactic-co-glycolic acid) (PLGA), polyethylene glycol–polylactic acid (PEG-PLA) copolymers, and chitosan, have gained popularity as multifunctional, biocompatible delivery systems for nucleic acids, proteins, and genome-editing technologies such as CRISPR/Cas9. FDA-approved biodegradable PLGA is frequently used to fabricate nanoparticles and microspheres owing to its favorable degradation profile and sustained release of entrapped therapeutics. Their enzymatic hydrolysis to lactic and glycolic acids allows for controlled release of drugs and low toxicity, rendering PLGA an ideal carrier for CRISPR components when enzymatic protection and controlled intracellular release are necessary. PLGA nanoparticles can be surface-engineered for increased cellular uptake and tissue targeting, broadening their application in gene editing and drug delivery.(Makadia & Siegel, 2011)

PEG-PLA copolymers combine the hydrophilic character of polyethylene glycol (PEG) with the biodegradable hydrophobic polylactic acid (PLA) to form amphiphilic block copolymers that self-assemble into micellar nanoparticles in water. The PEG corona provides steric stabilization, reduces opsonization, and prolongs circulation time by evading the immune system, while the PLA core encapsulates and protects nucleic acids or proteins. PEG-PLA nanoparticles have also shown improved biocompatibility, enhanced pharmacokinetics, and efficient intracellular delivery of CRISPR reagents, making them well-suited for systemic administration. The size and surface-chemistry tunability of the PEG-PLA micelles enable endosomal escape and targeted delivery, which are essential for achieving maximum genome-editing efficiency. (Sinha et al., 2024)

Chitosan, a cationic polysaccharide derived naturally from chitin, has specific advantages as a gene delivery vehicle due to its biocompatibility, biodegradability, and natural mucoadhesion. Its cationic property permits electrostatic complexation with anionic nucleic acids, forming stable nanoparticles that protect genetic material from degradation. Chitosan carriers facilitate increased cellular internalization through interactions with anionic cell membranes and endosomal release via the "proton sponge" effect. In addition, chitosan's chemical modifiability enables the incorporation of targeting moieties and stimulus-responsive functionalities to enhance specificity and controlled release. Chitosan nanoparticles are therefore extremely attractive for CRISPR system delivery via localized routes, such as intranasal and mucosal administration, where biocompatibility and minimal toxicity are critical. (Desai et al., 2023)

3.2.2. Biocompatibility and controlled release

Biocompatibility and controlled release are two fundamental features that define the success of any delivery system, especially for sensitive drugs like CRISPR/Cas9 components. Controlled release refers to the ability of a carrier to deliver its cargo in a sustained, consistent, and site-specific manner, thereby maintaining therapeutic levels for extended periods without inducing systemic toxicity. Attaining controlled release is crucial for genome editing applications, ensuring that CRISPR components remain functional at the target locus long enough to exert their biological consequences without being exposed for extended periods that may contribute to off-target activity or an immune response. (Dubey & Mostafavi, 2023)

Polymer carriers such as PLGA and PEG-PLA are better suited for controlled release because their degradation rates can be tuned by manipulating polymer structure, molecular weight, and particle size. For example, PLGA can degrade hydrolytically to lactic and glycolic acid with tunable rates from days to months, releasing a controlled diffusion of encapsulated CRISPR molecules. Similarly, PEG-PLA micelles offer a hydrophobic core that protects nucleic acids and releases them in a sustained fashion as the polymer matrix is hydrolyzed or in response to environmental signals such as pH changes. Controlled release also facilitates better intracellular delivery by enabling slow endosomal release and sustained cytoplasmic access to genome-editing tools, thereby promoting editing efficiency (J. Yang et al., 2024)

Biocompatibility ensures that the delivery system does not induce off-target immune response, toxicity, or inflammation that could endanger patient safety or therapeutic efficacy. Polymers such as PLGA, PEG-PLA, and chitosan have been extensively studied and are considered broadly safe due to their degradability into non-toxic metabolites and low immunogenicity. The biocompatibility of such polymers enables repeated dosing, should it be necessary —typically the case in chronic disease or regenerative medicine. Surface modifications, such as PEGylation, also reduce protein adsorption and immune recognition, extending circulation time and reducing phagocytic clearance. (Yadav et al., 2024) The mucoadhesive properties of chitosan and its positive charge enhance its biocompatibility with mucosal tissues, allowing for localized delivery with minimal irritation. Its nature also confers lower toxicity than that of synthetic polymers. Molecular weight, extent of deacetylation, and formulation parameters can influence biocompatibility, and careful optimization is needed for clinical applications.

3.3 Dendrimers and Gold Nanoparticles: New Platforms for CRISPR Delivery

Dendrimers and AuNPs are two of the most promising classes of nanomaterials under intense investigation for the delivery of gene-editing reagents, including CRISPR/Cas9 components. Both nanosystems have distinct structural and physicochemical properties that enable efficient loading of genes, intracellular delivery, and real-time tracking, a key for precise genome engineering. (Fateh et al., 2024)

Surface-engineered dendrimers are highly branched, monodisperse polymers with terminal groups that can be varied to accommodate a diverse range of functional components, thereby providing a versatile framework for gene loading. Poly(amidoamine) (PAMAM) dendrimers are one of the most widely used shapes that possess internal voids for encapsulation of nucleic acids and surface amine functionalities for electrostatic binding of negatively charged DNA, RNA, or Cas9 ribonucleoprotein complexes. Still, naked dendrimers are toxic due to their high cationic surface density. To prevent this, surface modification strategies such as PEGylation, acetylation, or conjugation to targeting ligands have been employed. These modifications not only reduce toxicity and immunogenicity but also enhance stability, circulation half-life, and cell-targeted delivery. Besides, dendrimers may be engineered with pH- or redox-sensitive spacers for controlled release in the intracellular environment, i.e., at acidic endosomes or reducing cytoplasmic milieus. All these characteristics make surface-engineered dendrimers an effective and customized platform for the safe and efficient delivery of CRISPR/Cas9 machinery. (Filipczak et al., 2021; Kharwade et al., 2020)

Gold nanoparticles (AuNPs), meanwhile, are effective nanocarriers for both targeted delivery and imaging of gene-editing reagents. AuNPs also exhibit unique optical properties, such as surface plasmon resonance, which can be used for imaging, biosensing, and photothermal therapy. Their ease of surface functionalization makes them a good candidate for conjugation to various biomolecules, such as DNA, RNA, peptides, and antibodies. For CRISPR-related applications, AuNPs were used effectively as delivery agents for guide RNAs and Cas9 proteins and as contrast agents to track biodistribution and intracellular transport. AuNPs ligand-coated are capable of targeting specific cell populations or tissues with increased efficiency and fewer off-target effects. They also have a high surface area-to-volume ratio, which enables multivalent interactions, allowing CRISPR cargo to be densely loaded. AuNPs can also be engineered to respond to stimuli such as light, pH, or temperature, providing controlled gene release and enhanced spatial-temporal regulation of genome editing(P.-T. Huang et al., 2024; Wu et al., 2025) (see Figure 02 for depictions of surface-modified dendrimers and gold nanoparticles as delivery vehicles for CRISPR/Cas9).

Figure 2. surface-modified dendrimers and gold nanoparticles (AuNPs) as delivery vehicles for CRISPR/Cas9 components.

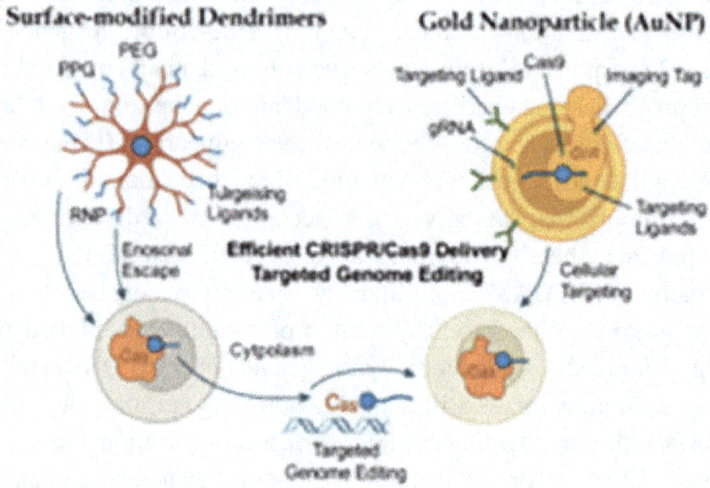

3.4 Exosomes and Biomimetic Systems for CRISPR Delivery

The application of exosomes and biomimetic systems as delivery tools for gene-editing tools like CRISPR/Cas9 has drawn a lot of attention due to their inherent characteristics of biological compatibility and unique delivery features. Exosomes are spontaneously secreted nano-scale extracellular vesicles (30–150 nm) from numerous cell types. They play a crucial role in intercellular communication by transporting bioactive molecules, such as proteins, lipids, and nucleic acids. Owing to these characteristics, they make an ideal delivery tool for gene-editing tools like CRISPR/Cas9. (Duan, Ouyang, Wang, et al., 2021)

Naturally occurring vesicles for CRISPR delivery offer several advantages over laboratory-made systems. Exosomes can carry CRISPR/Cas9 components, including plasmid DNA, mRNA, or ribonucleoprotein complexes (RNPs), and protect them from enzymatic degradation during systemic circulation. The lipid bilayer structure of exosomes is very similar to that of the native cell membrane and supports active fusion with target cells, followed by cargo release into the cytoplasm. Furthermore, exosomes can be engineered or obtained from single donor cells to express surface antigens that enable targeted delivery to intended tissues or cell types (X. Huang et al., 2023)

Low immunogenicity and targeting efficacy are two of the most significant characteristics of exosome-based drug delivery systems. As endogenous vesicles, exosomes are inherently biocompatible and induce minimal immune reactions. This renders them suitable for therapeutic applications. They do not integrate into the host genome like viral vectors, thereby reducing the likelihood of insertional mutagenesis. In addition, exosomes can be designed to incorporate targeting ligands or antibodies on the surface for specificity and to minimize off-target effects, a concern critical in genome editing therapies (J. Li et al., 2025)

In addition to endogenous exosomes, biomimetic particles such as exosome-mimicking liposomes or hybrid vesicles composed of both synthetic and biological elements have been developed. Such particles aim to replicate the desirable traits of native vesicles — i.e., biocompatibility and targeting — while also offering improvements in scalability, loading efficiency, and reproducibility. For instance, hybrid exosome-liposomes may be surface-functionalized with targeting peptides and loaded with CRISPR/Cas9 ribonucleoprotein complexes (RNPs) via electroporation or lipid fusion. These hybrid platforms provide a flexible and modular platform for gene editing applications in a range of cell types and tissues (Antimisiaris et al., 2018; Y.-J. Li et al., 2021) (see Table 01 for illustrative CRISPR/Cas9 delivery exosome and biomimetic vesicle platforms)

Table 1. Representative exosome and biomimetic vesicle platforms for CRISPR/Cas9 delivery.

System Type	Source	Loaded Cargo	Targeting Strategy	Key Advantages	Reference
Natural Exosomes	MSC-derived exosomes	Cas9 RNP	Surface display of tissue-homing peptides	Low immunogenicity; native membrane fusion	(Wan et al., 2022)
Dendritic Exosomes	Dendritic cell exosomes	Cas9 mRNA + sgRNA	DC-specific ligands (e.g., DEC-205 antibody)	Immune cell tropism; co-delivery of editing tools	(Balaraman et al., 2025)
RBC Exosomes	Red blood cell–derived exosomes	Cas9 RNP	Passive targeting (long circulation)	Ultra-low immunogenicity; abundant source	(L. Yang et al., 2022)
Exosome–Liposome Hybrid	Fusion of exosomal membranes with synthetic liposomes	Cas9 RNP	Ligand-PEGylation (e.g., folate)	Scalability, high loading efficiency	(Balaraman et al., 2025)

continued on following page

Table 1. Continued

System Type	Source	Loaded Cargo	Targeting Strategy	Key Advantages	Reference
Biomimetic Liposomes	Phospholipid vesicles functionalized with CD63 peptide	Cas9 mRNA + sgRNA	Peptide ligand for endothelium	Reproducible manufacture; customizable surface	(Sundaram & Aryal, 2025)
Cell Membrane-Coated NP	PLGA core coated with cancer cell membrane	Cas9 RNP	Homotypic targeting to tumor cells	Immune evasion; tumor-specific delivery	(Zeng et al., 2023)

- **MSC = mesenchymal stem cell**
- **RNP = ribonucleoprotein complex**
- **DEC-205 = dendritic cell receptor**

4. MECHANISMS OF CELLULAR UPTAKE AND ENDOSOMAL ESCAPE

Successful intracellular delivery of CRISPR/Cas9 reagents depends on the discovery and exploitation of cellular uptake processes and subsequent endosomal vesicle escape to reach the cytoplasm and nucleus. The primary pathways for cellular internalization of gene delivery vehicles include various forms of endocytosis, such as clathrin-mediated, caveolae-mediated, macropinocytosis, and clathrin/caveolae-independent processes. In such mechanisms, nanoparticles or complexes bind to cell-surface receptors or nonspecifically interact with the plasma membrane, leading to engulfment into endosomal membrane-bound vesicles. Certain delivery systems, however, such as lipid-based nanoparticles, can fuse directly with the plasma membrane, releasing their contents directly into the cytosol and bypassing entrapment in endosomes. (Durymanov & Reineke, 2018)

Following endocytosis, the most critical challenge is the escape of CRISPR reagents out of the acidic, enzyme-filled environment of endosomes before degradation. pH-sensitive carriers step in at this point. These carriers are engineered to respond to the acidic pH (~5-6) inside endosomes via conformational changes, membrane disruption, or protonation, thereby facilitating destabilization of the endosomal membrane. For example, "proton sponge" capacity polymers such as polyethyleneimine (PEI) or chemically modified chitosan buffer the acidic pH, provoking an ion and water influx that leads to swelling and lysis of endosomes. Similarly, pH-sensitive peptides or lipids incorporated in nanoparticle preparations become fusogenic at acidic pH, promoting endosomal release and delivery of genome editing machinery into the cytoplasm. (Brock et al., 2019)

Following entry into the cytoplasm, the CRISPR/Cas9 components must gain access to the nucleus to guide the genomic DNA. Facilitation of nuclear entry constitutes a crucial step, especially in the case of DNA or ribonucleoprotein (RNP) complexes, since nuclear membranes are a very effective barrier. Nuclear localization signals (NLS) can be conjugated to Cas9 proteins or delivery vehicles to induce active transport across nuclear pore complexes by binding to the cell's nuclear import apparatus. Moreover, synchronized delivery during cell division stages, when the nuclear envelope temporarily dissolves, can increase nuclear uptake. Advanced delivery systems also involve targeting ligands or microtubule-mediated transport for maximal intracellular trafficking to the nucleus (L. Li et al., 2018)

5. CASE STUDIES AND APPLICATIONS

5.1 Cancer Therapy

Cancer remains one of the hardest diseases to treat due to its genetic heterogeneity and complexity. The invention of CRISPR/Cas9 technology has opened numerous new avenues for cancer therapy targeting, due to its ability to remove or correct oncogenes and other cancer genes with enhanced specificity. Nano-CRISPR delivery systems, including the application of CRISPR/Cas9 components integrated with nanoparticles such as lipid nanoparticles, polymeric nanoparticles, and dendrimers, are a promising method to bypass biological barriers and increase the specificity and efficiency of gene editing in cancer cells. (Chehelgerdi et al., 2024)

One of the most important applications is the knockout of oncogenes that cause tumor initiation and growth. For example, KRAS, a frequently mutated pancreatic, colorectal, and lung cancer oncogene, has proven difficult to target by conventional therapies due to its "undruggable" nature. Nano-CRISPR platforms delivering Cas9 ribonucleoproteins or plasmids targeting mutant KRAS alleles demonstrated efficient gene disruption in preclinical models, leading to decreased tumor progression and enhanced chemotherapy sensitivity. Similarly, the epidermal growth factor receptor (EGFR) gene, overexpressed or mutated in glioblastoma and non-small cell lung cancer, has been targeted by lipid-based CRISPR delivery platforms to cause knockout of the gene that results in reduced invasiveness and growth of cancer cells. (Xu et al., 2021)

Another crucial target is BCL2, an anti-apoptotic gene overexpressed in most cancers that confers chemoresistance and tumor persistence. Inhibition of BCL2 via CRISPR-mediated delivery using polymeric nanoparticle carriers has been shown to induce apoptosis and enhance the efficacy of standard anticancer therapy. Such examples indicate the promise of nano-CRISPR reagents to selectively edit cancer-causing genes, thereby overcoming resistance determinants and improving cure rates. (Thomas et al., 2013)

5.2 *Genetic Diseases: Sickle Cell Disease and Duchenne Muscular Dystrophy—In Vivo Gene Correction Using Nanoparticles*

Sickle Cell Disease (SCD) and Duchenne Muscular Dystrophy (DMD) are genetic disorders caused by single-gene mutations leading to severe, life-threatening conditions. SCD is caused by a point mutation within the β-globin gene, leading to the production of dysfunctional hemoglobin S, sickle-shaped red cells, vaso-occlusive crises, hemolytic anemia, and organ destruction. DMD is an X-linked recessive disorder caused by mutations in the dystrophin gene, leading to the absence of functional dystrophin protein and progressive muscle degeneration. Traditional treatments for the disease are largely palliative or symptomatic, underscoring the need for curative approaches such as genome editing. (Inusa et al., 2019)

Gene correction with CRISPR/Cas9 has been a powerful approach for treating the primary genetic defects underpinning these diseases. Recent advances in nanoparticle-delivery systems have significantly improved the feasibility of in vivo genome editing by enabling safe, targeted delivery of CRISPR elements to target tissues. Lipid nanoparticles (LNPs) have shown great promise in this regard, as they can encapsulate and protect messenger RNA (mRNA) or ribonucleoprotein (RNP) complexes and deliver them into target cells. For instance, in preclinical models of SCD, LNPs have been used to deliver CRISPR/Cas9 systems targeting the BCL11A enhancer in hematopoietic stem cells, resulting in successful restoration of fetal hemoglobin synthesis and alleviation of disease severity. (Kolanu, 2024)

Similarly, in DMD, nanoparticles have facilitated the in vivo delivery of CRISPR/Cas9 components into skeletal and cardiac muscle tissues. Exon skipping and reading frame restoration of the dystrophin gene has been successfully done with polymeric nanoparticles or LNPs in research, leading to partial recovery of expression of dystrophin and improvement in muscle function. In vivo approaches are less invasive and avoid ex vivo stem cell manipulation, a more scalable and patient-friendly option. (Haque & Yokota, 2025)

5.3 Antiviral Approaches

The development of CRISPR/Cas9 as a gene-editing tool has shown potential avenues for the eradication of long-standing viral infections such as Hepatitis B virus (HBV) and Human Immunodeficiency Virus (HIV), which have defied curative interventions for decades. CRISPR/Cas9 can directly target and kill viral genomes, thereby eradicating latently infected reservoirs that are out of reach of conventional antiviral drugs. For HBV, CRISPR/Cas9 can be designed to target conserved regions of the virus covalently closed circular DNA (cccDNA), which is responsible for chronic infection and relapse after treatment. Several in vitro and in vivo studies have shown that HBV cccDNA cleavage by CRISPR results in significant reductions in viral replication and antigen expression, opening the door to a cure. (Stone et al., 2021)

Similarly, for HIV, CRISPR/Cas9 has been used to delete proviral DNA incorporated into host chromosomes, targeting key regions, such as long terminal repeats (LTRs), or essential viral genes, such as gag, pol, and env. Double or multiplexed guide RNA strategies have been developed to enhance efficiency and minimize the potential of viral escape. Interestingly, animal models and humanized mice studies have shown promising findings in reducing HIV burden and reactivation potential, reflecting the therapeutic promise of genome editing in antiviral therapy.(Das et al., 2019)

Nonetheless, the delivery of CRISPR components to infected cells remains a considerable challenge, primarily due to the presence of biological barriers such as the mucus layer, cell membranes, and systemic clearance pathways. Nanocarriers are emerging as a cornerstone solution to these issues by facilitating the safe and targeted delivery of genome-editing agents to infected tissues. Lipid nanoparticles (LNPs), polymeric nanoparticles, and dendrimer-based systems are under development for their potential to encapsulate and protect CRISPR plasmids, mRNA, or ribonucleoproteins (RNPs), facilitating efficient intracellular delivery while avoiding immune detection.(Sioson et al., 2021)

6. REGULATORY AND ETHICAL CONSIDERATIONS

The rapid progress in nanomedicine and gene-editing tools, particularly CRISPR/Cas9, has opened revolutionary possibilities for precision medicine. These advances, however, also raise fundamental regulatory and ethical challenges that must be addressed for them to be used safely, equitably, and responsibly. The integration of nanocarriers with gene-editing tools presents particular challenges to safety, long-term effects, and social acceptance, which regulatory agencies and ethical frameworks must jealously guard. (Kolanu, 2024)

Safety concerns in nanomedicine primarily focus on the pharmacokinetics, biodistribution, toxicity, and immunogenicity of nanoparticles. Unlike conventional drugs, nanocarriers are likely to exhibit complex in vivo behaviors controlled by size, shape, surface charge, and composition, leading to unforeseen accumulation in organs such as the liver, spleen, or lungs. Biocompatible excipients such as PLGA, PEG-PLA, and chitosan are considered safe. Still, their degradation products, interaction with the immune system, and long-term toxicity need a lot of preclinical and clinical investigation. Furthermore, surface-modified nanoparticles, including dendrimers and lipid-based formulations, need to be evaluated for off-target effects, cellular stress, and genotoxicity. The absence of harmonized characterization and testing methods of nanomedicines also complicates regulatory approval and underscores the need for international harmonization. (Havelikar et al., 2024)

The ethical issues of CRISPR are multifaceted and are strongest in the field of human genome editing. Somatic cell editing for therapy is generally accepted, but the potential for germline modification raises basic moral, social, and philosophical questions. Concerns include the risk of unintended genetic consequences, heritable alterations passed on to future generations, and the risk of creating "designer babies" with enhanced attributes. Access, equity, and consent issues must also be addressed to prevent genetic inequality or misuse of the technology. Ethical guidelines promote transparency, public engagement, and adherence to the precautionary principle, especially in the case of human embryos or fertility treatments.(Brokowski & Adli, 2019)

The regulatory framework for gene-edited nanomedicines is evolving rapidly, with institutions such as the U.S. Food and Drug Administration (FDA), the European Medicines Agency (EMA), and regulatory agencies in Asian and other countries setting the standards for evaluating these advanced products. Regulatory approval for CRISPR-based treatments delivered using nanocarriers involves examination of both the gene-editing process and the nanomaterial itself for delivery. Dual review in this manner requires unifying the regulatory frameworks of biologics, gene therapy, and nanomedicine. For instance, the FDA considers CRISPR systems to be biological products that require Investigational New Drug (IND) submissions, preclinical toxicology testing, and compliance with Good Manufacturing Practice (GMP) requirements. Furthermore, post-marketing surveillance must be conducted to monitor long-term effects and protect patients. (Parums, 2024)

7. FUTURE DIRECTIONS

Parallel to further improving CRISPR/Cas9 technology, the development and engineering of advanced nanocarrier systems are also advancing to meet the continuously growing demand for targeted, safe, and efficient gene delivery. Most potentially promising advancements involve new nanocarrier designs, particularly stimuli-responsive and multifunctional systems. These intelligent carriers are designed to respond to specific physiological or pathological signals, such as pH, temperature, redox potential, or enzymatic activity, thereby delivering CRISPR reagents directly to the site of interest with minimal systemic exposure. Multifunctional nanocarriers can be engineered to incorporate targeting ligands, imaging agents, and dual therapeutic payloads, enabling simultaneous gene editing, diagnosis, and therapeutic monitoring from a single platform. These new-generation nanocarriers are expected to enhance the safety and targeting of CRISPR-based therapies significantly. (Guzman Gonzalez et al., 2024)

A paradigm shift is also brought about by the use of artificial intelligence (AI) and machine learning algorithms of optimize delivery platforms. AI can analyze large datasets to predict optimal carrier combinations, surface chemistry, and formulation variables specific to a given tissue, disease model, or patient population. By integrating predictive modeling and high-throughput screening, AI facilitates rapid identification of the most effective and least harmful delivery methods. This process not only accelerates development but also reduces experimental costs and increases preclinical and clinical trial success rates. AI-driven platforms are currently widely used for the synthesis of custom nanocarriers with enhanced biodistribution, cell uptake, and gene-editing efficiency. (Jena et al., 2024)

8. CONCLUSION

A combination of nanodelivery systems with CRISPR/Cas9 technology represents a revolutionary step in the area of precision medicine, offering unparalleled opportunities for targeted, efficient, and safe genome editing. Nanocarriers such as lipid nanoparticles, polymeric nanoparticles (e.g., PLGA and PEG-PLA), chitosan-based systems, and surface-modified dendrimers have demonstrated great potential to address the inherent challenges of CRISPR delivery, including stability, specificity, controlled release, and immunogenicity. These nanoscale platforms shield CRISPR components from enzymatic cleavage, enhance cellular delivery and endosomal escape, and can be designed to target specific tissues, all of which are essential to maximize therapeutic potency while minimizing off-target toxicity.

The clinical advances made so far, most notably in monogenic disease treatments, inherited diseases, and cancer therapy, highlight the potential of nano-enabled CRISPR therapies. Technological advances in the field of nanotechnology have opened up avenues for safer ex vivo and in vivo delivery options, broadening the reach of gene editing beyond the classic viral vectors. Nevertheless, despite these positives, several challenges remain to be overcome before large-scale clinical use becomes possible. These include enhancing delivery efficacy, achieving prolonged safety, modulating immune responses, and addressing complex regulatory issues.

Future studies should aim to maximize nanocarrier design, investigate stimuli-responsive and multifunctional nanoparticles, and combine new genome-editing technologies such as base and prime editors. Interdisciplinary collaborations will be critical to tapping into the full therapeutic potential of CRISPR/Cas9 delivered via nanotechnology, enabling personalized treatment approaches tailored to unique genetic profiles. Nano-delivery systems are the solution to realizing CRISPR/Cas9's clinical and commercial potential in precision medicine, marking a new age of targeted, potent, and long-lasting genetic therapies that can transform health and enhance patient outcomes globally.

REFERENCES

Aljabali, A. A. A., El-Tanani, M., & Tambuwala, M. M. (2024). Principles of CRISPR-Cas9 technology: Advancements in genome editing and emerging trends in drug delivery. *Journal of Drug Delivery Science and Technology*, *92*, 105338. DOI: 10.1016/j.jddst.2024.105338

Almotiri, A., Abogosh, A., Abdelfattah, A., Alowaisy, D., & Rodrigues, N. P. (2025). Treating genetic blood disorders in the era of CRISPR-mediated genome editing. *Molecular Therapy : the Journal of the American Society of Gene Therapy*, *33*(6), 2645–2662. DOI: 10.1016/j.ymthe.2025.01.031 PMID: 39827371

Antimisiaris, S. G., Mourtas, S., & Marazioti, A. (2018). Exosomes and Exosome-Inspired Vesicles for Targeted Drug Delivery. *Pharmaceutics*, *10*(4), 218. DOI: 10.3390/pharmaceutics10040218 PMID: 30404188

Asmamaw Mengstie, M., Teshome Azezew, M., Asmamaw Dejenie, T., Teshome, A. A., Tadele Admasu, F., Behaile Teklemariam, A., Tilahun Mulu, A., Mekonnen Agidew, M., Adugna, D. G., Geremew, H., & Abebe, E. C. (2024). Recent Advancements in Reducing the Off-Target Effect of CRISPR-Cas9 Genome Editing. *Biologics*, *18*, 21–28. DOI: 10.2147/BTT.S429411 PMID: 38260716

Balaraman, A. K., Babu, M. A., Moglad, E., Mandaliya, V., Rekha, M. M., Gupta, S., Prasad, G. V. S., Kumari, M., Chauhan, A. S., Ali, H., & Goyal, K. (2025). Exosome-mediated delivery of CRISPR-Cas9: A revolutionary approach to cancer gene editing. *Pathology, Research and Practice*, *266*, 155785. DOI: 10.1016/j.prp.2024.155785 PMID: 39708520

Bareford, L., & Swaan, P. (2007). Endocytic mechanisms for targeted drug delivery☆. *Advanced Drug Delivery Reviews*, *59*(8), 748–758. DOI: 10.1016/j.addr.2007.06.008 PMID: 17659804

Brock, D. J., Kondow-McConaghy, H. M., Hager, E. C., & Pellois, J.-P. (2019). Endosomal Escape and Cytosolic Penetration of Macromolecules Mediated by Synthetic Delivery Agents. *Bioconjugate Chemistry*, *30*(2), 293–304. DOI: 10.1021/acs.bioconjchem.8b00799 PMID: 30462487

Brokowski, C., & Adli, M. (2019). CRISPR Ethics: Moral Considerations for Applications of a Powerful Tool. *Journal of Molecular Biology*, *431*(1), 88–101. DOI: 10.1016/j.jmb.2018.05.044 PMID: 29885329

Chehelgerdi, M., Chehelgerdi, M., Khorramian-Ghahfarokhi, M., Shafieizadeh, M., Mahmoudi, E., Eskandari, F., Rashidi, M., Arshi, A., & Mokhtari-Farsani, A. (2024). Comprehensive review of CRISPR-based gene editing: Mechanisms, challenges, and applications in cancer therapy. *Molecular Cancer*, *23*(1), 9. DOI: 10.1186/s12943-023-01925-5 PMID: 38195537

Das, A. T., Binda, C. S., & Berkhout, B. (2019). Elimination of infectious HIV DNA by CRISPR–Cas9. *Current Opinion in Virology*, *38*, 81–88. DOI: 10.1016/j.coviro.2019.07.001 PMID: 31450074

Desai, N., Rana, D., Salave, S., Gupta, R., Patel, P., Karunakaran, B., Sharma, A., Giri, J., Benival, D., & Kommineni, N. (2023). Chitosan: A Potential Biopolymer in Drug Delivery and Biomedical Applications. *Pharmaceutics*, *15*(4), 1313. DOI: 10.3390/pharmaceutics15041313 PMID: 37111795

Du, Y., Liu, Y., Hu, J., Peng, X., & Liu, Z. (2023). CRISPR/Cas9 systems: Delivery technologies and biomedical applications. *Asian Journal of Pharmaceutical Sciences*, *18*(6), 100854. DOI: 10.1016/j.ajps.2023.100854 PMID: 38089835

Duan, L., Ouyang, K., Wang, J., Xu, L., Xu, X., Wen, C., Xie, Y., Liang, Y., & Xia, J. (2021). Exosomes as Targeted Delivery Platform of CRISPR/Cas9 for Therapeutic Genome Editing. *ChemBioChem*, *22*(24), 3360–3368. DOI: 10.1002/cbic.202100359 PMID: 34418266

Duan, L., Ouyang, K., Xu, X., Xu, L., Wen, C., Zhou, X., Qin, Z., Xu, Z., Sun, W., & Liang, Y. (2021). Nanoparticle Delivery of CRISPR/Cas9 for Genome Editing. *Frontiers in Genetics*, *12*, 673286. DOI: 10.3389/fgene.2021.673286 PMID: 34054927

Dubey, A. K., & Mostafavi, E. (2023). Biomaterials-mediated CRISPR/Cas9 delivery: Recent challenges and opportunities in gene therapy. *Frontiers in Chemistry*, *11*, 1259435. DOI: 10.3389/fchem.2023.1259435 PMID: 37841202

Durymanov, M., & Reineke, J. (2018). Non-viral Delivery of Nucleic Acids: Insight Into Mechanisms of Overcoming Intracellular Barriers. *Frontiers in Pharmacology*, *9*, 971. DOI: 10.3389/fphar.2018.00971 PMID: 30186185

Elayaperumal, S., Sivamani, Y., Agarwal, P., & Srivastava, N. (2025). Plant-based nanotherapeutics: A new frontier in disease management and prevention. *Nano TransMed*, *4*, 100086. DOI: 10.1016/j.ntm.2025.100086

Fateh, S. T., Aghaii, A. H., Aminzade, Z., Shahriari, E., Roohpour, N., Koosha, F., & Dezfuli, A. S. (2024). Inorganic nanoparticle-cored dendrimers for biomedical applications: A review. *Heliyon*, *10*(9), e29726. DOI: 10.1016/j.heliyon.2024.e29726 PMID: 38694058

Filipczak, N., Yalamarty, S. S. K., Li, X., Parveen, F., & Torchilin, V. (2021). Developments in Treatment Methodologies Using Dendrimers for Infectious Diseases. *Molecules (Basel, Switzerland), 26*(11), 3304. DOI: 10.3390/molecules26113304 PMID: 34072765

Geszke-Moritz, M., & Moritz, M. (2024). Biodegradable Polymeric Nanoparticle-Based Drug Delivery Systems: Comprehensive Overview, Perspectives and Challenges. *Polymers, 16*(17), 2536. DOI: 10.3390/polym16172536 PMID: 39274168

Guzman Gonzalez, V., Grunenberger, A., Nicoud, O., Czuba, E., Vollaire, J., Josserand, V., Le Guével, X., Desai, N., Coll, J.-L., Divita, G., & Faure, V. (2024). Enhanced CRISPR-Cas9 RNA system delivery using cell penetrating peptides-based nanoparticles for efficient in vitro and in vivo applications. *Journal of Controlled Release, 376*, 1160–1175. DOI: 10.1016/j.jconrel.2024.11.008 PMID: 39521064

Haque, U. S., & Yokota, T. (2025). Gene Editing for Duchenne Muscular Dystrophy: From Experimental Models to Emerging Therapies. *Degenerative Neurological and Neuromuscular Disease, 15*, 17–40. DOI: 10.2147/DNND.S495536 PMID: 40241992

Havelikar, U., Ghorpade, K. B., Kumar, A., Patel, A., Singh, M., Banjare, N., & Gupta, P. N. (2024). Comprehensive insights into mechanism of nanotoxicity, assessment methods and regulatory challenges of nanomedicines. *Discover Nano, 19*(1), 165. DOI: 10.1186/s11671-024-04118-1 PMID: 39365367

Huang, P.-T., Chen, Y.-L., Lin, Y.-H., Wang, C.-C., & Chang, H.-T. (2024). Functional gold nanoparticles for analysis and delivery of nucleic acids. *Yao Wu Shi Pin Fen Xi, 32*(3), 252–273. DOI: 10.38212/2224-6614.3514 PMID: 39636773

Huang, X., Li, A., Xu, P., Yu, Y., Li, S., Hu, L., & Feng, S. (2023). Current and prospective strategies for advancing the targeted delivery of CRISPR/Cas system via extracellular vesicles. *Journal of Nanobiotechnology, 21*(1), 184. DOI: 10.1186/s12951-023-01952-w PMID: 37291577

Inusa, B., Hsu, L., Kohli, N., Patel, A., Ominu-Evbota, K., Anie, K., & Atoyebi, W. (2019). Sickle Cell Disease—Genetics, Pathophysiology, Clinical Presentation and Treatment. *International Journal of Neonatal Screening, 5*(2), 20. DOI: 10.3390/ijns5020020 PMID: 33072979

Jena, G. K., Patra, C. N., Jammula, S., Rana, R., & Chand, S. (2024). Artificial Intelligence and Machine Learning Implemented Drug Delivery Systems: A Paradigm Shift in the Pharmaceutical Industry. *Journal of Bio-X Research, 7*, 0016. DOI: 10.34133/jbioxresearch.0016

Jung, H. N., Lee, S.-Y., Lee, S., Youn, H., & Im, H.-J. (2022). Lipid nanoparticles for delivery of RNA therapeutics: Current status and the role of *in vivo* imaging. *Theranostics*, *12*(17), 7509–7531. DOI: 10.7150/thno.77259 PMID: 36438494

Kharwade, R., More, S., Warokar, A., Agrawal, P., & Mahajan, N. (2020). Starburst pamam dendrimers: Synthetic approaches, surface modifications, and biomedical applications. *Arabian Journal of Chemistry*, *13*(7), 6009–6039. DOI: 10.1016/j.arabjc.2020.05.002

Kolanu, N. D. (2024). CRISPR–Cas9 Gene Editing: Curing Genetic Diseases by Inherited Epigenetic Modifications. *Global Medical Genetics*, *11*(01), 113–122. DOI: 10.1055/s-0044-1785234 PMID: 38560484

Li, J., Wang, J., & Chen, Z. (2025). Emerging role of exosomes in cancer therapy: Progress and challenges. *Molecular Cancer*, *24*(1), 13. DOI: 10.1186/s12943-024-02215-4 PMID: 39806451

Li, L., Hu, S., & Chen, X. (2018). Non-viral delivery systems for CRISPR/Cas9-based genome editing: Challenges and opportunities. *Biomaterials*, *171*, 207–218. DOI: 10.1016/j.biomaterials.2018.04.031 PMID: 29704747

Li, T., Yang, Y., Qi, H., Cui, W., Zhang, L., Fu, X., He, X., Liu, M., Li, P., & Yu, T. (2023a). CRISPR/Cas9 therapeutics: Progress and prospects. *Signal Transduction and Targeted Therapy*, *8*(1), 36. DOI: 10.1038/s41392-023-01309-7 PMID: 36646687

Li, T., Yang, Y., Qi, H., Cui, W., Zhang, L., Fu, X., He, X., Liu, M., Li, P., & Yu, T. (2023b). CRISPR/Cas9 therapeutics: Progress and prospects. *Signal Transduction and Targeted Therapy*, *8*(1), 36. DOI: 10.1038/s41392-023-01309-7 PMID: 36646687

Li, Y.-J., Wu, J.-Y., Liu, J., Xu, W., Qiu, X., Huang, S., Hu, X.-B., & Xiang, D.-X. (2021). Artificial exosomes for translational nanomedicine. *Journal of Nanobiotechnology*, *19*(1), 242. DOI: 10.1186/s12951-021-00986-2 PMID: 34384440

Maeder, M. L., Stefanidakis, M., Wilson, C. J., Baral, R., Barrera, L. A., Bounoutas, G. S., Bumcrot, D., Chao, H., Ciulla, D. M., DaSilva, J. A., Dass, A., Dhanapal, V., Fennell, T. J., Friedland, A. E., Giannoukos, G., Gloskowski, S. W., Glucksmann, A., Gotta, G. M., Jayaram, H., & Jiang, H. (2019). Development of a gene-editing approach to restore vision loss in Leber congenital amaurosis type 10. *Nature Medicine*, *25*(2), 229–233. DOI: 10.1038/s41591-018-0327-9 PMID: 30664785

Makadia, H. K., & Siegel, S. J. (2011). Poly Lactic-co-Glycolic Acid (PLGA) as Biodegradable Controlled Drug Delivery Carrier. *Polymers*, *3*(3), 1377–1397. DOI: 10.3390/polym3031377 PMID: 22577513

Mehta, A., & Merkel, O. M. (2020). Immunogenicity of Cas9 Protein. *Journal of Pharmaceutical Sciences*, *109*(1), 62–67. DOI: 10.1016/j.xphs.2019.10.003 PMID: 31589876

Mendonça, M. C. P., Kont, A., Kowalski, P. S., & O'Driscoll, C. M. (2023). Design of lipid-based nanoparticles for delivery of therapeutic nucleic acids. *Drug Discovery Today*, *28*(3), 103505. DOI: 10.1016/j.drudis.2023.103505 PMID: 36708760

Mengstie, M. A., & Wondimu, B. Z. (2021). Mechanism and Applications of CRISPR/Cas-9-Mediated Genome Editing. *Biologics*, *15*, 353–361. DOI: 10.2147/BTT.S326422 PMID: 34456559

Parums, D. V. (2024). Editorial: First Regulatory Approvals for CRISPR-Cas9 Therapeutic Gene Editing for Sickle Cell Disease and Transfusion-Dependent β-Thalassemia. *Medical Science Monitor*, *30*. Advance online publication. DOI: 10.12659/MSM.944204 PMID: 38425279

Rauf, M. A., Rao, A., Sivasoorian, S. S., & Iyer, A. K. (2025). Nanotechnology-Based Delivery of CRISPR/Cas9 for Cancer Treatment: A Comprehensive Review. *Cells*, *14*(15), 1136. DOI: 10.3390/cells14151136 PMID: 40801569

Seijas, A., Cora, D., Novo, M., Al-Soufi, W., Sánchez, L., & Arana, Á. J. (2025). CRISPR/Cas9 Delivery Systems to Enhance Gene Editing Efficiency. *International Journal of Molecular Sciences*, *26*(9), 4420. DOI: 10.3390/ijms26094420 PMID: 40362657

Sinha, S., Tripathi, A. K., Pandey, A., Naik, P., Pandey, A., & Verma, V. S. (2024). Self-assembled PEGylated micelles for precise and targeted drug delivery: Current challenges and future directions. *Biocatalysis and Agricultural Biotechnology*, *60*, 103296. DOI: 10.1016/j.bcab.2024.103296

Sioson, V. A., Kim, M., & Joo, J. (2021). Challenges in delivery systems for CRISPR-based genome editing and opportunities of nanomedicine. *Biomedical Engineering Letters*, *11*(3), 217–233. DOI: 10.1007/s13534-021-00199-4 PMID: 34350049

Stone, D., Long, K. R., Loprieno, M. A., De Silva Feelixge, H. S., Kenkel, E. J., Liley, R. M., Rapp, S., Roychoudhury, P., Nguyen, T., Stensland, L., Colón-Thillet, R., Klouser, L. M., Weber, N. D., Le, C., Wagoner, J., Goecker, E. A., Li, A. Z., Eichholz, K., Corey, L., & Jerome, K. R. (2021). CRISPR-Cas9 gene editing of hepatitis B virus in chronically infected humanized mice. *Molecular Therapy. Methods & Clinical Development*, *20*, 258–275. DOI: 10.1016/j.omtm.2020.11.014 PMID: 33473359

Sundaram, V., & Aryal, S. (2025). Emerging Biomimetic Drug Delivery Nanoparticles Inspired by Extracellular Vesicles. *Wiley Interdisciplinary Reviews. Nanomedicine and Nanobiotechnology*, *17*(4), e70025. DOI: 10.1002/wnan.70025 PMID: 40840553

Sur, S., Rathore, A., Dave, V., Reddy, K. R., Chouhan, R. S., & Sadhu, V. (2019). Recent developments in functionalized polymer nanoparticles for efficient drug delivery system. *Nano-Structures & Nano-Objects*, *20*, 100397. DOI: 10.1016/j.nanoso.2019.100397

Thomas, S., Quinn, B. A., Das, S. K., Dash, R., Emdad, L., Dasgupta, S., Wang, X.-Y., Dent, P., Reed, J. C., Pellecchia, M., Sarkar, D., & Fisher, P. B. (2013). Targeting the Bcl-2 family for cancer therapy. *Expert Opinion on Therapeutic Targets*, *17*(1), 61–75. DOI: 10.1517/14728222.2013.733001 PMID: 23173842

Van Der Zanden, S. Y., Jongsma, M. L. M., Neefjes, A. C. M., Berlin, I., & Neefjes, J. (2023). Maintaining soluble protein homeostasis between nuclear and cytoplasmic compartments across mitosis. *Trends in Cell Biology*, *33*(1), 18–29. DOI: 10.1016/j.tcb.2022.06.002 PMID: 35778326

Vimal, S., Madar, I. H., Thirumani, L., Thangavelu, L., & Sivalingam, A. M. (2024). CRISPR/Cas9: Role of genome editing in cancer immunotherapy. *Oral Oncology Reports*, *10*, 100251. DOI: 10.1016/j.oor.2024.100251

Wan, T., Zhong, J., Pan, Q., Zhou, T., Ping, Y., & Liu, X. (2022). Exosome-mediated delivery of Cas9 ribonucleoprotein complexes for tissue-specific gene therapy of liver diseases. *Science Advances*, *8*(37), eabp9435. Advance online publication. DOI: 10.1126/sciadv.abp9435 PMID: 36103526

Wang, L., & Han, H. (2024). Strategies for improving the genome-editing efficiency of class 2 CRISPR/Cas system. *Heliyon*, *10*(19), e38588. DOI: 10.1016/j.heliyon.2024.e38588 PMID: 39397905

Wei, P.-S., Thota, N., John, G., Chang, E., Lee, S., Wang, Y., Ma, Z., Tsai, Y.-H., & Mei, K.-C. (2024). Enhancing RNA-lipid nanoparticle delivery: Organ- and cell-specificity and barcoding strategies. *Journal of Controlled Release*, *375*, 366–388. DOI: 10.1016/j.jconrel.2024.08.030 PMID: 39179112

Wu, J., Ko, S., Lee, E., Son, E., Kang, G., Hur, S., Lee, J.-H., Oh, J.-W., & Kim, Y. (2025). Gold nanoparticles in imaging: Advances, applications, and future perspectives. *Applied Spectroscopy Reviews*, *60*(9-10), 1–40. DOI: 10.1080/05704928.2025.2495022

Xu, X., Liu, C., Wang, Y., Koivisto, O., Zhou, J., Shu, Y., & Zhang, H. (2021). Nanotechnology-based delivery of CRISPR/Cas9 for cancer treatment. *Advanced Drug Delivery Reviews*, *176*, 113891. DOI: 10.1016/j.addr.2021.113891 PMID: 34324887

Yadav, H., Malviya, R., & Kaushik, N. (2024). Chitosan in biomedicine: A comprehensive review of recent developments. *Carbohydrate Polymer Technologies and Applications*, *8*, 100551. DOI: 10.1016/j.carpta.2024.100551

Yang, J., Zeng, H., Luo, Y., Chen, Y., Wang, M., Wu, C., & Hu, P. (2024). Recent Applications of PLGA in Drug Delivery Systems. *Polymers*, *16*(18), 2606. DOI: 10.3390/polym16182606 PMID: 39339068

Yang, L., Huang, S., Zhang, Z., Liu, Z., & Zhang, L. (2022). Roles and Applications of Red Blood Cell-Derived Extracellular Vesicles in Health and Diseases. *International Journal of Molecular Sciences*, *23*(11), 5927. DOI: 10.3390/ijms23115927 PMID: 35682606

Zeng, S., Tang, Q., Xiao, M., Tong, X., Yang, T., Yin, D., Lei, L., & Li, S. (2023). Cell membrane-coated nanomaterials for cancer therapy. *Materials Today. Bio*, *20*, 100633. DOI: 10.1016/j.mtbio.2023.100633 PMID: 37128288

Zhang, S., Shen, J., Li, D., & Cheng, Y. (2021). Strategies in the delivery of Cas9 ribonucleoprotein for CRISPR/Cas9 genome editing. *Theranostics*, *11*(2), 614–648. DOI: 10.7150/thno.47007 PMID: 33391496

Zhang, T., Yin, H., Li, Y., Yang, H., Ge, K., Zhang, J., Yuan, Q., Dai, X., Naeem, A., Weng, Y., Huang, Y., & Liang, X.-J. (2024). Optimized lipid nanoparticles (LNPs) for organ-selective nucleic acids delivery in vivo. *iScience*, *27*(6), 109804. DOI: 10.1016/j.isci.2024.109804 PMID: 38770138

Zhou, R., Geiger, R. C., & Dean, D. A. (2004). Intracellular trafficking of nucleic acids. *Expert Opinion on Drug Delivery*, *1*(1), 127–140. DOI: 10.1517/17425247.1.1.127 PMID: 16296725

Compilation of References

Abbasi, S., Totmaj, M. A., Abbasi, M., Hajazimian, S., Goleij, P., Behroozi, J., Shademan, B., Isazadeh, A., & Baradaran, B. (2023). Chimeric antigen receptor T (CAR-T) cells: Novel cell therapy for hematological malignancies. *Cancer Medicine*, *12*(7), 7844–7858. DOI: 10.1002/cam4.5551 PMID: 36583504

Abbott, T. R., Dhamdhere, G., Liu, Y., Lin, X., Goudy, L., Zeng, L., Chemparathy, A., Chmura, S., Heaton, N. S., Debs, R., Pande, T., Endy, D., La Russa, M. F., Lewis, D. B., & Qi, L. S. (2020). Development of CRISPR as an antiviral strategy to combat SARS-CoV-2 and influenza. *Cell*, *181*(4), 865–876. DOI: 10.1016/j.cell.2020.04.020 PMID: 32353252

Abou-el-Enein, M., Elsallab, M., Feldman, S. A., Fesnak, A. D., Heslop, H. E., Marks, P., Till, B. G., Bauer, G., & Savoldo, B. (2021). Scalable manufacturing of CAR T cells for cancer immunotherapy. *Blood Cancer Discovery*, *2*(5), 408–422. DOI: 10.1158/2643-3230.BCD-21-0084 PMID: 34568831

Abudayyeh, O. O., Gootenberg, J. S., Essletzbichler, P., Han, S., Joung, J., Belanto, J. J., Verdine, V., Cox, D. B. T., Kellner, M. J., Regev, A., Lander, E. S., Voytas, D. F., Ting, A. Y., & Zhang, F. (2017). RNA targeting with CRISPR–Cas13. *Nature*, *550*(7675), 280–284. DOI: 10.1038/nature24049 PMID: 28976959

Abusara, O. H., Agha, A. S. A. A., & Bardaweel, S. K. (2025). Advancements and innovations in liquid biopsy through microfluidic technology for cancer diagnosis. *Analyst*, *150*(9), 1711–1725. Advance online publication. DOI: 10.1039/D5AN00105F PMID: 40181713

Adegbola, P. I., & Adetutu, A. (2024). Genetic and epigenetic modulations in toxicity: The two-sided roles of heavy metals and polycyclic aromatic hydrocarbons from the environment. *Toxicology Reports*, *12*, 502–519. DOI: 10.1016/j.toxrep.2024.04.010 PMID: 38774476

Afrose, N., Chakraborty, R., Bhowmick, P., & Bhowmick, M. (2025). Advancing drug delivery systems through biomedical engineering: Innovations and future directions. In *Innovations in Biomedical Engineering* (pp. 1–32). Elsevier., DOI: 10.1016/B978-0-443-30146-9.00001-0

Afrose, N., Chakraborty, R., & Rajendran, K. (2025). Plasma-polymerized chitosan/agarose/chitosan-agarose composite coatings for antibacterial and antibiofilm wound dressings: A review. *International Journal of Biological Macromolecules, 320*, 145873. DOI: 10.1016/j.ijbiomac.2025.145873 PMID: 40645242

Aguilera, A., & García-Muse, T. (2013). Causes of Genome Instability. *Annual Review of Genetics, 47*(1), 1–32. DOI: 10.1146/annurev-genet-111212-133232 PMID: 23909437

Ahmed, R., Alghamdi, W. N., Alharbi, F. R., Alatawi, H. D., Alenezi, K. M., Alanazi, T. F., & Elsherbiny, N. M. (2025). CRISPR/Cas9 System as a Promising Therapy in Thalassemia and Sickle Cell Disease: A Systematic Review of Clinical Trials. *Molecular Biotechnology*. Advance online publication. DOI: 10.1007/s12033-025-01368-x PMID: 39794549

Akram, F., Tanveer, R., Andleeb, S., Shah, F. I., Ahmad, T., Shehzadi, S., Akhtar, A. M., & Syed, G. (2024). Deciphering the epigenetic symphony of cancer: Insights and epigenetic therapies implications. *Technology in Cancer Research & Treatment, 23*, 15330338241250317. DOI: 10.1177/15330338241250317 PMID: 38780251

Al Assi, A., Posty, S., Lamarche, F., Chebel, A., Guitton, J., Cottet-Rousselle, C., Prudent, R., Lafanechère, L., Giraud, S., Dallemagne, P., Suzanne, P., Verney, A., Genestier, L., Castets, M., Fontaine, E., Billaud, M., & Cordier-Bussat, M. (2024). A novel inhibitor of the mitochondrial respiratory complex I with uncoupling properties exerts potent antitumor activity. *Cell Death & Disease, 15*(5), 311. Advance online publication. DOI: 10.1038/s41419-024-06668-9 PMID: 38697987

Alagendran, S., Archunan, G., Prabhu, S. V., Orozco, B. E., & Guzman, R. G. (2010). Biochemical evaluation in human saliva with special reference to ovulation detection. *Indian Journal of Dental Research, 21*(2), 165–168. DOI: 10.4103/0970-9290.66625 PMID: 20657081

Al-Arafat, T., Mao, A., Katsube, T., & Wang, B. (2024). Exploring the role of p53 in radiosensitivity: A key player in cancer therapy. *Radiation, 4*(4), 309–324. DOI: 10.3390/radiation4040023

Ali, K. A., Choudhuri, S., & Roy, S. Kr., Reza, K. H., Mandal, D., Chakraborty, R., & Hossain, C. M. (2025). AI-Generated Nanoparticles in Immunological Research: Opportunities and Challenges. In J.-T. Chen (Ed.), *Advances in Computational Intelligence and Robotics* (pp. 295–322). IGI Global. DOI: 10.4018/979-8-3693-9725-1.ch011

Ali, K. A., Choudhuri, S., Bishal, A., Das, T., Nandy, S., & Hossain, C. M. (2025). Approaches of AI and Advanced Computational Biology for Combating HIV Infections. In J.-T. Chen, *AI and Precision Medicine in Infectious Disease Management* (1st ed., pp. 149–166). CRC Press. DOI: 10.1201/9781003615699-8

Ali, K. A., Choudhuri, S., Chakraborty, R., Pradhan, A., Kotal, A., & Shaharyar, M. A. (2025). AI Technology-Powered Precision Medicine and Smart Healthcare. In J.-T. Chen, *AI and Precision Medicine in Infectious Disease Management* (1st ed., pp. 51–70). CRC Press. DOI: 10.1201/9781003615699-3

Ali, K. A., Mohin, S., Mondal, P., Goswami, S., Ghosh, S., & Choudhuri, S. (2024). Influence of artificial intelligence in modern pharmaceutical formulation and drug development. *Future Journal of Pharmaceutical Sciences*, *10*(1), 53. DOI: 10.1186/s43094-024-00625-1

Aljabali, A. A., El-Tanani, M., & Tambuwala, M. M. (2024). Principles of CRISPR-Cas9 technology: Advancements in genome editing and emerging trends in drug delivery. *Journal of Drug Delivery Science and Technology*, *92*, 105338. DOI: 10.1016/j.jddst.2024.105338

Allan-Blitz, L. T., Adams, G., Sanders, G., Shah, P., Ramesh, K., Jarolimova, J., Ard, K. L., Branda, J. A., Klausner, J. D., Sabeti, P. C., & Lemieux, J. E. (2025). Preliminary clinical performance of a Cas13a-based lateral flow assay for detecting Neisseria gonorrhoeae in urine specimens. *MSphere*, *10*(1), e0067724. DOI: 10.1128/msphere.00677-24 PMID: 39688405

Allemailem, K. S., Almatroodi, S. A., Almatroudi, A., Alrumaihi, F., Al Abdulmonem, W., Al-Megrin, W. A. I., Aljamaan, A. N., Rahmani, A. H., & Khan, A. A. (2023). Recent advances in genome-editing technology with CRISPR/Cas9 variants and stimuli-responsive targeting approaches within tumor cells: A future perspective of cancer management. *International Journal of Molecular Sciences*, *24*(8), 7052. DOI: 10.3390/ijms24087052 PMID: 37108214

Allen, A., Cooper, B. H., Singh, J., Rohs, R., & Qin, P. Z. (2025). PAM-adjacent DNA flexibility tunes CRISPR-Cas12a off-target binding. *Scientific Reports*, *15*(1), 4930. DOI: 10.1038/s41598-025-87565-9 PMID: 39929897

Almotiri, A., Abogosh, A., Abdelfattah, A., Alowaisy, D., & Rodrigues, N. P. (2025). Treating genetic blood disorders in the era of CRISPR-mediated genome editing. *Molecular Therapy : the Journal of the American Society of Gene Therapy*, *33*(6), 2645–2662. DOI: 10.1016/j.ymthe.2025.01.031 PMID: 39827371

Almqvist, J., & Romano, P. R. C. (2020). The Regulation of Human Germline Genome Modification in Europe. In A. Boggio, C. P. R. Romano, & J. Almqvist (Eds.), *Human Germline Genome Modification and the Right to Science* (1st ed., pp. 155–216). Cambridge University Press. DOI: 10.1017/9781108759083.007

Alshammari, E., Zhang, Y., Sobota, J., & Yang, Z. (2022). Aberrant DNA methylation of tumor suppressor genes and oncogenes as cancer biomarkers. *Genomic and Epigenomic Biomarkers of Toxicology and Disease: Clinical and Therapeutic Actions*, 251-271.

Alves, E., Taifour, S., Dolcetti, R., Chee, J., Nowak, A. K., Gaudieri, S., & Blancafort, P. (2021). Reprogramming the anti-tumor immune response via CRISPR genetic and epigenetic editing. *Molecular Therapy. Methods & Clinical Development*, *21*, 592–606. DOI: 10.1016/j.omtm.2021.04.009 PMID: 34095343

Amabile, A., Migliara, A., Capasso, P., Biffi, M., Cittaro, D., Naldini, L., & Lombardo, A. (2016). Inheritable silencing of endogenous genes by hit-and-run targeted epigenetic editing. *Cell*, *167*(1), 219–232.e14. DOI: 10.1016/j.cell.2016.09.006 PMID: 27662090

Ambrosini, S., Mohammed, S. A., Costantino, S., & Paneni, F. (2020). Disentangling the epigenetic landscape in cardiovascular patients: A path toward personalized medicine. *Minerva Cardiology and Angiology*, *69*(3), 331–345. PMID: 32996305

Amoasii, L., Long, C., Li, H., Mireault, A. A., Shelton, J. M., Sanchez-Ortiz, E., & Olson, E. N. (2018). Single-cut genome editing restores dystrophin expression in a new mouse model of muscular dystrophy. *Science Translational Medicine*, *10*(437), eaap8415. PMID: 29187645

Ananthakumar, B., Nagaraja, B. G., Umesh, S. M., & Raghavan, R. (2025). Predictive Medicine. *Predictive Algorithms for Rehabilitation and Assistive Systems*, 291–314. DOI: 10.4018/979-8-3373-0194-5.ch011

Ananthakumar, B., Nanjundeshwara, N., Lakshmaiah, A., & Raghavan, R. (2024). Synthetic Gene Circuits as a Promising Approach in Cancer Immunotherapy. *Critical Developments in Cancer Immunotherapy*, 391–418. DOI: 10.4018/979-8-3693-3976-3.ch011

Ananthakumar, B., Nanjundeshwara, N., Lakshmaiah, A., & Raghavan, R. (2025). Nanocarriers for Delivering Nucleic Acids in Plants. *Nanocarriers in Plant Science and Agriculture*, 207–234. DOI: 10.4018/979-8-3693-9869-2.ch008

Ansori, A., Antonius, Y., Susilo, R. J. K., Hayaza, S., Kharisma, V. D., Parikesit, A. A., Zainul, R., Jakhmola, V., Saklani, T., Rebezov, M., Ullah, M. E., Maksimiuk, N., Derkho, M., & Burkov, P. (2023). Application of CRISPR-Cas9 genome editing technology in various fields: A review. *Narra J*, *3*(2), e184. DOI: 10.52225/narra.v3i2.184 PMID: 38450259

Antimisiaris, S. G., Mourtas, S., & Marazioti, A. (2018). Exosomes and Exosome-Inspired Vesicles for Targeted Drug Delivery. *Pharmaceutics*, *10*(4), 218. DOI: 10.3390/pharmaceutics10040218 PMID: 30404188

Antoniou, P., Miccio, A., & Brusson, M. (2021). Base and prime editing technologies for blood disorders. *Frontiers in Genome Editing*, *3*, 618406. DOI: 10.3389/fgeed.2021.618406 PMID: 34713251

Antón, Z., Mullally, G., Ford, H. C., van der Kamp, M. W., Szczelkun, M. D., & Lane, J. D. (2020). Mitochondrial import, health and mtDNA copy number variability seen when using type II and type V CRISPR effectors. *Journal of Cell Science*, *133*(18), jcs248468. Advance online publication. DOI: 10.1242/jcs.248468 PMID: 32843580

Anzalone, A. V., Koblan, L. W., & Liu, D. R. (2020). Genome editing with CRISPR–Cas nucleases, base editors, transposases and prime editors. *Nature Biotechnology*, *38*(7), 824–844. DOI: 10.1038/s41587-020-0561-9 PMID: 32572269

Anzalone, A. V., Randolph, P. B., Davis, J. R., Sousa, A. A., Koblan, L. W., Levy, J. M., Chen, P. J., Wilson, C., Newby, G. A., Raguram, A., & Liu, D. R. (2020). Search-and-replace genome editing without double-strand breaks or donor DNA. *Nature*, *576*(7785), 149–157. DOI: 10.1038/s41586-019-1711-4 PMID: 31634902

Aronova, A., Tosato, F., Naser, N., & Asare, Y. (2023). Innate immune pathways in atherosclerosis-From signalling to long-term epigenetic reprogramming. *Cells*, *12*(19), 2359. DOI: 10.3390/cells12192359 PMID: 37830572

Arora, R., Sinha, A., & Zhang, Y. (2022). Multi-omics integration in regenerative medicine: From data to therapies. *Trends in Biotechnology*, *40*(11), 1234–1248.

Arowosegbe, J. O. (2024). Before We Forget its Horrors: When Does Human Genome Editing Not Become Eugenics? *Mizan Law Review*, *18*(2), 241–270. DOI: 10.4314/mlr.v18i2.1

Arumugam, V. A., Thangavelu, S., Fathah, Z., Ravindran, P., Sanjeev, A. M. A., Babu, S., Arun, M., Mohd. Iqbal, Y., Khan, S., Ruchi, T., Megha Katare, P., Ranjit, S., Chandra, R., & Dhama, K. (2020). COVID-19 and the world with co-morbidities of heart disease, hypertension and diabetes. *Journal of Pure & Applied Microbiology*, *14*(3), 1623–1638. DOI: 10.22207/JPAM.14.3.01

Asif, M., Khan, W. J., Aslam, S., Aslam, A., & Chowdhury, M. A. (2024). The use of CRISPR-Cas9 genetic technology in cardiovascular disease: A comprehensive review of current progress and future prospective. *Cureus*, *16*(4). Advance online publication. DOI: 10.7759/cureus.57869 PMID: 38725755

Asmamaw Mengstie, M., Teshome Azezew, M., Asmamaw Dejenie, T., Teshome, A. A., Tadele Admasu, F., Behaile Teklemariam, A., Tilahun Mulu, A., Mekonnen Agidew, M., Adugna, D. G., Geremew, H., & Abebe, E. C. (2024). Recent Advancements in Reducing the Off-Target Effect of CRISPR-Cas9 Genome Editing. *Biologics*, *18*, 21–28. DOI: 10.2147/BTT.S429411 PMID: 38260716

Atkins, A., Chung, C.-H., Allen, A. G., Dampier, W., Gurrola, T. E., Sariyer, I. K., Nonnemacher, M. R., & Wigdahl, B. (2021). Off-Target Analysis in Gene Editing and Applications for Clinical Translation of CRISPR/Cas9 in HIV-1 Therapy. *Frontiers in Genome Editing*, *3*, 673022. DOI: 10.3389/fgeed.2021.673022 PMID: 34713260

Azizoglu, R. O. (2022). CRISPR-Cas systems in diagnostics: A comprehensive assessment of Cas effectors and biosensors. *Gene and Genome Editing*, *3*, 100019. https://doi.org/https://doi.org/10.1016/j.ggedit.2022.100019. DOI: 10.1016/j.ggedit.2022.100019

Baghban, R., Roshangar, L., Jahanban-Esfahlan, R., Seidi, K., Ebrahimi-Kalan, A., Jaymand, M., & Zarghami, N. (2024). A review of the literature on the use of CRISPR/Cas9 gene editing in cancer immunotherapy. *International Immunopharmacology*, *127*, 111741. DOI: 10.1016/j.intimp.2024.111741

Balaraman, A. K., Babu, M. A., Moglad, E., Mandaliya, V., Rekha, M. M., Gupta, S., Prasad, G. V. S., Kumari, M., Chauhan, A. S., Ali, H., & Goyal, K. (2025). Exosome-mediated delivery of CRISPR-Cas9: A revolutionary approach to cancer gene editing. *Pathology, Research and Practice*, *266*, 155785. DOI: 10.1016/j.prp.2024.155785 PMID: 39708520

Ball, M., van Bergen, N. J., Compton, A. G., Thorburn, D. R., Rahman, S., & Christodoulou, J. (2025). Therapies for Mitochondrial Disease: Past, Present, and Future. *Journal of Inherited Metabolic Disease*, *48*(4), e70065. DOI: 10.1002/jimd.70065 PMID: 40714961

Baltimore, D., Berg, P., Botchan, M., Carroll, D., Charo, R. A., Church, G., Corn, J. E., Daley, G. Q., Doudna, J. A., Fenner, M., Greely, H. T., Jinek, M., Martin, G. S., Penhoet, E., Puck, J., Sternberg, S. H., Weissman, J. S., & Yamamoto, K. R. (2015). A prudent path forward for genomic engineering and germline gene modification. *Science*, *348*(6230), 36–38. DOI: 10.1126/science.aab1028 PMID: 25791083

Balzano, L., Tumiatti, V., & Arnesano, F. (2021). Metallothioneins as multitasking mediators of metal homeostasis and redox balance. *Journal of Inorganic Biochemistry*, *225*, 111595. DOI: 10.1016/j.jinorgbio.2021.111595

Bandyopadhyay, A., Kancharla, N., Javalkote, V. S., Dasgupta, S., & Brutnell, T. P. (2020). CRISPR-Cas12a (Cpf1): A versatile tool in the plant genome editing tool box for agricultural advancement. *Frontiers in Plant Science*, *11*, 584151. DOI: 10.3389/fpls.2020.584151 PMID: 33214794

Barbi, V., De Martino, S., Aiello, A., Gottardi Zamperla, M., Negri, S., Cis, L., Pecci, V., Nanni, S., Farsetti, A., Martelli, F., Gaetano, C., & Atlante, S. (2025). Non-coding RNAs as novel biomarkers and therapeutic targets in breast cancer. *Oncology Reviews*, *19*, 1621144. DOI: 10.3389/or.2025.1621144 PMID: 40951838

Bareford, L., & Swaan, P. (2007). Endocytic mechanisms for targeted drug delivery☆. *Advanced Drug Delivery Reviews*, *59*(8), 748–758. DOI: 10.1016/j.addr.2007.06.008 PMID: 17659804

Barrangou, R., & Doudna, J. A. (2016). Applications of CRISPR technologies in research and beyond. *Nature Biotechnology*, *34*(9), 933–941. DOI: 10.1038/nbt.3659 PMID: 27606440

Barrangou, R., Fremaux, C., Deveau, H., Richards, M., Boyaval, P., Moineau, S., Romero, D. A., & Horvath, P. (2007). CRISPR provides acquired resistance against viruses in prokaryotes. *Science*, *315*(5819), 1709–1712. DOI: 10.1126/science.1138140 PMID: 17379808

Baylis, F., Darnovsky, M., Hasson, K., & Krahn, T. M. (2020). Human Germline and Heritable Genome Editing: The Global Policy Landscape. *The CRISPR Journal*, *3*(5), 365–377. DOI: 10.1089/crispr.2020.0082 PMID: 33095042

Ben Jehuda, R., Eisen, B., Shemer, Y., Mekies, L. N., Szantai, A., Reiter, I., & Binah, O. (2018). CRISPR correction of the PRKAG2 gene mutation in the patient's iPSC-derived cardiomyocytes eliminates the electrophysiological and structural abnormalities. *Heart Rhythm*, *15*(2), 267–276. DOI: 10.1016/j.hrthm.2017.09.024 PMID: 28917552

Bengtson, M., Bharadwaj, M., Franch, O., van der Torre, J., Meerdink, V., Schallig, H., & Dekker, C. (2022). CRISPR-dCas9 based DNA detection scheme for diagnostics in resource-limited settings. *Nanoscale*, *14*(5), 1885–1895. DOI: 10.1039/D1NR06557B PMID: 35044397

Ben-Mordechai, T., Palevski, D., Glucksam-Galnoy, Y., Elron-Gross, I., Margalit, R., & Leor, J. (2015). Targeting macrophage subsets for infarct repair. *Journal of Cardiovascular Pharmacology and Therapeutics*, *20*(1), 36–51. DOI: 10.1177/1074248414534916 PMID: 24938456

Bhardwaj, P., Kant, R., Behera, S. P., Dwivedi, G. R., & Singh, R. (2022). Next-generation diagnostic with CRISPR/Cas: Beyond nucleic acid detection. *International Journal of Molecular Sciences*, *23*(11), 6052. DOI: 10.3390/ijms23116052 PMID: 35682737

Bharucha, N., Arias, A., & Karakikes, I. (2022). The potential of CRISPR-Cas9 prime editing for cardiovascular disease research and therapy. *Current Opinion in Cardiology*, *37*(5), 413–418. DOI: 10.1097/HCO.0000000000000985 PMID: 35880456

Bharucha, N., Wong, C., Marti-Gutierrez, N., Nguyen, D., Puri, M., Wu, J., & Wu, J. C. (2021). Generation of AAVS1 integrated doxycycline-inducible prime editor into human iPSCs. *medRxiv*.

Bhotla, H. K., Balasubramanian, B., Meyyazhagan, A., Pushparaj, K., & Easwaran, M. (2021). Opportunistic mycoses in COVID-19 patients/survivors: Epidemic inside a pandemic. *Journal of Infection and Public Health*, *14*(11), 1720–1726. DOI: 10.1016/j.jiph.2021.10.010 PMID: 34700291

Bhuvana, G. N., Umesh, S. M., & Raghavan, R. (2025). Expanding the frontiers of Schistosomiasis Treatment: The promise of Nanoparticle-Based Therapies. *Journal of Vector Borne Diseases, 10*(4103/jvbd.jvbd_180_23). DOI: 10.4103/jvbd.jvbd_180_23

Bi, R., Li, Y., Xu, M., Zheng, Q., Zhang, D.-F., Li, X., Ma, G., Xiang, B., Zhu, X., Zhao, H., Huang, X., Zheng, P., & Yao, Y.-G. (2022). Direct evidence of CRISPR-Cas9-mediated mitochondrial genome editing. *Innovation (Cambridge (Mass.))*, *3*(6), 100329. DOI: 10.1016/j.xinn.2022.100329 PMID: 36275864

Bolotin, A., Quinquis, B., Sorokin, A., & Ehrlich, S. D. (2005). Clustered regularly interspaced short palindrome repeats (CRISPRs) have spacers of extrachromosomal origin. *Microbiology*, *151*(8), 2551–2561. DOI: 10.1099/mic.0.28048-0 PMID: 16079334

Bonowicz, K., Jerka, D., Piekarska, K., Olagbaju, J., Stapleton, L., Shobowale, M., & Gagat, M. (2025). CRISPR-Cas9 in cardiovascular medicine: Unlocking new potential for treatment. *Cells*, *14*(2), 131. DOI: 10.3390/cells14020131 PMID: 39851560

Bordin, D. L., Silva, T. B., & Pedrosa, F. O. (2023). Regulation and functional diversity of metallothionein isoforms across biological systems. *Environmental Toxicology and Pharmacology*, *104*, 104216. DOI: 10.1016/j.etap.2023.104216

Bower, J. J., Leonard, S. S., & Shi, X. L. (2005). Conference overview: Molecular mechanisms of metal toxicity and carcinogenesis. *Molecular and Cellular Biochemistry*, *279*(1–2), 3–15. https://doi.org/. DOI: 10.1007/s11010-005-8210-7 PMID: 16283510

Bowley, G., Kugler, E., Wilkinson, R., Lawrie, A., van Eeden, F., Chico, T. J., & Serbanovic-Canic, J. (2022). Zebrafish as a tractable model of human cardiovascular disease. *British Journal of Pharmacology*, *179*(5), 900–917. DOI: 10.1111/bph.15473 PMID: 33788282

Braddick, D., & Ramarohetra, R. F. (2020). Emergent challenges for CRISPR: biosafety, biosecurity, patenting, and regulatory issues. In *Genome engineering via CRISPR-Cas9 system* (pp. 281–307). Academic Press. DOI: 10.1016/B978-0-12-818140-9.00021-0

Bright, S. J., Flint, D. B., Martinus, D. K., Turner, B. X., Manandhar, M., Ben Kacem, M., & Sawakuchi, G. O. (2022). Targeted inhibition of DNA-PKcs, ATM, ATR, PARP, and Rad51 modulate response to X rays and protons. *Radiation Research*, *198*(4), 336–346. DOI: 10.1667/RADE-22-00040.1 PMID: 35939823

Brock, D. J., Kondow-McConaghy, H. M., Hager, E. C., & Pellois, J.-P. (2019). Endosomal Escape and Cytosolic Penetration of Macromolecules Mediated by Synthetic Delivery Agents. *Bioconjugate Chemistry*, *30*(2), 293–304. DOI: 10.1021/acs.bioconjchem.8b00799 PMID: 30462487

Brokowski, C., & Adli, M. (2019). CRISPR Ethics: Moral Considerations for Applications of a Powerful Tool. *Journal of Molecular Biology*, *431*(1), 88–101. DOI: 10.1016/j.jmb.2018.05.044 PMID: 29885329

Brossard, D., Scheufele, D. A., Kim, E., & Lewenstein, B. V. (2020). Public perceptions of gene-editing technologies: A cross-national survey. *Nature Biotechnology*, *38*(12), 1421–1430. PMID: 33273741

Broughton, J. P., Deng, X., Yu, G., Fasching, C. L., Servellita, V., Singh, J., Miao, X., Streithorst, J. A., Granados, A., Sotomayor-Gonzalez, A., Zorn, K., Gopez, A., Hsu, E., Gu, W., Miller, S., Pan, C.-Y., Guevara, H., Wadford, D. A., Chen, J. S., & Chiu, C. Y. (2020). CRISPR-Cas12-based detection of SARS-CoV-2. *Nature Biotechnology, 38*(7), 870–874. DOI: 10.1038/s41587-020-0513-4 PMID: 32300245

Buchanan, A., Borenstein, J., & Brown, A. (2021). Governance and oversight in human gene-editing technologies. *Journal of Law and the Biosciences, 8*(1), lsab003. PMID: 33981445

Bueren, J. A., Quintana-Bustamante, O., Almarza, E., Navarro, S., Río, P., Segovia, J. C., & Guenechea, G. (2020). Advances in the gene therapy of monogenic blood cell diseases. *Clinical Genetics, 97*(1), 89–102. DOI: 10.1111/cge.13593 PMID: 31231794

Butt, H., Mandava, M., & Jacobsohn, D. (2025). Advances in Gene Therapy for Sickle Cell Disease: From Preclinical Innovations to Clinical Implementation and Access Challenges. *The CRISPR Journal, 8*(3), 174–188. DOI: 10.1089/crispr.2024.0101 PMID: 40356202

Byalahunasishetter, B. I., N., V., P., S., C., M., S., S., & Raghavan, R. (2025). Spatial Omics and Bioprinting. *Innovations in Precision Medicine and Genomics*, 85–110. DOI: 10.4018/979-8-3693-5787-3.ch003

Cai, R., Lv, R., Shi, X., Yang, G., & Jin, J. (2023). CRISPR/dCas9 Tools: Epigenetic Mechanism and Application in Gene Transcriptional Regulation. *International Journal of Molecular Sciences, 24*(19), 14865. DOI: 10.3390/ijms241914865 PMID: 37834313

Cameron, D. E., Bashor, C. J., & Collins, J. J. (2020). Public engagement in synthetic biology and genome editing: Lessons for CRISPR. *Nature Reviews. Genetics, 21*(11), 721–733.

Capolla, S., Rasool, M., Toffoli, G., & Dal Bo, M. (2025). CAR-T Cell Manufacturing for Hematological and Solid Tumors: From the Preclinical to Clinical Point of View. *Cancer Medicine, 14*(5), e70726. DOI: 10.1002/cam4.70726 PMID: 40013750

Carbone, F., & Montecucco, F. (2024). Facing ethical concerns in the age of precise gene therapy: Outlook on inherited arrhythmias. *World Journal of Cardiology, 16*(2), 64–66. DOI: 10.4330/wjc.v16.i2.64 PMID: 38456071

Carinci, M., Vezzani, B., Patergnani, S., Ludewig, P., Lessmann, K., Magnus, T., Casetta, I., Pugliatti, M., Pinton, P., & Giorgi, C. (2021). Different Roles of Mitochondria in Cell Death and Inflammation: Focusing on Mitochondrial Quality Control in Ischemic Stroke and Reperfusion. *Biomedicines*, *9*(2), 169. DOI: 10.3390/biomedicines9020169 PMID: 33572080

Carroll, M. S., & Giacca, M. (2023). CRISPR activation and interference as investigative tools in the cardiovascular system. *The International Journal of Biochemistry & Cell Biology*, *155*, 106348. DOI: 10.1016/j.biocel.2022.106348 PMID: 36563996

Cattle, M. A., Aguado, L. C., Sze, S., Venkittu, S., Wang, Y., Papagiannakopoulos, T., Smith, S., Rice, C. M., Schneider, W. M., & Poirier, J. T. (2025). An enhanced Eco1 retron editor enables precision genome engineering in human cells without double-strand breaks. *Nucleic Acids Research*, *53*(14), gkaf716. Advance online publication. DOI: 10.1093/nar/gkaf716 PMID: 40744490

Chakraborty, R., Jana, B., Koner, S., Kanta Maiti, S., Afrose, N., Bhowmick, P., & Bhowmick, M. (2024). Exploring the transcriptomic landscape of rheumatoid arthritis using next-generation RNA sequencing. *INNOSC Theranostics & Pharmacological Sciences*, *8*(1), 14. DOI: 10.36922/itps.4449

Chakraborty, R., Kannan, S. P., Afrose, N., & Narayanasamy, D. (2025). Tumor microenvironment expressed enzymes (Matrix metalloproteinases, cathepsins, urokinase-type plasminogen activator) triggered polymersomes for liquid biopsy and cancer diagnostics: A review. *International Journal of Biological Macromolecules*, *321*, 146375. DOI: 10.1016/j.ijbiomac.2025.146375 PMID: 40738429

Changaei, M., Azimzadeh Tabrizi, Z., Karimi, M., Kashfi, S. A., Koochaki Chahardeh, T., Hashemi, S. M., & Soudi, S. (2025). From powerhouse to modulator: Regulating immune system responses through intracellular mitochondrial transfer. *Cell Communication and Signaling : CCS*, *23*(1), 232. DOI: 10.1186/s12964-025-02237-5 PMID: 40394666

Chang, W., Zhao, Y., Rayêe, D., Xie, Q., Suzuki, M., Zheng, D., & Cvekl, A. (2023). Dynamic changes in whole genome DNA methylation, chromatin and gene expression during mouse lens differentiation. *Epigenetics & Chromatin*, *16*(1), 4. DOI: 10.1186/s13072-023-00478-7 PMID: 36698218

Charlesworth, C. T., Deshpande, P. S., Dever, D. P., Dejene, B., Gomez-Ospina, N., Mantri, S., Pavel-Dinu, M., Camarena, J., Weinberg, K. I., & Porteus, M. H. (2019). Identification of preexisting adaptive immunity to Cas9 proteins in humans. *Nature Medicine*, *25*(2), 249–254. DOI: 10.1038/s41591-018-0326-x PMID: 30692695

Chatgilialoglu, C., Ferreri, C., Krokidis, M. G., Masi, A., & Terzidis, M. A. (2021). On the relevance of hydroxyl radical to purine DNA damage. *Free Radical Research*, *55*(4), 384–404. DOI: 10.1080/10715762.2021.1876855 PMID: 33494618

Chattopadhyay, R., Firdous, Z., & Bari, V. K. (2025). CRISPR/Cas9 and its derivatives to improve crop biotic stress resistance: Current status and prospects. *Physiological and Molecular Plant Pathology*, *135*, 102482. DOI: 10.1016/j.pmpp.2024.102482

Chaudhuri, O., Cooper, J. A., & Mooney, D. J. (2023). Multi-omics and CRISPR in tissue regeneration: Toward precision therapeutics. *Nature Reviews Bioengineering*, *1*(3), 211–225.

Chavez, A., Scheiman, J., Vora, S., Pruitt, B. W., Tuttle, M., & Iyer, P. R. (2015). Highly efficient Cas9-mediated transcriptional programming. *Nature Methods*, *12*(4), 326–328. DOI: 10.1038/nmeth.3312 PMID: 25730490

Chehelgerdi, M., Chehelgerdi, M., Allela, O. Q. B., Pecho, R. D. C., Jayasankar, N., Rao, D. P., & Amiri, M. (2024). Comprehensive review of CRISPR-based gene editing: Mechanisms, challenges, and applications in cancer therapy. *Molecular Cancer*, *23*(1), 9. DOI: 10.1186/s12943-023-01925-5 PMID: 38195537

Chemello, F., Chai, A. C., Li, H., Rodriguez-Caycedo, C., Sanchez-Ortiz, E., Atmanli, A., & Olson, E. N. (2021). Precise correction of dystrophin mutations in cardiomyocytes derived from Duchenne muscular dystrophy patients via base editing. *Science Advances*, *7*(12), eabg4910. DOI: 10.1126/sciadv.abg4910 PMID: 33931459

Chen, C., Tang, J., Xu, S., Zhang, W., & Jiang, H. (2020). Research Article miR-30a-5p Inhibits Proliferation and Migration of Lung Squamous Cell Carcinoma Cells by Targeting FOXD1.

Chen, J. S., & Doudna, J. A. (2017). The chemistry of Cas9 and its applications in genome editing. Nature Reviews Chemistry, 1(10), 0078.

Chen, Z., Lehertshuber, C., & Schunkert, H. (2022). Genome editing in dyslipidemia and atherosclerosis. *Genome Editing in Cardiovascular and Metabolic Diseases*, 139-156.

Chen, F., Alphonse, M., & Liu, Q. (2020). Strategies for nonviral nanoparticle-based delivery of CRISPR/Cas9 therapeutics. *Wiley Interdisciplinary Reviews. Nanomedicine and Nanobiotechnology*, *12*(3), e1609. DOI: 10.1002/wnan.1609 PMID: 31797562

Cheng, Jie, Duan, Yukan, & Cynthia, A Berlinicke, & Donald J Zack. (2023). CRISPR interference knockdown of Dynamin-Related Protein 1 (DRP1) restores mitochondrial homeostasis disrupted by inhibition of Optic Atrophy 1 (OPA1) expression. *Investigative Ophthalmology & Visual Science*, •••, 64.

Cheng, Q., Zhu, Y., Lv, S., Shi, J., Kuang, M., Wang, L., & Ji, X. (2026). Engineered bacteria and bacteria-derived nanomaterials for cancer therapy: Mechanisms, designs and advances. *Bioactive Materials*, *55*, 1–22. DOI: 10.1016/j.bioactmat.2025.09.012 PMID: 41368325

Chen, J., Liu, Y., Zhang, Y., & Xu, C. (2021). CRISPR-based modulation of PTEN in spinal cord injury models promotes axonal regeneration. *Nature Communications*, *12*(1), 1–13. PMID: 33397941

Chen, S., Sanjana, N. E., & Shalem, O. (2021). Dissecting neural and stem cell networks using CRISPR and single-cell transcriptomics. *Cell Stem Cell*, *28*(6), 1000–1014.

Chen, X., Torres, J. P. M., Li, Y., Lugones, L. G., & Wösten, H. A. (2023). Inheritable CRISPR based epigenetic modification in a fungus. *Microbiological Research*, *272*, 127397. DOI: 10.1016/j.micres.2023.127397 PMID: 37141850

Chen, X., Zhang, J., Gao, F., Liu, N., Du, H., Li, J., Li, Z., & Chen, R. (2025). Exercise therapy: An effective approach to mitigate the risk of cancer metastasis. *World Journal of Surgical Oncology*, *23*(1), 192. DOI: 10.1186/s12957-025-03846-7 PMID: 40380237

Chen, Y., Li, W., Zhang, H., & Xu, D. (2022). CRISPR/dCas9-mediated regulation of noncoding RNAs to overcome chemoresistance in cancer therapy. *Molecular Therapy. Nucleic Acids*, *28*, 1082–1095.

Chillón-Pino, D., Badonyi, M., Semple, C. A., & Marsh, J. A. (2024). Protein structural context of cancer mutations reveals molecular mechanisms and candidate driver genes. *Cell Reports*, *43*(11), 114905. DOI: 10.1016/j.celrep.2024.114905 PMID: 39441719

Chodnekar, S. R., Karsolia, R., & De, A. (2024). Systematic review of CRISPR clinical trials. *Gene Therapy*, *31*(11–12), 579–592. DOI: 10.1038/s41434-024-00456-3

Cho, H. M., Lee, K. H., Shen, Y. M., Shin, T. J., Ryu, P. D., Choi, M. C., & Cho, J. Y. (2020). Transplantation of hMSCs genome edited with LEF1 improves cardio-protective effects in myocardial infarction. *Molecular Therapy. Nucleic Acids*, *19*, 1186–1197. DOI: 10.1016/j.omtn.2020.01.007 PMID: 32069701

Choi, W., Cha, S., & Kim, K. (2024). Navigating the CRISPR/Cas landscape for enhanced diagnosis and treatment of Wilson's disease. *Cells*, *13*(14), 1214. Advance online publication. DOI: 10.3390/cells13141214 PMID: 39056796

Cho, S. I., Lim, K., Hong, S., Kim, J. S., & Bae, S. (2024). Engineering TALE-linked deaminases to facilitate precision adenine base editing in mitochondrial DNA. *Cell*, *187*(1), 95–109.e26. DOI: 10.1016/j.cell.2023.11.035 PMID: 38181745

Chuai, G., Ma, H., Yan, J., Chen, M., Hong, N., Xue, D., Zhou, C., Zhu, C., Chen, K., Duan, B., Gu, F., Qu, S., Huang, D., Wei, J., & Liu, Q. (2018). DeepCRISPR: Optimized CRISPR guide RNA design by deep learning. *Genome Biology*, *19*(1), 80. DOI: 10.1186/s13059-018-1459-4 PMID: 29945655

Chuang, Y.-F., Phipps, A. J., Lin, F.-L., Hecht, V., Hewitt, A. W., Wang, P.-Y., & Liu, G.-S. (2021). Approach for in vivo delivery of CRISPR/Cas system: A recent update and future prospect. *Cellular and Molecular Life Sciences : CMLS*, *78*(6), 2683–2708. DOI: 10.1007/s00018-020-03725-2 PMID: 33388855

Chung, H. K., Jacobs, C. L., Huo, Y., Yang, J., Krumm, S. A., Plemper, R. K., & Lin, M. Z. (2019). Tunable and reversible drug control of protein production via a self-excising degron. *Nature Chemical Biology*, *15*(5), 520–527. PMID: 26214256

Clemens, S. (2022). Metal transport and detoxification: Evolutionary insights and molecular mechanisms. *Annual Review of Plant Biology*, *73*, 129–159. DOI: 10.1146/annurev-arplant-102720-031359

Clift, C. L., Blaser, M. C., Gerrits, W., Turner, M. E., Sonawane, A., Pham, T., Andresen, J. L., Fenton, O. S., Grolman, J. M., Campedelli, A., Buffolo, F., Schoen, F. J., Hjortnaes, J., Muehlschlegel, J. D., Mooney, D. J., Aikawa, M., Singh, S. A., Langer, R., & Aikawa, E. (2024). Intracellular proteomics and extracellular vesiculomics as a metric of disease recapitulation in 3D-bioprinted aortic valve arrays. *Science Advances*, *10*(9), eadj9793. Advance online publication. DOI: 10.1126/sciadv.adj9793 PMID: 38416823

Cohen, I. G., & Adashi, E. Y. (2020). Access, equity, and the commercialization of CRISPR-based therapies. *Journal of the American Medical Association*, *324*(16), 1607–1608. PMID: 32970138

Coller, B. S. (2019). Ethics of Human Genome Editing. *Annual Review of Medicine*, *70*(1), 289–305. DOI: 10.1146/annurev-med-112717-094629 PMID: 30691366

Collias, D., & Beisel, C. L. (2021). CRISPR technologies and the search for the PAM-free nuclease. *Nature Communications*, *12*(1), 555. DOI: 10.1038/s41467-020-20633-y PMID: 33483498

Collins, L., Binder, P., Chen, H., & Wang, X. (2020). Regulation of long non-coding RNAs and microRNAs in heart disease: Insight into mechanisms and therapeutic approaches. *Frontiers in Physiology*, *11*, 798. DOI: 10.3389/fphys.2020.00798 PMID: 32754048

Cong, L., Ran, F. A., Cox, D., Lin, S., Barretto, R., Habib, N., Hsu, P. D., Wu, X., Jiang, W., Marraffini, L. A., & Zhang, F. (2013). Multiplex genome engineering using CRISPR/Cas systems. *Science*, *339*(6121), 819–823. DOI: 10.1126/science.1231143 PMID: 23287718

Costantino, S., Libby, P., Kishore, R., Tardif, J. C., El-Osta, A., & Paneni, F. (2018). Epigenetics and precision medicine in cardiovascular patients: From basic concepts to the clinical arena. *European Heart Journal*, *39*(47), 4150–4158. DOI: 10.1093/eurheartj/ehx568 PMID: 29069341

Cox, D. B. T., Platt, R. J., & Zhang, F. (2015). Therapeutic genome editing: Prospects and challenges. *Nature Medicine*, *21*(2), 121–131. DOI: 10.1038/nm.3793 PMID: 25654603

Cruz-Rodríguez, L., Guerra-Tschuschke, I., & González-Iglesias, H. (2024). Metallothioneins as biomarkers of metal-induced toxicity and cellular defense. *Toxicology Reports*, *11*, 210–223. DOI: 10.1016/j.toxrep.2024.02.015

Cyranoski, D. (2019). The CRISPR-baby scandal: What's next for human gene-editing. *Nature*, *566*(7745), 440–442. DOI: 10.1038/d41586-019-00673-1 PMID: 30809070

Dagar, G., Gupta, A., Masoodi, T., Nisar, S., Merhi, M., Hashem, S., Chauhan, R., Dagar, M., Mirza, S., Bagga, P., Kumar, R., Akil, A. S. A.-S., Macha, M. A., Haris, M., Uddin, S., Singh, M., & Bhat, A. A. (2023). Harnessing the potential of CAR-T cell therapy: Progress, challenges, and future directions in hematological and solid tumor treatments. *Journal of Translational Medicine*, *21*(1), 449. DOI: 10.1186/s12967-023-04292-3 PMID: 37420216

Dara, M., Shafieipour, N., Saffar, M., Dianatpour, M., Tabei, S.-M.-B., & Dastgheib, S.-A. (2025). Diagnosis of Infectious Diseases by CRISPR/Cas System. *OBM Genetics*, *9*(2), 1–39. DOI: 10.21926/obm.genet.2502289

Das, A. T., Binda, C. S., & Berkhout, B. (2019). Elimination of infectious HIV DNA by CRISPR–Cas9. *Current Opinion in Virology*, *38*, 81–88. DOI: 10.1016/j.coviro.2019.07.001 PMID: 31450074

Dasgupta, S., Gayen, S., Chakraborty, T., Afrose, N., Pal, R., Mahata, S., Nasare, V., & Roy, S. (2024). Potential role of immune cell therapy in gynecological cancer and future promises: A comprehensive review. *Medical Oncology (Northwood, London, England)*, *41*(5), 98. DOI: 10.1007/s12032-024-02337-1 PMID: 38536512

Das, S., Bano, S., Kapse, P., & Kundu, G. C. (2022). CRISPR based therapeutics: A new paradigm in cancer precision medicine. *Molecular Cancer*, *21*(1), 85. DOI: 10.1186/s12943-022-01552-6 PMID: 35337340

Datlinger, P., Rendeiro, A. F., Schmidl, C., Krausgruber, T., Traxler, P., Klughammer, J., & Bock, C. (2017). Pooled CRISPR screening with single-cell transcriptome readout. *Nature Methods*, *14*(3), 297–301. DOI: 10.1038/nmeth.4177 PMID: 28099430

Davis-Anderson, K., Micheva-Viteva, S., Solomon, E., Hovde, B., Cirigliano, E., Harris, J., Twary, S., & Iyer, R. (2023). CRISPR/Cas9 directed reprogramming of iPSC for accelerated motor neuron differentiation leads to dysregulation of neuronal fate patterning and function. *International Journal of Molecular Sciences*, *24*(22), 16161. DOI: 10.3390/ijms242216161 PMID: 38003351

de Vos, L., Grünwald, I., Bawden, E. G., Dietrich, J., Scheckenbach, K., Wiek, C., Zarbl, R., Bootz, F., Landsberg, J., & Dietrich, D. (2020). The landscape of CD28, CD80, CD86, CTLA4, and ICOS DNA methylation in head and neck squamous cell carcinomas. *Epigenetics*, *15*(11), 1195–1212. DOI: 10.1080/15592294.2020.1754675 PMID: 32281488

DeLuca, S., & Bursac, N. (2022). CRISPR library screening in cultured cardiomyocytes. In *Cardiac Tissue Engineering: Methods and Protocols* (pp. 1–13). Springer US. DOI: 10.1007/978-1-0716-2261-2_1

Deng, B., Kong, Y., Ma, Y., Zhan, Y., Sun, Y., Wang, R., Huang, P., & Liu, L. (2025). A novel carrier-free nanoparticle with stable distinctive three-dimensional structure for tumor-targeted precision chemoimmunotherapy. *Journal of Nanobiotechnology*, *23*(1), 1–15. DOI: 10.1186/s12951-025-03568-8 PMID: 40598510

Deng, S., Hu, Q., Zhang, H., Yang, F., Peng, C., & Huang, C. (2019). HDAC3 inhibition upregulates PD-L1 expression in B-cell lymphomas and augments the efficacy of anti–PD-L1 therapy. *Molecular Cancer Therapeutics*, *18*(5), 900–908. DOI: 10.1158/1535-7163.MCT-18-1068 PMID: 30824609

Deng, W., Chen, S., & Liu, Y. (2024). Genome-wide analysis of glutathione S-transferases and their roles in heavy metal detoxification. *Ecotoxicology and Environmental Safety*, *282*, 115919. DOI: 10.1016/j.ecoenv.2024.115919

Desai, N., Rana, D., Salave, S., Gupta, R., Patel, P., Karunakaran, B., Sharma, A., Giri, J., Benival, D., & Kommineni, N. (2023). Chitosan: A Potential Biopolymer in Drug Delivery and Biomedical Applications. *Pharmaceutics, 15*(4), 1313. DOI: 10.3390/pharmaceutics15041313 PMID: 37111795

Desouky, O., Ding, N., & Zhou, G. (2015). Targeted and non-targeted effects of ionizing radiation. *Journal of Radiation Research and Applied Sciences, 8*(2), 247–254. DOI: 10.1016/j.jrras.2015.03.003

Dhar, G. A., Saha, S., Mitra, P., & Nag Chaudhuri, R. (2021). DNA methylation and regulation of gene expression: Guardian of our health. *Nucleus (Austin, Tex.), 64*(3), 259–270. PMID: 34421129

Dhiflaoui, A., & Almawi, W. Y. (2025). Mechanisms and clinical implications of microRNA associations and signaling pathways in B-cell acute lymphoblastic leukemia. *Gene, 967*, 149730. DOI: 10.1016/j.gene.2025.149730 PMID: 40829668

Di Matteo, M., Belay, E., Chuah, M. K., & VandenDriessche, T. (2012). Recent developments in transposon-mediated gene therapy. *Expert Opinion on Biological Therapy, 12*(7), 841–858. DOI: 10.1517/14712598.2012.684875 PMID: 22679910

Dimitri, A., Herbst, F., & Fraietta, J. A. (2022). Engineering the next-generation of CAR T-cells with CRISPR-Cas9 gene editing. *Molecular Cancer, 21*(1), 78. DOI: 10.1186/s12943-022-01559-z PMID: 35303871

Dimond, R., & Stephens, N. (2018). Policy Work and Legitimacy at the Nuffield Council on Bioethics, the Human Fertilisation and Embryology Authority, and the Department of Health. In *Legalising Mitochondrial Donation: Enacting Ethical Futures in UK Biomedical Politics* (pp. 69–88). Springer International Publishing. DOI: 10.1007/978-3-319-74645-6_4

Dixit, S., Kumar, A., Srinivasan, K., Vincent, P. D. R., & Ramu Krishnan, N. (2024). Advancing genome editing with artificial intelligence: Opportunities, challenges, and future directions. *Frontiers in Bioengineering and Biotechnology, 11*, 1335901. DOI: 10.3389/fbioe.2023.1335901 PMID: 38260726

Dölarslan, M. (2023). CRISPR-Cas9 Mediated Gene Correction of CFTR Mutations in Cystic Fibrosis: Evaluating Efficacy, Safety, and Long-Term Outcomes in Patient-Derived Lung Organoids. *SHIFAA, 2023*, 41–47. DOI: 10.70470/SHIFAA/2023/005

Dominguez, A. A., Lim, W. A., & Qi, L. S. (2016). Beyond editing: Repurposing CRISPR–Cas9 for precision genome regulation and interrogation. *Nature Reviews. Molecular Cell Biology, 17*(1), 5–15. DOI: 10.1038/nrm.2015.2 PMID: 26670017

Donovan, D. A., Crandall, J. G., Banks, O. G., Jensvold, Z. D., Truong, V., Dinwiddie, D., McKnight, L. E., & McKnight, J. N. (2019). Engineered chromatin remodeling proteins for precise nucleosome positioning. *Cell Reports*, *29*(8), 2520–2535. DOI: 10.1016/j.celrep.2019.10.046 PMID: 31747617

Doudna, J. A. (2020). The promise and challenge of therapeutic genome editing. *Nature*, *578*(7794), 229–236. DOI: 10.1038/s41586-020-1978-5 PMID: 32051598

Doudna, J. A., & Charpentier, E. (2014). The new frontier of genome engineering with CRISPR-Cas9. *Science*, *346*(6213), 1258096. DOI: 10.1126/science.1258096 PMID: 25430774

Doudna, J. A., & Charpentier, E. (2023). The new frontier of genome engineering with CRISPR-Cas systems. *Science*, *380*(6640), 1020–1031.

Dou, H. Y., Huang, T. S., Wu, H. C., Hsu, C. H., Chen, F. J., & Liao, Y. C. (2025). Targeted sputum sequencing for rapid and broad drug resistance of Mycobacterium tuberculosis. *Infection*, *53*(4), 1413–1423. DOI: 10.1007/s15010-024-02463-y PMID: 39821740

Dräger, N. M., Sattler, S. M., Huang, C. T. L., Teter, O. M., Leng, K., Hashemi, S. H., Hong, J., Aviles, G., Clelland, C. D., Zhan, L., Udeochu, J. C., Kodama, L., Singleton, A. B., Nalls, M. A., Ichida, J., Ward, M. E., Faghri, F., Gan, L., & Kampmann, M. (2022). A CRISPRi/a platform in human iPSC-derived microglia uncovers regulators of disease states. *Nature Neuroscience*, *25*(9), 1149–1162. DOI: 10.1038/s41593-022-01131-4 PMID: 35953545

Dronina, J., Samukaite-Bubniene, U., & Ramanavicius, A. (2022). Towards application of CRISPR-Cas12a in the design of modern viral DNA detection tools [Review]. *Journal of Nanobiotechnology*, *20*(1), 41. DOI: 10.1186/s12951-022-01246-7 PMID: 35062978

Drost, J., van Jaarsveld, R. H., Ponsioen, B., Zimberlin, C., van Boxtel, R., Buijs, A., Sachs, N., Overmeer, R. M., Offerhaus, G. J., Begthel, H., Korving, J., van de Wetering, M., Schwank, G., Logtenberg, M., Cuppen, E., Snippert, H. J., Medema, J. P., Kops, G. J. P. L., & Clevers, H. (2015). Sequential cancer mutations in cultured human intestinal stem cells. *Nature*, *521*(7550), 43–47. DOI: 10.1038/nature14415 PMID: 25924068

Du, H., Xu, T., Yu, S., Wu, S., & Zhang, J. (2025). Mitochondrial metabolism and cancer therapeutic innovation. In *Signal Transduction and Targeted Therapy* (Vol. 10, Issue 1). Springer Nature. https://doi.org/DOI: 10.1038/s41392-025-02311-x

Duan, L., Ouyang, K., Wang, J., Xu, L., Xu, X., Wen, C., Xie, Y., Liang, Y., & Xia, J. (2021). Exosomes as Targeted Delivery Platform of CRISPR/Cas9 for Therapeutic Genome Editing. *ChemBioChem, 22*(24), 3360–3368. DOI: 10.1002/cbic.202100359 PMID: 34418266

Duan, L., Ouyang, K., Xu, X., Xu, L., Wen, C., Zhou, X., & Liang, Y. (2021). Nanoparticle delivery of CRISPR/Cas9 for genome editing. *Frontiers in Genetics, 12*, 673286. DOI: 10.3389/fgene.2021.673286 PMID: 34054927

Dubey, A. K., & Mostafavi, E. (2023). Biomaterials-mediated CRISPR/Cas9 delivery: Recent challenges and opportunities in gene therapy. *Frontiers in Chemistry, 11*, 1259435. DOI: 10.3389/fchem.2023.1259435 PMID: 37841202

Du, K., Zeng, Q., Jiang, M., Hu, Z., Zhou, M., & Xia, K. (2025). CRISPR/Cas12a-Based Biosensing: Advances in Mechanisms and Applications for Nucleic Acid Detection. *Biosensors (Basel), 15*(6), 360. DOI: 10.3390/bios15060360 PMID: 40558442

Durymanov, M., & Reineke, J. (2018). Non-viral Delivery of Nucleic Acids: Insight Into Mechanisms of Overcoming Intracellular Barriers. *Frontiers in Pharmacology, 9*, 971. DOI: 10.3389/fphar.2018.00971 PMID: 30186185

Du, Y., Ding, X., & Ye, Y. (2024). The spatial multi-omics revolution in cancer therapy: Precision redefined. *Cell Reports Medicine, 5*(9), 101740. DOI: 10.1016/j.xcrm.2024.101740 PMID: 39293393

Du, Y., Liu, Y., Hu, J., Peng, X., & Liu, Z. (2023). CRISPR/Cas9 systems: Delivery technologies and biomedical applications. *Asian Journal of Pharmaceutical Sciences, 18*(6), 100854. DOI: 10.1016/j.ajps.2023.100854 PMID: 38089835

Dzilic, E., Lahm, H., Dreßen, M., Deutsch, M. A., Lange, R., Wu, S. M., & Doppler, S. A. (2018). Genome editing redefines precision medicine in the cardiovascular field. *Stem Cells International, 2018*(1), 4136473. DOI: 10.1155/2018/4136473 PMID: 29731778

Eba, J., & Nakamura, K. (2022). Overview of the ethical guidelines for medical and biological research involving human subjects in Japan. *Japanese Journal of Clinical Oncology, 52*(6), 539–544. DOI: 10.1093/jjco/hyac034 PMID: 35349681

Eid, A., Alshareef, S., & Mahfouz, M. M. (2018). CRISPR base editors: Genome editing without double-stranded breaks. *Biochemical Journal, 475*(11), 1955–1964. DOI: 10.1042/BCJ20170793

Elayaperumal, S., Sivamani, Y., Agarwal, P., & Srivastava, N. (2025). Plant-based nanotherapeutics: A new frontier in disease management and prevention. *Nano TransMed*, *4*, 100086. DOI: 10.1016/j.ntm.2025.100086

Eyquem, J., Mansilla-Soto, J., Giavridis, T., Van Der Stegen, S. J., Hamieh, M., Cunanan, K. M., Odak, A., Gönen, M., & Sadelain, M. (2017). Targeting a CAR to the TRAC locus with CRISPR/Cas9 enhances tumour rejection. *Nature*, *543*(7643), 113–117. DOI: 10.1038/nature21405 PMID: 28225754

Fadul, S. M., Arshad, A., & Mehmood, R. (2023). CRISPR-based epigenome editing: Mechanisms and applications. *Epigenomics*, *15*(21), 1137–1155. DOI: 10.2217/epi-2023-0281 PMID: 37990877

Falandt, M., Bernal, P. N., Dudaryeva, O., Florczak, S., Größbacher, G., Schweiger, M., Longoni, A., Greant, C., Assunção, M., Nijssen, O., van Vlierberghe, S., Malda, J., Vermonden, T., & Levato, R. (2023). Spatial-Selective Volumetric 4D Printing and Single-Photon Grafting of Biomolecules within Centimeter-Scale Hydrogels via Tomographic Manufacturing. *Advanced Materials Technologies*, *8*(15), 2300026. DOI: 10.1002/admt.202300026 PMID: 37811162

Faltus, T., & De Miguel Beriain, I. (2023). The EU legal framework on clinical trials directed to therapeutic germline gene alteration: A critical and systematic analysis. *Medical Law International*, *23*(1), 44–70. DOI: 10.1177/09685332231154586

Fang, T., Cao, X., Ibnat, M., & Chen, G. (2022). Stimuli-responsive nanoformulations for CRISPR-Cas9 genome editing. *Journal of Nanobiotechnology*, *20*(1), 354. DOI: 10.1186/s12951-022-01570-y PMID: 35918694

Farrell, C. B. (2019). *"O brave new world" CRISPR Mediated Gene Editing in Somatic and Germ Line Contexts for Therapeutics and Enhancement of the Human Species – an Ethical and Legal Discussion.* 949365 Bytes. DOI: 10.25419/RCSI.10814045.V1

Fateh, S. T., Aghaii, A. H., Aminzade, Z., Shahriari, E., Roohpour, N., Koosha, F., & Dezfuli, A. S. (2024). Inorganic nanoparticle-cored dendrimers for biomedical applications: A review. *Heliyon*, *10*(9), e29726. DOI: 10.1016/j.heliyon.2024.e29726 PMID: 38694058

Feeney, O., Cockbain, J., & Sterckx, S. (2021). Ethics, Patents and Genome Editing: A Critical Assessment of Three Options of Technology Governance. *Frontiers in Political Science*, *3*, 731505. DOI: 10.3389/fpos.2021.731505

Feng, C., Shan, M., Liu, C., Xu, Y., & Wang, H. (2024). CRISPR/Cas9 technology for advancements in cancer immunotherapy. *International Immunopharmacology*, *139*, 112706. DOI: 10.1016/j.intimp.2024.112706 PMID: 39427211

Feoktistov, A. V., Georgieva, S. G., & Soshnikova, N. V. (2022). Role of the SWI/SNF chromatin remodeling complex in regulation of inflammation gene expression. *Molecular Biology*, *56*(2), 182–195. DOI: 10.1134/S0026893322020054 PMID: 35403618

Fernández-Borbolla, A., García-Hevia, L., & Fanarraga, M. L. (2024). Cell Membrane-Coated Nanoparticles for Precision Medicine: A Comprehensive Review of Coating Techniques for Tissue-Specific Therapeutics. *International Journal of Molecular Sciences*, *25*(4), 2071. DOI: 10.3390/ijms25042071 PMID: 38396747

Festa, R. A., & Thiele, D. J. (2023). Copper transport in health and disease. *Cell Metabolism*, *35*(4), 567–584. DOI: 10.1016/j.cmet.2023.03.007

Filipczak, N., Yalamarty, S. S. K., Li, X., Parveen, F., & Torchilin, V. (2021). Developments in Treatment Methodologies Using Dendrimers for Infectious Diseases. *Molecules (Basel, Switzerland)*, *26*(11), 3304. DOI: 10.3390/molecules26113304 PMID: 34072765

Fiumara, M., Ferrari, S., Omer-Javed, A., Beretta, S., Albano, L., Canarutto, D., Varesi, A., Gaddoni, C., Brombin, C., Cugnata, F., Zonari, E., Naldini, M. M., Barcella, M., Gentner, B., Merelli, I., & Naldini, L. (2024). Genotoxic effects of base and prime editing in human hematopoietic stem cells. *Nature Biotechnology*, *42*(6), 877–891. DOI: 10.1038/s41587-023-01915-4 PMID: 37679541

Fontana, M., Solomon, S. D., Kachadourian, J., Walsh, L., Rocha, R., Lebwohl, D., & Gillmore, J. D. (2024). CRISPR-Cas9 gene editing with nexiguran ziclumeran for ATTR cardiomyopathy. *The New England Journal of Medicine*, *391*(23), 2231–2241. DOI: 10.1056/NEJMoa2412309 PMID: 39555828

Forman, H. J. (2023). Glutathione: The master antioxidant—Mechanisms of synthesis and regulation. *Free Radical Biology & Medicine*, *197*, 87–100. DOI: 10.1016/j.freeradbiomed.2023.01.018

Frangoul, H., Altshuler, D., Cappellini, M. D., Chen, Y.-S., Domm, J., Eustace, B. K., Foell, J., de la Fuente, J., Grupp, S., Handgretinger, R., Ho, T. W., Kattamis, A., Kernytsky, A., Lekstrom-Himes, J., Li, A. M., Locatelli, F., Mapara, M. Y., de Montalembert, M., Rondelli, D., & Corbacioglu, S. (2021). CRISPR-Cas9 gene editing for sickle cell disease and β-thalassemia. *The New England Journal of Medicine*, *384*(3), 252–260. DOI: 10.1056/NEJMoa2031054 PMID: 33283989

Freyermuth, F., Rau, F., Kokunai, Y., Linke, T., Sellier, C., Nakamori, M., & Charlet-Berguerand, N. (2016). Splicing misregulation of SCN5A contributes to cardiac-conduction delay and heart arrhythmia in myotonic dystrophy. *Nature Communications*, *7*(1), 11067. DOI: 10.1038/ncomms11067 PMID: 27063795

Fu, Y., Foden, J. A., Khayter, C., Maeder, M. L., Reyon, D., Joung, J. K., & Sander, J. D. (2013). High-frequency off-target mutagenesis induced by CRISPR-Cas nucleases in human cells. *Nature Biotechnology*, *31*(9), 822–826. DOI: 10.1038/nbt.2623 PMID: 23792628

Gadicherla, R., Rai, N., Othayoth, R., & Kamma, S. (2024). Characterization of mitochondrial DNA mutations in colorectal cancer progression by in silico approach and use as potential biomarkers for diagnosis and prognosis. *The Egyptian Journal of Medical Human Genetics*, *25*(1), 126. DOI: 10.1186/s43042-024-00599-y

Galdos, F. X., Darsha, A. K., Paige, S. L., & Wu, S. M. (2020). Purification of pluripotent stem cell-derived cardiomyocytes using CRISPR/Cas9-mediated integration of fluorescent reporters. In *Cardiac Regeneration: Methods and Protocols* (pp. 223–240). Springer US.

Ganipineni, V. D. P., Gutlapalli, S. D., Danda, S., Garlapati, S. K. P., Fabian, D., Okorie, I., & Okorie, I. J. (2023). Clustered regularly interspaced short palindromic repeats (CRISPR) in cardiovascular disease: A comprehensive clinical review on dilated cardiomyopathy. *Cureus*, *15*(3), e35774. DOI: 10.7759/cureus.35774 PMID: 37025725

Gan, T., Yu, J., Deng, Z., & He, J. (2024). ERA-CRISPR/Cas12a system: A rapid, highly sensitive and specific assay for Mycobacterium tuberculosis. *Frontiers in Cellular and Infection Microbiology*, *14*, 1454076. DOI: 10.3389/fcimb.2024.1454076 PMID: 39233906

Gaudelli, N. M., Komor, A. C., Rees, H. A., Packer, M. S., Badran, A. H., Bryson, D. I., & Liu, D. R. (2017). Programmable base editing of A• T to G• C in genomic DNA without DNA cleavage. *Nature*, *551*(7681), 464–471. DOI: 10.1038/nature24644 PMID: 29160308

Genome-wide detection and analysis of CRISPR-Cas off-targets. (2021). In *Progress in Molecular Biology and Translational Science* (Vol. 181, pp. 31–43). Elsevier. DOI: 10.1016/bs.pmbts.2021.01.012

Geszke-Moritz, M., & Moritz, M. (2024). Biodegradable Polymeric Nanoparticle-Based Drug Delivery Systems: Comprehensive Overview, Perspectives and Challenges. *Polymers*, *16*(17), 2536. DOI: 10.3390/polym16172536 PMID: 39274168

Ghahremani, S., Kanwal, A., Pettinato, A., Ladha, F., Legere, N., Thakar, K., & Hinson, J. T. (2024). CRISPR activation reverses haploinsufficiency and functional deficits caused by TTN truncation variants. *Circulation*, *149*(16), 1285–1297. DOI: 10.1161/CIRCULATIONAHA.123.063972 PMID: 38235591

Giamarellos-Bourboulis, E. J., Aschenbrenner, A. C., Bauer, M., Bock, C., Calandra, T., Gat-Viks, I., Kyriazopoulou, E., Lupse, M., Monneret, G., Pickkers, P., Schultze, J. L., Van Der Poll, T., Van De Veerdonk, F. L., Vlaar, A. P. J., Weis, S., Wiersinga, W. J., & Netea, M. G. (2024). The pathophysiology of sepsis and precision-medicine-based immunotherapy. *Nature Immunology*, *25*(1), 19–28. DOI: 10.1038/s41590-023-01660-5 PMID: 38168953

Gilbert, L. A., Horlbeck, M. A., Adamson, B., Villalta, J. E., Chen, Y., Whitehead, E. H., Guimaraes, C., Panning, B., Ploegh, H. L., Bassik, M. C., Qi, L. S., Kampmann, M., & Weissman, J. S. (2014). Genome-scale CRISPR-mediated control of gene repression and activation. *Cell*, *159*(3), 647–661. DOI: 10.1016/j.cell.2014.09.029 PMID: 25307932

Gilbert, L. A., Larson, M. H., Morsut, L., Liu, Z., Brar, G. A., Torres, S. E., & Weissman, J. S. (2013). CRISPR-mediated modular RNA-guided regulation of transcription in eukaryotes. *Cell*, *154*(2), 442–451. DOI: 10.1016/j.cell.2013.06.044 PMID: 23849981

Gillmore, J. D., Gane, E., Taubel, J., Kao, J., Fontana, M., Maitland, M. L., Seitzer, J., O'Connell, D., Walsh, K. R., Wood, K., Phillips, J., Xu, Y., Amaral, A., Boyd, A. P., Cehelsky, J., McKee, M. D., Schiermeier, A., Harari, O., Murphy, A., & Lebwohl, D. (2021). CRISPR-Cas9 in vivo gene editing for transthyretin amyloidosis. *The New England Journal of Medicine*, *385*(6), 493–502. DOI: 10.1056/NEJMoa2107454 PMID: 34215024

Gong, H., Liu, M., Klomp, J., Merrill, B. J., Rehman, J., & Malik, A. B. (2017). Method for dual viral vector mediated CRISPR-Cas9 gene disruption in primary human endothelial cells. *Scientific Reports*, *7*(1), 42127. DOI: 10.1038/srep42127 PMID: 28198371

González Castro, N. G., Bjelic, J., Malhotra, G., Huang, C., & Alsaffar, S. H. (2021). Comparison of the Feasibility, Efficiency, and Safety of Genome Editing Technologies. *International Journal of Molecular Sciences*, *22*(19), 10355. DOI: 10.3390/ijms221910355 PMID: 34638696

González-Delgado, A., Lopez, S. C., Rojas-Montero, M., Fishman, C. B., & Shipman, S. L. (2024). Simultaneous multi-site editing of individual genomes using retron arrays. *Nature Chemical Biology*, *20*(11), 1482–1492. DOI: 10.1038/s41589-024-01665-7 PMID: 38982310

Goodwin, R. J. A., Platz, S. J., Reis-Filho, J. S., & Barry, S. T. (2024). Accelerating Drug Development Using Spatial Multi-omics. *Cancer Discovery*, *14*(4), 620–624. DOI: 10.1158/2159-8290.CD-24-0101 PMID: 38571424

Gootenberg, J. S., Abudayyeh, O. O., Kellner, M. J., Joung, J., Collins, J. J., & Zhang, F. (2017). Multiplexed and portable nucleic acid detection platform with Cas13, Cas12a, and Csm6. *Science*, *360*(6387), 439–444. DOI: 10.1126/science.aaq0179 PMID: 29449508

Grant, E. V. (2016). FDA Regulation of Clinical Applications of CRISPR-CAS Gene-Editing Technology. *Food and Drug Law Journal*, *71*(4), 608–633. PMID: 29140647

Greco, F., Cosentino, M., & Marino, F. (2024). The Italian breakthrough in CRISPR trials for rare diseases: A focus on beta-thalassemia and sickle cell disease treatment. *Frontiers in Medicine*, *11*, 1356578. DOI: 10.3389/fmed.2024.1356578 PMID: 38426160

Greenfield, A. (2021). Making sense of heritable human genome editing: Scientific and ethical considerations. In *Progress in Molecular Biology and Translational Science* (Vol. 182, pp. 1–28). Elsevier., DOI: 10.1016/bs.pmbts.2020.12.008

Grosch, M., Schraft, L., Chan, A., Küchenhoff, L., Rapti, K., Ferreira, A. M., & Steinmetz, L. M. (2023). Striated muscle-specific base editing enables correction of mutations causing dilated cardiomyopathy. *Nature Communications*, *14*(1), 3714. DOI: 10.1038/s41467-023-39352-1 PMID: 37349314

Gumer, J. M. (2019). The Wisdom of Germline Editing: An Ethical Analysis of the Use of CRISPR-Cas9 to Edit Human Embryos. *The New Bioethics : a Multidisciplinary Journal of Biotechnology and the Body*, *25*(2), 137–152. DOI: 10.1080/20502877.2019.1606151 PMID: 31130112

Gundappa, M., Arumugam, V. A., Hsieh, H. L., Balasubramanian, B., & Shanmugam, V. (2022). Expression of tissue factor and TF-mediated integrin regulation in HTR-8/SVneo trophoblast cells. *Journal of Reproductive Immunology*, *150*, 103473. DOI: 10.1016/j.jri.2022.103473 PMID: 35030354

Gupta, P. K., & Kumar, S. (2025). Third-generation novel technologies for gene editing. *Trends in Biotechnology, S0167779925002732*. Advance online publication. DOI: 10.1016/j.tibtech.2025.07.012 PMID: 40738762

Gupta, R. M., Schnitzler, G. R., Fang, S., Lee-Kim, V. S., & Barry, A. (2023). Multiomic analysis and CRISPR perturbation screens identify endothelial cell programs and novel therapeutic targets for coronary artery disease. *Arteriosclerosis, Thrombosis, and Vascular Biology*, *43*(5), 600–608. DOI: 10.1161/ATVBAHA.123.318328 PMID: 36994731

Gu, W., Crawford, E. D., O'Donovan, B. D., Wilson, M. R., Chow, E. D., Retallack, H., & DeRisi, J. L. (2016). Depletion of Abundant Sequences by Hybridization (DASH): Using Cas9 to remove unwanted high-abundance species in sequencing libraries and molecular counting applications. *Genome Biology*, *17*(1), 41. DOI: 10.1186/s13059-016-0904-5 PMID: 26944702

Guzman Gonzalez, V., Grunenberger, A., Nicoud, O., Czuba, E., Vollaire, J., Josserand, V., Le Guével, X., Desai, N., Coll, J.-L., Divita, G., & Faure, V. (2024). Enhanced CRISPR-Cas9 RNA system delivery using cell penetrating peptides-based nanoparticles for efficient in vitro and in vivo applications. *Journal of Controlled Release*, *376*, 1160–1175. DOI: 10.1016/j.jconrel.2024.11.008 PMID: 39521064

GVS, A. C., Ananthakumar, B., & Raghavan, R. (2025). Enhancing Assistive Technologies With Neuromorphic Computing. *Revolutionizing AI with Brain-Inspired Technology: Neuromorphic Computing*, 207–232. DOI: 10.4018/979-8-3693-6303-4.ch009

Habbane, M., Montoya, J., Rhouda, T., Sbaoui, Y., Radallah, D., & Emperador, S. (2021). Human Mitochondrial DNA: Particularities and Diseases. *Biomedicines*, *9*(10), 1364. DOI: 10.3390/biomedicines9101364 PMID: 34680481

Haft, D. H., Selengut, J., Mongodin, E. F., & Nelson, K. E. (2005). A guild of 45 CRISPR-associated protein families and multiple CRISPR/Cas subtypes exist in prokaryotic genomes. *PLoS Computational Biology, 1*(6), 0474–0483. https://doi.org/DOI: 10.1371/journal.pcbi.0010060

Hall, T., Gulati, S., Sang, R., Jia, Z., Mckinnirey, F., Vesey, G., Goldys, E., & Deng, F. (2025). CRISPR biosensing with lateral flow assays for point of care Diagnostics: Overcoming commercial development challenges. *Trends in Analytical Chemistry*, *118275*, 118275. https://doi.org/https://doi.org/10.1016/j.trac.2025.118275. DOI: 10.1016/j.trac.2025.118275

Han, J. L., & Entcheva, E. (2023). Gene modulation with CRISPR-based tools in human iPSC-cardiomyocytes. *Stem Cell Reviews and Reports*, *19*(4), 886–905. DOI: 10.1007/s12015-023-10506-4 PMID: 36656467

Han, J. L., Heinson, Y. W., Chua, C. J., Liu, W., & Entcheva, E. (2023). CRISPRi gene modulation and all-optical electrophysiology in post-differentiated human iPSC-cardiomyocytes. *Communications Biology*, *6*(1), 1236. DOI: 10.1038/s42003-023-05627-y PMID: 38062109

Hanna, R., Frangoul, H., McKinney, C., Pineiro, L., Mapara, M., Dalal, J., Chang, K.-H., Jaskolka, M., Kim, K., Farrington, D. L., Wally, M., Mei, B., Lawal, A., Afonja, O. O., & Walters, M. C. (2023). AsCas12a Gene Editing of *HBG1/2* Promoters with EDIT-301 Results in Rapid and Sustained Normalization of Hemoglobin and Increased Fetal Hemoglobin in Patients with Severe Sickle Cell Disease and Transfusion-Dependent Beta-Thalassemia. *Blood*, *142*(Supplement 1), 4996–4996. DOI: 10.1182/blood-2023-187397

Han, S., Xu, Q., Du, Y., Tang, C., Cui, H., Xia, X., & Shang, H. (2024). Single-cell spatial transcriptomics in cardiovascular development, disease, and medicine. *Genes & Diseases*, *11*(6), 101163. DOI: 10.1016/j.gendis.2023.101163 PMID: 39224111

Hanscom, T., & McVey, M. (2020). Regulation of error-prone DNA double-strand break repair and its impact on genome evolution. *Cells*, *9*(7), 1657. DOI: 10.3390/cells9071657 PMID: 32660124

Haque, U. S., & Yokota, T. (2025). Gene Editing for Duchenne Muscular Dystrophy: From Experimental Models to Emerging Therapies. *Degenerative Neurological and Neuromuscular Disease*, *15*, 17–40. DOI: 10.2147/DNND.S495536 PMID: 40241992

Havelikar, U., Ghorpade, K. B., Kumar, A., Patel, A., Singh, M., Banjare, N., & Gupta, P. N. (2024). Comprehensive insights into mechanism of nanotoxicity, assessment methods and regulatory challenges of nanomedicines. *Discover Nano*, *19*(1), 165. DOI: 10.1186/s11671-024-04118-1 PMID: 39365367

Hayes, B. H., Zhu, H., Andrechak, J. C., & Discher, D. E. (2022). Titrating CD47 by mismatch CRISPRi reveals incomplete repression can eliminate IgG-opsonized tumors but CD47 heterogeneity limits induction of anti-tumor IgG. bioRxiv, 2022-09.

Heller, E. A., Cates, H. M., Peña, C. J., Sun, H., Shao, N., Feng, J., & Nestler, E. J. (2020). Locus-specific epigenetic remodeling controls addiction- and depression-related behaviors. *Nature Neuroscience*, *23*(5), 572–583. PMID: 25347353

Hendriks, D., Hageman, G. J., & Woltjen, K. (2020). Genetic programming of human iPSCs using CRISPR tools: Towards tissue engineering and regenerative medicine. *Cell Stem Cell.*

He, X., Liang, J., Zhang, J., Fang, W., Liu, J., Zhang, M., Wang, L., & Song, C. (2025). CRISPR/Cas13a triggered-DNA walker amplified SERS sensor for ultrasensitive detection of cancer-related exosomal miRNA. *Biosensors & Bioelectronics, 289*, 117924. DOI: 10.1016/j.bios.2025.117924 PMID: 40886435

Hillary, V. E., & Ceasar, S. A. (2023). A Review on the Mechanism and Applications of CRISPR/Cas9/Cas12/Cas13/Cas14 Proteins Utilized for Genome Engineering. *Molecular Biotechnology, 65*(3), 311–325. DOI: 10.1007/s12033-022-00567-0 PMID: 36163606

Hillary, V. E., & Ceasar, S. A. (2024). CRISPR/Cas system-mediated base editing in crops: Recent developments and future prospects. *Plant Cell Reports, 43*(11), 271. DOI: 10.1007/s00299-024-03346-0 PMID: 39453560

Ho, M. K., & Ho, J. W. K. (2025). Mitochondrial DNA common deletion in cancer: Emerging evidence and methodological challenges. *Computational and Structural Biotechnology Journal, 27*, 3929–3940. DOI: 10.1016/j.csbj.2025.09.010

Hossain, M. A. (2021). CRISPR-Cas9: A fascinating journey from bacterial immune system to human gene editing. *Progress in Molecular Biology and Translational Science, 178*, 63–83. DOI: 10.1016/bs.pmbts.2021.01.001 PMID: 33685600

Hou, C., Xiao, L., Ren, X., Cheng, L., Guo, B., Zhang, M., & Yan, N. (2022). EZH2-mediated H3K27me3 is a predictive biomarker and therapeutic target in uveal melanoma. *Frontiers in Genetics, 13*, 1013475. DOI: 10.3389/fgene.2022.1013475 PMID: 36276954

Hsu, P. D., Lander, E. S., & Zhang, F. (2014). Development and applications of CRISPR-Cas9 for genome engineering. *Cell, 157*(6), 1262–1278. DOI: 10.1016/j.cell.2014.05.010 PMID: 24906146

Huang, B., Guo, L., Yin, H., Wu, Y., Zeng, Z., Xu, S., Lou, Y., Ai, Z., Zhang, W., Kan, X., Yu, Q., Du, S., Li, C., Wu, L., Huang, X., Wang, S., & Wang, X. (2024). Deep learning enhancing guide RNA design for CRISPR/Cas12a-based diagnostics. *iMeta, 3*(4), e214. DOI: 10.1002/imt2.214 PMID: 39135699

Huang, J., & Fussenegger, M. (2025). Programming mammalian cell behaviors by physical cues. *Trends in Biotechnology, 43*(1), 16–42. DOI: 10.1016/j.tibtech.2024.07.014 PMID: 39179464

Huang, P.-T., Chen, Y.-L., Lin, Y.-H., Wang, C.-C., & Chang, H.-T. (2024). Functional gold nanoparticles for analysis and delivery of nucleic acids. *Yao Wu Shi Pin Fen Xi*, *32*(3), 252–273. DOI: 10.38212/2224-6614.3514 PMID: 39636773

Huang, Q., Li, Y., Huang, Y., Wu, J., Bao, W., Xue, C., Li, X., Dong, S., Dong, Z., & Hu, S. (2025). Advances in molecular pathology and therapy of non-small cell lung cancer. *Signal Transduction and Targeted Therapy*, *10*(1), 186. DOI: 10.1038/s41392-025-02243-6 PMID: 40517166

Huang, X., Li, A., Xu, P., Yu, Y., Li, S., Hu, L., & Feng, S. (2023). Current and prospective strategies for advancing the targeted delivery of CRISPR/Cas system via extracellular vesicles. *Journal of Nanobiotechnology*, *21*(1), 184. DOI: 10.1186/s12951-023-01952-w PMID: 37291577

Huang, X., Zhang, L., & Chen, Y. (2022). CRISPR/dCas9-mediated epigenetic editing reverses drug resistance in cancer: Mechanisms and therapeutic potential. *Molecular Cancer*, *21*(1), 143. PMID: 35820907

Huang, X., Zhou, G., Wu, W., Duan, Y., Ma, G., Song, J., & Lei, H. (2017). Genome editing abrogates angiogenesis in vivo. *Nature Communications*, *8*(1), 112. DOI: 10.1038/s41467-017-00140-3 PMID: 28740073

Hu, C., Shi, X., Guo, B., Yang, Z., Zhou, J., & Wang, F. (2025). Toehold-Based CRISPR-dCas9 Transcriptional Activation Platform for Spatiotemporally Controllable Gene Therapy in Tumor and Diabetic Mouse Models. *ACS Nano*, *19*(12), 12277–12287. DOI: 10.1021/acsnano.5c01078 PMID: 40123515

Hu, J. H., Miller, S. M., Geurts, M. H., Tang, W., Chen, L., Sun, N., Zeina, C. M., Gao, X., Rees, H. A., Lin, Z., & Liu, D. R. (2018). Evolved Cas9 variants with broad PAM compatibility and high DNA specificity. *Nature*, *556*(7699), 57–63. DOI: 10.1038/nature26155 PMID: 29512652

Hussen, B. M., Najmadden, Z. B., Abdullah, S. R., Rasul, M. F., Mustafa, S. A., Ghafouri-Fard, S., & Taheri, M. (2024). CRISPR/Cas9 gene editing: A novel strategy for fighting drug resistance in respiratory disorders. *Cell Communication and Signaling : CCS*, *22*(1), 329. DOI: 10.1186/s12964-024-01713-8 PMID: 38877530

Hwang, H.-Y., Lee, M., Yi, H., Seok, C., Lim, K., Na, Y. R., Kang, J.-S., Park, J.-H., & Kim, D. (2025). Engineered Sdd7 cytosine base editors with enhanced specificity. *Nature Communications*, *16*(1), 5881. DOI: 10.1038/s41467-025-60789-z PMID: 40593661

Ibraheim, R., Song, C. Q., Mir, A., Amrani, N., Xue, W., & Sontheimer, E. J. (2018). All-in-one adeno-associated virus delivery and genome editing by Neisseria meningitidis Cas9 in vivo. *Genome Biology*, *19*(1), 137. DOI: 10.1186/s13059-018-1515-0 PMID: 30231914

Ieki, H., Ito, K., Zhang, S., Koyama, S., Miyazawa, K., Yoshida, H., & Komuro, I. (2024). Machine learning uncovers the contribution of rare genetic variants and enhances risk prediction for coronary artery disease in the japanese population. *Circulation*, *150*(Suppl_1), A4136501–A4136501. DOI: 10.1161/circ.150.suppl_1.4136501

Institute, I. G. (2025). *CRISPR Clinical Trials: A 2024 Update*. https://innovativegenomics.org/news/crispr-clinical-trials-2024/

Inusa, B., Hsu, L., Kohli, N., Patel, A., Ominu-Evbota, K., Anie, K., & Atoyebi, W. (2019). Sickle Cell Disease—Genetics, Pathophysiology, Clinical Presentation and Treatment. *International Journal of Neonatal Screening*, *5*(2), 20. DOI: 10.3390/ijns5020020 PMID: 33072979

Isasi, R., Knoppers, B. M., & Abouelhoda, M. (2021). Ensuring equitable access to genome editing therapies: Global perspectives and policy considerations. *Human Gene Therapy*, *32*(15-16), 913–925.

Ishii, T. (2021). Human genome editing: Ethical and regulatory perspectives. *Trends in Molecular Medicine*, *27*(2), 125–134.

Ishino, Y., Shinagawa, H., Makino, K., Amemura, M., & Nakata, A. (1987). Nucleotide sequence of the *iap* gene, responsible for alkaline phosphatase isozyme conversion in *Escherichia coli*, and identification of the gene product. *Journal of Bacteriology*, *169*(12), 5429–5433. DOI: 10.1128/jb.169.12.5429-5433.1987 PMID: 3316184

Iyemere, V. P., Proudfoot, D., Weissberg, P. L., & Shanahan, C. M. (2006). Vascular smooth muscle cell phenotypic plasticity and the regulation of vascular calcification. *Journal of Internal Medicine*, *260*(3), 192–210. DOI: 10.1111/j.1365-2796.2006.01692.x PMID: 16918817

Jansen, R., Embden, J. D. V., Gaastra, W., & Schouls, L. M. (2002). Identification of genes that are associated with DNA repeats in prokaryotes. *Molecular Microbiology*, *43*(6), 1565–1575. DOI: 10.1046/j.1365-2958.2002.02839.x PMID: 11952905

Jansen, R., Van Embden, J. D. A., Gaastra, W., & Schouls, L. M. (2002). Identification of a novel family of sequence repeats among prokaryotes. *OMICS: A Journal of Integrative Biology*, *6*(1), 23–33. DOI: 10.1089/15362310252780816 PMID: 11883425

Jayarajan, D., Kumar Dutta, A., Paripuranam, T. D., Krishnaveni, M., & Bhavya, S. (2025). Recent Advancements in CRISPR-Cas9 Technology for Precision Gene Editing and Therapeutic Applications. *Journal of Neonatal Surgery, 14*(7), 818–827. DOI: 10.63682/jns.v14i7.6384

Jena, G. K., Patra, C. N., Jammula, S., Rana, R., & Chand, S. (2024). Artificial Intelligence and Machine Learning Implemented Drug Delivery Systems: A Paradigm Shift in the Pharmaceutical Industry. *Journal of Bio-X Research, 7*, 0016. DOI: 10.34133/jbioxresearch.0016

Jiang, F. Q., Liu, K., Chen, J. X., Cao, Y., Chen, W. Y., Zhao, W. L., & Qi, X. F. (2022). Mettl3-mediated m6A modification of Fgf16 restricts cardiomyocyte proliferation during heart regeneration. *eLife, 11*, e77014. DOI: 10.7554/eLife.77014 PMID: 36399125

Jiang, F., & Doudna, J. A. (2017). CRISPR-Cas9 structures and mechanisms. *Annual Review of Biophysics, 46*(1), 505–529. Advance online publication. DOI: 10.1146/annurev-biophys-062215-010822 PMID: 28375731

Jiménez-Boland, D., Robles-Fernández, A., Martín-Rodríguez, A., Cuadros, M. Á., Traverso, J. Á., Sánchez-Moreno, P., & Bramini, M. (2025). Breaking Barriers in Glioblastoma Targeting through Advanced Nanoparticle Cell Membrane Coating. *ACS Applied Materials & Interfaces, 17*(24), 35288–35303. DOI: 10.1021/acsami.5c07306 PMID: 40488690

Jinek, M., Chylinski, K., Fonfara, I., Hauer, M., Doudna, J. A., & Charpentier, E. (2012). A programmable dual-RNA–guided DNA endonuclease in adaptive bacterial immunity. *science, 337*(6096), 816-821.

Jinek, M., Chylinski, K., Fonfara, I., Hauer, M., Doudna, J. A., & Charpentier, E. (2012). A programmable dual-RNA–guided DNA endonuclease in adaptive bacterial immunity. *Science, 337*(6096), 816–821. DOI: 10.1126/science.1225829 PMID: 22745249

Joneidi, Z., Mortazavi, Y., Memari, F., Roointan, A., Chahardouli, B., & Rostami, S. (2019). The impact of genetic variation on metabolism of heavy metals. *Biomedicine and Pharmacotherapy, 113*, 108642. DOI: 10.1016/j.biopha.2019.108642 PMID: 30849640

Joshi, S., & Mishra, S. (2022). Recent advances in biofuel production through metabolic engineering. *Bioresource Technology, 352*, 127037. DOI: 10.1016/j.biortech.2022.127037 PMID: 35318143

Jung, H. N., Lee, S.-Y., Lee, S., Youn, H., & Im, H.-J. (2022). Lipid nanoparticles for delivery of RNA therapeutics: Current status and the role of *in vivo* imaging. *Theranostics*, *12*(17), 7509–7531. DOI: 10.7150/thno.77259 PMID: 36438494

Jung, W. J., Park, S.-J., Cha, S., & Kim, K. (2024). Factors affecting the cleavage efficiency of the CRISPR-Cas9 system. *Animal Cells and Systems*, *28*(1), 75–83. DOI: 10.1080/19768354.2024.2322054 PMID: 38440123

Kaljanac, M., & Abken, H. (2023). Do treg speed up with CARs? Chimeric antigen receptor treg engineered to induce transplant tolerance. *Transplantation*, *107*(1), 74–85. DOI: 10.1097/TP.0000000000004316 PMID: 36226849

Kalter, N., Fuster-Garcia, C., Silva, A., Ronco-Diaz, V., Roncelli, S., Turchiano, G., Gorodkin, J., Cathomen, T., Benabdellah, K., Lee, C., & Hendel, A. (2025). Off-target effects in CRISPR-Cas genome editing for human therapeutics: Progress and challenges. *Molecular Therapy. Nucleic Acids*, *36*(3), 102636. DOI: 10.1016/j.omtn.2025.102636 PMID: 40777742

Kampouri, E., Reynolds, G., Teh, B. W., & Hill, J. A. (2024). Chimeric antigen receptor-T-cell therapies going viral: Latent and incidental viral infections. *Current Opinion in Infectious Diseases*, *37*(6), 526–535. DOI: 10.1097/QCO.0000000000001066 PMID: 39361275

Kanazhevskaya, L. Y., Zhdanova, P. V., Chernonosov, A. A., & Koval, V. V. (2025). Off-target interactions in CRISPR-Cas9 machinery. *Biochemistry and Biophysics Reports*, *43*, 102134. Advance online publication. DOI: 10.1016/j.bbrep.2025.102134 PMID: 40688512

Kar, B., Castillo, S. R., Sabharwal, A., Clark, K. J., & Ekker, S. C. (2023). Mitochondrial base editing: Recent advances toward therapeutic opportunities. *International Journal of Molecular Sciences*, *24*(6), 5798. DOI: 10.3390/ijms24065798 PMID: 36982871

Karimi, M. A., Paryan, M., Behrouzian Fard, G., Sadeghian, H., Zarrinfar, H., & Hosseini Bafghi, M. (2025). Challenges and Opportunities in the Application of CRISPR-Cas9: A Review on Genomic Editing and Therapeutic Potentials. *Medical Principles and Practice : International Journal of the Kuwait University, Health Science Centre*, •••, 1–17. DOI: 10.1159/000547334 PMID: 40675140

Kaur, N., & Pati, P. K. (2024). Retron Library Recombineering: Next Powerful Tool for Genome Editing after CRISPR/Cas. *ACS Synthetic Biology*, *13*(4), 1019–1025. DOI: 10.1021/acssynbio.3c00667 PMID: 38480006

Kaur, R., Gupta, S., Chauhan, A., Mishra, V., Sharma, M. K., & Singh, J. (2024). Harnessing the power of clustered regularly interspaced short palindromic repeats (CRISPR) based microfluidics for next-generation molecular diagnostics. *Molecular Biology Reports*, *51*(1), 896. DOI: 10.1007/s11033-024-09840-8 PMID: 39115550

Kazemian, P., Yu, S. Y., Thomson, S. B., Birkenshaw, A., Leavitt, B. R., & Ross, C. J. (2022). Lipid-nanoparticle-based delivery of CRISPR/Cas9 genome-editing components. *Molecular Pharmaceutics*, *19*(6), 1669–1686. DOI: 10.1021/acs.molpharmaceut.1c00916 PMID: 35594500

Keepers, B. P., Qian, L., Liu, J., & Shutt, S. (2022). Abstract P1134: Activation of endogenous transcription factors by crispra mediates cardiac reprogramming. *Circulation Research*, *131*(Suppl_1), AP1134–AP1134. DOI: 10.1161/res.131.suppl_1.P1134

Khan, S. A., Nurulain, S. M., Qureshi, R. N., Zafar, A., Riaz, Z., Shoukat, A., & Shah, S. T. A. (2024). Exposure to heavy metals, antioxidant status, and immune response. *Toxicology Reports*, *11*, 108642. Advance online publication. DOI: 10.1016/j.toxrep.2024.108642

Khan, S. S., Rangraze, I. R., Wali, A. F., Jhancy, M., Attia, R. A., Elshamly, H. A. H., Adam, S., & Elbeshbeishy, R. A. M. (2025). Repurposing Metformin in Precision Oncology: Mechanistic Insights, Biomarker-Guided Strategies, and Translational Imperatives. *Medicina*, *61*(9), 1577. DOI: 10.3390/medicina61091577 PMID: 41010968

Kharwade, R., More, S., Warokar, A., Agrawal, P., & Mahajan, N. (2020). Starburst pamam dendrimers: Synthetic approaches, surface modifications, and biomedical applications. *Arabian Journal of Chemistry*, *13*(7), 6009–6039. DOI: 10.1016/j.arabjc.2020.05.002

Khatri, D. K., Sun, W., Singh, S. B., Gugulothu, D., Srivastava, S., & Vora, L. (2023). CRISPR/Cas9 assisted stem cell therapy in Parkinson's disease. *Biomaterials Research*, *27*(1), 46. DOI: 10.1186/s40824-023-00381-y PMID: 37194005

Khobragade, S. S., Deshmukh, M., Vyas, U., & Ingle, R. G. (2025). Innovative Approaches in Bone Tissue Engineering: Strategies for Cancer Treatment and Recovery. *International Journal of Molecular Sciences*, *26*(9), 3937. DOI: 10.3390/ijms26093937 PMID: 40362178

Khoshandam, M., Soltaninejad, H., Hosseinkhani, S., Saltanatpour, Z., Goudarzi, M. T. H., & Hamidieh, A. A. (2024). CRISPR/Cas and artificial intelligence to improve precision medicine: Future perspectives and potential limitations. *Human Gene*, *42*, 201356. DOI: 10.1016/j.humgen.2024.201356

Khoshandam, M., Soltaninejad, H., Mousazadeh, M., Hamidieh, A. A., & Hosseinkhani, S. (2024). Clinical applications of the CRISPR/Cas9 genome-editing system: Delivery options and challenges in precision medicine. *Genes & Diseases*, *11*(1), 268–282. DOI: 10.1016/j.gendis.2023.02.027 PMID: 37588217

Khurana, A., Sayed, N., Singh, V., Khurana, I., Allawadhi, P., Rawat, P. S., Navik, U., Pasumarthi, S. K., Bharani, K. K., & Weiskirchen, R. (2022). A comprehensive overview of CRISPR/Cas 9 technology and application thereof in drug discovery. *Journal of Cellular Biochemistry*, *123*(10), 1674–1698. https://doi.org/https://doi.org/10.1002/jcb.30329. DOI: 10.1002/jcb.30329 PMID: 36128934

Kim, H. J., Lee, S. Y., & Park, J. H. (2021). CRISPR/dCas9-mediated epigenetic reprogramming to overcome chemoresistance in cancer cells. *Journal of Experimental & Clinical Cancer Research : CR*, *40*, 300.

Kim, H. K., Lee, S., Kim, Y., Park, J., Min, S., Choi, J. W., Huang, T. P., Yoon, S., Liu, D. R., & Kim, H. H. (2020). High-throughput analysis of the activities of xCas9, SpCas9-NG and SpCas9 at matched and mismatched target sequences in human cells. *Nature Biomedical Engineering*, *4*(1), 111–124. DOI: 10.1038/s41551-019-0505-1 PMID: 31937939

Kim, H. K., Min, S., Song, M., Jung, S., Choi, J. W., & Kim, H. (2021). Deep learning improves prediction of CRISPR–Cpf1 guide RNA activity. *Nature Biotechnology*, *39*(2), 170–178. PMID: 32958957

Kim, H., Choi, S., Heo, H., Cho, S. H., Lee, Y., Kim, D., & Rhee, S. (2025). Applications of single-cell omics technologies for induced pluripotent stem cell-based cardiovascular research. *International Journal of Stem Cells*, *18*(1), 37–48. DOI: 10.15283/ijsc23183 PMID: 39129179

Kim, H.-K., Cheong, H., Kim, M.-Y., & Jin, H.-E. (2025). Therapeutic Targeting in Ovarian Cancer: Nano-Enhanced CRISPR/Cas9 Gene Editing and Drug Combination Therapy. *International Journal of Nanomedicine*, *20*, 3907–3931. DOI: 10.2147/IJN.S507688 PMID: 40191042

Kim, H., Kim, S., & Kim, J.-S. (2021). Delivery strategies for CRISPR/Cas systems in vitro and in vivo. *Experimental & Molecular Medicine*, *53*, 1220–1235.

Kim, H., Lee, W. J., Oh, Y., Kang, S. H., Hur, J. K., Lee, H., Song, W. J., Lim, K.-S., Park, Y.-H., Song, B.-S., Jin, Y. B., Jun, B.-H., Jung, C., Lee, D.-S., Kim, S.-U., & Lee, S. H. (2020). Enhancement of target specificity of CRISPR–Cas12a by using a chimeric DNA–RNA guide. *Nucleic Acids Research*, *48*(15), 8601–8616. DOI: 10.1093/nar/gkaa605 PMID: 32687187

Kim, M., Hwang, Y., Lim, S., Jang, H. K., & Kim, H. O. (2024). Advances in nanoparticles as non-viral vectors for efficient delivery of CRISPR/Cas9. *Pharmaceutics, 16*(9), 1197. DOI: 10.3390/pharmaceutics16091197 PMID: 39339233

Kim, S., Kim, D., Cho, S. W., Kim, J., & Kim, J. S. (2020). Highly efficient RNA-guided genome editing in human cells via delivery of purified Cas9 ribonucleoproteins. Genome Research, 30(4), 517–526. Li, Z., Wang, Y., & Xu, X. (2022). Engineering exosomes for targeted CRISPR-Cas9 delivery. *Theranostics*, *12*(1), 131–149.

Kingwell, K. (2023). First CRISPR therapy seeks landmark approval. *Nature Reviews. Drug Discovery*, *22*(5), 339–341. DOI: 10.1038/d41573-023-00050-8 PMID: 37012339

Kirkpatrick, R., & Stevens, H. (2019). Public ethics and genome editing: Navigating moral boundaries. *Bioethics*, *33*(7), 845–856.

Klaihmon, P., Lorthongpanich, C., Kheolamai, P., Luanpitpong, S., & Issaragrisil, S. (2022). Distinctive roles of yap and taz in human endothelial progenitor cells growth and functions. *Biomedicines*, *10*(1), 147. DOI: 10.3390/biomedicines10010147 PMID: 35052826

Klasfeld, S. J., Knutson, K. A., Miller, M. R., Fauman, E. B., Berghout, J., Moccia, R., & Kim, H. I. (2025). Common genetic modifiers influence cardiomyopathy susceptibility among the carriers of rare pathogenic variants. *HGG Advances*, *6*(3), 100460. DOI: 10.1016/j.xhgg.2025.100460 PMID: 40411145

Kleinstiver, B. P., Pattanayak, V., Prew, M. S., Tsai, S. Q., Nguyen, N. T., Zheng, Z., & Joung, J. K. (2016). High-fidelity CRISPR–Cas9 nucleases with no detectable genome-wide off-target effects. *Nature*, *529*(7587), 490–495. DOI: 10.1038/nature16526 PMID: 26735016

Kobayashi, T., Sato, M., & Nishikawa, S. (2023). Omics-guided regenerative medicine: From CRISPR editing to patient-specific therapy. *Science Translational Medicine*, *15*(678), eabn1234.

Kok, C. L., Koh, Y. Y., Ho, C. K., Thanh, N. T. C., & Teo, T. H. (2024). Artificial intelligence in cardiology. In *2024 IEEE 17th International Symposium on Embedded Multicore/Many-core Systems-on-Chip (MCSoC)* (pp. 462-466). IEEE. DOI: 10.1109/MCSoC64144.2024.00082

Kolanu, N. D. (2024). CRISPR–Cas9 Gene Editing: Curing Genetic Diseases by Inherited Epigenetic Modifications. *Global Medical Genetics*, *11*(01), 113–122. DOI: 10.1055/s-0044-1785234 PMID: 38560484

Komor, A. C., Badran, A. H., & Liu, D. R. (2017). CRISPR-based technologies for the manipulation of eukaryotic genomes. *Cell, 169*(3), 559–563. DOI: 10.1016/j.cell.2017.04.005 PMID: 28431253

Komor, A. C., Kim, Y. B., Packer, M. S., Zuris, J. A., & Liu, D. R. (2017). Programmable editing of a target base in genomic DNA without double-stranded DNA cleavage. *Nature, 533*(7603), 420–424. DOI: 10.1038/nature17946 PMID: 27096365

Konstantinov, I. E., King, G., & Porrello, E. R. (2022). From genome editing to blastocyst complementation: A new horizon in heart transplantation? *JTCVS Techniques, 12*, 177–184. DOI: 10.1016/j.xjtc.2022.01.012 PMID: 35403039

Koo, B., Kim, D. E., Kweon, J., Jin, C. E., Kim, S. H., Kim, Y., & Shin, Y. (2018). CRISPR/dCas9-mediated biosensor for detection of tick-borne diseases. *Sensors and Actuators. B, Chemical, 273*, 316–321. DOI: 10.1016/j.snb.2018.06.069 PMID: 32288252

Kordyś, M., Sen, R., & Warkocki, Z. (2022). Applications of the versatile CRISPR-Cas13 RNA targeting system. *Wiley Interdisciplinary Reviews. RNA, 13*(3), e1694. DOI: 10.1002/wrna.1694 PMID: 34553495

Korrapati, C., Yaraladdi, A. B., Vaddi, H., & Raghavan, R. (2024). Applications of the Saffron European Kitchen Spice Crocus sativus. *Therapeutic and Pharmacological Applications of Ethnobotany*, 98–127. DOI: 10.4018/979-8-3693-1986-4.ch005

Kostyushev, D., Brezgin, S., Kostyusheva, A., Zarifyan, D., Goptar, I., & Chulanov, V. (2019). Orthologous CRISPR/Cas9 systems for specific and efficient degradation of covalently closed circular DNA of hepatitis B virus. *Cellular and Molecular Life Sciences : CMLS, 76*(9), 1779–1794. DOI: 10.1007/s00018-019-03021-8 PMID: 30673820

Kotit, S. (2023). Lessons from the first-in-human in vivo CRISPR/Cas9 editing of the TTR gene by NTLA-2001 trial in patients with transthyretin amyloidosis with cardiomyopathy. *Global Cardiology Science & Practice, 2023*(1), e202304. DOI: 10.21542/gcsp.2023.4 PMID: 37928601

Kuchi Bhotla, H., Kaul, T., Balasubramanian, B., Easwaran, M., Arumugam, V. A., Pappusamy, M., Muthupandian, S., & Meyyazhagan, A. (2020). Platelets to surrogate lung inflammation in COVID-19 patients. *Medical Hypotheses, 143*, 110098. DOI: 10.1016/j.mehy.2020.110098 PMID: 32688185

Kulcsár, P. I., Tálas, A., Huszár, K., Ligeti, Z., Tóth, E., & Welker, E. (2017). Crossing enhanced and high-fidelity SpCas9 nucleases to optimize specificity and cleavage. *Genome Biology, 18*(1), 190. DOI: 10.1186/s13059-017-1318-8 PMID: 28985763

Kumar, R., Bargoti, T., Sengar, S., Singh, D., & Nain, V. (2025). A Decade of Genome Editing: Comparative Review of ZFN, TALEN, and CRISPR/Cas9. International Journal of Innovative Science and Research Technology, 3708–3717. https://doi.org/DOI: 10.38124/ijisrt/25apr2221

Kumar, S., Goyal, S., & Tiwari, R. (2020). Regulatory considerations for CRISPR and epigenetic therapies: Global perspectives. *Regenerative Medicine*, *15*(7), 2031–2045.

Kweon, J., Park, S., Jeon, M. Y., Lim, K., Jang, G., Jang, A.-H., Lee, M., Seok, C., Lee, C., Park, S., Ahn, J., Jang, J., Kim, N., Sung, Y. H., Kim, D., & Kim, Y. (2025). High-efficiency base editing for nuclear and mitochondrial DNA with an optimized DYW-like deaminase. *Molecular Therapy : the Journal of the American Society of Gene Therapy*, *33*(11), 5611–5623. Advance online publication. DOI: 10.1016/j.ymthe.2025.08.007 PMID: 40781772

Kwon, D. Y., & Lim, Y. (2023). Delivery strategies for CRISPR/dCas9-based epigenome editing in cancer therapy. *Pharmaceutics*, *15*(6), 1752. PMID: 37376200

Kwon, D. Y., Zhao, Y. T., Lamonica, J. M., & Zhou, Z. (2021). Locus-specific epigenetic remodeling controls neural differentiation. *Nature Communications*, *12*(1), 2931. PMID: 34006841

Kwon, J., Kim, M., Hwang, W., Jo, A., Hwang, G.-H., Jung, M., Kim, U. G., Cui, G., Kim, H., Eom, J.-H., Hur, J. K., Lee, J., Kim, Y., Kim, J., Bae, S., & Lee, J. K. (2023). Extru-seq: A method for predicting genome-wide Cas9 off-target sites with advantages of both cell-based and in vitro approaches. *Genome Biology*, *24*(1), 4. DOI: 10.1186/s13059-022-02842-4 PMID: 36627653

Lafi, Z., Ata, T., & Asha, S. (2025). CRISPR in clinical diagnostics: Bridging the gap between research and practice. *Bioanalysis*, *17*(4), 281–290. DOI: 10.1080/17576180.2025.2459520 PMID: 39902531

Lakshmaiah, A., Korrapati, C., Suresh, C., & Raghavan, R. (2025). Synthetic Biology and AI. *Spatially Variable Genes in Cancer*, 265–290. DOI: 10.4018/979-8-3693-7728-4.ch010

Lancaster, M. A., Renner, M., Martin, C. A., Wenzel, D., Bicknell, L. S., Hurles, M. E., Homfray, T., Penninger, J. M., Jackson, A. P., & Knoblich, J. A. (2013). Cerebral organoids model human brain development and microcephaly. *Nature*, *501*(7467), 373–379. DOI: 10.1038/nature12517 PMID: 23995685

Lanphier, E., Urnov, F., Haecker, S. E., Werner, M., & Smolenski, J. (2015). Don't edit the human germ line. *Nature*, *519*(7544), 410–411. DOI: 10.1038/519410a PMID: 25810189

Lanza, R., Russell, D. W., & Nagy, A. (2022). Engineering iPSCs with CRISPR for regenerative medicine. *Nature Reviews. Molecular Cell Biology*, *23*(4), 244–261.

Lau, C. H., Li, X., Liang, Q. L., Guo, R., Ren, X., Xu, Z., Zhu, Y., Lai, Y., Liu, G., Huang, Y., Wu, W., Zhu, H., Chen, J., & Zhang, X. (2025). CRISPR technology for diagnosis and treatment of human brucellosis. *Journal of Microbiological Methods*, *240*, 107339. DOI: 10.1016/j.mimet.2025.107339 PMID: 41274609

Lau, C.-H., & Suh, Y. (2017). Genome and Epigenome Editing in Mechanistic Studies of Human Aging and Aging-Related Disease. *Gerontology*, *63*(2), 103–117. DOI: 10.1159/000452972 PMID: 27974723

Laufer, B. I., & Singh, S. M. (2015). Strategies for precision modulation of gene expression by epigenome editing: An overview. *Epigenetics & Chromatin*, *8*(1), 34. DOI: 10.1186/s13072-015-0023-7 PMID: 26388942

Lecluze, E., Belhaj, H., Diaz, C., Lo, C. S., Lettre, G., & Lo, K. S. (2024). CRISPRi screening for functional characterization of blood pressure (BP) and coronary artery disease (CAD)-associated genes in endothelial cells. *protocols.io*.https://dx.doi.org/ DOI: 10.17504/protocols.io.5qpvo9ybxv4o/v1

Lee, E., Ozigbo, A. A., Varon, J., Halma, M., Laezzo, M., Ang, S. P., & Iglesias, J. (2025). Mitochondrial Reactive Oxygen Species: A Unifying Mechanism in Long COVID and Spike Protein-Associated Injury: A Narrative Review. *Biomolecules*, *15*(9), 1339. DOI: 10.3390/biom15091339 PMID: 41008646

Lee, H., Kim, B., Park, J., Park, S., Yoo, G., Yum, S., Kang, W., Lee, J.-M., Youn, H., & Youn, B. (2025). Cancer stem cells: Landscape, challenges and emerging therapeutic innovations. *Signal Transduction and Targeted Therapy*, *10*(1), 248. DOI: 10.1038/s41392-025-02360-2 PMID: 40759634

Lee, H., Rho, W. Y., Kim, Y. H., Chang, H., & Jun, B. H. (2025). CRISPR-Cas9 gene therapy: Non-viral delivery and stimuli-responsive nanoformulations. *Molecules (Basel, Switzerland)*, *30*(3), 542. DOI: 10.3390/molecules30030542 PMID: 39942646

Lee, J., Lim, K., Kim, A., Mok, Y. G., Chung, E., Cho, S. I., Lee, J. M., & Kim, J. S. (2023). Prime editing with genuine Cas9 nickases minimizes unwanted indels. *Nature Communications*, *14*(1), 1786. DOI: 10.1038/s41467-023-37507-8 PMID: 36997524

Lee, R. G., Mazzola, A. M., Braun, M. C., Platt, C., Vafai, S. B., Kathiresan, S., & Khera, A. V. (2023). Efficacy and safety of an investigational single-course CRISPR base-editing therapy targeting PCSK9 in nonhuman primate and mouse models. *Circulation*, *147*(3), 242–253. DOI: 10.1161/CIRCULATIONAHA.122.062132 PMID: 36314243

Lee, S., Kim, T., Lee, E., Lee, C., Kim, H., Rhee, H., Park, S. Y., Son, H.-J., Yu, S., Park, J. W., Choo, E. J., Park, S., Loeb, M., & Kim, T. H. (2020). Clinical course and molecular viral shedding among asymptomatic and symptomatic patients with SARS-CoV-2 infection in a community treatment center in the Republic of Korea. *JAMA Internal Medicine*, *180*(11), 1447–1452. DOI: 10.1001/jamainternmed.2020.3862 PMID: 32780793

Lee, T. H., Liu, P. S., Wang, S. J., Tsai, M. M., Shanmugam, V., & Hsieh, H. L. (2021). Bradykinin, as a reprogramming factor, induces transdifferentiation of brain astrocytes into neuron-like cells. *Biomedicines*, *9*(8), 923. DOI: 10.3390/biomedicines9080923 PMID: 34440126

Lee, T. L., & Sawai, T. (2024). Navigating equity in global access to genome therapy expanding access to potentially transformative therapies and benefiting those in need requires global policy changes. *Frontiers in Genetics*, *15*, 1381172. DOI: 10.3389/fgene.2024.1381172 PMID: 38638119

Lee, W., Zamudio-Ochoa, A., Buchel, G., Podlesniy, P., Gutierrez, N. M., Puigròs, M., Calderon, A., Tang, H.-Y., Li, L., Mikhalchenko, A., Koski, A., Trullas, R., Mitalipov, S., & Temiakov, D. (2023). Molecular Basis for Maternal Inheritance of Human Mitochondrial DNA. *Nature Genetics*, *55*(10), 1632–1639. DOI: 10.1038/s41588-023-01505-9 PMID: 37723262

Lei, J., Jiang, X., Huang, D., Jing, Y., Yang, S., Geng, L., & Qu, J. (2024). Human ESC-derived vascular cells promote vascular regeneration in a HIF-1α dependent manner. *Protein & Cell*, *15*(1), 36–51. DOI: 10.1093/procel/pwad027 PMID: 37158785

Lei, R., & Qiu, R. (2020). Ethical and regulatory issues in human gene editing: Chinese perspective. *Biotechnology and Applied Biochemistry*, *67*(6), 880–891. DOI: 10.1002/bab.2032 PMID: 33011985

Lei, T., Wang, Y., Zhang, Y., Yang, Y., Cao, J., Huang, J., Chen, J., Chen, H., Zhang, J., Wang, L., Xu, X., Gale, R. P., & Wang, L. (2024). Leveraging CRISPR gene editing technology to optimize the efficacy, safety and accessibility of CAR T-cell therapy. *Leukemia*, *38*(12), 2517–2543. DOI: 10.1038/s41375-024-02444-y PMID: 39455854

Liang, J., & Wang, Y. G. (2022). In vivo Reprogramming of Mouse Fibroblasts into Cardiac Progenitor Cells for Myocardial Infarction Treatment. *Circulation, 146*(Suppl_1), A13281–A13281. DOI: 10.1161/circ.146.suppl_1.13281

Liang, P., Xu, Y., Zhang, X., Ding, C., Huang, J., Zhang, Z., & Zhou, C. (2022). CRISPR-engineered iPSC-derived retinal cells for treating degenerative blindness. *Cell Stem Cell, 29*(3), 420–435.

Liang, P., Xu, Y., Zhang, X., Ding, C., Huang, R., & Zhang, Z.. (2020). CRISPR/Cas9-mediated gene editing in human embryos: Long-term effects and safety assessment. *Cell Research, 30*(3), 212–225.

Li, C., Zhang, H., & Wu, X. (2023). Functional characterization of GST gene families. *Plant Physiology and Biochemistry, 199*, 107952. DOI: 10.1016/j.plaphy.2023.107952

Li, C., Zhan, W., Jin, Y., Wang, X., & Jin, Y. (2023). Prime editing: A powerful genome editing tool for human genetic diseases. *Translational Medicine (Sunnyvale, Calif.), 21*, 36.

Li, J., Wang, J., & Chen, Z. (2025). Emerging role of exosomes in cancer therapy: Progress and challenges. *Molecular Cancer, 24*(1), 13. DOI: 10.1186/s12943-024-02215-4 PMID: 39806451

Li, L., Hu, S., & Chen, X. (2018). Non-viral delivery systems for CRISPR/Cas9-based genome editing: Challenges and opportunities. *Biomaterials, 171*, 207–218. DOI: 10.1016/j.biomaterials.2018.04.031 PMID: 29704747

Lillestøl, R. K., Redder, P., Garrett, R. A., & Brügger, K. (2006). Putative viral defence mechanism in archaeal cells. *Archaea : an International Microbiological Journal, 2*(1), 59–72. DOI: 10.1155/2006/542818 PMID: 16877322

Lima, A., & Maddalo, D. (2021). SEMMs: Somatically engineered mouse models. A new tool for in vivo disease modeling for basic and translational research. *Frontiers in Oncology, 11*, 667189. DOI: 10.3389/fonc.2021.667189 PMID: 33968774

Lino, C. A., Harper, J. C., Carney, J. P., & Timlin, J. A. (2018). Delivering CRISPR: A review of the challenges and approaches. *Drug Delivery, 25*(1), 1234–1257. DOI: 10.1080/10717544.2018.1474964 PMID: 29801422

Lin, Y., & Sun, Z. (2022). Klotho deficiency-induced arterial calcification involves osteoblastic transition of VSMCs and activation of BMP signalling. *Journal of Cellular Physiology, 237*(1), 720–729. DOI: 10.1002/jcp.30541 PMID: 34368951

Lin, Y., Zhu, Y., Cui, Y., Chen, R., Chen, Z., Li, G., Fan, M., Chen, J., Li, Y., Guo, X., Zheng, X., Chen, L., & Wang, F. (2021). Derepression of specific miRNA-target genes in rice using CRISPR/Cas9. *Journal of Experimental Botany*, *72*(20), 7067–7077. DOI: 10.1093/jxb/erab336 PMID: 34283216

Li, P., Xu, Q., & Wu, Y. (2021). CRISPR/Cas9-based targeting of non-coding RNAs in drug-resistant cancers: From mechanisms to therapeutic opportunities. *Cancer Letters*, *522*, 202–213.

Li, T., Li, S., Kang, Y., Zhou, J., & Yi, M. (2024). Harnessing the evolving CRISPR/Cas9 for precision oncology. *Journal of Translational Medicine*, *22*(1), 749. DOI: 10.1186/s12967-024-05570-4 PMID: 39118151

Li, T., Yang, Y., Qi, H., Cui, W., Zhang, L., Fu, X., He, X., Liu, M., Li, P., & Yu, T. (2023a). CRISPR/Cas9 therapeutics: Progress and prospects. *Signal Transduction and Targeted Therapy*, *8*(1), 36. DOI: 10.1038/s41392-023-01309-7 PMID: 36646687

Liu, C., Gu, L., Deng, W., Meng, Q., Li, N., Dai, G., & Fang, H. (2022). N6-methyladenosine RNA methylation in cardiovascular diseases. *Frontiers in Cardiovascular Medicine*, *9*, 887838. DOI: 10.3389/fcvm.2022.887838 PMID: 35571209

Liu, C., Shen, M., Manhas, A., Zhao, S., Belbachir, N., Nishiga, M., & Wu, J. (2024). Abstract Or121: Druggable Genome CRISPRi/a screens in iPSC-derived cardiomyocytes identify therapeutic targets for doxorubicin-induced cardiotoxicity. *Circulation Research*, *135*(Suppl_1), AOr121–AOr121. DOI: 10.1161/res.135.suppl_1.Or121

Liu, J., Qiao, S., Li, S., Jiang, Y., Zhang, S., & Li, Y. (2026). Integrating Cu2O nanocubes-enhanced BiVO4 photoanodes and CRISPR/Cas12a for ultra-sensitive miRNA-21 detection in cancer diagnostics. *Sensors and Actuators. B, Chemical*, *446*, 138637. DOI: 10.1016/j.snb.2025.138637

Liu, J., Wang, H., & Zhao, M. (2023). Genome editing for regenerative medicine: CRISPR-based approaches in stem cells and tissue repair. *Advanced Drug Delivery Reviews*, *195*, 114726.

Liu, M., Rehman, S., Tang, X., Gu, K., Fan, Q., Chen, D., & Ma, W. (2019). Methodologies for improving HDR efficiency. *Frontiers in Genetics*, *9*, 691. Advance online publication. DOI: 10.3389/fgene.2018.00691 PMID: 30687381

Liu, N., & Olson, E. N. (2022). CRISPR modelling and correction of cardiovascular disease. *Circulation Research*, *130*(12), 1827–1850. DOI: 10.1161/CIRCRESAHA.122.320496 PMID: 35679361

Liu, X. S., Wu, H., & Ji, X. (2016). Targeted DNA methylation in vivo using an engineered dCas9-MQ1 fusion protein. *Nature Communications*, *7*, 13593.

Liu, X. S., Wu, H., Ji, X., Stelzer, Y., Wu, X., Czauderna, S., Shu, J., Dadon, D., Young, R. A., & Jaenisch, R. (2016). Editing DNA methylation in the mammalian genome. *Cell, 167*(1), 233–247.e17. DOI: 10.1016/j.cell.2016.08.056 PMID: 27662091

Liu, X., Gao, Y., & Xie, Z. (2021). Multiplexed CRISPR-based epigenome editing for combinatorial gene regulation. *Nature Communications*, *12*, 3407.

Liu, X., Wu, H., Wang, J., & Zhang, C. (2023). CRISPR-based epigenome editing in cancer research and therapy. *Cancer Letters*, *564*, 216186.

Liu, Y., Lim, J., Teoh, S. H., & Zhang, Y. (2019). Engineering of osteoinductive signals in biodegradable scaffolds for bone tissue engineering using CRISPR. *Biomaterials*, *183*, 232–244.

Liu, Y., & Shanahan, C. M. (2011). Signalling pathways and vascular calcification. *Frontiers in Bioscience*, *16*(1302), 14. PMID: 21196233

Liu, Y., Yu, C., Daley, T. P., Wang, F., Cao, W. S., Bhate, S., & Qi, L. S. (2021). CRISPR activation and interference screens in primary human T cells decode cytokine regulation. *Cell*, *184*(2), 403–420.e17.

Liu, Y., Yu, C., Daley, T. P., Wang, F., Cao, W. S., Bhate, S., & Qi, L. S. (2022). CRISPR-based chromatin and epigenetic editing. *Nature Reviews. Methods Primers*, *2*, 6.

Li, W., Zhu, X., Xu, Y., Chen, J., Zhang, H., Yang, Z., Qi, Y., Hong, J., Li, Y., Wang, G., Shen, J., & Qian, C. (2022). Simultaneous editing of TCR, HLA-I/II and HLA-E resulted in enhanced universal CAR-T resistance to allo-rejection. *Frontiers in Immunology*, *13*, 1052717. DOI: 10.3389/fimmu.2022.1052717 PMID: 36532006

Li, X., Wang, H., Qi, X., Ji, Y., Li, F., Chen, X., Li, K., & Li, L. (2024). PCR Independent Strategy-Based Biosensors for RNA Detection. *Biosensors (Basel)*, *14*(4), 200. DOI: 10.3390/bios14040200 PMID: 38667193

Li, X., Zhang, W., & Chen, Z. (2022). Gene editing strategies for promoting axonal regeneration after spinal cord injury. *Stem Cell Research & Therapy*, *13*(1), 1–12. PMID: 34998430

Li, Y.-J., Wu, J.-Y., Liu, J., Xu, W., Qiu, X., Huang, S., Hu, X.-B., & Xiang, D.-X. (2021). Artificial exosomes for translational nanomedicine. *Journal of Nanobiotechnology*, *19*(1), 242. DOI: 10.1186/s12951-021-00986-2 PMID: 34384440

Li, Y., Li, D., You, L., Deng, T., Pang, Q., Meng, X., & Zhu, B. (2023). dCas9-Based PDGFR–β Activation ADSCs Accelerate Wound Healing in Diabetic Mice through Angiogenesis and ECM Remodeling. *International Journal of Molecular Sciences*, *24*(6), 5949. DOI: 10.3390/ijms24065949 PMID: 36983022

Li, Y., Ma, G., Weintraub, N., & Tang, Y. (2015). Enhancing transplanted cardiac progenitor cells homing and survival in ischemic heart by CRISPR-dcas9/sgRNA-based endogenous CXCR4 promoter activation. *Circulation*, *132*(suppl_3), A16620–A16620. DOI: 10.1161/circ.132.suppl_3.16620

Li, Y.-R., Fang, Y., Niu, S., Chen, Y., Lyu, Z., & Yang, L. (2025). Managing allorejection in off-the-shelf CAR-engineered cell therapies. *Molecular Therapy : the Journal of the American Society of Gene Therapy*, *33*(6), 2368–2390. DOI: 10.1016/j.ymthe.2024.11.035 PMID: 39600090

Li, Z. H., Wang, J., Xu, J. P., Wang, J., & Yang, X. (2023). Recent advances in CRISPR-based genome editing technology and its applications in cardiovascular research. *Military Medical Research*, *10*(1), 12. DOI: 10.1186/s40779-023-00447-x PMID: 36895064

Li, Z., & Zhang, Z. (2025). A tale of two strands: Decoding chromatin replication through strand-specific sequencing. *Molecular Cell*, *85*(2), 238–261. DOI: 10.1016/j.molcel.2024.10.035 PMID: 39824166

Loganathan, T., & Doss, C. G. P. (2023). Non-coding RNAs in human health and disease: Potential function as biomarkers and therapeutic targets. *Functional & Integrative Genomics*, *23*(1), 33. DOI: 10.1007/s10142-022-00947-4 PMID: 36625940

Lopes, J., Lopes, D., Motallebi, M., Ye, M., Xue, Y., Vieira, A. C. F., Singh, S. K., Dua, K., Veiga, F., Sethi, G., Paiva-Santos, A. C., & Makvandi, P. (2025). Biomembrane-coated nanosystems as next-generation delivery systems for the treatment of gastrointestinal cancers. *Bioengineering & Translational Medicine*, *10*(4), e70006. DOI: 10.1002/btm2.70006 PMID: 40708975

Lopes, R., & Prasad, M. K. (2024). Beyond the promise: Evaluating and mitigating off-target effects in CRISPR gene editing for safer therapeutics. *Frontiers in Bioengineering and Biotechnology*, *11*, 1339189. DOI: 10.3389/fbioe.2023.1339189 PMID: 38390600

Lovell-Badge, R., Anthony, E., Barker, R. A., Bubela, T., Brivanlou, A. H., Carpenter, M., Charo, R. A., Clark, A., Clayton, E., Cong, Y., Daley, G. Q., Fu, J., Fujita, M., Greenfield, A., Goldman, S. A., Hill, L., Hyun, I., Isasi, R., Kahn, J., & Zhai, X. (2021). ISSCR guidelines for stem cell research and clinical translation: The 2021 update. *Stem Cell Reports*, *16*(6), 1398–1408. DOI: 10.1016/j.stemcr.2021.05.012 PMID: 34048692

Lučanský, V., Holubeková, V., Kolková, Z., Halašová, E., Samec, M., & Golubnitschaja, O. (2023). Multi-faceted CRISPR/Cas technological innovation aspects in the framework of 3P medicine. *The EPMA Journal*, *14*(2), 201–217. DOI: 10.1007/s13167-023-00324-6 PMID: 37275547

Lukjanov, V., Koutná, I., & Šimara, P. (2021). CAR T-Cell Production Using Nonviral Approaches. *Journal of Immunology Research*, *2021*, 1–9. DOI: 10.1155/2021/6644685 PMID: 33855089

Luo, X., Weidinger, E., Burghardt, T., Höhn, M., & Wagner, E. (2025). CRISPR/Cas9 Ribonucleoprotein Delivery Enhanced by Lipo-Xenopeptide Carriers and Homology-Directed Repair Modulators: Insights from Reporter Cell Lines. *International Journal of Molecular Sciences*, *26*(9), 4361. DOI: 10.3390/ijms26094361 PMID: 40362595

Lupan, I., Bintintan, V., Deleanu, D., & Samasca, G. (2024). Epigenetic Regulation of DNA Methylation and RNA Interference in Gastric Cancer: A 2024 Update. *Biomedicines*, *12*(9), 2001. DOI: 10.3390/biomedicines12092001 PMID: 39335515

Lu, Q., Chen, R., Du, S., Chen, C., Pan, Y., Luan, X., Yang, J., Zeng, F., He, B., Han, X., & Song, Y. (2022). Activation of the cGAS-STING pathway combined with CRISPR-Cas9 gene editing triggering long-term immunotherapy. *Biomaterials*, *291*, 121871. DOI: 10.1016/j.biomaterials.2022.121871 PMID: 36323073

Lu, X., Zhang, M., Li, G., Zhang, S., Zhang, J., Fu, X., & Sun, F. (2023). Applications and Research Advances in the Delivery of CRISPR/Cas9 Systems for the Treatment of Inherited Diseases. *International Journal of Molecular Sciences*, *24*(17), 13202. DOI: 10.3390/ijms241713202 PMID: 37686009

Lu, Y., Xue, J., Deng, T., Zhou, X., Yu, K., Deng, L., & You, Y. (2017). Safety and feasibility of CRISPR-edited T cells in patients with refractory non-small-cell lung cancer. *Nature Medicine*, *26*(5), 732–740. DOI: 10.1038/s41591-020-0840-5 PMID: 32341578

Lv, W., Shao, Y., Hoshino, A., Arany, Z., Musunuru, K., & McDermott-Roe, C. (2022). CRISPR/Cas9 screens implicate RARA and SPNS1 in doxorubicin cardiotoxicity. bioRxiv, 2022-08.

Lv, Y., Shu, F., Luo, D., Jia, R., Huang, Y., & Xu, C. (2025). Targeting tumor angiogenesis and metabolism: A new perspective in pediatric thoracic tumor therapy. *Frontiers in Cell and Developmental Biology*, *13*, 1558403. DOI: 10.3389/fcell.2025.1558403 PMID: 40213391

Maan, K., Baghel, R., Dhariwal, S., Sharma, A., Bakhshi, R., & Rana, P. (2023). Multi-omics integration reveals radiation-induced altered pathways. *NPJ Systems Biology and Applications*, *9*(1), 42. DOI: 10.1038/s41540-023-00305-5 PMID: 37689794

Mabsor-Zgandaoui, S., Boulmane, H., Rachmoune, K., & Zgandaoui, Y. (2026b). Activation of Enzymes by Radiation: A Frontier for the Discovery of Neuroprotective Drugs. In Tiwari, P., Kadiri, S., & Minocha, N. (Eds.), *Recent Breakthroughs in Neurodegenerative Disorders and Enzyme Therapy* (pp. 107–128). IGI Global Scientific Publishing., DOI: 10.4018/979-8-3373-4627-4.ch005

Mabsor-Zgandaoui, S., Moussaif, A., El Hessni, A., Mesfioui, A., El Mzibri, M., Iddar, A., & Moutaouakkil, A. (2025a). Ionising and Ultraviolet Radiation Induce the Secretion of Melanin in *Tetrahymena Pyriformis. The International Journal of Environmental Studies*, *82*(1), 1–14. DOI: 10.1080/00207233.2024.2412455

Mabsor-Zgandaoui, S., Rachmoune, K., Aftais, I., Elamrani, F. E., Amradi, I., El Housseini, A., Hamdan, Y. A., Zgandaoui, Y., Iddar, A., El Mzibri, M., Moutaouakkil, A., El Hessni, A., & Mesfioui, A. (2026a). Biomedical Polymers: Foundations and Future Innovations in Medicine. In Erişen, D., Khan, F., Chang, C., & Yan, K. (Eds.), *Biomedical Applications and Toxicity of Polymers, Nanoparticles, Biomaterials, and Metal Ions* (pp. 1–30). IGI Global Scientific Publishing., DOI: 10.4018/979-8-3373-0055-9.ch001

Mabsor-Zgandaoui, S., Rachmoune, K., El Hessni, A., El Mzibri, M., Iddar, A., & Moutaouakkil, A. (2025b). Effect of Hydrogen Peroxide Pre-Treatment on Enhancing the Cellular Resistance of *Tetrahymena Pyriformis* Exposed to Ionizing Radiation. *Toxicological and Environmental Chemistry*, *107*(2), 294–309. DOI: 10.1080/02772248.2025.2460804

Maeder, M. L., & Gersbach, C. A. (2016). Genome-editing technologies for gene and cell therapy. *Molecular Therapy : the Journal of the American Society of Gene Therapy*, *24*(3), 430–446. DOI: 10.1038/mt.2016.10 PMID: 26755333

Maeder, M. L., Stefanidakis, M., Wilson, C. J., Baral, R., Barrera, L. A., Bounoutas, G. S., & Rebar, E. J. (2019). Development of a CRISPR-Cas9 gene-editing therapy for inherited retinal disease. *Nature Medicine*, *25*(2), 229–233. DOI: 10.1038/s41591-018-0327-9 PMID: 30664785

Maganti, H. B., Jrade, H., Cafuir, L. A., Pegliasco, E. M., Nalagatla, T., Abou-El-Enein, M., & Kanashova, T. (2022). Use of CRISPR/Cas9 gene editing to improve chimeric antigen receptor T cell therapy: A systematic review and meta-analysis. *Frontiers in Immunology*, *13*, 867888. DOI: 10.3389/fimmu.2022.867888 PMID: 35039239

Mahfouz, M. (2024). Revolutionizing Point-of-Care Diagnostics via CRISPR Systems. *ACS Synthetic Biology*, *13*(2), 411–412. DOI: 10.1021/acssynbio.3c00763 PMID: 38197893

Makadia, H. K., & Siegel, S. J. (2011). Poly Lactic-co-Glycolic Acid (PLGA) as Biodegradable Controlled Drug Delivery Carrier. *Polymers*, *3*(3), 1377–1397. DOI: 10.3390/polym3031377 PMID: 22577513

Makarova, K. S., Haft, D. H., Barrangou, R., Brouns, S. J., Charpentier, E., Horvath, P., Moineau, S., Mojica, F. J. M., Wolf, Y. I., Yakunin, A. F., van der Oost, J., & Koonin, E. V. (2011). Evolution and classification of the CRISPR–Cas systems. *Nature Reviews. Microbiology*, *9*(6), 467–477. DOI: 10.1038/nrmicro2577 PMID: 21552286

Makarova, K. S., Wolf, Y. I., Alkhnbashi, O. S., Costa, F., Shah, S. A., Saunders, S. J., Barrangou, R., Brouns, S. J. J., Charpentier, E., Haft, D. H., Horvath, P., Moineau, S., Mojica, F. J. M., Terns, R. M., Terns, M. P., White, M. F., Yakunin, A. F., Garrett, R. A., van der Oost, J., & Koonin, E. V. (2015). Updated evolutionary classification of CRISPR-Cas systems. *Nature Reviews. Microbiology*, *13*(11), 722–736. DOI: 10.1038/nrmicro3569 PMID: 26411297

Makarova, K. S., Wolf, Y. I., Iranzo, J., Shmakov, S. A., Alkhnbashi, O. S., Brouns, S. J. J., Charpentier, E., Cheng, D., Haft, D. H., Horvath, P., Moineau, S., Mojica, F. J. M., Scott, D., Shah, S. A., Siksnys, V., Terns, M. P., Venclovas, Č., White, M. F., Yakunin, A. F., & Koonin, E. V. (2020). Evolutionary classification of CRISPR–Cas systems. *Nature Reviews. Microbiology*, *18*(2), 67–83. Advance online publication. DOI: 10.1038/s41579-019-0299-x PMID: 31857715

Malekshoar, M., Azimi, S. A., Kaki, A., Mousazadeh, L., Motaei, J., & Vatankhah, M. (2023). CRISPR-Cas9 targeted enrichment and next-generation sequencing for mutation detection. *The Journal of Molecular Diagnostics : JMD*, *25*(5), 249–262. DOI: 10.1016/j.jmoldx.2023.01.010 PMID: 36841425

Manfrin, A., Tabata, Y., Paquet, E. R., Vuaridel, A. R., Rivest, F. R., Naef, F., & Lutolf, M. P. (2019). Engineered signaling centers for the spatially controlled patterning of human pluripotent stem cells. *Nature Methods*, *16*(7), 640–648. DOI: 10.1038/s41592-019-0455-2 PMID: 31249412

Manghwar, H., Li, B., Ding, X., Hussain, A., Lindsey, K., Zhang, X., & Jin, S. (2020). CRISPR/Cas Systems in Genome Editing: Methodologies and Tools for sgRNA Design, Off-Target Evaluation, and Strategies to Mitigate Off-Target Effects. *Advancement of Science*, *7*(6), 1902312. DOI: 10.1002/advs.201902312 PMID: 32195078

Mao, X., Fu, P., Wang, L., & Xu, J. (2019). CRISPR-Cas9-mediated gene editing in osteoarthritis: Targeting MMP13 for cartilage regeneration. *Annals of the Rheumatic Diseases*, *78*(2), 163–171.

Marchant, G. E. (2021). Global Governance of Human Genome Editing: What Are the Rules? *Annual Review of Genomics and Human Genetics*, *22*(1), 385–405. DOI: 10.1146/annurev-genom-111320-091930 PMID: 33667117

Marina, R. J., Brannan, K. W., Dong, K. D., Yee, B. A., & Yeo, G. W. (2020). Evaluation of engineered CRISPR-Cas-mediated systems for site-specific RNA editing. *Cell Reports*, *33*(5), 108350. DOI: 10.1016/j.celrep.2020.108350 PMID: 33147453

Marín-García, J., & Akhmedov, A. T. (2015). Epigenetics of the failing heart. *Heart Failure Reviews*, *20*(4), 435–459. DOI: 10.1007/s10741-015-9483-x PMID: 25847519

Martinez-Lage, M., Torres-Ruiz, R., Puig-Serra, P., Moreno-Gaona, P., Martin, M. C., Moya, F. J., & Rodriguez-Perales, S. (2020). In vivo CRISPR/Cas9 targeting of fusion oncogenes for selective elimination of cancer cells. *Nature Communications*, *11*(1), 5060. DOI: 10.1038/s41467-020-18875-x PMID: 33033246

Matsumoto, D., Matsugi, E., Kishi, K., Inoue, Y., Nigorikawa, K., & Nomura, W. (2024). SpCas9-HF1 enhances accuracy of cell cycle-dependent genome editing by increasing HDR efficiency, and by reducing off-target effects and indel rates. *Molecular Therapy. Nucleic Acids*, *35*(1), 102124. DOI: 10.1016/j.omtn.2024.102124 PMID: 38328481

McArthur, K. M., Kay, A. M., Mosier, J. A., Grant, J. N., Stewart, J. A., & Simpson, C. L. (2017). Manipulating the plasticity of smooth muscle cells to regulate vascular calcification. *AIMS Cell Tissue Eng*, *1*(2), 165–179. DOI: 10.3934/celltissue.2017.3.165

McCutcheon, S. R., Rohm, D., Iglesias, N., & Gersbach, C. A. (2024). Epigenome editing technologies for discovery and medicine. *Nature Biotechnology*, *42*(8), 1199–1217. DOI: 10.1038/s41587-024-02320-1 PMID: 39075148

Medvedev, S. P., & Zakian, S. M. (2015). The CRISPR/Cas9 system-a tool for studying hereditary cardiovascular diseases. *Blood Circulation Pathology and Cardiac Surgery, 19* (S4-2), 113-117.

Mehta, A., & Merkel, O. M. (2020). Immunogenicity of Cas9 Protein. *Journal of Pharmaceutical Sciences*, *109*(1), 62–67. DOI: 10.1016/j.xphs.2019.10.003 PMID: 31589876

Mendell, J. R., Al-Zaidy, S. A., Rodino-Klapac, L. R., Goodspeed, K., Gray, S. J., Kay, C. N., Boye, S. L., Boye, S. E., George, L. A., Salabarria, S., Corti, M., Byrne, B. J., & Tremblay, J. P. (2021). Current clinical applications of in vivo gene therapy with AAVs. *Molecular Therapy : the Journal of the American Society of Gene Therapy*, *29*(2), 464–488. DOI: 10.1016/j.ymthe.2020.12.007 PMID: 33309881

Mendell, J. R., Al-Zaidy, S., Rodino-Klapac, L. R., Goodspeed, K., Gray, S. J., Kay, C. N., & Chamberlain, J. R. (2021). Current clinical landscape of AAV-based therapies. *Nature Reviews. Drug Discovery*, *20*(1), 35–55.

Mendonça, M. C. P., Kont, A., Kowalski, P. S., & O'Driscoll, C. M. (2023). Design of lipid-based nanoparticles for delivery of therapeutic nucleic acids. *Drug Discovery Today*, *28*(3), 103505. DOI: 10.1016/j.drudis.2023.103505 PMID: 36708760

Mengstie, M. A., & Wondimu, B. Z. (2021). Mechanism and applications of CRISPR/Cas9. *Biologics*, *15*, 353–361. Advance online publication. DOI: 10.2147/BTT.S326422

Meng, Z., Dong, J., Guo, L., Su, F., Liu, J., Wang, G., & Zhao, Z. (2026). Now and future of artificial intelligence-based signet ring cell diagnosis: A survey. *Expert Systems with Applications*, *296*, 129188. DOI: 10.1016/j.eswa.2025.129188

Meyyazhagan, A., Balasubramanian, B., Easwaran, M., Alagamuthu, K. K., Shanmugam, S., Kuchi Bhotla, H., Pappusamy, M., Arumugam, V. A., Thangaraj, A., Kaul, T., Keshavarao, S., & Cacabelos, R. (2020). Biomarker study of the biological parameter and neurotransmitter levels in autistics. *Molecular and Cellular Biochemistry*, *474*(1-2), 277–284. DOI: 10.1007/s11010-020-03851-2 PMID: 32740790

Meyyazhagan, A., Pushparaj, K., Balasubramanian, B., Kuchi Bhotla, H., Pappusamy, M., Arumugam, V. A., Easwaran, M., Pottail, L., Mani, P., Tsibizova, V., & Di Renzo, G. C. (2022). COVID-19 in pregnant women and children: Insights on clinical manifestations, complexities, and pathogenesis. *International Journal of Gynaecology and Obstetrics: the Official Organ of the International Federation of Gynaecology and Obstetrics*, *156*(2), 216–224. DOI: 10.1002/ijgo.14007 PMID: 34735717

Miguel-dos-Santos, R., & Cingolani, E. (2023). The hunt for novel AAV capsids with improved cardiac tropism. *Molecular Therapy. Methods & Clinical Development*, *31*, 31. DOI: 10.1016/j.omtm.2023.101124 PMID: 37886605

Mikhaylova, E. V., Kuluev, B. R., Gerashchenkov, G. A., Chemeris, D. A., Garafutdinov, R. R., Kuluev, A. R., Baymiev, A., Baymiev, A. K., & Chemeris, A. V. (2024). Prime-editing methods and pegRNA design programs. *Molekuliarnaia Biologiia*, *58*(1), 22–39. DOI: 10.31857/S0026898424010023 PMID: 38943578

Minczuk, M., Silva-Pinheiro, P., Jackson, C. B., & Gammage, P. A. (2022). In vivo mitochondrial base editing via adeno-associated viral delivery to mouse post-mitotic tissue. *Nature Communications*, *13*(1), 750. DOI: 10.1038/s41467-022-28358-w PMID: 35136065

Minnesota Medical School. U. (2025). New gene-editing therapy shows early success in fighting advanced GI cancers. *Medical News*. https://med.umn.edu/news/new-gene-editing-therapy-shows-early-success-fighting-advanced-gi-cancers

Mishra, S., Dwivedi, S. P., & Singh, R. B. (2010). Epigenetic effect of heavy metals. *The Open Nutraceuticals Journal*, *3*, 188–193. DOI: 10.2174/1876396001003010018 8

Mladenov, E., Kallies, M., Stuschke, M., Gkika, E., & Iliakis, G. (2025). CRISPR/Cas9-generated DSB clusters mimic complex lesions. *Scientific Reports*, *15*(1), 36480. DOI: 10.1038/s41598-025-22945-9 PMID: 41115973

Mohanraju, P., Saha, C., van Baarlen, P., Louwen, R., Staals, R. H. J., & van der Oost, J. (2022). Alternative functions of CRISPR-Cas systems in the evolutionary arms race. *Nature Reviews. Microbiology*, *20*(6), 351–364. DOI: 10.1038/s41579-021-00663-z PMID: 34992260

Mohd, Y., Balasubramanian, B., Meyyazhagan, A., Kuchi Bhotla, H., Shanmugam, S.K., Ramesh Kumar, M.K., Pappusamy, M., Alagamuthu, K.K., Keshavarao, S., & Arumugam, V.A. (2021). Extricating the association between the prognostic factors of colorectal cancer. *J Gastrointest Cancer, 52*(3), 022-1028.

Mohd, Y., Kumar, P., Kuchi Bhotla, H., Meyyazhagan, A., Balasubramanian, B., Ramesh Kumar, M. K., Pappusamy, M., Alagamuthu, K. K., Orlacchio, A., Keshavarao, S., Sampathkumar, P., & Arumugam, V. A. (2021). Transmission jeopardy of adenomatosis polyposis coli and methylenetetrahydrofolate reductase in colorectal cancer. *Journal of the Renin-Angiotensin-Aldosterone System : JRAAS*, *2021*, 2021. DOI: 10.1155/2021/7010706 PMID: 34956401

Mojica, F. J., Díez-Villaseñor, C., García-Martínez, J., & Soria, E. (2005). Intervening sequences of regularly spaced prokaryotic repeats derive from foreign genetic elements. *Journal of Molecular Evolution*, *60*(2), 174–182. DOI: 10.1007/s00239-004-0046-3 PMID: 15791728

Mojica, F. J., & Montoliu, L. (2016). On the origin of CRISPR-Cas technology: From prokaryotes to mammals. *Trends in Microbiology*, *24*(10), 811–820. DOI: 10.1016/j.tim.2016.06.005 PMID: 27401123

Mojtahedi, H., Yazdanpanah, N., & Rezaei, N. (2021). Chronic myeloid leukemia stem cells: Targeting therapeutic implications. *Stem Cell Research & Therapy*, *12*(1), 603. DOI: 10.1186/s13287-021-02659-1 PMID: 34922630

Molinaro, C., Scalise, M., Leo, I., Salerno, L., Sabatino, J., Salerno, N., & Marino, F. (2023). Polarizing macrophage functional phenotype to foster cardiac regeneration. *International Journal of Molecular Sciences*, *24*(13), 10747. DOI: 10.3390/ijms241310747 PMID: 37445929

Moloudi, K., & Haghdoost, S. (2025). Ionizing radiation enhances proliferation and differentiation via NRF2. *Antioxidants*, *14*(8), 986. DOI: 10.3390/antiox14080986 PMID: 40867882

Moon, J., & Liu, C. (2023). Asymmetric CRISPR enabling cascade signal amplification for nucleic acid detection by competitive crRNA. *Nature Communications*, *14*(1), 7504. DOI: 10.1038/s41467-023-43389-7 PMID: 37980404

Moore, O. M., Aguilar-Sanchez, Y., Lahiri, S. K., Hulsurkar, M. M., Navarro-Garcia, J. A., Word, T. A., & Wehrens, X. H. (2024). Long-term efficacy and safety of cardiac genome editing for catecholaminergic polymorphic ventricular tachycardia. *The Journal of Cardiovascular Aging*, *4*(1), 8. DOI: 10.20517/jca.2023.42 PMID: 38464671

Moreno, A. M., Fu, X., Zhu, J., Katrekar, D., Shih, Y. R. V., Marlett, J., Cabotaje, J., Tat, J., Naughton, J., Lisowski, L., Varghese, S., Zhang, K., & Mali, P. (2018). In situ gene therapy via AAV-CRISPR-Cas9-mediated targeted gene regulation. *Molecular Therapy : the Journal of the American Society of Gene Therapy*, *26*(7), 1818–1827. DOI: 10.1016/j.ymthe.2018.04.017 PMID: 29754775

Moreno, A., Fernández, C., & Domínguez, O. (2021). CRISPR technology in stem cell therapy: Strategies, challenges, and opportunities. *Frontiers in Bioengineering and Biotechnology*, *9*, 679763.

Morita, S., Noguchi, H., Horii, T., Nakabayashi, K., Kimura, M., & Okamura, K.. (2016). Targeted DNA demethylation for epigenetic therapy. *Nature Chemical Biology*, *12*(12), 1090–1097.

Morita, S., Noguchi, H., Horii, T., Nakabayashi, K., Kimura, M., Okamura, K., Sakai, A., Nakashima, H., Hata, K., Nakashima, K., & Hatada, I. (2016). Targeted DNA demethylation in vivo using dCas9–peptide repeat and scFv–TET1 catalytic domain fusions. *Nature Biotechnology*, *34*(10), 1060–1065. DOI: 10.1038/nbt.3658 PMID: 27571369

Moses, C., Hodgetts, S. I., Nugent, F., Ben-Ary, G., Park, K. K., Blancafort, P., & Harvey, A. R. (2020). Transcriptional repression of PTEN in neural cells using CRISPR/dCas9 epigenetic editing. *Scientific Reports*, *10*(1), 11393. DOI: 10.1038/s41598-020-68257-y PMID: 32647121

Motta, B. M., Pramstaller, P. P., Hicks, A. A., & Rossini, A. (2017). The impact of CRISPR/Cas9 technology on cardiac research: From disease modelling to therapeutic approaches. *Stem Cells International*, *2017*(1), 8960236. DOI: 10.1155/2017/8960236 PMID: 29434642

Moya, I. M., & Halder, G. (2019). Hippo-YAP/TAZ signalling in organ regeneration and regenerative medicine. *Nature Reviews. Molecular Cell Biology*, *20*(4), 211–226. DOI: 10.1038/s41580-018-0086-y PMID: 30546055

Musunuru, K. (2022). Moving toward genome-editing therapies for cardiovascular diseases. *The Journal of Clinical Investigation*, *132*(1), e148555. DOI: 10.1172/JCI148555 PMID: 34981785

Musunuru, K. (2023). CRISPR and cardiovascular diseases. *Cardiovascular Research*, *119*(1), 79–93. DOI: 10.1093/cvr/cvac048 PMID: 35388882

Naeem, M., Hoque, M. Z., Ovais, M., Basheer, C., & Ahmad, I. (2021). Stimulus-responsive smart nanoparticles-based CRISPR-cas delivery for therapeutic genome editing. *International Journal of Molecular Sciences*, *22*(20), 11300. DOI: 10.3390/ijms222011300 PMID: 34681959

Naeem, M., Majeed, S., Hoque, M. Z., & Ahmad, I. (2020). Latest Developed Strategies to Minimize the Off-Target Effects in CRISPR-Cas-Mediated Genome Editing. *Cells*, *9*(7), 1608. DOI: 10.3390/cells9071608 PMID: 32630835

Nakamura, M., Srinivasan, P., Chavez, M., Carter, M. A., Dominguez, A. A., La Russa, M., & Qi, L. S. (2021). Anti-CRISPR-mediated control of gene editing and synthetic circuits in eukaryotic cells. *Nature Communications*, *12*(1), 530. PMID: 33483501

Naseri, G., & Koffas, M. A. G. (2020). Application of combinatorial optimization strategies in synthetic biology. *Nature Communications*, *11*(1), 2446. DOI: 10.1038/s41467-020-16175-y PMID: 32415065

Navarro Quiroz, R., Villarreal Camacho, J., Zarate Peñata, E., Bello Lemus, Y., López-Fernández, C., Gomez Escorcia, L., Fernández-Ponce, C., Rebolledo Cobos, M., Fandiño Moreno, J., Fiorillo-Moreno, O., & Navarro Quiroz, E. (2025). Multiscale information processing in the immune system. *Frontiers in Immunology, 16*, 1563992. Advance online publication. DOI: 10.3389/fimmu.2025.1563992 PMID: 40761781

NCT02793856. (2016). *PD-1 Knockout Engineered T Cells for Metastatic Non-small Cell Lung Cancer. ClinicalTrials.gov*. https://clinicaltrials.gov/study/NCT02793856

NCT04637763. (2020). *CRISPR-Edited Allogeneic Anti-CD19 CAR-T Cell Therapy. ClinicalTrials.gov*. https://clinicaltrials.gov/study/NCT04637763

NCT05643742. (2022). *A Safety and Efficacy Study Evaluating CTX112 in Subjects with Relapsed or Refractory B-cell Malignancies. ClinicalTrials.gov*. https://clinicaltrials.gov/study/NCT05643742

Nedorezova, D. D., Dubovichenko, M. V., Belyaeva, E. P., Grigorieva, E. D., Peresadina, A. V., & Kolpashchikov, D. M. (2022). Specificity of oligonucleotide gene therapy (OGT) agents. *Theranostics, 12*(16), 7132–7157. DOI: 10.7150/thno.77830 PMID: 36276652

Neiger, H. E., Siegler, E. L., & Shi, Y. (2021). Breast cancer predisposition genes and synthetic lethality. *International Journal of Molecular Sciences, 22*(11), 5614. DOI: 10.3390/ijms22115614 PMID: 34070674

Nelson, C. E., Wu, Y., Gemberling, M. P., Oliver, M. L., Waller, M. A., Bohning, J. D., Robinson-Hamm, J. N., Bulaklak, K., Castellanos Rivera, R. M., Collier, J. H., Asokan, A., & Gersbach, C. A. (2019). Long-term evaluation of AAV-CRISPR genome editing for Duchenne muscular dystrophy. *Nature Medicine, 25*(3), 427–432. DOI: 10.1038/s41591-019-0344-3 PMID: 30778238

Neslund, J., Carroll, J., & Bryant, R. (2022). Metallothionein expression as prognostic biomarker. *Cancer Biomarkers, 35*(2), 233–245. DOI: 10.3233/CBM-210536

Ngo, V., & Duennwald, M. L. (2022). Nrf2 and oxidative stress overview. *Antioxidants, 11*(12), 2345. DOI: 10.3390/antiox11122345 PMID: 36552553

Niazi, S. K. (2023). The Dawn of In Vivo Gene Editing Era: A Revolution in the Making. *Biologics (Basel, Switzerland), 3*(4), 253–295. DOI: 10.3390/biologics3040014

Nie, J., Han, Y., Jin, Z., Hang, W., Shu, H., Wen, Z., & Wang, D. W. (2023). Homology-directed repair of an MYBPC3 gene mutation in a rat model of hypertrophic cardiomyopathy. *Gene Therapy, 30*(6), 520–527. DOI: 10.1038/s41434-023-00384-3 PMID: 36765144

Nielsen, A. A. K., Der, B. S., Shin, J., Vaidyanathan, P., Paralanov, V., Strychalski, E. A., & Voigt, C. A. (2016). Genetic circuit design automation. Science, 352(6281), Nissim, L., Perli, S. D., Fridkin, A., Perez-Pinera, P., & Lu, T. K. (2014). Multiplexed and programmable regulation of gene networks with an integrated RNA and CRISPR/Cas toolkit in human cells. *Molecular Cell, 54*(4), 698–710.

Nielsen, A. A., & Keasling, J. D. (2016). Engineering cellular metabolism. *Cell, 164*(6), 1185–1197. DOI: 10.1016/j.cell.2016.02.004 PMID: 26967285

Nihongaki, Y., Kawano, F., Nakajima, T., & Sato, M. (2015). Photoactivatable CRISPR-Cas9 for optogenetic genome editing. *Nature Biotechnology, 33*(7), 755–760. DOI: 10.1038/nbt.3245 PMID: 26076431

Nikfarjam, S., & Singh, K. K. (2023). DDR signaling — Link between cancer and CVD. *Cancer Medicine, 12*(4), 4380–4404. DOI: 10.1002/cam4.5274 PMID: 36156462

Nikitchina, N., Heckel, A.-M., Shebanov, N., Mazunin, I., Tarassov, I., & Entelis, N. (2025). Targeted deletions in human mitochondrial DNA engineered by Type V CRISPR–Cas12a system. *NAR Molecular Medicine, 2*(2), ugaf021. Advance online publication. DOI: 10.1093/narmme/ugaf021 PMID: 41255706

Nishiga, M., Qi, L. S., & Wu, J. C. (2021). CRISPRi/a screening with human iPSCs. In *Pluripotent Stem-Cell Derived Cardiomyocytes* (pp. 261–281). Springer US. DOI: 10.1007/978-1-0716-1484-6_23

Nishikura, K. (2010). Functions and Regulation of RNA Editing by ADAR Deaminases. *Annual Review of Biochemistry, 79*(1), 321–349. DOI: 10.1146/annurev-biochem-060208-105251 PMID: 20192758

Nishimasu, H., Ran, F. A., Hsu, P. D., Konermann, S., Shehata, S. I., Dohmae, N., Ishitani, R., Zhang, F., & Nureki, O. (2014). Crystal structure of Cas9. *Cell, 156*(5), 935–949. DOI: 10.1016/j.cell.2014.02.001 PMID: 24529477

Nishiyama, T., Zhang, Y., Mcanally, J., Xu, L., Liu, N., Bassel-Duby, R. S., & Olson, E. N. (2022). Precise genomic correction of RBM20 mutations rescues dilated cardiomyopathy. *Circulation, 146*(Suppl_1), A11164–A11164. DOI: 10.1161/circ.146.suppl_1.11164

Nissanka, N., & Moraes, C. T. (2020). Mitochondrial DNA heteroplasmy in disease and targeted nuclease-based therapeutic approaches. *EMBO Reports, 21*(3), e49612. DOI: 10.15252/embr.201949612 PMID: 32073748

Norota, K., Ishizuka, S., Hirose, M., Sato, Y., Maeki, M., Tokeshi, M., Ibrahim, S. M., Harashima, H., & Yamada, Y. (2025). Lipid nanoparticle delivery of the CRISPR/Cas9 system directly into the mitochondria of cells carrying m.7778G>T mutation in MtDNA (mt-Atp8). *Scientific Reports*, *15*(1), 18717. Advance online publication. DOI: 10.1038/s41598-025-03671-8 PMID: 40537532

Nuñez, J. K., Chen, J., Pommier, G., Cogan, J. Z., Replogle, J. M., Adriaens, C., Ramalingam, S., Kweon, J., Takahashi, C., & Weissman, J. S. (2021). Genome-wide programmable transcriptional memory by CRISPR-based epigenome editing. *Cell*, *184*(6), 2503–2519.e17. DOI: 10.1016/j.cell.2021.03.025 PMID: 33838111

O'Geen, H., Henry, I. M., Bhakta, M. S., Meckler, J. F., & Segal, D. J. (2019). A genome-wide analysis of Cas9 binding specificity using ChIP-seq and targeted sequence capture. *Nucleic Acids Research*, *47*(9), 4881–4895. PMID: 25712100

Okesanya, O. J., Ahmed, M. M., Ogaya, J. B., Amisu, B. O., Ukoaka, B. M., Adigun, O. A., Manirambona, E., Adebusuyi, O., Othman, Z. K., Oluwakemi, O. G., Ayando, O. D., Tan, M. I. R. S., Idris, N. B., Kayode, H. H., Oso, T. A., Ahmed, M., Kouwenhoven, M. B. N., Ibrahim, A. M., & Lucero-Prisno, D. E.III. (2025). Reinvigorating AMR resilience: Leveraging CRISPR-Cas technology potentials to combat the 2024 WHO bacterial priority pathogens for enhanced global health security-a systematic review. *Tropical Medicine and Health*, *53*(1), 43. DOI: 10.1186/s41182-025-00728-2 PMID: 40176174

Osborn, M. J., Starker, C. G., McElroy, A. N., Webber, B. R., & Moriarity, B. S. (2021). Base editing-mediated correction of COL7A1 nonsense mutations in dystrophic epidermolysis bullosa. *Molecular Therapy. Nucleic Acids*, *26*, 333–342.

Ostrovska, S. S., Abramov, S. V., Dychko, E. N., Vyselko, A. D., Konovalova, O. S., & Danilchenko, A. K. (2023). Epigenetic effects of heavy metals. *Bulletin of Problems in Biology and Medicine*, *1*(168), 36–43. DOI: 10.29254/2077-4214-2023-1-168-36-43

Ou, Z., Niu, X., He, W., Chen, Y., Song, B., Xian, Y., Fan, D., Tang, D., & Sun, X. (2016). The combination of CRISPR/Cas9 and iPSC technologies in the gene therapy of human β-thalassemia in mice. *Scientific Reports*, *6*(1), 32463. DOI: 10.1038/srep32463 PMID: 27581487

Owen, M. C., & Kopecky, B. J. (2024). Targeting macrophages in organ transplantation: A step toward personalized medicine. *Transplantation*, *108*(10), 2045–2056. DOI: 10.1097/TP.0000000000004978 PMID: 38467591

Ozden, F., & Minary, P. (2024). Learning to quantify uncertainty in off-target activity for CRISPR guide RNAs. *Nucleic Acids Research, 52*(18), e87. DOI: 10.1093/nar/gkae759 PMID: 39275984

P., A., D., K. M., M., J., P., S., & Raghavan, R. (2025). The Cancer Exposome. *Cancer Exposomics and Environmental Influences on Carcinogenesis*, 161–192. DOI: 10.4018/979-8-3373-2165-3.ch004

Pache, M. M., Rutuja, R. P., Nikam, S. V., Rode, R. B., & Darekar, A. B. (2025). CRISPR-Cas9 in Pharmaceutical Research: Applications, Challenges, Ethical Considerations and Future Directions. *Asian Journal of Pharmacy and Technology., 15*(3), 296–4.

Pallares, R. M., Barmin, R. A., Wang, A., Kiessling, F., & Lammers, T. (2025). Clinical cancer nanomedicines. *Journal of Controlled Release, 385*, 113991. DOI: 10.1016/j.jconrel.2025.113991 PMID: 40582644

Paranitharan, N., Kataria, S., Arumugam, V. A., Hsieh, H., Muthukrishnan, S., & Velayuthaprabhu, S. (2025). Integrin α1 upregulation by tf:fviia complex promotes cervical cancer migration through par2-dependent mek1/2 activation. *Biochemical and Biophysical Research Communications, 742*, 151151. DOI: 10.1016/j.bbrc.2024.151151 PMID: 39657349

Parhi, S., Afrose, N., Rajendran, K., & Narayanasamy, D. (2025). Central composite design (CCD) based formulation, optimization, *in-vitro* and *ex-vivo* characterization of 5-fluorouracil-loaded emulgel for enhanced dermal penetration and psoriasis management. *Drug Development and Industrial Pharmacy, 51*(3), 244–261. DOI: 10.1080/03639045.2025.2464782 PMID: 39932381

Parisi, G. F., Terlizzi, V., Manti, S., Papale, M., Pecora, G., Presti, S., Tosto, M., & Leonardi, S. (2025). Cutting-Edge Advances in Cystic Fibrosis: From Gene Therapy to Personalized Medicine and Holistic Management. *Genes, 16*(4), 402. DOI: 10.3390/genes16040402 PMID: 40282362

Park, B.-S., Lee, M., Kim, J., & Kim, T. (2025). Perturbomics: CRISPR–Cas screening-based functional genomics approach for drug target discovery. *Experimental & Molecular Medicine, 57*(7), 1443–1454. DOI: 10.1038/s12276-025-01487-0 PMID: 40588529

Park, H., Shin, J., Kim, Y., Saito, T., Saido, T. C., & Kim, J. (2022). CRISPR/dCas9-Dnmt3a-mediated targeted DNA methylation of APP rescues brain pathology in a mouse model of Alzheimer's disease. *Translational Neurodegeneration, 11*(1), 41. DOI: 10.1186/s40035-022-00314-0 PMID: 36109806

Park, S. H., & Bao, G. (2021). CRISPR/Cas9 gene editing for curing sickle cell disease. *Transfusion and Apheresis Science : Official Journal of the World Apheresis Association : Official Journal of the European Society for Haemapheresis, 60*(1), 103060. DOI: 10.1016/j.transci.2021.103060 PMID: 33455878

Parums, D. V. (2024). Editorial: First Regulatory Approvals for CRISPR-Cas9 Therapeutic Gene Editing for Sickle Cell Disease and Transfusion-Dependent β-Thalassemia. *Medical Science Monitor, 30*. Advance online publication. DOI: 10.12659/MSM.944204 PMID: 38425279

Patel, R., Singh, N., & Kumar, A. (2023). Translational applications of CRISPR/dCas9-mediated epigenetic reprogramming in overcoming cancer drug resistance. *Cancer Research, 83*(14), 2875–2887.

Patel, S., Kim, J., Herrera, M., Mukherjee, A., Kabanov, A. V., & Sahay, G. (2021). Development of CRISPR-Cas9 delivery vehicles for in vivo gene editing. *Advanced Drug Delivery Reviews, 168*, 28–47.

Pausch, P., Al-Shayeb, B., Bisom-Rapp, E., Tsuchida, C. A., Li, Z., Cress, B. F., Knott, G. J., Jacobsen, S. E., Banfield, J. F., & Doudna, J. A. (2020). CRISPR–CasΦ from huge phages is a hypercompact genome editor. *Science, 369*(6501), 333–337. DOI: 10.1126/science.abb1400 PMID: 32675376

Peng, N., Song, S. J., Hui, V. W.-K., Lai, J. C.-T., Wong, G. L.-H., Wong, V. W.-S., & Yip, T. C.-F. (2026). Foundations of Artificial Intelligence in Hepatology: What a Clinician Needs to Know. *Journal of Clinical and Experimental Hepatology, 16*(1), 103183. Advance online publication. DOI: 10.1016/j.jceh.2025.103183 PMID: 41048886

Pérez-Amado, C. J., Bazan-Cordoba, A., Hidalgo-Miranda, A., & Jiménez-Morales, S. (2021). Mitochondrial Heteroplasmy Shifting as a Potential Biomarker of Cancer Progression. *International Journal of Molecular Sciences, 22*(14), 7369. DOI: 10.3390/ijms22147369 PMID: 34298989

Pfeiffer, L. S., & Stafforst, T. (2023). Precision RNA base editing with engineered and endogenous effectors. *Nature Biotechnology, 41*(11), 1526–1542. DOI: 10.1038/s41587-023-01927-0 PMID: 37735261

Pickar-Oliver, A., & Gersbach, C. A. (2019). The next generation of CRISPR–Cas technologies and applications. *Nature Reviews. Molecular Cell Biology, 20*(8), 490–507. DOI: 10.1038/s41580-019-0131-5 PMID: 31147612

Plasschaert, L. W., MacDonald, K. D., & Moffit, J. S. (2024). Current landscape of cystic fibrosis gene therapy. *Frontiers in Pharmacology, 15*, 1476331. DOI: 10.3389/fphar.2024.1476331 PMID: 39439894

Platt, R. J., Chen, S., Zhou, Y., Yim, M. J., Swiech, L., Kempton, H. R., & Zhang, F. (2014). CRISPR-Cas9 knockin mice for genome editing and cancer modeling. *Cell, 159*(2), 440–455. DOI: 10.1016/j.cell.2014.09.014 PMID: 25263330

Pollanen, S., & Syan, R. (2024). Gene-based targeting for treatment of hypercholesterolemia and prevention of atherosclerotic cardiovascular disease. *Qapsule: Queen's Undergraduate. Health Science Journal, 1*(2).

Polstein, L. R., & Gersbach, C. A. (2015). A light-inducible CRISPR-Cas9 system for control of endogenous gene activation. *Nature Chemical Biology, 11*(3), 198–200. DOI: 10.1038/nchembio.1753 PMID: 25664691

Pourcel, C., Salvignol, G., & Vergnaud, G. (2005). CRISPR elements in *Yersinia pestis. Microbiology, 151*(3), 653–663. DOI: 10.1099/mic.0.27437-0 PMID: 15758212

Prasad, K. M., Xu, Y., Yang, Z., Acton, S. T., & French, B. A. (2011). Robust cardiomyocyte-specific gene expression following systemic injection of AAV: In vivo gene delivery follows a Poisson distribution. *Gene Therapy, 18*(1), 43–52. DOI: 10.1038/gt.2010.105 PMID: 20703310

Prezza, G., Heckel, T., Dietrich, S., Homberger, C., Westermann, A. J., & Vogel, J. (2020). Improved bacterial RNA-seq by Cas9-based depletion of ribosomal RNA reads. *RNA (New York, N.Y.), 26*(8), 1069–1078. DOI: 10.1261/rna.075945.120 PMID: 32345633

Priya, S., Boyd, A. S., Torres-Bautista, A., Sawadkar, P., Mosahebi, A., Narayan Rath, S., & Kalaskar, D. M. (2024). Accelerating vascular graft development: Adipose-derived stem cells and PODS® (Polyhedrin delivery system with tissue-specific growth factors) – Enhanced 3D bioprinting for functional blood vessels. *Chemical Engineering Journal, 494*, 152934. DOI: 10.1016/j.cej.2024.152934

Pushparaj, K., Kuchi Bhotla, H., Arumugam, V. A., Pappusamy, M., Easwaran, M., Liu, W. C., Issara, U., Rengasamy, K. R. R., Meyyazhagan, A., & Balasubramanian, B. (2022). Mucormycosis (black fungus) ensuing COVID-19 and comorbidity meets - magnifying global pandemic grieve and catastrophe begins. *The Science of the Total Environment, 805*, 150355. DOI: 10.1016/j.scitotenv.2021.150355 PMID: 34818767

Qi, L. S., Larson, M. H., Gilbert, L. A., Doudna, J. A., Weissman, J. S., Arkin, A. P., & Lim, W. A. (2013). Repurposing CRISPR as an RNA-guided platform for sequence-specific control of gene expression. *Cell*, *152*(5), 1173–1183. DOI: 10.1016/j.cell.2013.02.022 PMID: 23452860

Qi, M., Ma, S., Liu, J., Liu, X., Wei, J., Lu, W. J., & Lan, F. (2024). In vivo base editing of Scn5a rescues type 3 long QT syndrome in mice. *Circulation*, *149*(4), 317–329. DOI: 10.1161/CIRCULATIONAHA.123.065624 PMID: 37965733

Qiu, M., & Li, P. (2021). CRISPR/Cas-based diagnostics and gene therapy. *BIO Integration.*, *2*(3), 121. DOI: 10.15212/bioi-2020-0048

Quan, J., Langelier, C., Kuchta, A., Batson, J., Teyssier, N., Lyden, A., Caldera, S., McGeever, A., Dimitrov, B., King, R., Wilheim, J., Murphy, M., Ares, L. P., Travisano, K. A., Sit, R., Amato, R., Mumbengegwi, D. R., Smith, J. L., Bennett, A., & Crawford, E. D. (2019). FLASH: A next-generation CRISPR diagnostic for multiplexed detection of antimicrobial resistance sequences. *Nucleic Acids Research*, *47*(14), e83. DOI: 10.1093/nar/gkz418 PMID: 31114866

Rafeeq, H., Afsheen, N., Rafique, S., Arshad, A., Intisar, M., Hussain, A., Bilal, M., & Iqbal, H. M. (2023). Genetically engineered microorganisms for environmental remediation. *Chemosphere*, *310*, 136751. DOI: 10.1016/j.chemosphere.2022.136751 PMID: 36209847

Raghavan, R., Moger, S. S., Umesh, S., & Bhuvana, G. N. (2025). Revolutionizing Pain Management. *Unveiling Technological Advancements and Interdisciplinary Solutions for Pain Care*, 129–164. DOI: 10.4018/979-8-3693-9501-1.ch005

Raghunath, A., Krishna, T., Nandan, P. K., & Sivaraman, J. (2025). A Paradigm Shift in Cystic Fibrosis: Insights into Molecular Diagnosis, Newborn Screening, and Nanotechnology-Based Drug Delivery for CFTreatment. *Current Respiratory Medicine Reviews*, *21*(5), 485–499. Advance online publication. DOI: 10.2174/01 1573398X359962250527113134

Rajanathadurai, J., Perumal, E., & Sindya, J. (2024). Advances in targeting cancer epigenetics using CRISPR-dCas9 technology: A comprehensive review and future prospects. *Functional & Integrative Genomics*, *24*(5), 164. DOI: 10.1007/s10142-024-01455-3 PMID: 39292321

Raj, B., Wagner, D. E., McKenna, A., Pandey, S., Klein, A. M., Shendure, J., Gagnon, J. A., & Schier, A. F. (2018). Simultaneous single-cell profiling of lineages and cell types in the vertebrate brain. *Nature Biotechnology*, *36*(5), 442–450. DOI: 10.1038/nbt.4103 PMID: 29608178

Ramakrishnan, M., Satish, L., Sharma, A., Kurungara Vinod, K., Emamverdian, A., Zhou, M., & Wei, Q. (2022). Transposable elements in plants: Recent advancements, tools and prospects. *Plant Molecular Biology Reporter*, *40*(4), 628–645. DOI: 10.1007/s11105-022-01342-w

Ramya, S., Poornima, P., Jananisri, A., Geofferina, I. P., Bavyataa, V., Divya, M., Priyanga, P., Vadivukarasi, J., Sujitha, S., Elamathi, S., Anand, A. V., & Balamuralikrishnan, B. (2023). Role of hormones and the potential impact of multiple stresses on infertility. *Stresses*, *3*(2), 454–474. DOI: 10.3390/stresses3020033

Rana, D., Rangel, V. R., Padmanaban, P., Trikalitis, V. D., Kandar, A., Kim, H.-W., & Rouwkema, J. (2025). Bioprinting of Aptamer-Based Programmable Bioinks to Modulate Multiscale Microvascular Morphogenesis in 4D. *Advanced Healthcare Materials*, *14*(1), 2402302. DOI: 10.1002/adhm.202402302 PMID: 39487611

Rasheed, A., Gill, R. A., Hassan, M. U., Mahmood, A., Qari, S., Zaman, Q. U., Ilyas, M., Aamer, M., Batool, M., Li, H., & Wu, Z. (2021). A critical review: Recent advancements in the use of CRISPR/Cas9 technology to enhance crops and alleviate global food crises. *Current Issues in Molecular Biology*, *43*(3), 1950–1976. DOI: 10.3390/cimb43030135 PMID: 34889892

Rauf, M. A., Rao, A., Sivasoorian, S. S., & Iyer, A. K. (2025). Nanotechnology-Based Delivery of CRISPR/Cas9 for Cancer Treatment: A Comprehensive Review. *Cells*, *14*(15), 1136. DOI: 10.3390/cells14151136 PMID: 40801569

Ravichandran, M., & Maddalo, D. (2023). Applications of CRISPR-Cas9 for advancing precision medicine in oncology: From target discovery to disease modeling. *Frontiers in Genetics*, *14*, 1273994. DOI: 10.3389/fgene.2023.1273994 PMID: 37908590

Reddy, P., Vilella, F., Izpisua Belmonte, J. C., & Simón, C. (2020). Use of Customizable Nucleases for Gene Editing and Other Novel Applications. *Genes*, *11*(9), 976. DOI: 10.3390/genes11090976 PMID: 32842577

Rees, H. A., & Liu, D. R. (2018). Base editing: Precision chemistry on the genome and transcriptome of living cells. *Nature Reviews. Genetics*, *19*(12), 770–788. DOI: 10.1038/s41576-018-0059-1 PMID: 30323312

RegMedNet. (2022, February 11). In vivo gene editing of mitochondrial DNA shows potential to treat mitochondrial disorders. *Regenerative Medicine Net*.

Replogle, J. M., Norman, T. M., Xu, A., Hussmann, J. A., Chen, J., & Weissman, J. S. (2020). Combinatorial single-cell CRISPR screens by direct guide RNA capture and targeted sequencing. *Nature Biotechnology*, *38*(8), 954–961. DOI: 10.1038/s41587-020-0470-y PMID: 32231336

Richhariya, S., Shin, D., Schlichting, M., & Rosbash, M. (2025). Metabolic rewiring prevents neurodegeneration caused by chronic mitochondrial dysfunction. *Current Biology*, *35*(22), 5443–5459.e5. DOI: 10.1016/j.cub.2025.09.063 PMID: 41151583

Riley, J. S., & Tait, S. W. (2020). Mitochondrial DNA in inflammation and immunity. *EMBO Reports*, *21*(4), e49799. DOI: 10.15252/embr.201949799 PMID: 32202065

Risato, G., Brañas Casas, R., Cason, M., Bueno Marinas, M., Pinci, S., De Gaspari, M., & Celeghin, R. (2024). In vivo approaches to understand arrhythmogenic cardiomyopathy: Perspectives on animal models. *Cells*, *13*(15), 1264. DOI: 10.3390/cells13151264 PMID: 39120296

Rocheleau, G., Clarke, S. L., Auguste, G., Hasbani, N. R., Morrison, A. C., Heath, A. S., & Do, R. (2024). Rare variant contribution to the heritability of coronary artery disease. *Nature Communications*, *15*(1), 8741. DOI: 10.1038/s41467-024-52939-6 PMID: 39384761

Rogg, E. M., & Dimmeler, S. (2019). Therapeutic potential of non-coding RNAs for acute and chronic myocardial ischemia. *Current Cardiology*, *8*(03), 223–229.

Rooney, K., & Sadikovic, B. (2022). DNA methylation episignatures in neurodevelopmental disorders associated with large structural copy number variants: Clinical implications. *International Journal of Molecular Sciences*, *23*(14), 7862. DOI: 10.3390/ijms23147862 PMID: 35887210

Rós, F. A., Couto, S. C. F., Milhomens, J., Ovider, I., Maio, K. T., Jennifer, V., Ramos, R. N., Picanço-Castro, V., Kashima, S., Calado, R. T., Barros, L. R. C., & Rocha, V. (2023). A systematic review of clinical trials for gene therapies for β-hemoglobinopathy around the world. *Cytotherapy*, *25*(12), 1300–1306. DOI: 10.1016/j.jcyt.2023.05.006 PMID: 37318395

Roshanravan, N., Tutunchi, H., Najafipour, F., Dastouri, M., Ghaffari, S., & Jebeli, A. (2022). A glance at the application of CRISPR/Cas9 gene-editing technology in cardiovascular diseases. *Journal of Cardiovascular and Thoracic Research*, *14*(2), 77–83. DOI: 10.34172/jcvtr.2022.14 PMID: 35935390

Roth, R., Kim, S., Kim, J., & Rhee, S. (2020). Single-cell and spatial transcriptomics approaches of cardiovascular development and disease. *BMB Reports*, *53*(8), 393–399. DOI: 10.5483/BMBRep.2020.53.8.130 PMID: 32684243

Rubeis, G. (2018). Risks and benefits of human germline genome editing. *Frontiers in Bioengineering and Biotechnology*, *6*, 171. PMID: 33717282

Ryu, H. W., Lee, D. H., & Won, H. R.. (2015). Toxic metals and epigenetic regulation. *Toxicological Research*, *31*(1), 1–9. DOI: 10.5487/TR.2015.31.1.001 PMID: 25874027

S., G. V., C., A., Biradar, S., & Raghavan, R. (2025). Neuroscience-Driven Solutions for Speech Impairment. *Advancing Medical Research Through Neuroscience*, 119–146. DOI: 10.4018/979-8-3693-5464-3.ch006

Saeed, A., Ahmad, A., & Husnain, A. (2024). Harnessing AI for advancements in cardiovascular disease management and drug discovery. *International Journal of Science and Research Archive*, *13*(1), 244–249. DOI: 10.30574/ijsra.2024.13.1.1623

Sage, C. L., Lawo, S., Panckier, P., Scales, T. M., Rahman, S. A., Little, A. S., McCarthy, N. J., Moore, J. D., & Cross, B. C. (2018). Dual-direction CRISPR transcriptional regulation screening uncovers gene networks driving drug resistance. *Cancer Research*, *78*(13, Supplement), 409–409. DOI: 10.1158/1538-7445.AM2018-409

Sakari, M., Suzuki, T., Yano, S., & Tsukahara, T. (2020). Effective RNA knockdown using CRISPR-Cas13a and molecular targeting of the EML4-ALK transcript in H3122 lung cancer cells. *International Journal of Molecular Sciences*, *21*(23), 8904. DOI: 10.3390/ijms21238904 PMID: 33255340

Sandoval-Villegas, N., Nurieva, W., Amberger, M., & Ivics, Z. (2021). Contemporary Transposon Tools: A Review and Guide through Mechanisms and Applications of Sleeping Beauty, piggyBac and Tol2 for Genome Engineering. *International Journal of Molecular Sciences*, *22*(10), 5084. DOI: 10.3390/ijms22105084 PMID: 34064900

Sangeetha, T., Anand, A. V., & Begum, T. N. (2022). Assessment of inter-relationship between anemia and COPD in accordance with altitude. *The Open Respiratory Medicine Journal*, *16*(1), e187430642206270. DOI: 10.2174/18743064-v16-e2206270 PMID: 37273961

Sangeetha, T., Nargis Begum, T., Balamuralikrishnan, B., Arun, M., Rengasamy, K. R. R., Senthilkumar, N., Velayuthaprabhu, S., Saradhadevi, M., Sampathkumar, P., & Vijaya Anand, A. (2022). Influence of SERPINA1 gene polymorphisms on anemia and chronic obstructive pulmonary disease. *Journal of the Renin-Angiotensin-Aldosterone System : JRAAS*, *2022*, 2022. DOI: 10.1155/2022/2238320 PMID: 36320441

Sanzo, K. M. (2022). Regulation of Genome Editing in Human iPS Cells: United States. In Dederer, H.-G., & Frenken, G. (Eds.), *Regulation of Genome Editing in Human iPS Cells* (pp. 199–222). Springer International Publishing., DOI: 10.1007/978-3-030-93023-3_6

Sapozhnikov, D. M., & Szyf, M. (2021). Unraveling the functional role of DNA demethylation at specific promoters by targeted steric blockage of DNA methyltransferase with CRISPR/dCas9. *Nature Communications*, *12*(1), 5711. DOI: 10.1038/s41467-021-25991-9 PMID: 34588447

Sapp, V., Aguirre, A., Mainkar, G., Ding, J., Adler, E., Liao, R., & Jain, M. (2021). Genome-wide CRISPR/Cas9 screening in human iPS derived cardiomyocytes uncovers novel mediators of doxorubicin cardiotoxicity. *Scientific Reports*, *11*(1), 13866. DOI: 10.1038/s41598-021-92988-1 PMID: 34230586

Schary, Y., Rotem, I., Caller, T., Lewis, N., Shaihov-Teper, O., Brzezinski, R. Y., & Leor, J. (2023). CRISPR-Cas9 editing of TLR4 to improve the outcome of cardiac cell therapy. *Scientific Reports*, *13*(1), 4481. DOI: 10.1038/s41598-023-31286-4 PMID: 36934130

Scheufele, D. A., Xenos, M. A., Howell, E. A., Rose, K. M., Brossard, D., & Hardy, B. W. (2017). U.S. attitudes on human genome editing. *Science*, *357*(6351), 553–554. DOI: 10.1126/science.aan3708 PMID: 28798120

Schicktanz, S., Schweda, M., & Wynne, B. (2019). The ethics of genome editing in human embryos: International perspectives. *European Journal of Human Genetics : EJHG*, *27*(5), 545–553.

Schmidt, R., Steinhart, Z., Layeghi, M., Freimer, J. W., Nguyen, V. Q., Blaeschke, F., & Marson, A. (2021). CRISPR activation and interference screens in primary human T cells decode cytokine regulation. bioRxiv, 2021-05.

Schmidt, T. J. N., Berarducci, B., Konstantinidou, S., & Raffa, V. (2023). CRISPR/Cas9 in nanomedicine. *Drug Discovery Today*, *28*(1), 103375. DOI: 10.1016/j.drudis.2022.103375 PMID: 36174966

Schoger, E., Lelek, S., Panáková, D., & Zelarayán, L. C. (2022). Tailoring cardiac synthetic transcriptional modulation towards precision medicine. *Frontiers in Cardiovascular Medicine*, *8*, 783072. DOI: 10.3389/fcvm.2021.783072 PMID: 35097003

Schreurs, J., Sacchetto, C., Colpaert, R. M., Vitiello, L., Rampazzo, A., & Calore, M. (2021). Recent advances in CRISPR/Cas9-based genome editing tools for cardiac diseases. *International Journal of Molecular Sciences*, *22*(20), 10985. DOI: 10.3390/ijms222010985 PMID: 34681646

Schubert, M. S., Thommandru, B., Woodley, J., Turk, R., Yan, S., Kurgan, G., McNeill, M. S., & Rettig, G. R. (2021). Optimized design parameters for CRISPR Cas9 and Cas12a homology-directed repair. *Scientific Reports*, *11*(1), 19482. DOI: 10.1038/s41598-021-98965-y PMID: 34593942

Schwank, G., Koo, B. K., Sasselli, V., Dekkers, J. F., Heo, I., Demircan, T., Sasaki, N., Boymans, S., Cuppen, E., van der Ent, C. K., Nieuwenhuis, E. E. S., Beekman, J. M., & Clevers, H. (2013). Functional repair of CFTR by CRISPR/Cas9 in intestinal stem cell organoids of cystic fibrosis patients. *Cell Stem Cell*, *13*(6), 653–658. DOI: 10.1016/j.stem.2013.11.002 PMID: 24315439

Schweiggert, J., Habeck, G., Hinkel, R., Kehr, D., Cimen, K., Kramp, X., & Busch, M. (2023). Engineered cardiac-specific AAV capsids from a novel next-generation platform (next-cap) with cross-species selection for cardiac-targeted gene therapy. *Circulation*, *148*(Suppl_1), A14481–A14481. DOI: 10.1161/circ.148.suppl_1.14481

Seem, K., Kaur, S., Kumar, S., & Mohapatra, T. (2024). Epigenome editing for targeted DNA (de) methylation: A new perspective in modulating gene expression. *Critical Reviews in Biochemistry and Molecular Biology*, *59*(1-2), 69–98. DOI: 10.1080/10409238.2024.2320659 PMID: 38440883

Segers, S. (2023). Heritable genome editing: Ethical aspects of a developing domain. *Human Reproduction (Oxford, England)*, *38*(11), 2055–2061. DOI: 10.1093/humrep/dead167 PMID: 37581898

Seijas, A., Cora, D., Novo, M., Al-Soufi, W., Sánchez, L., & Arana, Á. J. (2025). CRISPR/Cas9 Delivery Systems to Enhance Gene Editing Efficiency. *International Journal of Molecular Sciences*, *26*(9), 4420. DOI: 10.3390/ijms26094420 PMID: 40362657

Sen, T., & Thummer, R. P. (2022). CRISPR and iPSCs: Recent developments and future perspectives in neurodegenerative disease modelling, research, and therapeutics. *Neurotoxicity Research*, *40*(5), 1597–1623. DOI: 10.1007/s12640-022-00564-w PMID: 36044181

Sessions, D. T., Kim, K. B., Kashatus, J. A., Churchill, N., Park, K. S., Mayo, M. W., Sesaki, H., & Kashatus, D. F. (2022). Opa1 and Drp1 reciprocally regulate cristae morphology, ETC function, and NAD+ regeneration in KRas-mutant lung adenocarcinoma. *Cell Reports*, *41*(11), 111818. Advance online publication. DOI: 10.1016/j.celrep.2022.111818 PMID: 36516772

Shah, A. H., Shetty, O., & Joshi, A. (2024). CRISPR/Cas9 applications in breast and cervical cancers. *Cancer Gene Therapy*, *31*(3), 425–435. DOI: 10.1038/s41417-024-00712-w

Shah, S. H., & Gerszten, R. E. (2024). Sifting for gold in terabytes of data: Illuminating cardiovascular biology in the 'omics age. *Circulation, 149*(17), 1323–1325. DOI: 10.1161/CIRCULATIONAHA.123.066988 PMID: 38648274

Shan, H., & Fei, T. (2023). CRISPR screening in cardiovascular research. *Frontiers in Cell and Developmental Biology, 11*, 1175849. DOI: 10.3389/fcell.2023.1175849 PMID: 37123412

Shao, M., Xu, T. R., & Chen, C. S. (2016). Development of CRISPR/Cas9 in animal models. *Zoological Research.* Advance online publication. DOI: 10.13918/j.issn.2095-8137.2016.4.191 PMID: 27469250

Sharma, B., Chauhan, I., Kumar, G., Bhardwaj, K., & Tiwari, R. K. (2024). Nanoparticle carriers: A new era of precise crispr/cas9 gene editing. *MicroRNA (Shariqah, United Arab Emirates), 14*(2), 101–111. DOI: 10.2174/0122115366319848241022092805 PMID: 39484758

Sharma, G., Sharma, A. R., Bhattacharya, M., Lee, S. S., & Chakraborty, C. (2021). CRISPR-Cas9: A Preclinical and Clinical Perspective for the Treatment of Human Diseases. *Molecular Therapy : the Journal of the American Society of Gene Therapy, 29*(2), 571–586. DOI: 10.1016/j.ymthe.2020.09.028 PMID: 33238136

Shelton, S. D., House, S., Martins Nascentes Melo, L., Ramesh, V., Chen, Z., Wei, T., Wang, X., Llamas, C. B., Venigalla, S. S. K., Menezes, C. J., Allies, G., Krystkiewicz, J., Rösler, J., Meckelmann, S. W., Zhao, P., Rambow, F., Schadendorf, D., Zhao, Z., Gill, J. G., & Mishra, P. (2024). Pathogenic mitochondrial DNA mutations inhibit melanoma metastasis. *Science Advances, 10*(44), eadk8801. Advance online publication. DOI: 10.1126/sciadv.adk8801 PMID: 39485847

Sherif, Z. A., Ogunwobi, O. O., & Ressom, H. W. (2025). Mechanisms and technologies in cancer epigenetics. *Frontiers in Oncology, 14*, 1513654. Advance online publication. DOI: 10.3389/fonc.2024.1513654 PMID: 39839798

Shi, J.-Y., Li, Z.-W., Yao, Z.-L., Han, G.-M., Li, M.-Y., Cai, Q.-L., & Kong, D.-M. (2025). 5′-End Engineering of CRISPR/Cas12a Activators: A Versatile Platform for Multiple Biomarker Analysis and Clinical Cancer Tissue Identification. *Analytical Chemistry, 97*(33), 18188–18198. DOI: 10.1021/acs.analchem.5c03168 PMID: 40811645

Shi, L., Tang, X., & Tang, G. (2016). GUIDE-Seq to Detect Genome-wide Double-Stranded Breaks in Plants. *Trends in Plant Science, 21*(10), 815–818. DOI: 10.1016/j.tplants.2016.08.005 PMID: 27593568

Shimada, M., Matsumoto, Y., & Otsuka, K. (2025). Live-cell imaging of DNA damage. *The Journal of Biological Chemistry, 110328*(7), 110328. Advance online publication. DOI: 10.1016/j.jbc.2025.110328 PMID: 40473209

Shim, Y. J. (2025). Therapeutic targeting of DNA damage response in TP53/ATM tumors. *Brain Tumor Research and Treatment, 13*(3), 73–80. DOI: 10.14791/btrt.2025.0017 PMID: 40759474

Shiridokht, F., Kehtari, P., Eskandani, M., Farajollahi, A., & Vandghanooni, S. (2025). Harnessing radiosensitizers and nanotechnology. *International Journal of Pharmaceutics: X, 100419*, 100419. Advance online publication. DOI: 10.1016/j.ijpx.2025.100419 PMID: 41189671

Shirk, B. D., Rodriguez, C. Z., Pacheco, M. O., McTyer, J. B., Shirk, P. D., & Stoppel, W. L. (2026). Enhancing CRISPR homology directed repair in IAL-PiD2 insect cells via reagent delivery optimization and cell synchronization. *Biochemical Engineering Journal, 225*, 109906. DOI: 10.1016/j.bej.2025.109906

Shmakov, S., Smargon, A., Scott, D., Cox, D. B. T., Pyzocha, N., Yan, W., Abudayyeh, O. O., Gootenberg, J. S., Makarova, K. S., Wolf, Y. I., Severinov, K., Zhang, F., & Koonin, E. V. (2017). Diversity and evolution of class 2 CRISPR-Cas systems. *Nature Reviews. Microbiology, 15*(3), 169–182. DOI: 10.1038/nrmicro.2016.184 PMID: 28111461

Shu, J., Xie, X., Wang, S., Du, Z., Huang, P., Chen, Y., & He, Z. (2025). CRISPR/Cas-edited iPSCs and mesenchymal stem cells: A concise review of their potential in thalassemia therapy. *Frontiers in Cell and Developmental Biology, 13*, 1595897. Advance online publication. DOI: 10.3389/fcell.2025.1595897 PMID: 40970094

Shuo, Y., Feng, X., Zhao, C., Zhang, J., Wang, Y., Wang, J., Ding, Y., Shi, X., Li, J., Song, X., & Wang, J. (2025). A dual enhancement strategy for ultrasensitive detection of norovirus based on catalytic hairpin assembly-assisted CRISPR/Cas12a sensing system. *Journal of Dairy Science, 108*(5), 4726–4733. DOI: 10.3168/jds.2024-26150 PMID: 40139379

Silva-Pinheiro, P., Gammage, P. A., & Minczuk, M. (2022). Mitochondrial DNA editing: Recent advances toward therapeutic translation. *Trends in Molecular Medicine, 28*(6), 514–527.

Simon, A. J., Ellington, A. D., & Finkelstein, I. J. (2019). Retrons and their applications in genome engineering. *Nucleic Acids Research, 47*(21), 11007–11019. DOI: 10.1093/nar/gkz865 PMID: 31598685

Singh, D. (2024). Revolutionizing Lung Cancer Treatment: Innovative CRISPR-Cas9 Delivery Strategies. *AAPS PharmSciTech*, *25*(5), 129. DOI: 10.1208/s12249-024-02834-6 PMID: 38844700

Singh, M., Bindal, G., Misra, C. S., & Rath, D. (2022). The era of Cas12 and Cas13 CRISPR-based disease diagnosis. *Critical Reviews in Microbiology*, *48*(6), 714–729. DOI: 10.1080/1040841X.2021.2025041 PMID: 35164636

Sinha, S., Tripathi, A. K., Pandey, A., Naik, P., Pandey, A., & Verma, V. S. (2024). Self-assembled PEGylated micelles for precise and targeted drug delivery: Current challenges and future directions. *Biocatalysis and Agricultural Biotechnology*, *60*, 103296. DOI: 10.1016/j.bcab.2024.103296

Sioson, V. A., Kim, M., & Joo, J. (2021). Challenges in delivery systems for CRISPR-based genome editing and opportunities of nanomedicine. *Biomedical Engineering Letters*, *11*(3), 217–233. DOI: 10.1007/s13534-021-00199-4 PMID: 34350049

Sivagnanam, S., Das, K., Pan, I., Stewart, A., Barik, A., Maity, B., & Das, P. (2024). Engineered triphenylphosphonium-based, mitochondrial-targeted liposomal drug delivery system facilitates cancer cell killing actions of chemotherapeutics. *RSC Chemical Biology*, *5*(3), 236–248. DOI: 10.1039/D3CB00219E PMID: 38456034

Sivagourounadin, K., Ravichandran, M., & Rajendran, P. (2021). National guidelines for gene therapy product (2019): A road-map to gene therapy products development and clinical trials. *Perspectives in Clinical Research*, *12*(3), 118–125. DOI: 10.4103/picr.PICR_189_20 PMID: 34386375

Slattery, J. R., Naung, N. Y., Kalinna, B. H., & Pal, M. (2025). CRISPR-Powered Liquid Biopsies in Cancer Diagnostics. *Cells*, *14*(19), 1539. DOI: 10.3390/cells14191539 PMID: 41090767

Slaymaker, I. M., Gao, L., Zetsche, B., Scott, D. A., Yan, W. X., & Zhang, F. (2016). Rationally engineered Cas9 nucleases with improved specificity. *Science*, *351*(6268), 84–88. DOI: 10.1126/science.aad5227 PMID: 26628643

Smeir, M., Chumala, P., Katselis, G. S., & Liu, L. (2024). Lymphocyte-Specific protein 1 regulates expression and stability of endothelial nitric oxide synthase. *Biomolecules*, *14*(1), 111. DOI: 10.3390/biom14010111 PMID: 38254711

Smith, C., Abalde-Atristain, L., He, C., Brodsky, B. R., Braunstein, S., & Chaudhari, P.. (2020). Efficient and allele-specific genome editing of disease loci in human iPSCs. *Molecular Therapy : the Journal of the American Society of Gene Therapy*, *28*(2), 422–434. PMID: 31843447

Smith, C., Abalde-Cela, S., & Abell, C. (2022). Delivery technologies for genome editing. *Nature Reviews. Materials*, *7*(12), 991–1009.

Socha, M. W., Flis, W., & Wartęga, M. (2024). Epigenetic genome modifications during pregnancy: The impact of essential nutritional supplements on DNA methylation. *Nutrients*, *16*(5), 678. DOI: 10.3390/nu16050678 PMID: 38474806

Sokka, J. E., Hendriks, D., Jussila, M., Tuusa, J., Guo, Y., Huhtala, M., & Weltner, J. (2022). CRISPR activation enables high-fidelity reprogramming into human pluripotent stem cells. *Stem Cell Reports*, *17*(3), 620–634. DOI: 10.1016/j.stemcr.2021.12.017 PMID: 35063129

Song, R. J., Han, L., Sun, H. F., & Shen, B. (2023). Advances in mitochondrial DNA base editing technology. *Yi Chuan*, *45*(8), 632–642. PMID: 37609815

Song, Y., Zheng, Z., & Lian, J. (2022). Deciphering common long QT syndrome using CRISPR/Cas9 in human-induced pluripotent stem cell-derived cardiomyocytes. *Frontiers in Cardiovascular Medicine*, *9*, 889519. DOI: 10.3389/fcvm.2022.889519 PMID: 35647048

Srinivasan, D., Balakrishnan, R., Chauhan, A., Kumar, J., Girija, D. M., Shrestha, R., Shrestha, R., & Subbarayan, R. (2025). Epithelial–Mesenchymal Transition in Cancer: Insights Into Therapeutic Targets and Clinical Implications. *MedComm*, *6*(9), e70333. DOI: 10.1002/mco2.70333 PMID: 40895189

Srivastava, R. A. K. (2023). New opportunities in the management and treatment of refractory hypercholesterolemia using in vivo CRISPR-mediated genome/base editing. *Nutrition, Metabolism, and Cardiovascular Diseases*, *33*(12), 2317–2325. DOI: 10.1016/j.numecd.2023.08.010 PMID: 37805309

Srivastava, S., Upadhyay, D. J., & Srivastava, A. (2020). Next-generation molecular diagnostics development by CRISPR/Cas tool: Rapid detection and surveillance of viral disease outbreaks. *Frontiers in Molecular Biosciences*, *7*, 582499. DOI: 10.3389/fmolb.2020.582499 PMID: 33425987

Stadtmauer, E. A., Fraietta, J. A., Davis, M. M., Cohen, A. D., Weber, K. L., Lancaster, E., & June, C. H. (2020). CRISPR-engineered T cells in patients with refractory cancer. *Science*, *367*(6481), 7365. DOI: 10.1126/science.aba7365 PMID: 32029687

Sternberg, S. H., Redding, S., Jinek, M., Greene, E. C., & Doudna, J. A. (2014). DNA interrogation by the CRISPR RNA-guided endonuclease Cas9. *Nature*, *507*(7490), 62–67. DOI: 10.1038/nature13011 PMID: 24476820

Stevens, A. J., de Jong, L., & Kennedy, M. A. (2022). The dynamic regulation of G-quadruplex DNA structures by cytosine methylation. *International Journal of Molecular Sciences, 23*(5), 2407. DOI: 10.3390/ijms23052407 PMID: 35269551

Stinson, B. M., & Loparo, J. J. (2021). DSB repair by NHEJ. *Annual Review of Biochemistry, 90*, 137–164. DOI: 10.1146/annurev-biochem-080320-110356 PMID: 33556282

Stone, D., Long, K. R., Loprieno, M. A., De Silva Feelixge, H. S., Kenkel, E. J., Liley, R. M., Rapp, S., Roychoudhury, P., Nguyen, T., Stensland, L., Colón-Thillet, R., Klouser, L. M., Weber, N. D., Le, C., Wagoner, J., Goecker, E. A., Li, A. Z., Eichholz, K., Corey, L., & Jerome, K. R. (2021). CRISPR-Cas9 gene editing of hepatitis B virus in chronically infected humanized mice. *Molecular Therapy. Methods & Clinical Development, 20*, 258–275. DOI: 10.1016/j.omtm.2020.11.014 PMID: 33473359

Stuart, W. D., Fink-Baldauf, I. M., Tomoshige, K., Guo, M., & Maeda, Y. (2021). CRISPRi-mediated functional analysis of NKX2-1-binding sites in the lung. *Communications Biology, 4*(1), 568. DOI: 10.1038/s42003-021-02083-4 PMID: 33980985

Subica, A. M. (2023). CRISPR in Public Health: The Health Equity Implications and Role of Community in Gene-Editing Research and Applications. *American Journal of Public Health, 113*(8), 874–882. DOI: 10.2105/AJPH.2023.307315 PMID: 37200601

Sundaram, V., & Aryal, S. (2025). Emerging Biomimetic Drug Delivery Nanoparticles Inspired by Extracellular Vesicles. *Wiley Interdisciplinary Reviews. Nanomedicine and Nanobiotechnology, 17*(4), e70025. DOI: 10.1002/wnan.70025 PMID: 40840553

Sun, X., Dong, M., Li, J., Sun, Y., Gao, Y., Wang, Y., Du, L., Liu, Y., Ji, K., He, N., Wang, J., Zhang, M., Song, H., Xu, C., & Liu, Q. (2024). NRF2 promotes radiation resistance. *Theranostics, 14*(2), 681–698. DOI: 10.7150/thno.88899 PMID: 38169561

Sun, Y., Guo, Z., & Li, R. (2023). Epigenetic reprogramming via CRISPR/Cas9 to overcome multidrug resistance in tumor therapy. *Frontiers in Oncology, 13*, 1187653.

Sun, Y., Wen, T., Zhang, P., Wang, M., & Xu, Y. (2024). Recent Advances in the CRISPR/Cas-Based Nucleic Acid Biosensor for Food Analysis: A Review. *Foods, 13*(20), 3222. DOI: 10.3390/foods13203222 PMID: 39456285

Sur, S., Rathore, A., Dave, V., Reddy, K. R., Chouhan, R. S., & Sadhu, V. (2019). Recent developments in functionalized polymer nanoparticles for efficient drug delivery system. *Nano-Structures & Nano-Objects, 20*, 100397. DOI: 10.1016/j.nanoso.2019.100397

Sutherland, H. G., Jenkins, B., & Griffiths, L. R. (2024). Genetics of migraine: Complexity, implications, and potential clinical applications. *Lancet Neurology*, *23*(4), 429–446. DOI: 10.1016/S1474-4422(24)00026-7 PMID: 38508838

Suwito, B. E., Adji, A. S., Wardani, V. A. K., Widjaja, J. S., Angel, S. C. S., & Rahman, F. S. (2022). A review of CRISPR Cas9 for ASCVD: Treatment strategies and could target PSCK9 gene using CRISPR cas9 prevent the patient from atherosclerotic vascular disease? *Bali Medical Journal*, *11*(2), 985–993. DOI: 10.15562/bmj.v11i2.3414

Swarts, D. C., & Jinek, M. (2018). Cas9 versus Cas12a/Cpf1: Structure–function comparisons and implications for genome editing. *Wiley Interdisciplinary Reviews. RNA*, *9*(5), e1481. DOI: 10.1002/wrna.1481 PMID: 29790280

Tada, H., Usui, S., Sakata, K., Takamura, M., & Kawashiri, M. A. (2021). Challenges of precision medicine for atherosclerotic cardiovascular disease based on human genome information. *Journal of Atherosclerosis and Thrombosis*, *28*(4), 305–313. DOI: 10.5551/jat.60087 PMID: 33229815

Tait, J., Meslin, E., & Sheehan, M. (2019). Public engagement in genome editing: Models and outcomes. *Journal of Community Genetics*, *10*(2), 123–135.

Takara Bio. (n.d.). *Introduction to the CRISPR–Cas9 system* [Illustration]. https://www.takarabio.com/learning-centers/gene-function/gene-editing/gene-editing-tools-and-information/introduction-to-the-crisprcas9-system

Takebe, T., Sekine, K., Enomura, M., Koike, H., Kimura, M., Ogaeri, T., ... & Taniguchi, H. (2015). Vascularized and functional human liver from an iPSC-derived organ bud transplant.

Tamaddondoust, R. N., Wong, A., Chandrashekhar, M., Azzam, E. I., Alain, T., & Wang, Y. (2022). Novel regulators of radiosensitivity. *International Journal of Molecular Sciences*, *23*(15), 8774. DOI: 10.3390/ijms23158774 PMID: 35955908

Tang, J., & Du, K. (2025). Mitochondrial base editing: From principle, optimization to application. *Cell & Bioscience*, *15*(1), 9. DOI: 10.1186/s13578-025-01351-8 PMID: 39856740

Tang, J., Zhang, J., Yang, R., Chen, H., Yu, X., Peng, W., & Zeng, P. (2025). The causal relationships between mitochondria and six types of cancer: A Mendelian randomization study. *BMC Cancer*, *25*(1), 794. DOI: 10.1186/s12885-025-14201-0 PMID: 40295943

Tang, R., Long, T., Lui, K. O., Chen, Y., & Huang, Z. P. (2020). A roadmap for fixing the heart: RNA regulatory networks in cardiac disease. *Molecular Therapy. Nucleic Acids*, *20*, 673–686. DOI: 10.1016/j.omtn.2020.04.007 PMID: 32380417

Tang, S., Chen, X., Tong, X., & Zhu, L. (2025). Overcoming the Delivery Challenges in CRISPR/Cas9 Gene Editing for Effective Cancer Treatment: A Review of Delivery Systems. *International Journal of Medical Sciences*, *22*(14), 3625–3649. DOI: 10.7150/ijms.112724 PMID: 40959562

Tang, T. H., Bachellerie, J. P., Rozhdestvensky, T., Bortolin, M. L., Huber, H., Drungowski, M., Elge, T., Brosius, J., & Hüttenhofer, A. (2002). Identification of 86 candidates for small non-messenger RNAs from the archaeon *Archaeoglobus fulgidus*. *Proceedings of the National Academy of Sciences of the United States of America*, *99*(11), 7536–7541. DOI: 10.1073/pnas.112047299 PMID: 12032318

Tao, J., Bauer, D. E., & Chiarle, R. (2023). Assessing and advancing the safety of CRISPR-Cas tools: From DNA to RNA editing. *Nature Communications*, *14*(1), 212. DOI: 10.1038/s41467-023-35886-6 PMID: 36639728

Tao, R., Han, X., Bai, X., Yu, J., Ma, Y., Chen, W., Zhang, D., & Li, Z. (2024). Revolutionizing cancer treatment: Enhancing CAR-T cell therapy with CRISPR/Cas9 gene editing technology. *Frontiers in Immunology*, *15*, 1354825. DOI: 10.3389/fimmu.2024.1354825 PMID: 38449862

Taylor, A. D., Hathaway, Q. A., Kunovac, A., Pinti, M. V., Newman, M. S., Cook, C. C., Cramer, E. R., Starcovic, S. A., Winters, M. T., Westemeier-Rice, E. S., Fink, G. K., Durr, A. J., Rizwan, S., Shepherd, D. L., Robart, A. R., Martinez, I., & Hollander, J. M. (2024). Mitochondrial sequencing identifies long noncoding RNA features that promote binding to PNPase. *American Journal of Physiology. Cell Physiology*, *327*(2), C221–C236. DOI: 10.1152/ajpcell.00648.2023 PMID: 38826135

Tei, C., Hata, S., Mabuchi, A., Okuda, S., Ito, K. K., Genova, M., Fukuyama, M., Yamamoto, S., Chinen, T., Toyoda, A., & Kitagawa, D. (2025). Comparative analysis of multiple DNA double-strand break repair pathways in CRISPR-mediated endogenous tagging. *Communications Biology*, *8*(1), 749. DOI: 10.1038/s42003-025-08187-5 PMID: 40360740

Tejedor, J. R., Peñarroya, A., Gancedo-Verdejo, J., Santamarina-Ojeda, P., Pérez, R. F., López-Tamargo, S., Díez-Borge, A., Alba-Linares, J. J., González-del-Rey, N., Urdinguio, R. G., Mangas, C., Roberti, A., López, V., Morales-Ruiz, T., Ariza, R. R., Roldán-Arjona, T., Meijón, M., Valledor, L., Cañal, M. J., & Fraga, M. F. (2023). CRISPR/dCAS9-mediated DNA demethylation screen identifies functional epigenetic determinants of colorectal cancer. *Clinical Epigenetics*, *15*(1), 133. DOI: 10.1186/s13148-023-01546-1 PMID: 37612734

Testa, L. C., Musunuru, K., & Wang, X. (2023). Epigenome editing of the AGT gene as a therapeutic approach for cardiovascular disorders. *Arteriosclerosis, Thrombosis, and Vascular Biology*, *43*(Suppl_1), A430–A430. DOI: 10.1161/atvb.43.suppl_1.430

Thakore, P. I., Black, J. B., Hilton, I. B., & Gersbach, C. A. (2015). Editing the epigenome: Technologies for programmable transcription and epigenetic modulation. *Nature Methods*, *12*(10), 933–936. PMID: 26820547

Thakore, P. I., D'ippolito, A. M., Song, L., Safi, A., Shivakumar, N. K., Kabadi, A. M., Reddy, T. E., Crawford, G. E., & Gersbach, C. A. (2015). Highly specific epigenome editing by CRISPR-Cas9 repressors for silencing of distal regulatory elements. *Nature Methods*, *12*(12), 1143–1149. DOI: 10.1038/nmeth.3630 PMID: 26501517

Theardy, M. S., Takeda, M., Sorokin, A., Chen, S., Yang, Z., Wang, X., Kanikarla, P., Coker, O., Nguyen, P., Wei, Y., Yao, J., Yan, L., Jin, Y., Cai, Y., Paku, M., Chen, Z., Li, K. Z., Citron, F., Tomihara, H., & Yao, W. (2025). Therapeutic targeting of syndecan-1 axis overcomes acquired resistance to KRAS-targeted therapy in gastrointestinal cancers. *Cell Reports Medicine*, *6*(8), 102253. DOI: 10.1016/j.xcrm.2025.102253 PMID: 40713971

Thomas, S., Quinn, B. A., Das, S. K., Dash, R., Emdad, L., Dasgupta, S., Wang, X.-Y., Dent, P., Reed, J. C., Pellecchia, M., Sarkar, D., & Fisher, P. B. (2013). Targeting the Bcl-2 family for cancer therapy. *Expert Opinion on Therapeutic Targets*, *17*(1), 61–75. DOI: 10.1517/14728222.2013.733001 PMID: 23173842

Thomsen, E. A., Mich, J. K., Yao, Z., Hodge, R. D., Doyle, A. M., Jang, S., & Zeng, H. (2022). Implementation of CRISPR/Cas9 Genome Editing to Generate Murine Lung Cancer Models That Depict the Mutational Landscape of Human Disease. *Frontiers in Cell and Developmental Biology*, *9*, 641618. DOI: 10.3389/fcell.2021.641618 PMID: 33738287

Tian, R., Cao, C., He, D., Dong, D., Sun, L., Liu, J., Chen, Y., Wang, Y., Huang, Z., Li, L., Jin, Z., Huang, Z., Xie, H., Zhao, T., Zhong, C., Hong, Y., & Hu, Z. (2023). Massively parallel CRISPR off-target detection enables rapid off-target prediction model building. *Med (New York, N.Y.)*, *4*(7), 478–492.e6. DOI: 10.1016/j.medj.2023.05.005 PMID: 37279759

Tiwari, A., Saraf, S., Verma, A., Panda, P. K., & Patel, B. K. (2023). CRISPR/Cas9 as a therapeutic tool for triple negative breast cancer: From bench to clinics. *Frontiers in Molecular Biosciences*, *10*, 1214489. DOI: 10.3389/fmolb.2023.1214489 PMID: 37469704

Toma, L., Barbălată, T., Sanda, G. M., Niculescu, L. S., Sima, A. V., & Stancu, C. S. (2021). CRISPR/dCas9 Transcriptional Activation of Endogenous Apolipoprotein AI and Paraoxonase 1 in Enterocytes Alleviates Endothelial Cell Dysfunction. *Biomolecules*, *11*(12), 1769. DOI: 10.3390/biom11121769 PMID: 34944413

Tram, T. T. B., Ha, V. T. N., Trieu, L. P. T., Ashton, P. M., Crawford, E. D., Thu, D. D. A., Quang, N. L., Thwaites, G. E., Walker, T. M., Anscombe, C., & Thuong, N. T. T. (2023). FLASH-TB: An Application of Next-Generation CRISPR to Detect Drug Resistant Tuberculosis from Direct Sputum. *Journal of Clinical Microbiology*, *61*(4), e0163422. DOI: 10.1128/jcm.01634-22 PMID: 37010411

Tripathi, S., Kharkwal, G., Mishra, R., & Singh, G. (2024). NRF2 signaling in metal-induced oxidative stress. *Heliyon*, *10*(18). Advance online publication. DOI: 10.1016/j.heliyon.2024.e37545 PMID: 39309893

Tripathi, S., Khatri, P., Fatima, Z., Pandey, R. P., & Hameed, S. (2022). A Landscape of CRISPR/Cas Technique for Emerging Viral Disease Diagnostics and Therapeutics: Progress and Prospects. *Pathogens (Basel, Switzerland)*, *12*(1), 56. DOI: 10.3390/pathogens12010056 PMID: 36678404

Tsai, S. Q., Nguyen, N. T., Malagon-Lopez, J., Topkar, V. V., Aryee, M. J., & Joung, J. K. (2017). CIRCLE-seq: A highly sensitive in vitro screen for genome-wide CRISPR–Cas9 nuclease off-targets. *Nature Methods*, *14*(6), 607–614. DOI: 10.1038/nmeth.4278 PMID: 28459458

Tufail, M., Jiang, C.-H., & Li, N. (2024). Altered metabolism in cancer: Insights into energy pathways and therapeutic targets. *Molecular Cancer*, *23*(1), 203. DOI: 10.1186/s12943-024-02119-3 PMID: 39294640

Tufail, M., Jiang, C.-H., & Li, N. (2025). Wnt signaling in cancer: From biomarkers to targeted therapies and clinical translation. *Molecular Cancer*, *24*(1), 107. DOI: 10.1186/s12943-025-02306-w PMID: 40170063

Tycko, J., Wainberg, M., Marinov, G. K., Ursu, O., Hess, G. T., Ego, B. K., Aradhana, , Li, A., Truong, A., Trevino, A. E., Spees, K., Yao, D., Kaplow, I. M., Greenside, P. G., Morgens, D. W., Phanstiel, D. H., Snyder, M. P., Bintu, L., Greenleaf, W. J., & Bassik, M. C. (2019). Mitigation of off-target toxicity in CRISPR-Cas9 screens for essential non-coding elements. *Nature Communications*, *10*(1), 4063. DOI: 10.1038/s41467-019-11955-7 PMID: 31492858

Tyson, J., Bundy, K., Roach, C., Douglas, H., Ventura, V., Segars, M. F., & Simpson, C. L. (2020). Mechanisms of the osteogenic switch of smooth muscle cells in vascular calcification: WNT signalling, BMPs, mechanotransduction, and EndMT. *Bioengineering (Basel, Switzerland)*, *7*(3), 88. DOI: 10.3390/bioengineering7030088 PMID: 32781528

Umeh-Garcia, M., O'Geen, H., Simion, C., Gephart, M. H., Segal, D. J., & Sweeney, C. A. (2022). Aberrant promoter methylation contributes to LRIG1 silencing in basal/triple-negative breast cancer. *British Journal of Cancer*, *127*(3), 436–448. DOI: 10.1038/s41416-022-01812-8 PMID: 35440669

Valente, D., Gentileschi, M. P., Guerrisi, A., Bruzzaniti, V., Morrone, A., Soddu, S., & Verdina, A. (2022). Factors to consider for the correct use of γH2AX in the evaluation of DNA double-strand breaks caused by ionizing radiation. *Cancers (Basel)*, *14*(24), 6204. DOI: 10.3390/cancers14246204 PMID: 36551689

Van Der Zanden, S. Y., Jongsma, M. L. M., Neefjes, A. C. M., Berlin, I., & Neefjes, J. (2023). Maintaining soluble protein homeostasis between nuclear and cytoplasmic compartments across mitosis. *Trends in Cell Biology*, *33*(1), 18–29. DOI: 10.1016/j.tcb.2022.06.002 PMID: 35778326

Van Opijnen, T., & Levin, H. L. (2020). Transposon Insertion Sequencing, a Global Measure of Gene Function. *Annual Review of Genetics*, *54*(1), 337–365. DOI: 10.1146/annurev-genet-112618-043838 PMID: 32886545

van Staalduinen, J., van Staveren, T., Grosveld, F., & Wendt, K. S. (2023). Live-cell imaging of chromatin contacts opens a new window into chromatin dynamics. *Epigenetics & Chromatin*, *16*(1), 27. DOI: 10.1186/s13072-023-00503-9 PMID: 37349773

Velayuthaprabhu, S., & Archunan, G. (2005). Evaluation of anticardiolipin antibodies and antiphosphatidylserine antibodies in women with recurrent abortion. *Indian Journal of Medical Sciences*, *59*(8), 347–352. DOI: 10.4103/0019-5359.16651 PMID: 16129928

Velayuthaprabhu, S., Archunan, G., & Balakrishnan, K. (2007). Placental thrombosis in experimental anticardiolipin antibodies-mediated intrauterine fetal death. *American Journal of Reproductive Immunology (New York, N.Y.)*, *57*(4), 270–276. DOI: 10.1111/j.1600-0897.2007.00474.x PMID: 17362388

Velayuthaprabhu, S., Matsubayashi, H., & Archunan, G. (2016). Beta-2 GPI induced tissue factor and placental apoptosis for the pathophysiology of pregnancy loss in antiphospholipid syndrome. *International Journal of Research in Medical Sciences*, *4*(8), 3109–3113. DOI: 10.18203/2320-6012.ijrms20162267

Velayuthaprabhu, S., Matsubayashi, H., Sugi, T., Nakamura, M., Ohnishi, Y., Ogura, T., & Archunan, G. (2013). Expression of apoptosis in placenta of experimental antiphospholipid syndrome (APS) mouse. *American Journal of Reproductive Immunology (New York, N.Y.)*, *69*(5), 486–494. DOI: 10.1111/aji.12084 PMID: 23398188

Velayuthaprabhu, S., Matsubayashi, H., Sugi, T., Nakamura, M., Ohnishi, Y., Ogura, T., Tomiyama, T., & Archunan, G. (2011). A unique preliminary study on fetal resorption and placental apoptosis in mice with passive immunization of anti-phosphatidylethanolamine antibodies and anti-Factor XII antibodies. *American Journal of Reproductive Immunology (New York, N.Y.)*, *66*(5), 373–384. DOI: 10.1111/j.1600-0897.2011.01008.x PMID: 21623987

Verma, M. K., Roychowdhury, S., Sahu, B. D., Mishra, A., & Sethi, K. K. (2022). CRISPR-based point-of-care diagnostics incorporating Cas9, Cas12, and Cas13 enzymes advanced for SARS-CoV-2 detection. *Journal of Biochemical and Molecular Toxicology*, *36*(8), e23113. DOI: 10.1002/jbt.23113 PMID: 35642647

Vermersch, E., Jouve, C., & Hulot, J. S. (2020). CRISPR/Cas9 gene-editing strategies in cardiovascular cells. *Cardiovascular Research*, *116*(5), 894–907. DOI: 10.1093/cvr/cvz250 PMID: 31584620

Viktorsson, K., Rieckmann, T., Fleischmann, M., Diefenhardt, M., Hehlgans, S., & Rödel, F. (2023). Advances in molecular targeted therapies to increase efficacy of (chemo)radiation therapy. *Strahlentherapie und Onkologie*, *199*(12), 1091–1109. DOI: 10.1007/s00066-023-02064-y PMID: 37041372

Vimal, S., Madar, I. H., Thirumani, L., Thangavelu, L., & Sivalingam, A. M. (2024). CRISPR/Cas9: Role of genome editing in cancer immunotherapy. *Oral Oncology Reports*, *10*, 100251. DOI: 10.1016/j.oor.2024.100251

Vlassis, A., Beattie, G. M., Sosa, E., Park, T. S., & Cohen, A. R. (2023). Integration of CRISPR-MAD7 constructs into genomic safe harbor sites enables stable gene expression in iPSC-derived lineages. *Frontiers in Genetics*, *14*, 10682145.

Volodina, O., & Smirnikhina, S. (2025). The Future of Gene Therapy: A Review of In Vivo and Ex Vivo Delivery Methods for Genome Editing-Based Therapies. *Molecular Biotechnology*, *67*(2), 425–437. DOI: 10.1007/s12033-024-01070-4 PMID: 38363528

von Holdt, B. M., Kartzinel, R. Y., van Oers, K., Verhoeven, K. J., & Ouyang, J. Q. (2023). Changes in the rearing environment cause reorganization of molecular networks associated with DNA methylation. *Journal of Animal Ecology*, *92*(3), 648–664. DOI: 10.1111/1365-2656.13878 PMID: 36567635

Wallace, D. C. (2012). Mitochondria and cancer. *Nature Reviews. Cancer*, *12*(10), 685–698. DOI: 10.1038/nrc3365 PMID: 23001348

Wang, H.-Z., & He, Z.-X. (2024). Recent advances in the treatment of β-thalassemia: From traditional therapies to gene therapy. *New Discovery*, 1–15. DOI: 10.61958/NDQD9971

Wang, J., Jonker, T., Cervera I Barea, A., Boender, A. R., Klerk, M., Yang, Y., & Boink, G. J. J. (2024). Adeno-associated virus (AAV) 6 provides efficient gene transfer upon intramyocardial injection in vivo. *Cardiovascular Research, 120*(Supplement_1), cvae088-073.

Wang, S., Xiang, S., Wu, X., Li, N., & Qi, D. (2016). Regulation of macrophage polarization during atherosclerosis. *Lipid Cardiovasc Res, 2*, 05-12.

Wang, D., Tai, P. W. L., & Gao, G. (2020). Adeno-associated virus vector as a platform for gene therapy delivery. *Nature Reviews. Drug Discovery*, *19*(5), 325–350. PMID: 30710128

Wang, D., Zhang, F., & Gao, G. (2020). CRISPR-based therapeutic genome editing: Strategies and in vivo delivery by AAV vectors. *Cell*, *181*(1), 136–150. DOI: 10.1016/j.cell.2020.03.023 PMID: 32243786

Wang, G. H., Wang, C. M., Wu, X. J., Chu, T., Huang, D. W., & Li, J. (2024). The Development of SpCas9 Variants with High Specificity and Efficiency Based on the HH Theory. *Molecular Biology*, *58*(1), 133–146. DOI: 10.1134/S0026893324010187 PMID: 38943587

Wang, G., & Li, J. (2021). Review, analysis, and optimization of the CRISPR Streptococcus pyogenes Cas9 system. *Medicine in Drug Discovery*, *9*, 100080. DOI: 10.1016/j.medidd.2021.100080

Wang, H., La Russa, M., & Qi, L. S. (2022). CRISPR-based epigenome editing in vivo: Delivery methods and monitoring strategies. *Trends in Biotechnology*, *40*(8), 920–935.

Wang, H., Zou, W., & Cao, Y. (2025). Radiation-induced senescence and adaptive response. *Radiation Medicine and Protection*. Advance online publication. DOI: 10.1016/j.radmp.2025.05.002

Wang, J., Wang, K., Deng, Z., Zhong, Z., Sun, G., Mei, Q., Zhou, F., Deng, Z., & Sun, Y. (2024). Engineered cytosine base editor enabling broad-scope and high-fidelity gene editing in Streptomyces. *Nature Communications*, *15*(1), 5687. DOI: 10.1038/s41467-024-49987-3 PMID: 38971862

Wang, K. C., Zheng, T., & Hubbard, B. P. (2025). CRISPR/Cas technologies for cancer drug discovery and treatment. *Trends in Pharmacological Sciences*, *46*(5), 437–452. DOI: 10.1016/j.tips.2025.02.009 PMID: 40133194

Wang, L., & Han, H. (2024). Strategies for improving the genome-editing efficiency of class 2 CRISPR/Cas system. *Heliyon*, *10*(19), e38588. DOI: 10.1016/j.heliyon.2024.e38588 PMID: 39397905

Wang, M., Wang, H., Li, K., Li, X., Wang, X., & Wang, Z. (2023). Review of CRISPR/Cas Systems on Detection of Nucleotide Sequences. *Foods*, *12*(3), 477. DOI: 10.3390/foods12030477 PMID: 36766007

Wang, P., Liu, Z., Zhang, X., Li, J., Sun, L., Ju, Z., & Qu, J. (2018). CRISPR/Cas9-mediated gene knockout reveals a guardian role of NF-κB/RelA in maintaining the homeostasis of human vascular cells. *Protein & Cell*, *9*(11), 945–965. DOI: 10.1007/s13238-018-0560-5 PMID: 29968158

Wang, Q., Ma, C., Mao, H., & Wang, J. (2024). Epigenome editing by a CRISPR-Cas9-based acetyltransferase activates the ZNF334 gene to inhibit the growth of colorectal cancer. *International Journal of Biological Macromolecules*, *277*, 134580. DOI: 10.1016/j.ijbiomac.2024.134580 PMID: 39122070

Wang, X., Whittaker, M. N., Jindal, I., Cortez-Alvarado, S. V., Qu, P., & Musunuru, K. (2022). Reversible epigenome editing of PCSK9 as a therapeutic strategy. *Arteriosclerosis, Thrombosis, and Vascular Biology*, *42*(Suppl_1), A125–A125. DOI: 10.1161/atvb.42.suppl_1.125

Wang, Y., Lin, R.-Z., Harris, M., Lavayen, B., Diwanji, N., McCreedy, B., Hofmeister, R., & Getts, D. (2025). CRISPR-Enabled Autonomous Transposable Element (CREATE) for RNA-based gene editing and delivery. *EMBO Reports*, *26*(4), 1062–1083. DOI: 10.1038/s44319-024-00364-7 PMID: 39789389

Wang, Y., Wang, F., & Zhang, L. (2021). CRISPR-based regenerative medicine: Target discovery and therapeutic design. *Cell Reports Medicine*, *2*(9), 100389. PMID: 34622226

Wang, Y., Zhang, H., & Lin, C. (2022). CRISPR-Cas9-based therapeutic approaches in osteoarthritis. *Bone Research*, *10*(1), 1–14. PMID: 34975148

Wang, Z., Chen, Y., & Liu, J. (2023). CRISPR-corrected iPSCs in translational regenerative medicine. *Science Translational Medicine*, *15*(685), eabo6851.

Wan, T., Zhong, J., Pan, Q., Zhou, T., Ping, Y., & Liu, X. (2022). Exosome-mediated delivery of Cas9 ribonucleoprotein complexes for tissue-specific gene therapy of liver diseases. *Science Advances*, *8*(37), eabp9435. Advance online publication. DOI: 10.1126/sciadv.abp9435 PMID: 36103526

Wei, P.-S., Thota, N., John, G., Chang, E., Lee, S., Wang, Y., Ma, Z., Tsai, Y.-H., & Mei, K.-C. (2024). Enhancing RNA-lipid nanoparticle delivery: Organ- and cell-specificity and barcoding strategies. *Journal of Controlled Release*, *375*, 366–388. DOI: 10.1016/j.jconrel.2024.08.030 PMID: 39179112

Weiss, K. H., Stremmel, W., & Gotthardt, D. N. (2023). ATP7B dysfunction in Wilson disease. *Journal of Hepatology*, *78*(4), 789–805. DOI: 10.1016/j.jhep.2022.12.017

Wei, Y., Fu, Y., Li, C., Chen, S., Xie, L., & Chen, M. (2024). Ultrasensitive detection of circulating tumor DNA using a CRISPR/Cas9 nickase-driven 3D DNA walker based on a COF-AuNPs sensing platform. *Microchimica Acta*, *191*(11), 671. DOI: 10.1007/s00604-024-06749-8 PMID: 39404875

Welch, D. R., Foster, C., & Rigoutsos, I. (2022). Roles of mitochondrial genetics in cancer metastasis. *Trends in Cancer*, *8*(12), 1002–1018. DOI: 10.1016/j.trecan.2022.07.004 PMID: 35915015

Weng, Z., You, Z., Yang, J., Mohammad, N., Lin, M., Wei, Q., Gao, X., & Zhang, Y. (2023). CRISPR-Cas biochemistry and CRISPR-based molecular diagnostics. *Angewandte Chemie International Edition*, *62*(17), e202214987. DOI: 10.1002/anie.202214987 PMID: 36710268

Wen, H., Deng, H., Li, B., Chen, J., Zhu, J., Zhang, X., Yoshida, S., & Zhou, Y. (2025). Mitochondrial diseases: From molecular mechanisms to therapeutic advances. *Signal Transduction and Targeted Therapy*, *10*(1), 9. DOI: 10.1038/s41392-024-02044-3 PMID: 39788934

Westerman, K., Fernández-Sanlés, A., Patil, P., Sebastiani, P., Jacques, P., Starr, J. M., & Ordovás, J. M. (2020). Epigenomic assessment of cardiovascular disease risk and interactions with traditional risk metrics. *Journal of the American Heart Association*, *9*(8), e015299. DOI: 10.1161/JAHA.119.015299 PMID: 32308120

Whittaker, M. N., Testa, L. C., Quigley, A., Jindal, I., Cortez-Alvarado, S. V., Qu, P., Yang, Y., Alameh, M.-G., Musunuru, K., & Wang, X. (2023). Epigenome editing durability varies widely across cardiovascular disease target genes. *Arteriosclerosis, Thrombosis, and Vascular Biology*, *43*(10), 2075–2077. DOI: 10.1161/ATVBAHA.123.319748 PMID: 37589141

Wiley, L., Cheek, M., LaFar, E., Ma, X., Sekowski, J., Tanguturi, N., & Iltis, A. (2025, June). The ethics of human embryo editing via CRISPR-Cas9 technology: A systematic review of ethical arguments, reasons, and concerns. []). Dordrecht: Springer Netherlands.]. *HEC Forum*, *37*(2), 267–303. DOI: 10.1007/s10730-024-09538-1 PMID: 39302534

Williams, K., & Sobol, R. W. (2013). Mutation research/fundamental and molecular mechanisms of mutagenesis. *Mutation Research. Fundamental and Molecular Mechanisms of Mutagenesis*, *743–744*, 1–3. DOI: 10.1016/j.mrfmmm.2013.04.009 PMID: 23688353

Wilton-Clark, H., & Yokota, T. (2023). CRISPR-Cas9-mediated exon skipping as a cardioprotective strategy in duchenne muscular dystrophy. *Molecular Therapy. Methods & Clinical Development*, *30*, 500–501. DOI: 10.1016/j.omtm.2023.08.013 PMID: 37693945

World Health Organization (WHO). (2021). *Human genome editing: Recommendations*. WHO.

Wroblewska, L., Kitada, T., Endo, K., Siciliano, V., Stillo, B., Saito, H., & Weiss, R. (2015). Mammalian synthetic circuits with RNA binding proteins for RNA-only delivery. *Nature Biotechnology*, *33*(9), 839–841. DOI: 10.1038/nbt.3301 PMID: 26237515

Wu, H.-p., & Yang, Z.-l. Advances in the applications of CRISPR/Cas system for tumor molecular diagnostics. LabMed Discovery. 2024;1(1):100011. https://doi.org/https://doi.org/10.1016/j.lmd.2024.100011

Wu, D., Chen, Q., Chen, X., Han, F., Chen, Z., & Wang, Y. (2023). The blood–brain barrier: Structure, regulation and drug delivery. *Signal Transduction and Targeted Therapy*, *8*(1), 217. DOI: 10.1038/s41392-023-01481-w PMID: 37231000

Wu, J., He, S., Song, Z., Chen, S., Lin, X., Sun, H., & Liu, X. (2023). Macrophage polarization states in atherosclerosis. *Frontiers in Immunology*, *14*, 1185587. DOI: 10.3389/fimmu.2023.1185587 PMID: 37207214

Wu, J., Ko, S., Lee, E., Son, E., Kang, G., Hur, S., Lee, J.-H., Oh, J.-W., & Kim, Y. (2025). Gold nanoparticles in imaging: Advances, applications, and future perspectives. *Applied Spectroscopy Reviews*, *60*(9-10), 1–40. DOI: 10.1080/05704928.2025.2495022

Wu, P. Y., Chandramohan, R., Phan, J. H., Mahle, W. T., Gaynor, J. W., Maher, K. O., & Wang, M. D. (2014). Cardiovascular transcriptomics and epigenomics using next-generation sequencing: Challenges, progress, and opportunities. *Circulation: Cardiovascular Genetics*, *7*(5), 701–710. DOI: 10.1161/CIRCGENETICS.113.000129 PMID: 25518043

Wu, W., Shan, J., Li, Y., Luo, L., Sun, G., Zhou, Y., & Feng, L. (2012). Adoptive transfusion of tolerance dendritic cells prolongs the survival of cardiac allograft: A systematic review of 44 basic studies in mice. *Journal of Evidence-Based Medicine*, *5*(3), 139–153. DOI: 10.1111/j.1756-5391.2012.01191.x PMID: 23672221

Wu, Y., Chen, Z., Chen, X., Hu, Y., & Ma, Y. (2025). Recent advances in the investigation of the regulatory network underlying reactive nitrogen species-mediated tumorigenesis: Molecular mechanisms and targeted therapeutic strategies. *Redox Report : Communications in Free Radical Research*, *30*(1), 2564593. DOI: 10.1080/13510002.2025.2564593 PMID: 41036604

Wu, Y., Kong, X. J., Ji, Y. Y., Fan, J., Chen, X. M., Ji, C. C., & Wu, S. H. (2024). Correction of Ito in human induced pluripotent stem Cell-derived cardiomyocyte carrying DPP6 mutation in early repolarization syndrome by CRISPR/Cas9 genome editing. *Experimental Cell Research*, *435*(2), 113929. DOI: 10.1016/j.yexcr.2024.113929 PMID: 38272106

Xiang, X., Corsi, G. I., Anthon, C., Qu, K., Pan, X., Liang, X., Han, P., Dong, Z., Liu, L., Zhong, J., Ma, T., Wang, J., Zhang, X., Jiang, H., Xu, F., Liu, X., Xu, X., Wang, J., Yang, H., & Luo, Y. (2021). Enhancing CRISPR-Cas9 gRNA efficiency prediction by data integration and deep learning. *Nature Communications*, *12*(1), 3238. DOI: 10.1038/s41467-021-23576-0 PMID: 34050182

Xiao, R., Zhou, M., Wang, P., Zeng, B., Wu, L., Hu, Z., & Liang, D. (2022). Full-length dystrophin restoration via targeted exon addition in DMD-patient specific iPSCs and cardiomyocytes. *International Journal of Molecular Sciences*, *23*(16), 9176. DOI: 10.3390/ijms23169176 PMID: 36012442

Xie, R., Wang, Y., & Gong, S. (2021). External stimuli-responsive nanoparticles for spatially and temporally controlled delivery of CRISPR-Cas genome editors. *Biomaterials Science*, *9*(18), 6012–6022. DOI: 10.1039/D1BM00558H PMID: 34286726

Xie, S., Yue, Y., & Yang, F. (2024). Recent Advances in CRISPR/Cas System-Based Biosensors for the Detection of Foodborne Pathogenic Microorganisms. *Micromachines*, *15*(11), 1329. DOI: 10.3390/mi15111329 PMID: 39597141

Xing, H., & Meng, L. H. (2020). CRISPR-Cas9: A powerful tool towards precision medicine in cancer treatment. *Acta Pharmacologica Sinica*, *41*(5), 583–587. DOI: 10.1038/s41401-019-0322-9 PMID: 31792341

Xu, C., Qi, L. S., & Zhao, H. (2020). Machine learning in CRISPR guide RNA design: Advances and opportunities. *Trends in Biotechnology*, *38*(12), 1292–1304. DOI: 10.1016/j.tibtech.2020.03.012 PMID: 32307119

Xue, C., & Greene, E. C. (2021). Pathway choices in CRISPR genome editing. *Trends in Genetics*, *37*(7), 639–656. DOI: 10.1016/j.tig.2021.02.008 PMID: 33896583

Xue, M., Gonzalez, D. H., Osikpa, E., Gao, X., & Lillehoj, P. B. (2025). Rapid and automated interpretation of CRISPR-Cas13-based lateral flow assay test results using machine learning. *Sensors & Diagnostics*, *4*(2), 171–181. DOI: 10.1039/D4SD00314D PMID: 39817182

Xu, H., Xiao, T., Chen, C. H., Li, W., Meyer, C. A., Wu, Q., & Liu, X. S. (2022). Machine learning–based personalized CRISPR guide RNA design for therapeutic genome editing. *Nature Communications*, *13*(1), 4556. PMID: 35961959

Xu, S., Shiomi, H., Yamashita, Y., Koyama, S., Horie, T., Baba, O., & Ono, K. (2024). CRISPR-Cas9-guided amplification-free genomic diagnosis for familial hypercholesterolemia using nanopore sequencing. *PLoS One*, *19*(3), e0297231. DOI: 10.1371/journal.pone.0297231 PMID: 38507394

Xu, W., Wang, J., & Wang, D. (2023). Advances in CRISPR-iPSC-derived cell therapies: Opportunities and challenges. *Nature Biotechnology*, *41*(7), 950–963.

Xu, X., Liu, C., Wang, Y., Koivisto, O., Zhou, J., Shu, Y., & Zhang, H. (2021). Nanotechnology-based delivery of CRISPR/Cas9 for cancer treatment. *Advanced Drug Delivery Reviews*, *176*, 113891. DOI: 10.1016/j.addr.2021.113891 PMID: 34324887

Xu, X., Wan, T., Xin, H., Li, D., Pan, H., Wu, J., Tang, R., & Ping, Y. (2021). Delivery of CRISPR-Cas9 for therapeutic genome editing. *Journal of Controlled Release*, *329*, 114–124.

Yadav, H., Malviya, R., & Kaushik, N. (2024). Chitosan in biomedicine: A comprehensive review of recent developments. *Carbohydrate Polymer Technologies and Applications*, *8*, 100551. DOI: 10.1016/j.carpta.2024.100551

Yamamoto, M., & Kensler, T. W. (2024). Nrf2 signaling in oxidative and xenobiotic stress: Mechanisms and therapeutic modulation. *Nature Reviews. Molecular Cell Biology*, *25*, 145–161. DOI: 10.1038/s41580-023-00630-7

Yandrapalli, N. (2025). Bottom-up development of lipid-based synthetic cells for practical applications. *Trends in Biotechnology*, *0*(0), 2150–2169. Advance online publication. DOI: 10.1016/j.tibtech.2025.03.009 PMID: 40263003

Yang, H., Ren, S., Yu, S., Pan, H., Li, T., Ge, S., Zhang, J., & Xia, N. (2020). Methods favoring homology-directed repair choice in response to CRISPR/Cas9-induced double-strand breaks. *International Journal of Molecular Sciences*, *21*(18), 6461. DOI: 10.3390/ijms21186461 PMID: 32899704

Yang, J., Zeng, H., Luo, Y., Chen, Y., Wang, M., Wu, C., & Hu, P. (2024). Recent Applications of PLGA in Drug Delivery Systems. *Polymers*, *16*(18), 2606. DOI: 10.3390/polym16182606 PMID: 39339068

Yang, L., Huang, S., Zhang, Z., Liu, Z., & Zhang, L. (2022). Roles and Applications of Red Blood Cell-Derived Extracellular Vesicles in Health and Diseases. *International Journal of Molecular Sciences*, *23*(11), 5927. DOI: 10.3390/ijms23115927 PMID: 35682606

Yang, N., He, Z., & Lang, T. (2025). Drug Delivery Systems Based on Metal–Organic Frameworks for Tumor Immunotherapy. *Pharmaceutics*, *17*(2), 225. DOI: 10.3390/pharmaceutics17020225 PMID: 40006592

Yang, P., Zhang, S., Hu, D., Li, X., Guo, Y., Guo, H., Zhang, L., & Ding, X. (2024). Research Progress on the Mechanism and Application of the Type I CRISPR-Cas System. *International Journal of Molecular Sciences*, *25*(23), 12544. DOI: 10.3390/ijms252312544 PMID: 39684256

Yang, S., Chang, R., Yang, H., Zhao, T., Hong, Y., Kong, H. E., Sun, X., Qin, Z., Jin, P., Li, S., & Li, X. J. (2017). CRISPR/Cas9-mediated gene editing ameliorates neurotoxicity in mouse model of Huntington's disease. *The Journal of Clinical Investigation*, *127*(7), 2719–2724. DOI: 10.1172/JCI92087 PMID: 28628038

Yang, X., Bui, T. A., Mei, H., Aksoy, Y. A., Deng, F., Hutvagner, G., & Deng, W. (2024). Exploring the potential and challenges of CRISPR delivery and therapeutics for genetic disease treatment. *Advanced Functional Materials*, *34*(38), 2402630. DOI: 10.1002/adfm.202402630

Yang, Z. X., Fu, Y. W., Zhao, J. J., Zhang, F., Li, S. A., Zhao, M., Wen, W., Zhang, L., Cheng, T., Zhang, J.-P., & Zhang, X. B. (2023). Superior fidelity and distinct editing outcomes of SaCas9 compared with SpCas9 in genome editing. *Genomics, Proteomics & Bioinformatics*, *21*(6), 1206–1220. DOI: 10.1016/j.gpb.2022.12.003 PMID: 36549468

Yan, X., Han, D., Chen, Z., Han, C., Dong, W., Han, L., Zou, L., Zhang, J., Liu, Y., & Chai, J. (2020). RUNX2 interacts with BRG1 to target CD44 for promoting invasion and migration of colorectal cancer cells. *Cancer Cell International*, *20*(1), 505. DOI: 10.1186/s12935-020-01544-w PMID: 33071648

Ye, R., Chen, X., & Li, Y. (2023). CRISPR/Cas systems for therapeutic purposes: Applications in disease prevention, mechanism study, and treatment. *Highlights in Science. Engineering and Technology*, *54*, 1–7.

Yildirim, D. T., Baki Yildirim, A., Salzet, M., Bertelli, M., Beccari, T., Prakash, S., Pascucci, L., & Dundar, M. (2025). Messaging malignancy: Tumour-derived exosomes at the nexus of immune escape, vascular remodelling and metastatic competence. *The EuroBiotech Journal*, *9*(3), 216–237. DOI: 10.2478/ebtj-2025-0018

Yin, H., Kauffman, K. J., & Anderson, D. G. (2017). Delivery technologies for genome editing. *Nature Reviews. Drug Discovery*, *16*(6), 387–399. DOI: 10.1038/nrd.2016.280 PMID: 28337020

Yin, H., Xue, W., Chen, S., Bogorad, R. L., Benedetti, E., Grompe, M., Koteliansky, V., Sharp, P. A., Jacks, T., & Anderson, D. G. (2014). Genome editing with Cas9 in adult mice corrects a disease mutation and phenotype. *Nature Biotechnology*, *32*(6), 551–553. DOI: 10.1038/nbt.2884 PMID: 24681508

Yin, T., Luo, J., Huang, D., & Li, H. (2022). Current Progress of Mitochondrial Genome Editing by CRISPR. *Frontiers in Physiology*, *13*, 883459. Advance online publication. DOI: 10.3389/fphys.2022.883459 PMID: 35586709

Yuan, J., Ma, Y., Huang, T., Chen, Y., Peng, Y., Li, B., & Li, C. (2021). Efficient base editing for disease modeling in human pluripotent stem cells. *Cell Regeneration (London, England)*, *10*(1), 6. PMID: 33644818

Yu, L., Peng, Y., Sheng, M., Wang, Q., Huang, J., & Yang, X. (2023). Sensitive and Amplification-Free Electrochemiluminescence Biosensor for HPV-16 Detection Based on CRISPR/Cas12a and DNA Tetrahedron Nanostructures. *ACS Sensors*, *8*(7), 2852–2858. DOI: 10.1021/acssensors.3c00806 PMID: 37402133

Yusa, K., Rashid, S. T., Strick-Marchand, H., Varela, I., Liu, P. Q., Paschon, D. E., & Lacaud, G. (2011). Targeted gene correction of α1-antitrypsin deficiency in induced pluripotent stem cells. *Nature*, *478*(7369), 391–394. DOI: 10.1038/nature10424 PMID: 21993621

Zaidi, A. A., Wilton, P. R., Su, M. S.-W., Paul, I. M., Arbeithuber, B., Anthony, K., Nekrutenko, A., Nielsen, R., & Makova, K. D. (2019). Bottleneck and selection in the germline and maternal age influence transmission of mitochondrial DNA in human pedigrees. *Proceedings of the National Academy of Sciences of the United States of America*, *116*(50), 25172–25178. DOI: 10.1073/pnas.1906331116 PMID: 31757848

Zeng, S., Lei, S., Qu, C., Wang, Y., Teng, S., & Huang, P. (2023). CRISPR/Cas-based gene editing in therapeutic strategies for beta-thalassemia. *Human Genetics*, *142*(12), 1677–1703. DOI: 10.1007/s00439-023-02610-9 PMID: 37878144

Zeng, S., Tang, Q., Xiao, M., Tong, X., Yang, T., Yin, D., Lei, L., & Li, S. (2023). Cell membrane-coated nanomaterials for cancer therapy. *Materials Today. Bio*, *20*, 100633. DOI: 10.1016/j.mtbio.2023.100633 PMID: 37128288

Zhang, B., Li, J., & Yu, W. (2025). Integration of CRISPR/dCas9-Based methylation editing with guide positioning sequencing identifies dynamic changes of mrDEGs in breast cancer progression. *Cellular and Molecular Life Sciences : CMLS*, *82*(1), 46. DOI: 10.1007/s00018-024-05562-z PMID: 39833630

Zhang, F. (2019). Development of CRISPR-Cas systems for genome editing and beyond. *Quarterly Reviews of Biophysics*, *52*, e6. DOI: 10.1017/S0033583519000052

Zhang, H., Qin, C., An, C., Zheng, X., Wen, S., Chen, W., Liu, X., Lv, Z., Yang, P., Xu, W., Gao, W., & Wu, Y. (2021). Application of the CRISPR/Cas9-based gene editing technique in basic research, diagnosis, and therapy of cancer. *Molecular Cancer*, *20*(1), 126. DOI: 10.1186/s12943-021-01431-6 PMID: 34598686

Zhang, J., & He, P. (2025). In silico and in vitro assessment of TP53, ATM, RAD51, and BAX genes in gastric cancer and their contribution to radiotherapy resistance. *Hereditas*, *162*(1), 1–20. DOI: 10.1186/s41065-025-00496-3 PMID: 40652278

Zhang, L., Zhao, D., Wei, Z., Zhu, X., Sha, T., Tang, W., Bi, C., & Zhang, X. (2025). The Advancement of Prime Editing Technology. *ChemBioChem*, *26*(12), e202500193. DOI: 10.1002/cbic.202500193 PMID: 40350377

Zhang, R., Li, Y., Fu, H., Zhao, C., Li, X., Wang, Y., Sun, Y., & Li, Y. (2025). Nanomedicine strategies for cuproptosis: Metabolic reprogramming and tumor immunotherapy. *Acta Pharmaceutica Sinica. B*, *15*(9), 4582–4613. Advance online publication. DOI: 10.1016/j.apsb.2025.07.007 PMID: 41049741

Zhang, S., Shen, J., Li, D., & Cheng, Y. (2021). Strategies in the delivery of Cas9 ribonucleoprotein for CRISPR/Cas9 genome editing. *Theranostics*, *11*(2), 614–648. DOI: 10.7150/thno.47007 PMID: 33391496

Zhang, T., Yin, H., Li, Y., Yang, H., Ge, K., Zhang, J., Yuan, Q., Dai, X., Naeem, A., Weng, Y., Huang, Y., & Liang, X.-J. (2024). Optimized lipid nanoparticles (LNPs) for organ-selective nucleic acids delivery in vivo. *iScience*, *27*(6), 109804. DOI: 10.1016/j.isci.2024.109804 PMID: 38770138

Zhang, X. H., Tee, L. Y., Wang, X. G., Huang, Q. S., & Yang, S. H. (2020). Off-target effects in CRISPR/Cas9-mediated genome engineering. *Molecular Therapy. Nucleic Acids*, *19*, 1–12. DOI: 10.1016/j.omtn.2019.10.033 PMID: 31790971

Zhang, X., Cai, H., Xu, H., Dong, S., & Ma, H. (2023). Critical roles of m6A methylation in cardiovascular diseases. *Frontiers in Cardiovascular Medicine*, *10*, 1187514. DOI: 10.3389/fcvm.2023.1187514 PMID: 37273867

Zhang, X., Li, L., Xu, J., Wang, Z., & Wang, H. (2021). Lipid nanoparticles for genome editing: Progress and prospects. *Advanced Drug Delivery Reviews*, *168*, 137–151.

Zhang, Y., Ren, Q., Tang, X., Liu, S., Malzahn, A. A., Zhou, J., Wang, J., Yin, D., Pan, C., Yuan, M., Huang, L., Yang, H., Zhao, Y., Fang, Q., Zheng, X., Tian, L., Cheng, Y., Le, Y., McCoy, B., & Qi, Y. (2021). Expanding the scope of plant genome engineering with Cas12a orthologs and highly multiplexable editing systems. *Nature Communications*, *12*(1), 1944. DOI: 10.1038/s41467-021-22330-w PMID: 33782402

Zhang, Y., Zhang, Z., & Ding, Y. (2021). CRISPR/Cas9-mediated genome editing in cancer therapy. *Biomarker Research*, *9*, 69. DOI: 10.1186/s40364-021-00323-x

Zhang, Z., Li, J., & Huang, Y. (2023). Omics-informed CRISPR therapeutics: Precision targeting for regeneration. *Frontiers in Cell and Developmental Biology*, *11*, 1189.

Zhang, Z., Ren, F., & Hohenstein, A. (2023). Structural basis of ZIP family transport. *Science Advances*, *9*(12), eadg8721. Advance online publication. DOI: 10.1126/sciadv.adg8721

Zhang, Z., Zhang, W., Blakes, R., Sundby, L. J., Shi, Z., Rockey, D. C., & Nam, Y. J. (2022). Fibroblast fate determination during cardiac reprogramming by remodelling of actin filaments. *Stem Cell Reports*, *17*(7), 1604–1619. DOI: 10.1016/j.stemcr.2022.05.004 PMID: 35688153

Zhan, Y., Cao, C., Li, A., Mei, H., & Liu, Y. (2023). Enhanced RNA knockdown efficiency with engineered fusion guide RNAs that function with both CRISPR-CasRx and hammerhead ribozyme. *Genome Biology*, *24*(1), 9. DOI: 10.1186/s13059-023-02852-w PMID: 36650600

Zhao, C., Cheng, Y., Huang, P., Wang, C., Wang, W., Wang, M., & Deng, H. (2023). X-ray-guided in situ genetic engineering of macrophages for sustained cancer immunotherapy. *Advanced Materials*, *35*(14), 2208059. DOI: 10.1002/adma.202208059 PMID: 36527738

Zhao, H., Li, Y., He, L., Pu, W., Yu, W., Li, Y., & Zhou, B. (2020). In vivo AAV-CRISPR/Cas9-mediated gene editing ameliorates atherosclerosis in familial hypercholesterolemia. *Circulation*, *141*(1), 67–79. DOI: 10.1161/CIRCULATIONAHA.119.042476 PMID: 31779484

Zhao, L., Wang, J., & Chen, S. (2022). CRISPR/dCas9-mediated epigenetic regulation to overcome chemoresistance in cancer therapy. *Epigenetics*, *17*(6), 563–577.

Zhao, R., Long, J., & Chen, P. (2024). Nrf2 activation protects against heavy metal–induced oxidative stress in mammalian cells. *Redox Biology*, *68*, 102853. DOI: 10.1016/j.redox.2024.102853

Zheng, Y., Nie, L., & Han, G. (2022). Integrating CRISPR perturbations with single-cell and spatial epigenomics for functional genomics. *Nature Reviews. Genetics*, *23*, 555–573.

Zhou, H., Ye, P., Xiong, W., Duan, X., Jing, S., He, Y., Zeng, Z., Wei, Y., & Ye, Q. (2024). Genome-scale CRISPR-Cas9 screening in stem cells: Theories, applications and challenges. *Stem Cell Research & Therapy*, *15*(1), 218. DOI: 10.1186/s13287-024-03831-z PMID: 39026343

Zhou, J., Li, Z., Seun Olajide, J., & Wang, G. (2024). CRISPR/Cas-based nucleic acid detection strategies: Trends and challenges. *Heliyon*, *10*(4), e26179. DOI: 10.1016/j.heliyon.2024.e26179 PMID: 38390187

Zhou, L., & Simonian, A. L. (2024). CRISPR/Cas Technology: The Unique Synthetic Biology Genome-Editing Tool Shifting the Paradigm in Viral Diagnostics, Defense, and Therapeutics. *Annual Review of Biomedical Engineering*, *26*(1), 247–272. DOI: 10.1146/annurev-bioeng-081723-013033 PMID: 38346278

Zhou, Q., Cao, T., Li, F., Zhang, M., Li, X., Zhao, H., & Zhou, Y. (2024). Mitochondria: A new intervention target for tumor invasion and metastasis. *Molecular Medicine (Cambridge, Mass.)*, *30*(1), 129. DOI: 10.1186/s10020-024-00899-4 PMID: 39179991

Zhou, R., Geiger, R. C., & Dean, D. A. (2004). Intracellular trafficking of nucleic acids. *Expert Opinion on Drug Delivery*, *1*(1), 127–140. DOI: 10.1517/17425247.1.1.127 PMID: 16296725

Zhou, W., Wang, C., Chang, J., Huang, Y., Xue, Q., Miao, C., & Wu, P. (2021). RNA methylations in cardiovascular diseases, molecular structure, biological functions and regulatory roles in cardiovascular diseases. *Frontiers in Pharmacology*, *12*, 722728. DOI: 10.3389/fphar.2021.722728 PMID: 34489709

Zhou, Y., Liang, P., & Li, S. (2022). CRISPR-Cas9 in stem cell therapy: Applications in disease modeling and regenerative medicine. *Stem Cell Research & Therapy*, *13*, 457.

Zhou, Y., Zhu, S., Cai, C., Yuan, P., Li, C., Huang, Y., & Huang, Y. (2019). Induction of muscle progenitor cells from fibroblasts by CRISPR-mediated activation of endogenous MyoD1. *Cell Reports*, *29*(3), 598–610.e4.

Zhou, Z., Chen, G., Shen, M., Li, J., Liu, K., Liu, M., Shi, S., Yang, D., Chen, W., Chen, S., Yin, Y., Qin, Y., Su, X., Chen, W., & Kang, M. (2023). Genome-scale CRISPR-Cas9 knockout screening in nasopharyngeal carcinoma for radiosensitive and radioresistant genes. *Translational Oncology*, *30*, 101625. DOI: 10.1016/j.tranon.2023.101625 PMID: 36739730

Zhu, Y. (2022). Advances in CRISPR/Cas9. *BioMed Research International*, *2022*(1), 9978571. DOI: 10.1155/2022/9978571 PMID: 36193328

Zou, R. S., Liu, Y., Gaido, O. E. R., Konig, M. F., Mog, B. J., Shen, L. L., Aviles-Vazquez, F., Marin-Gonzalez, A., & Ha, T. (2023). Improving the sensitivity of in vivo CRISPR off-target detection with DISCOVER-Seq+. *Nature Methods*, *20*(5), 706–713. DOI: 10.1038/s41592-023-01840-z PMID: 37024653

Zuccaro, M. V., Xu, J., Mitchell, C., Marin, D., Zimmerman, R., Rana, B., Weinstein, E., King, R. T., Palmerola, K. L., Smith, M. E., Tsang, S. H., Goland, R., Jasin, M., Lobo, R., Treff, N., & Egli, D. (2020). Allele-specific chromosome removal after Cas9 cleavage in human embryos. *Cell*, *183*(6), 1650–1664.e15. DOI: 10.1016/j.cell.2020.10.025 PMID: 33125898

About the Contributors

Jen-Tsung Chen is a professor of cell biology at the National University of Kaohsiung in Taiwan. He teaches cell biology, genomics, proteomics, plant physiology, and plant biotechnology. Dr. Chen's research interests include bioactive compounds, chromatography techniques, plant molecular biology, plant biotechnology, bioinformatics, and systems pharmacology. He is an active editor of academic books and international journals to advance the exploration of multidisciplinary knowledge involving plant physiology, plant biotechnology, nanotechnology, materials science, ethnopharmacology, and systems biology. He serves as an associate editor, editorial board member, and guest editor in reputed journals. Dr. Chen published books in collaboration with international publishers, and he is handling book projects on diverse topics such as AI technologies, drug discovery, drug development, herbal medicine, medicinal biotechnology, nanotechnology, bioengineering, plant functional genomics, plant speed breeding, epigenetics, functional RNAs, and CRISPR-based genome editing. Dr. Chen is a productive author in academic publications and was recognized as one of the World's Top 2% Scientists 2023, 2024, and 2025 by Elsevier and Stanford University. Dr. Chen received the Springer Nature Editor of Distinction Award in 2025.

***

Naureen Afrose is an Assistant Professor of Pharmaceutics at the Bengal College of Pharmaceutical Sciences and Research in Durgapur, India, Ms. Afrose is driven by a passion to revolutionize drug delivery systems. Her research delves into a variety of areas, including traditional Solid formulations, but her true focus lies in pushing boundaries with Novel drug delivery formulations, Nanoformulations, and utilizing Transdermal and Film-based methods for more efficient drug delivery. She's keenly interested in the potential of AI-related systems and their role in shaping the future of drug delivery, actively exploring how these advancements can further optimize drug effectiveness and patient experience.

Deepak Uttamrao Bhusare working as an Assistant professor, Department of Microbiology, Digambarrao Bindu ACS college bhokar, i have published more than 30 research articles in national and international reputed journal, i am research supervisor, currently guide to 3 research scholars, i have actively did MRP sanctioned by SRTMU, Nanded, Maharashtra, India.

Savitha C is Masters Student in the Department of Life Sciences, Sri Sathya Sai University for Human Excllence, India.

Rideb Chakraborty currently works as an Assistant Professor of Pharmaceutics at Bengal College of Pharmaceutical Sciences and Research (BCPSR) in Durgapur, West Bengal. He attained his postgraduate degree with a focus on Pharmaceutics from the Department of Pharmaceutical Technology at Jadavpur University, Kolkata. His research spans various domains including Oral Solid Dosage (OSD) formulation, Nanoformulation, Transdermal drug delivery systems, and Novel Drug Delivery Systems (NDDS). Additionally, he holds a strong interest in leveraging novel artificial intelligence for drug discovery and development.

Soumia Ed-Day is a neuroscientist with a PhD in Physiology, Neuroscience, and Health from Ibn Tofail University, Morocco. Her research focuses on neurotoxicology, neurodegenerative diseases, and phytotherapy.

Anne Sithuvili Hettiarachchige Done's academic and research interests focus on biomedical sciences, with particular emphasis on clinical microbiology, diagnostic medicine, and translational research. Currently a biomedical sciences student at Coventry University, she has developed strong foundations in microbiology, virology, immunology, and clinical biochemistry. Her work explores laboratory-based diagnostic approaches, antimicrobial resistance, and public awareness of antibiotic misuse, especially within low- and middle-income healthcare settings. She has gained exposure to regulated laboratory environments and quality-focused testing frameworks. Sithuvili is interested in innovative biomedical technologies, including novel therapeutics and vaccine development, and aims to bridge scientific research with practical healthcare applications to generate meaningful societal impact.

C M Hossain, did his B.Pharm degree from Department of Pharmaceutical Technology, Jadavpur University. He is a recipient of University Gold Medal in Master of Pharmacy programme. He did his Ph.D. Degree in Pharmacology from Department of Pharmaceutical Technology, Jadavpur University. Currently He is a Professor & HoD, Department of Pharmaceutical Science & Technology, MAKAUT, WB. He is a recipient of most prestigious DST-Inspire Fellowship from Department of Science and Technology, Govt. of India to pursue PhD degree. His doctoral research is on Chemoprevention and drug discovery of Diabetes and Liver cancer. Dr. Hossain is former Postdoctoral Fellow in School of Medicine, University of Florida, USA. He worked on the Industry project on Drug Discovery with Foxchase Chemical Diversity Centre, United States. He is also engaged in Consultancy with Dept. of Ayush, Govt of India. Dr. Hossain worked as a Principal in Bengal School of Technology-College of Pharmacy under MAKAUT and the College got NAAC and NBA accreditation under his active Leadership. He has experience of facing many AICTE and PCI inspections and its compliances. He has wide experience on Administrative, Developmental, Academics and R&D works. Dr. Hossain has expertise on drug discovery research on Infectious Diseases, diabetes and liver cancer. Currently he is working on the involvement of host defense peptides (HDPs) such as beta-defensin in the Innate Immunity against infectious diseases and trying to explore the effect of Defensin Peptide mimetics against the pathogenesis of microbial infection. His goal is to develop safe and efficacious novel therapeutic compounds for treating life threatening and debilitating ailments. He has published number of research papers in peer reviewed journals of International repute such as Nature group, Elsevier, Wiley, Springer etc. He is an editor for a couple of International journals and contributed as a co-author of several books and book chapters on relevant topics. He has guided number of UG and PG students in USA & India, and currently guiding Ph.D students. He is a recipient of â€œTeacher of the Year-2017â€ award by MAKAUT for outstanding performance in Teaching-Learning and Research. He has delivered number of Invited Talks in National and International Conferences in India and USA. He is also a reviewer of many journals of International repute. He has visited Harvard University and MIT, Boston during his work in US. He is a Member of Indian Pharmaceutical Association (IPA) and Association of Pharmaceutical Teachers of India (APTI). Dr. Hossain is a member several prestigious International organization such as American Association of Pharmaceutical Scientists (AAPS), American Association of Microbiology (ASM), American Association of Immunologists (AAI) and American Medical Writerâ€™s Association (AMWA) etc.

Hettige Kaveesha Sandani Dinudya's academic and research interests encompass biomedical sciences and pharmaceutical studies, with a particular focus on clinical microbiology, diagnostic medicine, and translational research. As a biomedical sciences student at Coventry University, she has developed strong foundations in microbiology, virology, immunology, and clinical biochemistry, while also expanding her expertise in pharmacology, drug mechanisms, and therapeutic applications. Her work explores laboratory-based diagnostic approaches, antimicrobial resistance, and public awareness of antibiotic misuse, especially within low- and middle-income healthcare systems. She has gained valuable experience in regulated laboratory environments, applying quality-driven testing frameworks to ensure accuracy and reliability.

Poonam Kodag has completed her graduation from Mspm's Deogiri College Aurangabad and post graduation in botany from department of Botany, Dr. B.A.M. University Aurangabad. She is currently working as an assistant professor in Rashtriya College Of Pharmacy Hatnoor. She has more than 3 year experience in field of environmental science and microbiology. She has published 2 reviewed & 1 research article in National and international journal and published 1 book chapter. Miss.kodag has presented poster in 'Shodh A National Level Research Convention'. She has attended number of national conference and state level seminar.

Kudhekar A.R. (M.Pharm) working as a Assistant Professor, Nandkumar Shinde College of Pharmacy, Aghur Tq. Vaijapur Dist. Chh. Sambhajinagar (MS) India. It is his passion to design, synthesize & characterize synthetic derivative as well as design molecules by computer Aided Drug Design Methods. He has authored 3 research papers and 5 review article published in national and international journals. He has attended several national and international conferences, workshops, and faculty development programs.

Manish Kumar, M. Pharm, Ph.D. (Pharmaceutics) is presently working as a Professor and Head in Dept. of Pharmaceutics, ISF College of Pharmacy, Moga, Punjab-142001, India, He has completed B. Pharm, from DCS'S A.R.A. College of Pharmacy, Dhule, Maharashtra in 2003 and M.Pharm from Bapu ji Pharmacy College, Davangere, Karnataka in 2005 and Ph.D. from Karpagam University Coimbatore in 2014, Tamil Nadu. He has a total teaching and research experience of 20 years out of which, 08 years administrative experience as Faculty Head, and Dean. He has guided 24 M Pharm dissertations and 6 PhD research scholars. He has published 175 papers in various national and international journals and 8 books and 20 book chapters through various international publishers such as Elsevier, Informa, etc. He also has 31 papers in various national and international conferences and has 25 patents to his credit. His h-index is 23. The areas of research interest include nanotechnology for drug delivery, herbal drug delivery systems, Nanoparticulate drug delivery system for ocular, nasal, topical etc. He is Life-member of various organizations including APTI, IPGA, SPER. He has received various Awards which include SPER eminent scientist Awards (2019), He has received Travel Grant of 35000 INR from ICMR New Delhi for attending International Conference in Hong Kong in 2016. He has worked on many Industry consultancy projects. He has been Member Secretary/ Chairperson of a number of management and policy boards including Board of Studies, Department Research Committee (DRC), Faculty

Research Committee (FRC), Faculty Appraisal Committee in various universities. Working as Editorial Board Member of many national/ international journals. He is also working as Reviewer for many national / international journals published by Bentham, Elsevier, Springer-Nature, Informa. He has organized more than 20 workshops / conferences / Webinars as Chairperson / Convener / Organization Committee member and attended 15 conferences in India and abroad as Resource person / Session Chair.

Saloua Mabsor-Zgandaoui holds a PhD in Cell Biology, Biotechnology, and Health from the Faculty of Sciences at Ibn Tofail University, Kenitra. Her doctoral research was conducted in collaboration with the National Centre for Nuclear Energy, Science, and Technology (CNESTEN), where she investigated the enzymatic and molecular mechanisms of cellular response to ionizing radiation, contributing to advances in radiobiology and antioxidant defense strategies. She has authored several peer-reviewed publications and book chapters in internationally indexed platforms. Dr. Mabsor-Zgandaoui is currently an associate member of the Biology and Health Laboratory, Department of Biology, Faculty of Sciences, Ibn Tofail University, where she continues to develop research at the interface of Cellular and Molecular Biology, Biotechnology, and Cellular Stress Responses.

A. A. P. S. Manamperi is a Senior Professor in the Molecular Medicine Unit at the Faculty of Medicine, University of Kelaniya, Sri Lanka. Her expertise lies in molecular medicine with a focus on infectious diseases, molecular epidemiology, and the development of novel in vitro diagnostic platforms. She has an extensive publication record in peer-reviewed journals and has contributed to foundational research on infectious disease mechanisms, genetic diversity of pathogens, and public health genomics. Professor Manamperi has also authored and contributed to multiple academic textbooks and manuals on molecular biology and diagnostics, and has received numerous research awards and recognitions from national scientific bodies. She continues to mentor students and lead research that bridges molecular science with clinical and public health applications.

Khawla Rachmoune is specialized in biology and biochemistry, with a strong interdisciplinary background in biomaterials science. Her research focuses on the development and characterization of radioprotective biomaterials within multidisciplinary projects that integrate biology, biochemistry, and materials engineering. She is the first author and co-author of several scientific articles published in peer-reviewed international journals. Passionate and highly dedicated to research, she is particularly interested in the applications of biopolymers in human health, environmental protection, and radiation shielding. Her work aims to explore innovative, sustainable, and biologically inspired solutions to mitigate the harmful effects of ionizing radiation, with potential applications in medicine, industry, and environmental safety.

Ramya Raghavan, Assistant Professor, Sri Sathya Sai University for Human Excellence, Karnataka, India. Further investigated neural markers, molecular mechanisms associated with severe speech impairment. Her recent editorial work on a book dedicated to advancing cancer treatment through spatial omics approaches underscores her commitment to innovation. As an experienced editor specializing in medical and scientific publications, Dr. Raghavan possesses a unique ability to bridge complex scientific concepts with broader societal impact.

M. P. Sawani N. Rodrigo is a molecular pathology postgraduate, researcher, and lecturer affiliated with the International Institute of Health Sciences (IIHS), Sri Lanka. She holds a BSc with a major in Genetics, Biotechnology, and Biochemistry, and has hands-on experience in molecular biology, gene editing, and translational biomedical research. Her research interests focus on CRISPR/Cas9 technologies, mitochondrial biology, cancer metabolism, and precision medicine. She is particularly interested in how genome-editing tools can be applied to modulate immune responses and target mitochondrial dysfunction in cancer. Her work bridges fundamental molecular mechanisms with emerging therapeutic strategies in cancer and genetic diseases.

Meenambiga S S is an Assistant Professor who completed 6 years of teaching experience and 4 years of research experience in the field of Biotechnology. Passionate about teaching and have trained many young students in the field of Mycology, Natural Product Research, Nanoparticle synthesis, Drug Discovery from Plants and fungal sources and in-silico analysis of drug targets using modern Bio-informatics tools. Skilled in scientific writing covering the following aspects: Proposal submission, research articles suitable for national and international publications,drafting, proof reading and editing research thesis and expertise in critically reviewing the research manuscript for indexed journals. Member of Board of Studies for B.Tech Biotechnology curriculum in VISTAS. Guided more than 20 undergraduate students projects and served as Principal Investigator in the student's project funded by Tamilnadu State Council for Science and Technology. Published more than 30 research and review articles in national and international indexed journal and has volunteered as reviewer for many reputed journals. Life member of Mycological Society of India, BRSI, IARA, IAENG and many other professional organizations. Completed

Y. I. Nilmini Silva Gunawardene is a Senior Professor and Head of the Molecular Medicine Unit at the Faculty of Medicine, University of Kelaniya, Sri Lanka. Her expertise spans molecular biology, recombinant gene technology, medical entomology, and molecular medicine, with a strong focus on understanding disease mechanisms at a molecular level. Prof. Gunawardene has made significant contributions in medical entomology and mosquito control through genetic studies and mentoring numerous students. Her work has been recognised internationally, including receiving the SOMA Lifetime Achievement Award for her contributions to medical arthropodology. She maintains an active research programme and has published extensively on parasitic infection, vector biology, and molecular diagnostics, advancing both scientific knowledge and public health in Sri Lanka and beyond.

Pallavi VN is a Graduate Student in Life Sciences Department in Sri Sathya Sai University for Human Excellence, India

Suresh Waghamare specializes in employing bioactivity-driven techniques for extracting and identifying phytochemicals, ensuring the consistency of drugs and herbal remedies, and analyzing medicinal plants and their tissue cultures for bioactive compounds. Additionally, he has experience with various laboratory instruments such as UV spectrophotometer, Optical Microscope, Soxhlet Apparatus, Clevenger Apparatus, Sonicator, Rota evaporator, IR and techniques like Thin Layer and Column Chromatography. He focuses on formulating and assessing herbal products for their antimicrobial properties. Also, he filed for two patents. He is credited with nine publications in reputable national and international magazines, four books, and one book chapter.

Index

E

G

T

www.ingramcontent.com/pod-product-compliance
Lightning Source LLC
LaVergne TN
LVHW061525290926
849162LV00004B/662